PROTEASES II
Potential Role in Health and Disease

ADVANCES IN EXPERIMENTAL MEDICINE AND BIOLOGY

Recent Volumes in this Series

A Continuation Order Plan is available for this series. A continuation order will bring delivery of each new volume immediately upon publication. Volumes are billed only upon actual shipment. For further information please contact the publisher.

PROTEASES II
Potential Role in Health and Disease

Edited by

Walter H. Hörl
University of Freiburg
Freiburg, Federal Republic of Germany

and

August Heidland
University of Würzburg
Würzburg, Federal Republic of Germany

PLENUM PRESS • NEW YORK AND LONDON

Library of Congress Cataloging in Publication Data

International Symposium on Proteases: Potential Role in Health and Disease (2nd: 1987: Rothenburg ob der Tauber, Germany)
    Proteases II: potential role in health and disease.

    (Advances in experimental medicine and biology; v. 240)
    "Proceedings of the Second International Symposium on Proteases: Potential Role in Health and Disease, held May 17-20, 1987 in Rothenburg ob der Tauber, Federal Republic of Germany"—T.p. verso.
    Includes bibliographies and index.
    1. Proteolytic enzymes—Congresses. 2. Proteolytic enzymes—Pathophysiology—Congresses. 3. Proteolytic enzyme inhibitors—Congresses. I. Hörl, Walter H. II. Heidland, August. III. Title. IV. Series. [DNLM: 1. Peptide Hydrolases—congresses. 2. Protease Inhibitors—congresses. W1 AD559 v.240 / QU 136 I61 1987]
QP609.P78I57 1987   612'.01516   88-25549
ISBN-13: 978-1-4612-8313-3   e-ISBN-13: 978-1-4613-1057-0
DOI: 10.1007/978-1-4613-1057-0

Proceedings of the Second International Symposium on Proteases:
Potential Role in Health and Disease, held May 17-20, 1987, in
Rothenburg ob der Tauber, Federal Republic of Germany

© 1988 Plenum Press, New York
Softcover reprint of the hardcover 1st edition 1988
A Division of Plenum Publishing Corporation
233 Spring Street, New York, N.Y. 10013

TO OUR WIVES

Ursula Hörl
Gundula Heidland

AND TO OUR CHILDREN

Matthias, Johannes, and
Lukas Hörl

Ulrich and Jochen Heidland

PREFACE

We are pleased to present to our readers the Proceedings of the Second International Symposium "Proteases: Potential Role in Health and Disease" which was held in Rothenburg ob der Tauber (FRG) during May 17-20, 1987.

The topics discussed included those dealing with the physiology and pathophysiology of proteases and their inhibitors, the inter-actions of proteases and hormones, the kallikrein-kinin, com-plement and coagulation system, the function of proteases in arthritis, malignoma, pancreatitis, intestinal tract, lung and kidney disease as well as in hypercatabolic states (acute renal failure, multiple trauma and septicemia). Furthermore some reports dealt with the role of proteases during extracorporeal circulation.

The papers presented answered many questions, but raised many more concerning the significance of proteases and their in-hibitors in clinical medicine. It was unfortunately impossible in this volume, to include the extended, lively and stimulating discussions which were enjoyed by the participiants during the conference.

The meeting has provided a unique framework for close inter-action between scientists from various disciplines, including molecular biology, biochemistry, physiology, surgery, anaesthe-siology, endocrinology, hematology, pneumatology and nephrology.

We would like to express our gratitude and appreciation for all those who have stimulated, encouraged and supported us to hold the symposium in Rothenburg. This endeavour could not have been possible without the generous financial support of Asid-

Bonz (Böblingen), Bayer AG (Leverkusen), Behring AG (Marburg), Biotest Pharma (Dreieich), Boehringer (Ingelheim), Boehringer GmbH (Mannheim), Braun (Melsungen), Ciba-Geigy (Wehr/Baden), Cyanamid GmbH (Wolfratshausen), Diamed (Köln), Fresenius AG (Oberursel), Gambro (Martinsried), Gry-Pharma (Kirchzarten), von Heyden (München), Kabi Vitrum (München), Knoll AG (Ludwigs-hafen), Merck AG (Darmstadt), Nephropharma (Bad Aibling), Pfrimmer & Co. (Erlangen), Röhm-Pharma (Darmstadt), Sandoz AG (Nürnberg), Schwarz (Monheim), Travenol GmbH (München), Tropon Werke GmbH (Köln), and Zyma GmbH (München). In particular the financial support of the Paul-Martini-Stiftung (Mainz) and the "Verein zur Bekämpfung der Hochdruck- und Nierenkrankheiten" is appreciated.

We also are indepted to Mrs. I. Hevendehl, Mrs. E. Hammer, Miss C. Winter, Mrs. M. Röder, Miss B. Schäfer for their invaluable assistance and help in the organization of the meeting. It is a pleasure to acknowledge the excellent secretarial help of Christiane Ehret for preparing the manuscripts.

August Heidland
Walter H. Hörl

CONTENTS

II PROTEASES AND LUNG

III PROTEASES AND LIVER

IV MUSCLE PROTEIN DEGRADATION

V PROTEASES, KIDNEY AND UREMIA

VIII PROTEASES AND MALIGNOMA

ASPARTIC PROTEINASES AND INHIBITORS FOR THEIR CONTROL

IN HEALTH AND DISEASE

J.Kay[1] R.A.Jupp[1] C.G.Norey[1] A.D.Richards[1] W.A.Reid[2]

R.T.Taggart[3] I.M.Samloff[3] and B.M.Dunn[4]

[1]Dept. of Biochemistry, University College, P.O.Box 78
        Cardiff, CF1 1XL. Wales, U.K.

[2]Dept. of Pathology, University of Leeds, Leeds
        LS2 9JT, U.K.

[3]V.A.Medical Centre, Sepulveda, Ca 91343, U.S.A.

[4]Dept. of Biochemistry & Mol.Biology, J.Hillis Miller
        Health Centre, University of Florida
        Gainesville, Fl. 32610, U.S.A.

Aspartic proteinases are produced by a number of cells and tissues within the human body (for a review - see 1). Many are secretory proteins that are deliberately released into extracellular spaces while some exert their function primarily within the cell of origin. Most are active in the acidic pH range.

In the adult stomach, two different types of gastric enzyme predominate, the pepsins and gastricsins (2), each of which is produced in zymogen form in the chief cells (Fig.1). The genes encoding these proteins appear to be expressed from an early stage in foetal development in that both enzymes have been shown to be present from as early as 17 weeks of gestation (3). In gastric carcinomas, only a small percentage (<6%) of the tumour cells produce pepsinogen (4) whereas (pro)gastricsin is present in a much higher proportion(>30%) of cases (5). In addition, local lymph note metastases of the positive cases showed positive staining for gastricsin (>80%) in the neoplastic cells so that gastricsin may have some diagnostic value as a tumour marker. Its potential in this role somewhat diminished by its presence in acinar epithelium of the prostate and in the tumour cells of prostatic carcinomas (6). The prostate would thus appear to be the source of the aspartic proteinase (zymogen) found in seminal fluid (7).

Two further aspartic proteinases are found in the stomachs of a number of species. Chymosin(rennin) is one of the world's most valuable enzymes in monetary terms in its commercial role of cheese manufacture (8) and recombinant enzyme has been produced by gene technology for such purposes (9). The biological role of chymosin within the young ruminant has been discussed (10) but it probably has very little significance in the gastrointestinal tract of humans.

The least well-characterised member of the aspartic proteinase family present in the stomach of humans (and rats, mice and dogs) has been termed slow moving proteinase because of its electrophoretic mobility. The distinct nature of this enzyme has been reported recently, however, and it is virtually certain that slow moving proteinase in man is identical with the cathepsin E of animal tissues (11). The enzyme has a strategic location within the cytoplasm of surface epithelial cells in the human stomach. It appears to be present also, however, in tissues other than gastric mucosa e.g. small intestine, colon, brain, white cells, red cell membranes (12). It has been found in about 50% of gastric cancers (13) and the enzyme purified from cells grown in culture from explants of one such tumour has been shown to be identical with the enzyme isolated from healthy gastric mucosa.

Considerable uncertainty has existed in the past as to whether slow moving proteinase/cathepsin E was an enzyme distinct from a further aspartic proteinase, cathepsin D. This has now been demonstrated unequivocally (11,12,14). Cathepsin D is a ubiquitous enzyme found in the lysosomes of many cell types such as macrophages and other connective tissue cells and in epithelium (15). It is present in spleen, liver, lung, brain, lymph nodes, muscle, stomach and just as with the other aspartic proteinases, cathepsin D is found in neoplasms arising in those tissues in which it is produced normally. Because of this wide distribution,it would appear to have no specificity as a tumour marker. It has, however, been postulated to have a role in producing cachexia, possibly in enhanced proteolysis in muscle and other tissues (16).

The other aspartic proteinase whose action has a most significant effect within the human body is renin. This is synthesised mainly within the juxtaglomerular cells of the kidney but is produced also in other locations e.g. in chorionic cells, in brain and in the sub-maxillary gland of mice. Its action upon its substrate, angiotensinogen, results in the generation of angiotensin I which is further converted by an unrelated enzyme to the effective hormone, angiotensin II. This molecule has many effects concerned with the control of electrolyte balance and the regulation of blood pressure. The control of renin activity in relation to hypertension has been the subject of intense scientific endeavour(reviewed in 16).

In addition to their importance in mammalian tissues, aspartic proteinases have been characterised from a number of microbial sources. Included among these are the well-documented proteinases from Endothia parasitica, Penicillium janthinellum, Rhizopus chinensis, yeast, Mucor pusillus and Mucor miehei. Recently, it has been recognised that retroviruses including the AIDS virus, H.I.V.-1 (18) and H.I.V.-2 (19) encode a proteolytic enzyme within the genomic RNA (in the pol region within the AIDS virus). The predicted amino acid sequences for the putative proteinases show them to have more than a fortuitous resemblance to aspartic proteinases. The sequence homology is high at what is proposed

2

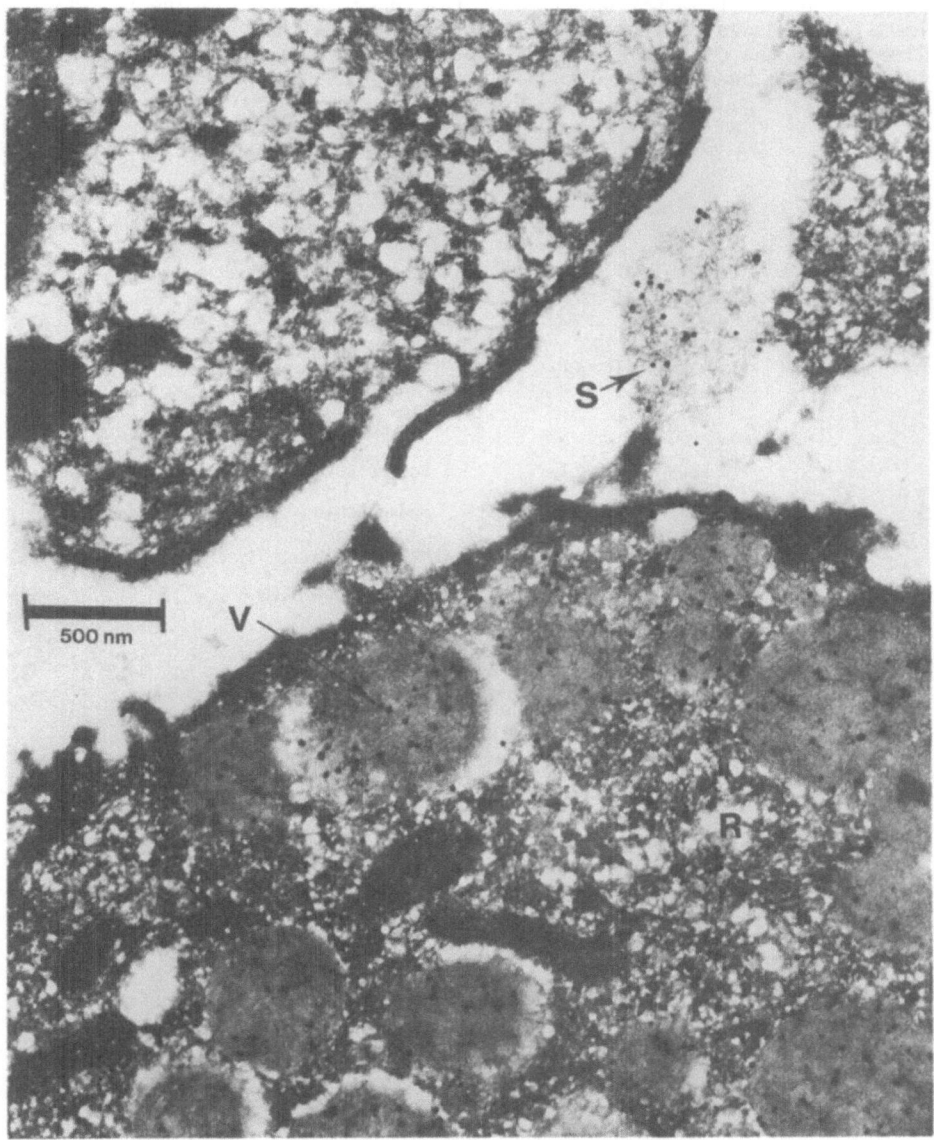

Fig.1. Electron microscopic immunocytochemical localisation of
progastricsin in human stomach. Parts of a chief cell (lower)
and parietal cell (upper) are shown. Progastricsin was
localised by the immunogold method using 20nm gold particles in
secretory vesicles (V) and rough endoplasmic reticulum (R) of
the chief cell and in a luminal secretion (S). The parietal
cell is negative.

to be an active site aspartic acid residue in the viral proteinases (Fig.2). Particularly convincing further lines of evidence for the likelihood that these enzymes do belong to the aspartic proteinase family comes a) from the presence of the threonine and glycine residues immediately downstream from the aspartic acid. Although not involved directly in the catalytic mechanism of aspartic proteinases, these residues have been proposed to be absolutely essential for the correct

215	
Ile-Val-ASP-Thr-Gly-	Human,Pig,Cow,Chicken, Monkey Pepsins Pig, Human Cathepsin D, Pig,Monkey Gastricsins
Ile-Leu-ASP-Thr-Gly-	Bovine Chymosin
Val-Val-ASP-Thr-Gly-	Mouse Renin
Leu-Val-ASP-Thr-Gly-	Human Renin
Ala-Ile-ASP-Thr-Gly-	Yeast Proteinase A
Ile-Leu-ASP-Thr-Gly-	Rhizopuspepsin
Ile-Ala-ASP-Thr-Gly-	Endothia parasitica proteinase,Penicillo -pepsin
Leu-Leu-ASP-Thr-Gly-	H.I.V.-1
Leu-Leu-ASP-Thr-Gly-	H.I.V.-2

Fig.2. Alignment of a section of the predicted amino acid sequences of the pol-proteinases from human immunedeficiency (H.I.V.) viruses -1 and -2 with the sequences of residues adjacent to one of the catalytic aspartic acid residues (no 215, pig pepsin numbering) in a variety of archetypal mammalian and microbial aspartic proteinases.

positioning of the catalytic aspartic acid residues by means of a 'fireman's grip' (20); b) from the relative ease with which much of the predicted amino acid sequences of the viral enzymes (particularly for H.I.V.-1) can be aligned to produce significant homology with the C-terminal half of the sequences of almost 20 conventional aspartic proteinases with c) the need for very few inserted stretches of sequence and d) deletions that can be projected by computer graphics modelling techniques to occur predominantly on the exterior of the postulated three-dimensional structure (21) of the molecule in loops and turns connecting the strands of β-sheet in the polypeptide.

The viral proteinases differ from archetypal aspartic proteinases in that their molecular weights must lie in the size range of 10-20,000 whereas the typical enzymes have molecular masses between 35,000 and 40,000 daltons (depending on the state of glycosylation) with the sole exception of slow moving proteinase/cathepsin E which is composed of two subunits of mol.wts 41,000 and 43,000 respectively that are disulphide-linked (11) to give an intact mol.wt. approaching 90,000. Whereas most aspartic proteinases are single polypeptide chains e.g. human pepsin and rat cathepsin D, others such as renin and human cathepsin D consist of light and heavy chains as a result of having undergone internal cleavage. From X-Ray crystallographic analyses, detailed structures at fine resolution have been determined for pepsin (22) and its precursor, pepsinogen (23), penicillopepsin (24) and the proteinases from Rhizopus chinensis (25) and Endothia parasitica (26,27). All are broadly similar and it is clear that these proteins are composed of two very similar domains that must have arisen through evolution from a single domain by duplication and fusion of an ancestral gene (28). Thus, while archetypal aspartic proteinases function by folding a single polypeptide chain (or 2 chains-S-S-bridged) in such a way as to place the two catalytically active aspartic acid residues in close juxtaposition to one another, in the case of the retroviral enzymes, the generation of activity may require two physically separate subunits (of mol.wt. 10-20K) to interact via the aforementioned threonine residues to provide and orientate two aspartic acid residues in a catalytically active 2-domain entity. Modern methods of recombinant DNA technology may be expected to provide rapid answers to such problems by generating sufficient amounts of recombinant viral enzyme for study via cloning and expressing the appropriate region of the viral genome.

The homologies in primary sequences and the similar 3-dimensional structures suggest that all of the members of the aspartic proteinase family are likely to resemble one another very closely. However, each type of enzyme must have evolved subtle distinctions in structure and therefore activity from the others in order to carry out its specific physiological function. The active site cleft of these enzymes is a long deep groove that extends for about 30 A° across the junction between the NH_2 and COOH-terminal lobes and which is essentially unoccluded except for one highly-conserved section of the polypeptide chain (residues 73-80). This overlies the cleft as an extended β-hairpin or flap and is capable of enclosing molecules that are bound within the active site cleft (29). This extended cleft can accommodate at least seven (and perhaps as many as nine) amino acid residues of a substrate or inhibitor in the S_4-S_3'sub-sites with a distinct preference for cleavage to occur between hydrophobic residues occupying the S_1-S_1' sites (30,31).

The differences in specificity and activity may be explained by discrete alterations in some or all of the sub-sites in the various enzymes. Considerable information has been obtained recently on the nature and identity of the amino acid residues which create the topography and which dictate the selectivity of binding within individual sub-sites (32,33,34). Much of this insight has resulted from interactive biochemical and crystallographic studies. Thus, data derived from kinetic analysis of the interaction of individual enzymes with systematic series of synthetic

Table 1

Sequences of K-casein (residues 101-112) and of two synthetic statine-containing peptides (I & II). K_i values for the inhibition of calf chymosin at pH 6.0 by each of these compounds were determined.

	101				105	106					111	
K-casein:	∿Pro-His-Leu-Ser-Phe-Met-Ala-Ile-Pro-Pro-Lys-Lys∿

I   R—His-Leu-Ser ——Sta—— Ala-Ile-Pro-Pro-Lys-Lys

II   R—Leu-Ser ——Sta—— Ala-Ile-Pro-Pro-Lys-Lys

Peptide	K_i (nM)
I	160
II	11

(Data taken from ref. 37).

chromogenic substrates or inhibitors may be interpreted at the molecular level on the basis of known (or modelled) structures produced for free enzymes or enzyme-inhibitor complexes. Reciprocally, interpretation of 3-Dimensional structural features has to be compatible with the data obtained in solution and a logical basis is thus created for inhibitor design (27,29,34). Detailed discussion of the topography of a number of sub-sites in several enzymes is beyond the scope of this article but it can be seen that the molecular architecture of sub-sites in individual aspartic proteinases can be established by this approach and the contribution of key amino acid residues can be verified through site-directed mutagenesis experiments. Interactions occurring at these distant sub-sites in the active site cleft can begin to offer a molecular explanation for earlier observations of "secondary specificity" wherein substantial increases in catalytic rate (kcat) were noted with substrates of increasing chain length with relatively little variation in the corresponding K_m values (35). With specific docking of substrate residues into most of the sub-sites along the length of the active site crevice (27,34). the exquisite specificity displayed by enzymes such as renin can be explained.

Knowledge of the details of sub-sites in each of the enzymes has permitted, in turn, the design of inhibitors that are specifically targetted towards individual enzymes. The rationale behind such 'designer' drugs has often been based on the modification of sequences of amino acids that occur naturally in the normal substrate. For example, inhibitors have been generated by synthesis of a peptide of an appropriate length containing the amino acid residues known to be present in the substrate but with the introduction of statine (4-amino-3-hydroxy-6-methylheptanoic acid) in place of the two residues that contribute the scissile peptide bond of the substrate (36). Statine acts as a non-hydrolysable dipeptide analogue so that a peptide (inhibitor) resistant

Table 2

Sequences of angiotensinogen (residues 6-13) and of two synthetic inhibitors derived therefrom (L-363,564 and H-142). K_i values were determined for the inhibition of renin at neutral pH and for the other enzymes at pH 3.1 as detailed in ref.27.

	6	10 11	13
Angiotensinogen:	∿His-Pro-Phe-His-Leu-Val-Ile-His∿		
L-363,564	Boc-His-Pro-Phe-His——Sta——Leu-Phe-NH$_2$		
H-142	Pro-His-Pro-Phe-His-Leu-Val-Ile-His-Lys (R over Leu-Val)		

	L-363,564	H-142
	K_i(nM)	
Human renin	3	10
Human pepsin	130	>40,000
Human gastricsin	200	15,000
Human cathepsin D	110	>40,000

(Data taken from ref.27; R = -CH$_2$-NH-)

to attack by the proteinase is thus generated. As an illustration of this strategy, the sequence of the residues immediately adjacent to the Phe-Met bond in K-casein that is cleaved by chymosin to initiate the cascade reaction of milk-clotting is shown in Table 1. By modification of this sequence, synthetic statine-containing derivatives were produced (Table 1) which were effective as inhibitors of chymosin (both naturally-occurring and recombinant enzymes - 37). Such inhibitors (in which R=progesterone) have been incorporated into a novel assay system for use by farmers for the detection of the onset of oestrus in dairy cattle in order to permit timely artificial insemination. The estimated market for this diagnostic kit is approx. $50 million per annum so that synthetic inhibitors of aspartic proteinases are clearly of significant value for diagnostic purposes.

Similar strategies of incorporation of non-hydrolysable analogues (of the tetrahedral transition-state formed during peptide bond hydrolysis)into partial sequences of angiotensinogen have also been used to produce effective inhibitors of human renin (36,38). These have contained either statine (Table 2) or purpose-made dipeptide analogues in which the scissile peptide bond (-CONH-) between residues P_1-P_1' of the substrate is replaced (38) by the synthetic secondary amine (e.g. -CH$_2$-NH- in H-142-Table 2). H-142 is effective in lowering blood pressure in humans (39) and it has no significant effect on the other aspartic proteinases in the human body (Table 2). With established sales of anti-hypertensive agents somewhere in excess of $100 million per annum, it is perhaps not surprising, therefore, that there has been a keen interest within the pharmaceutical industry to modify such compounds in order to develop specific, orally active inhibitors of renin for therapeutic purposes.

With such precedents thus established, attention may become diversified to include other enzymes in the family with a view to development of selective inhibitors for their control under pathophysiological conditions e.g. inhibitors of pepsin/gastricsin for use in the treatment of erosions/ulcers in the gastrointestinal tract (40), inhibitors of cathepsin D in cancer progression (16). Similarly, as more information becomes available on the 'half' enzymes of the retroviruses (which appear to have an unusual specificity directed towards -X-Pro-bonds where X is an aromatic amino acid), it may well be possible to develop selective inhibitors of the HIV-pol-proteinase that may have therapeutic potential as anti-viral agents for the treatment of AIDS. The sensitivity of aspartic proteinases to modulation by synthetic inhibitors administered exogenously may be facilitated somewhat by the fact that naturally-occurring inhibitors of mammalian aspartic proteinases are relatively uncommon (41) and appear to have little physiological significance.

Thus, considerable headway has been made in elucidating the structure, activity and significance of the aspartic proteinases so that it should now be possible to capitalise on the information thus amassed to expedite future investigations into the involvement of these enzymes in processes of clinical and commercial significance.

Acknowledgements

This work was sponsored by grants from the Science and Engineering Research Council, the Yorkshire Regional Health Authority, the National Institutes of Health and by direct support from Boots-Celltech Diagnostics Ltd., Celltech Ltd., Glaxo Group Research Ltd (all U.K.)& Merck, Sharp & Dohme, U.S.A. We are very grateful to our many colleagues in other laboratories, both academic and industrial, who have contributed to the work described in ways too numerous to describe in detail. It is also a pleasure to acknowledge the vital contribution of Barbara Power in producing this manuscript in typed form.

References

1. J.Kay, Aspartic proteinases and their inhibitors, in: "Aspartic Proteinases and their Inhibitors" V.Kostka, ed. pp 1-17, de Gruyter, Berlin (1985).
2. I.M.Samloff, R.T.Taggart and K.L.Hengels, in: "Aspartic Proteinases and their Inhibitors" V.Kostka, ed. pp 79-95, de Gruyter, Berlin (1985).
3. W.A.Reid, K.McGechaen, T.Branch, H.D.A.Gray, W.D.Thompson, E.S.Gray and J.Kay. Immunolocalisation of aspartic proteinases in the developing human stomach, m/s in submission (1987).
4. W.A.Reid, W.D.Thompson and J.Kay, Pepsinogen in gastric carcinoma cells, J.Clin.Path. 36: 137-139 (1983).
5. W.A.Reid and J.Kay, Gastricsin in the cells of gastric carcinomas, Disease Markers 1:263-269 (1983).
6. W.A.Reid, C.N.Liddle, J.Svasti and J.Kay, Gastricsin in the benign and malignant prostate, J.Clin.Pathol. 38: 639-643 (1985).

7. W.A.Reid, L.Vongsorasak, J.Svasti, M.J.Valler and J.Kay. Identification of the acid proteinase in human seminal fluid as a gastricsin originating in the prostate, Cell Tissue Res. 236: 597-600 (1984).

8. M.K.Harboe, Commercial aspects of aspartic proteinases,in: "Aspartic Proteinases and their Inhibitors" V.Kostka, ed. pp 537-550, de Gruyter, Berlin (1985).

9. F.A.O. Marston, P.A.Lowe, M.T. Doel, J.M.Schoemaker, S.White and S.Angal, Purification of calf prochymosin synthesised in E.coli, Biotechnology 2: 800-804 (1984).

10. B.Foltmann and N.H.Axelsen, Gastric proteinases and their zymogens. Phylogenetic and developmental aspects, Febs. Proc. 60: 271-280 (1980)

11. I.M.Samloff, R.T.Taggart, T.Shiraishi, T.Branch, W.A.Reid R.Heath, R.W.Lewis, M.J.Valler and J.Kay, Slow moving proteinase. Isolation, characterisation and immunohistochemical localisation in gastric mucosa, Gastroenterology, 93: in press (1987).

12. N.I.Tarasova, P.B.Szecsi and B.Foltmann. An aspartic proteinase from human erythrocytes is immunochemically indistinguishable from slow moving proteinase from gastric mucosa, Biochim. Biophys.Acta. 880: 96-100 (1986).

13. T.Shiraishi, I.M.Samloff, R.T.Taggart and G.N. Stemmermann, Slow moving proteinase in gastric cancer and its relationship to pepsinogen and progastricsin, Gastroenterology,m/s in submission (1987).

14. S.Yonezawa, T. Tanaka, N. Muto and S.Tani, Immunochemical similarity between a gastric mucosa non-pepsin acid proteinase and neutrophil cathepsin E of the rat, Biochem.Biophys.Res.Commun, 144: 1251-1256.(1987).

15. W.A.Reid, M.J.Valler and J.Kay, Immunolocalisation of cathepsin D in normal and neoplastic human tissues, J.Clin.Path. 39: 1323-1330 (1986).

16. L.M.Greenbaum and J.H.R.Sutherland, Host cathepsin D response to tumour in the normal and pepstatin-treated mouse. Cancer Res. 43:2584-7 (1983).

17. J.E.Hall, H.L.Mizelle and L.L.Woods, The renin-angiotensin system and the long-term regulation of arterial pressure, J.Hypertension 4: 387-397.(1986).

18. L.Ratner, W.Haseltine, R.Patarca, K.J.Livak, B.Starcich, S.F.Josephs, E.R.Doran, J.A.Rafalski, E.A.Whitehorn, K.Baumeister, L.Ivanoff, S.R.Petteway, M.L.Pearson, J.A.Lautenberger, T.S.Papas, J.Ghrayeb, N.T.Chang, R.C.Gallo and F.Wong-Staal, Complete nucleotide sequence of the AIDS virus, HTLV-III, Nature 313: 277-284 (1985).

19. M.Guyader, M.Emerman, P.Sonigo, F.Clavel, L.Montagnier and M.Alizon, Genome organisation and transactivation of the human immunodeficiency virus type 2, Nature 326, 662-669 (1987).

20. L.H.Pearl and T.L.Blundell, The active site of aspartic proteinases, FEBS Lett. 174:96-101 (1984).

21. L.H.Pearl and W.R.Taylor, A structural model for the retroviral proteinases, m/s in submission (1987).

22. N.S.Andreeva, A.S. Zdanov, E.E.Gustchina and I.A. Fedorov, Structure of ethanol-inhibited porcine pepsin at 2A° resolution, J.Biol.Chem, 259:11353-11365 (1984).

23. M.N.G.James and A.R.Sielecki, Molecular structure of an aspartic proteinase zymogen, pig pepsinogen at 1.8A° resolution, Nature 319:33-39 (1986).
24. M.N.G.James and A.R.Sielecki, Structure and refinement of penicillopepsin at 1.8A° resolution, J.Mol.Biol. 163:299-361(1983).
25. R.Bott, E.Subramanian and D.R.Davies, Three-dimensional structure of the complex of Rhizopus chinensis proteinase and pepstatin at 2.5A° resolution, Biochemistry 21:6956-6962 (1982).
26. T.L.Blundell, J.Jenkins, L.H.Pearl, T.Sewell and V.Pedersen, The high-resolution structure of Endothiapepsin, in: "Aspartic Proteinases and their Inhibitors" V.Kostka, ed. pp 151-161, de Gruyter,Berlin (1985).
27. S.I.Foundling, J.Cooper, F.E.Watson, A.Cleasby, L.H.Pearl, B.L.Sibanda, A.Hemmings, S.P.Wood, T.L.Blundell, M.J.Valler, C.G.Norey, J.Kay, J.S.Boger, B.M.Dunn, B.J.Leckie, D.M.Jones, B.Atrash, A.Hallett and M.Szelke, High resolution X-ray analyses of renin inhibitor-aspartic proteinase complexes, Nature 327: 349-353 (1987).
28. J.Tang, M.N.G.James, I-N.Hsu, J.A.Jenkins and T.L.Blundell, Structural evidence for gene duplication in the evolution of the acid proteinases, Nature 271:618-621 (1978).
29. A.Hallett, D.M.Jones, B.Atrash, M.Szelke, B.J.Leckie, S.Beattie, B.M.Dunn, M.J.Valler, C.E.Rolph, J.Kay, S.I.Foundling, S.P.Wood, L.H.Pearl, F.E.Watson and T.L.Blundell, Inhibition of aspartic proteinases by transition state analogues, in: "Aspartic Proteinases and their Inhibitors" V.Kostka, ed pp 467-478, de Gruyter, Berlin (1985).
30. J.C.Powers, A.D.Harley and D.V.Myers, Subsite specificity of porcine pepsin, in: Acid Proteases" J.Tang, ed.pp 141-157 Plenum Press, New York (1977).
31. T.Imoto, K.Okazaki, H.Koga and H.Yamada, Specificity of rat liver cathepsin D, J.Biochem. 101:575-580 (1987).
32. T.Hofmann, R.S.Hodges and M.N.G.James, Effect of pH on the activities of penicillopepsin and rhizopuspepsin and a proposal for the productive substrate binding mode in penicillopepsin, Biochemistry 23:635-643 (1984).
33. B.M.Dunn, M.Jimenez, B.F.Parten, M.J.Valler, C.E.Rolph and J.Kay, A systematic series of synthetic chromophoric substrates for aspartic proteinases, Biochem.J. 237: 899-906 (1986).
34. B.M.Dunn, M.J.Valler, C.E.Rolph, S.I.Foundling, M.Jimenez and J.Kay, The pH dependence of the hydrolysis of chromogenic substrates by selected aspartic proteinases: Evidence for specific interactions in subsites S_3 and S_2, Biochim.Biophys Acta, in press (1987).
35. M.Blum, A.Cunningham, M.Bendiner and T.Hofmann, Penicillopepsin substrate-binding effects and intermediates in transpeptidation reactions, Biochem.Soc. Trans.13: 1044-6 (1985).
36. J.S.Boger, N.S.Lohr, E.Ulm, M.Poe, E.H.Blaine, G.M.Fanelli, T-Y Lin, L.S.Payne, T.W.Schorn, B.I.Lamont, T.C.Vassil, I.I.Stabilito, D.F.Veber, D.H.Rich, and A.S.Bopari, Novel renin inhibitors containing the amino acid, statine, Nature 303: 81-84 (1983).

37. M.J.Powell, R.J.Holdsworth, T.S.Baker, R.C.Titmas, C.C.Bose, A.Phipps, M.Eaton, C.E.Rolph, M.J.Valler and J.Kay, Design and synthesis of statine-containing inhibitors of chymosin, in: "Aspartic Proteinases and their Inhibitors" V.Kostka,ed pp 479-483, de Gruyter, Berlin (1985).
38. M.Szelke, B.J.Leckie, A.Hallett, D.M.Jones, J.Sueiras-Diaz, B.Atrash and A.F.Lever, Potent new inhibitors of human renin, Nature 299: 555-557 (1982).
39. D.J.Webb, A.M.M.Cumming, B.J.Leckie, A.F.Lever, J.J.Morton, J.I.S.Robertson, M.Szelke & B.Donovan, Reductions of blood pressure in man with H-142, a potent new renin inhibitor, Lancet 2: 1486-7 (1983).
40. T.F.Ford, D.A.W.Grant, B.M.Austen and J.Herman-Taylor, Intramucosal activation of pepsinogens in the pathogenesis of acute gastric erosions and their prevention by pepstatinyl-Gly-Gly-Lys-Lys, Clin.Chim.Acta 145:37-47 (1985).
41. J.Kay, M.J.Valler and B.M.Dunn, Naturally occurring inhibitors of aspartic proteinases, in:Proteinase Inhibitors, N.Katunuma, H.Umezawa, H.Holzer, eds. pp 153-163, Springer-Verlag, Berlin (1983).

HUMAN NEUTRAL ENDOPEPTIDASE 24.11 (NEP, ENKEPHALINASE);

FUNCTION, DISTRIBUTION AND RELEASE

Ervin G. Erdös and Randal A. Skidgel

Departments of Anesthesiology and Pharmacology
University of Illinois College of Medicine
Chicago, IL

Neutral endopeptidase 24.11 (NEP), although discovered in 1974 (1), received much attention recently because of its very wide distribution and many potential functions in the body (2). NEP is also known as enkephalinase (3). This name wrongfully indicates that it only cleaves enkephalins, whereas it actually hydrolyzes a variety of vaso- and CNS-active peptides (2-7). NEP, a metalloendopeptidase, was first detected in the brush border of animal kidney by Kerr and Kenny (1), who used the B-chain of insulin as substrate. The name "enkephalinase" was given later to an enzyme present in brain which cleaved enkephalins at the same bond as the angiotensin I converting enzyme (kininase II or ACE) (8). Schwartz et al (3) suggested that this "enkephalinase" was a second peptidyl dipeptidase present in the striatum of mouse brain. Because the Km of enkephalins with the latter enzyme was lower than with ACE, the second peptidyl dipeptidase was named the true "enkephalinase". Both enzymes are membrane-bound and contain zinc as a cofactor, but human NEP is not a second peptidyl dipeptidase. It differs from ACE in its substrate specificity; for example, it releases a tetrapeptide from angiotensin II and inactivates angiotensin I by cleaving the C-terminal tripeptide instead of activating it as ACE does by releasing His-Leu (4,7). It does not crossreact with antiserum to human ACE (9), and is inhibited by phosphoramidon and thiorphan (2,5) but not by captopril or enalapril. Human NEP can be classified as a thermolysin-type metalloendo-peptidase (2,10,11).

We purified human NEP from the same membrane fraction of human kidney from which we extracted and purified ACE. NEP was extracted with Triton X-100 and purified to homogeneity. The purification was eleven hundred-fold with a 14% yield (5).

Human NEP is a membrane-bound glycoprotein with a molecular weight of about 90,000 containing 10% neutral sugars. It is present in the brush border of the proximal tubules in very high concentration, as is ACE (9,12). In addition, NEP is in the brain (2,3,13), intestinal (13), and placental brush border (6). NEP of the placenta (6) and prostate gland is

13

richer in sialic acid than the renal enzyme (6,9).

Although NEP is present in many of the same microvillar structures as ACE, the distribution of the two enzymes differs in the vasculature. ACE is mainly in the endothelial cells, while NEP is in subendothelial structures such as fibroblasts (14). The presence of NEP in high concentrations in male genital tract (in the prostates, epididimydes and seminal plasma) is puzzling (9), although the reason for the large amounts of ACE in the human male genital tract is also waiting for a clarification.

In the human kidney, NEP, besides being on the luminal microvilli of the proximal tubules, is present also in the basal infoldings but it seems to be absent from distal tubules. This was shown by using immunogold techniques in ultrastructural studies (12). In addition, NEP is released into the urine (12).

NEP cleaves peptide substrates at the amino side of

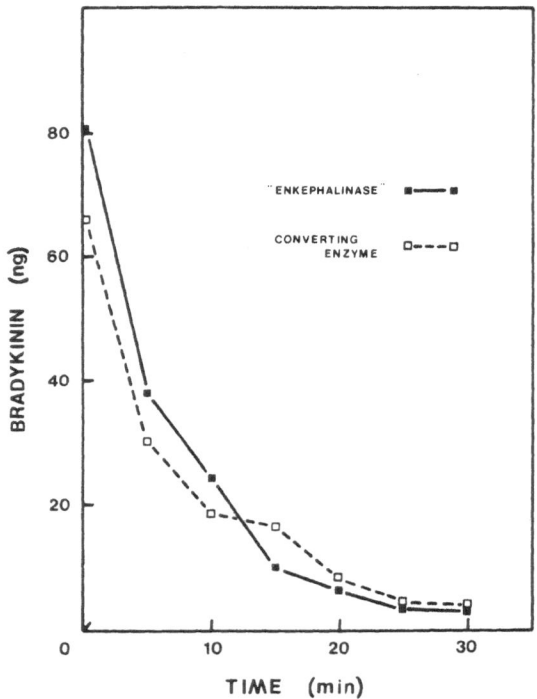

Fig. 1. Inactivation of bradykinin by human ACE and NEP measured on the isolated rat uterus. Equal concentrations of the enzymes were incubated at 37°C in the presence of 2 μM bradykinin and aliquots removed and injected into the tissue bath at regular time intervals. (From Gafford et al., Human Enkephalinase, in: "Degradation of Endogenous Opioids: Its Relevance in Human Pathology and Therapy," S. Ehrenpreis and F. Sicuteri, ed., pp. 43-49, Raven Press, New York, 1983.)

Table 1. Some Active Peptides Cleaved by Human NEP.

Peptide	Sequence
Bradykinin	Arg-Pro-Pro-Gly↓Phe-Ser-Pro↓Phe-Arg
Enkephalin	Tyr-Gly-Gly↑Phe-(Met/Leu)
Angiotensin I	Asp-Arg↓Val-Tyr↓Ile-His-Pro↑Phe-His-Leu
Angiotensin II	Asp-Arg↓Val-Tyr↓Ile-His-Pro-Phe
Substance P	Arg-Pro-Lys-Pro-Gln-Gln↓Phe↓Phe-Gly↓Leu-Met-NH$_2$
Oxytocin	Cys-Tyr-Ile-Gln-Asn-Cys-Pro↓Leu-Gly-NH$_2$
Neurotensin	<Glu-Leu-Tyr-Glu-Asn-Lys-Pro-Arg-Arg-Pro↓Tyr↓Ile-Leu
Chemotactic Peptide	fMet↓Leu-Phe

Arrows denote the sites of primary (↑) and secondary (⋮) cleavage.
(For references see # 5-8, 19).

hydrophobic amino acids such as Phe, Leu or Ile (Table 1). We studied the cleavage of biologically active peptides by the human enzyme. First, bradykinin was the substrate and we compared its cleavage by the two homogeneous human enzymes; NEP and ACE (which is also called kininase II). Both enzymes hydrolyzed bradykinin by cleaving the C-terminal Phe-Arg dipeptide (5). Although human NEP has a higher turnover number (k_{cat}) than ACE (4770 vs. 500 min^{-1}), ACE has a more favorable K_m, hence the k_{cat}/K_m of ACE is higher (500 vs. 40 $min^{-1} \mu M^{-1}$). In spite of the high K_m (120 μM), NEP is a very effective kininase and we named it, tongue-in-cheek, kininase-two- an'-a-half. As shown in Fig. 1, at a concentration of 2 μM, bradykinin is cleaved by the two enzymes at an equal rate. The inactivation by ACE is completely inhibited by captopril (1μM) while phosphoramidon (1μM) inhibited NEP. NEP is also the major kininase in rat urine even at 10 nM concentration of bradykinin (15). The rest of the kininase activity in rat urine was due to a kininase I-type carboxypeptidase and to ACE.

The specificity constant (k_{cat}/K_m) of enkephalin hydrolysis by human NEP is only slightly better than that for bradykinin (5). When Turner et al. compared the hydrolysis of various substrates by purified hog NEP, substance P had the highest specificity constant, and this decreased in the order Met^5-enkephalin-Arg^6-Phe^7, bradykinin, Leu^5-enkephalin, CCK-8, dynorphin(1-9), neurotensin and LHRH (4). Recent studies focused on the cleavage of the atrial natriuretic factor (ANF) by NEP (16). This peptide is cleaved probably at the Cys^7-Phe^8 bond within the ring formed by the S-S bridge of the two cysteine residues (17).

The finding of the hydrolysis of the endogenous opioids, enkephalins, by NEP in the CNS and the possible therapeutic benefit of the inhibition of this reaction in vivo generated a great deal of research work. NEP is inhibited by the thermolysin inhibitor phosphoramidon and by thiorphan and other synthetic inhibitors as well (2,3). Acetorphan is a lipophilic derivative of thiorphan, and can be administered parenterally (18). This inhibitor acts on NEP present on synaptic membranes in the CNS as a prime target. The CNS enzyme is identical with the enzyme purified from renal brush border (2).

NEP as a membrane-bound enzyme is anchored to the plasma membrane of cells by hydrophobic amino acids in an anchor peptide. The fact that the level of NEP in normal human blood is very low stimulated our studies of possible disease states where the enzyme would be released into the circulation, especially from the lung. There are several potential sites from which NEP could be released into the blood. Fibroblasts in lung are one possible source of circulating NEP. In fibroblasts, the half-life of NEP is estimated to be 3 to 7 days (Lorkowski et al., to be published). Other possible sources for NEP in human plasma include the neutrophils (19) or the lymph nodes (20). As the blood level of NEP is elevated in sarcoidosis (21), lymph nodes could be a source in that condition. We studied NEP in leukocytes because these cells are sequestered in lung, especially in shock. We found significant NEP activity on neutrophil cell membranes (19), indicating that these cells could be another possible source for circulating NEP.

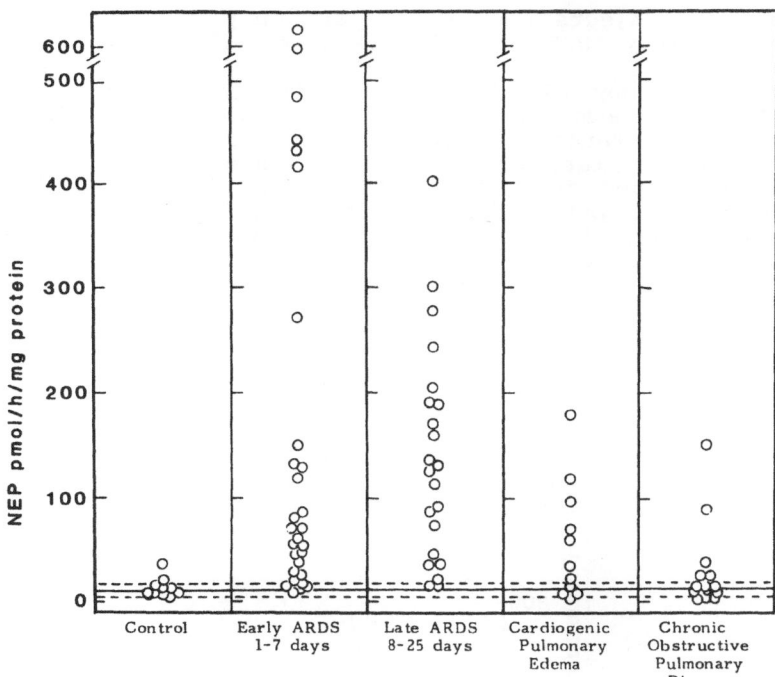

Fig. 2. Neutral endopeptidase levels in serum samples from patients with early or late ARDS, cardiogenic pulmonary edema, or chronic obstructive pulmonary disease. The specific activity (pmol/h/mg) is given on the ordinate. The individual values are indicated by the open triangles in each group. The heavy line is the mean value for the control (normal) group, and the dashed lines indicate 1 standard deviation from the control mean. (Modified from Johnson, et al., reference # 22).

We measured serum levels of NEP in patients with adult respiratory distress syndrome (ARDS) and found that the level of NEP could rise 50-60 fold (22). Fig. 2 shows the variations in the level of NEP in sera of normal individuals and in patients suffering from various lung diseases. The enzyme released into the circulation in these patients appears to be identical with the human renal enzyme; phosphoramidon and antiserum to renal NEP inhibited the activity. The NEP values in 13 control serum samples ranged from 2 to 35 pmol/h/mg protein, with a mean ± SEM of 11.6 ± 2.5. In both early and late ARDS groups, serum NEP values were considerably higher than in the control group. In patients studied within 1 to 7 days after admission, the specific activity of serum NEP ranged from 7 to 635 pmol/h/mg. Only 5 values were within 1 standard deviation of the mean of the control group, and 11 patients had values at least 10 times higher than normal. All of the patients with values above 1 standard deviation from the control subjects had sepsis, and only 6 survived in this group with early ARDS (1 to 7 days after admission).

In the group of 22 patients studied more than 7 days after hospital admission, 20 of them had serum NEP values higher than 1 standard deviation of the mean for the control groups. In the late ARDS group, the specific activities of the enzyme ranged from 20 to 410 pmol/h/mg. The activity of serum NEP in 13 patients was 10-fold higher than in control subjects. Only 2 patients with values higher than 100 pmol/h/mg did survive.

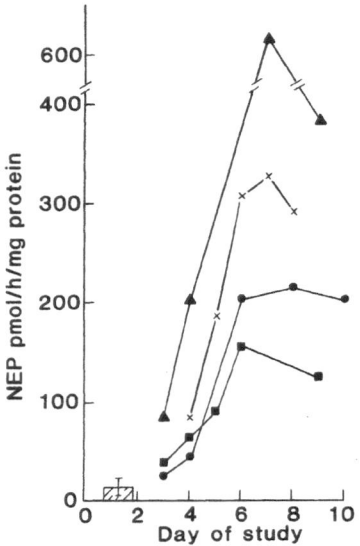

Fig. 3. Changes in serum NEP over several days in 4 patients. Four patients with early ARDS were studied on sequential days. The specific activity of serum NEP is given on the ordinate while the day of the study is given on the abscissa. For comparison, the hatched bar indicates the mean value ± SD of NEP for the control subjects. (Modified from Johnson, et al., reference # 22).

Patients with cardiogenic pulmonary edema and chronic obstructive lung disease also had elevated levels of serum NEP. In the relatively small sample of 11 patients with cardiogenic pulmonary edema, 7 had serum NEP values beyond 1 standard deviation of the normal value. In contrast, 10 of the 15 patients with chronic obstructive lung disease had normal or low serum values of NEP (fig. 2).

Serial serum samples were collected from several patients with early ARDS. The NEP activity measured 3 days after admission were 2 to 7 times higher than in the healthy control subjects, and rose progressively in subsequent samples. The peak activities (days 6 to 8) were more than 10-fold higher than normal and in 1 patient it was 60 times higher (Fig. 3). When 13 patients with pneumonia complicated by sepsis were compared with 8 who had pneumonia without sepsis, the specific activities of serum NEP were significantly higher (p<0.01) in the septic patients. Similarly, 14 septic patients who had ARDS from nonpulmonary causes had greater NEP activities than nonseptic patients (p<0.01). Mortality was greater in both septic groups; 11 of the 13 patients with pneumonia died within 3 to 5 days of blood sampling, and all 14 septic patients with nonpulmonary causes of ARDS died. In the group with pneumonia uncomplicated by sepsis, 3 of the 8 patients died (22).

In conclusion, NEP released into the circulation appears to be a useful marker for lung injury. In the lung, NEP is not associated with vascular endothelium as ACE is, but with fibroblasts (22) and airway epithelial cells. The source of NEP appearing in the circulation is not known. It is possible that during lung injury, neutrophils sequestered in the lung release NEP from the cell membrane by some yet unknown mechanism.

ACKNOWLEDGEMENTS

These studies were supported by Grants HL36473 and HL36082 from the National Institutes of Health.

REFERENCES

1. M.A. Kerr, and A.J. Kenny, The purification and specificity of a neutral endopeptidase from rabbit kidney brush border, Biochem. J. 137:477 (1974).
2. A.J. Turner, Metabolism of enkephalins, ISI Atlas Sci.: Pharmacol. 1:74 (1987).
3. J.-C. Schwartz, B. Malfroy, and S. De La Baume, Biological inactivation of enkephalins and the role of enkephalin-dipeptidyl-carboxypeptidase ("enkephalinase") as neuropeptidase, Life Sci. 29:1715 (1981).
4. R. Matsas, A.J. Kenny, and A.J. Turner, The metabolism of neuropeptides: The hydrolysis of peptides, including enkephalins, tachykinins and their analogues, by endopeptidase-24.11, Biochem. J. 223:433 (1984).
5. J.T. Gafford, R.A. Skidgel, E.G. Erdös, and L.B. Hersh, Human kidney "enkephalinase", a neutral metallo-endopeptidase that cleaves active peptides, Biochemisty 22:3265 (1983).
6. A.R. Johnson, R.A. Skidgel, J.T. Gafford, and E.G. Erdös, Enzymes in placental microvilli: angiotensin I converting enzyme, angiotensinase A, carboxypeptidase, and neutral endopeptidase ("enkephalinase"), Peptides 5:789 (1984).

7. R.A. Skidgel, S. Engelbrecht, A.R. Johnson, and E.G. Erdös, Hydrolysis of substance P and neurotensin by converting enzyme and neutral endopeptidase, <u>Peptides</u> 5:769 (1984).

8. E.G. Erdös, A.R. Johnson, and N.T. Boyden, Hydrolysis of enkephalin by cultured human endothelial cells and by purified peptidyl dipeptidase, <u>Biochem. Pharmacol</u>. 27:843 (1978).

9. E.G. Erdös, W.W. Schulz, J.T. Gafford, and R. Defendini, Neutral metalloendopeptidase in human male genital tract: comparison to angiotensin I-converting enzyme, <u>Lab</u>. <u>Invest</u>. 52:437 (1985).

10. S. Blumberg, Z. Vogel, and M. Altstein, Inhibition of enkephalin-degrading enzymes from rat brain and of thermolysin by amino acid hydroxamates, <u>Life</u> <u>Sci</u>. 28:301 (1981).

11. J. Almenoff, S. Wilk, and M. Orlowski, Membrane bound pituitary metalloendopeptidase: apparent identity to enkephalinase, <u>Biochem</u>. <u>Biophys</u>. <u>Res</u>. <u>Commun</u>. 102:206 (1981).

12. R.A. Skidgel, W.W. Schulz, L.-T. Tam, and E. G. Erdös, Human renal angiotensin I converting enzyme and neutral endopeptidase, <u>Kidney</u> <u>Int</u>. (In press, 1987).

13. R. Matsas, A.J. Kenny, and A.J. Turner, An immunohisto-chemical study of endopeptidase-24.11 ("enkepha-linase") in the pig nervous system, <u>Neurosci</u>. 18:991 (1986).

14. A.R. Johnson, J. Ashton, W.W. Schulz, and E.G. Erdös, Neutral metalloendopeptidase in human lung tissue and cultured cells, <u>Am</u>. <u>Rev</u>. <u>Respir</u>. <u>Dis</u>. 132:564 (1985).

15. N. Ura, O.A. Carretero, and E.G. Erdös, The role of renal endopeptidase 24.11 in kinin metabolism, <u>Kidney</u> <u>Int</u>. (In press).

16. G.M. Olins, K.L. Spear, N.R. Siegel, H.A. Zurcher-Neely, and C.E. Smith, Proteolytic degradation of atrio-peptin III by rabbit kidney brush border membranes, <u>Fed</u>. <u>Proc</u>. 45:427 (1986).

17. N.G. Delaney, D.W. Cushman, M.B. Rom, M.M. Asaad, J.L. Bergey, and A.A. Seymour, Neutral endopeptidase cleaves ANP 103-126 to ring-opened products with diminished biological activity, <u>Fed</u>. <u>Proc</u>. 46:1296 (1987)

18. J.-M. Lecomte, J. Costentin, A. Vlaiculescu, P. Chaillet, H. Marcais-Collado, C. Llorens-Cortes, M. Leboyer and J.-C. Schwartz, Pharmacological properties of acetorphan, a parenterally active "enkephalinase" inhibitor, <u>J</u>. <u>Pharmacol</u>. <u>Exper</u>. <u>Therap</u>. 237:937 (1986).

19. J.C. Connelly, R.A. Skidgel, W.W. Schulz, A.R. Johnson, and E.G. Erdös, Neutral endopeptidase 24.11 in human neutrophils: Cleavage of chemotactic peptide, <u>Proc</u>. <u>Natl</u>. <u>Acad</u>. <u>Sci</u>. 82:8737 (1985).

20. M.A. Bowes, and A.J. Kenny, Endopeptidase-24.11 in pig lymph nodes. Purification and immunocytochemical localization in reticular cells, <u>Biochem</u>. <u>J</u>. 236:801 (1986).

21. J. Almenoff, A.S. Teirstein, J.C. Thornton, and M. Orlowski, Identification of a thermolysin-like metalloendopeptidase in serum: activity in normal subjects and in patients with sarcoidosis, <u>J</u>. <u>Lab</u>.

<u>Clin</u>. <u>Med</u>. 103:420 (1984).

22. A.R. Johnson, J.J. Coalson, J. Ashton, M. Larumbide, and E.G. Erdös, Neutral endopeptidase in serum samples from patients with adult respiratory distress syndrome: comparison with angiotensin-converting enzyme, <u>Am</u>. <u>Rev</u>. <u>Respir</u>. <u>Dis</u>. 132:1262 (1985).

NEUTROPHIL ELASTASE AND CATHEPSIN G:

STRUCTURE, FUNCTION, AND BIOLOGICAL CONTROL

Wieslaw Watorek, David Farley, Guy Salvesen
and James Travis

Department of Biochemistry
University of Georgia
Athens, Georgia
30602

INTRODUCTION

Human neutrophils utilize a variety of destructive enzymes during the process of phagocytosis. These include proteinases, phosphatases, glycosidases, nucleases, and oxidases. The major enzymes have been determined to be elastase, cathepsin G, myeloperoxidase, and lysozyme (1), while minor proteins include collagenase, lactoferrin, alkaline phosphatase, and a myriad of other proteins.

For several years (2,3) it has been suggested that elastase and cathepsin G might play significant roles in the abnormal degradation of healthy tissue, as a result of their release during either phagocytosis or cell death. Since the combination of these two enzymes can result in the degradation of a variety of connective tissue proteins, including elastin, collagen, and proteoglycan (4), both have been implicated as having major roles in the development of diseases such as pulmonary emphysema, rheumatoid arthritis, and glomerulo-nephritis.

Both elastase and cathepsin G are under the control of plasma proteinase inhibitors (5) which can penetrate into various tissues or, alternatively, by inhibitors synthesized directly by the organ under attack. For this reason one must not only discuss α-1-proteinase inhibitor (α-1-PI) and α-1-antichymotrypsin (α-1-Achy) (6), which apparently regulate the activity of elastase and cathepsin G, respectively, but also the bronchial mucous proteinase inhibitor (BMPI) (7) which can inactivate either enzyme and, more importantly, can be found in a large variety of tissues (8).

Because of the almost certain role of elastase and cathepsin G in both normal and pathogenic states we have investigated both enzymes in some detail. In this paper we review much of the data which has been gathered not only on the structure and function of both enzymes but also with regard to their control.

STRUCTURAL FEATURES OF NEUTROPHIL ELASTASE AND CATHEPSIN G

Elastase and cathepsin G are typical serine proteinases, being rapidly inactivated by di-isopropyl fluorophosphate, synthetic acylating inhibitors (9), and plasma serine proteinase inhibitors. However, they differ in many other aspects both from other serine proteinases and from each other. Some of these differences are summarized in Table 1.

Table 1. Properties of Neutrophil Elastase and Cathepsin G

Property	Elastase	Cathepsin G
Isozyme	3 forms	4 forms
Carbohydrate	3 sites	1 site
Isoelectric Point	11.0	near 13.0
Zymogen	?	Predicted
Carboxy Terminal Extension	20 residues	12 residues
Disulfide Bonds	4	3
Amino Acid Specificity	Valine	Leucine
Controlling Inhibitor	α-1-PI	α-1-Achy

These properties will be discussed in more detail in the following sections.

Amino Acid Sequence Analysis

Both neutrophil elastase and cathepsin G have recently been sequenced (10,11) using traditional protein sequencing methodology for the former and DNA sequencing of a cDNA clone for the latter. In fact, both enzymes are synthesized by U-937 monocytes (12,13) so that it has been possible to also isolate a partial clone for elastase, as well. The sequence of each indicates a carboxy terminal extension when comparison is made with mature enzyme. These sequences are presumably removed during post-translational modification of the enzyme, although it is not likely that they are in any way responsible for the isozyme character of either enzyme. These extensions are shown in Table 2.

Table 2. Carboxy Terminal Extensions on Elastase and Cathepsin G

Elastase (mature)	I I Q
Elastase (cDNA)	I I Q R S E D N P C P H D R D F D P A S R T H
Cathepsin G (mature)	I R T T M R
Cathepsin G (cDNA)	I R T T M R S F K L L D Q M E T P L

Examination of the primary structure of each enzyme indicates that both contain the expected catalytic triad involving the interactions of His-57, Asp-102, and Ser-195 (chymotrypsinogen numbering). However, alignment of the primary structure of each enzyme with other serine proteinases suggests significant differences in putative binding sites which are probably responsible for some of their unique properties. These are amplified in the following sections.

Carbohydrate Analysis

The primary structure of each of the proteinases indicates the potentiality for carbohydrate attachment sites. In the case of elastase there are three possible sites, two of which are apparently utilized. Cathepsin G has one site which has not, as yet, been completely defined. Analysis of each of the isozyme forms of elastase clearly suggests that in the major form (previously referred to as E-4) (14) one glycosylation site is utilized, while in the other forms it is likely that two or possibly all three are involved. It is of some significance that treatment of a mixture of the elastase isozymes or individual isozymes with endoglycosidase F results in the conversion of all to a single form, strongly suggesting that the isozyme character of these proteinases is carbohydrate derived. A similar explanation is likely for the isozyme character of cathepsin G.

Preliminary carbohydrate analysis of the major form of neutrophil elastase indicates that it has retained a traditional complex carbohydrate structure, with only one of the three potential sites being used. Other forms have considerably more carbohydrate and at least two sites have so far been confirmed as having been glycosylated. Cathepsin G, on the other hand, contains only small amounts of carbohydrate, and it remains to be established as to whether the single site is utilized by any or all of the isozyme forms.

X-Ray Crystal Structure Analysis of Neutrophil Elastase

Recently (15), a preliminary investigation of the crystal structure of neutrophil elastase was reported. This involved the utilization of crystals prepared from a complex of neutrophil elastase with the third domain of turkey ovomucoid. The results indicated that this enzyme did share some common structural features with porcine pancreatic elastase, its most homologous counterpart, although there were significant differences. These are summarized as follows:

1. Nearly all of the arginine residues of the neutrophil enzyme are positioned to one side of the enzyme surface so that they can properly interact with acidic polysaccharides present in the matrix of the azurophil granule.

2. Both carbohydrate components are attached on opposite sides of the enzyme and away from the reactive site. It should be pointed out that crystals were made from a mixture of the isozymes so that the finding of two carbohydrate components indicates that the minor forms of elastase are more heavily glycosylated than the major form in which only one site is apparently utilized.

3. The area in which the binding and catalytic sites reside is similar to that found in the pancreatic enzyme. However, there are significant differences inside the binding pocket. These include the replacement of several residues believed to be involved in substrate binding for the pancreatic enzyme. This results in the pocket of the neutrophil enzyme becoming very hydrophobic. In addition, one also finds an acidic residue at the bottom of the pocket whose function appears to be in the binding of water molecules as well as in hydrogen binding to other amino acid residues near the pocket. In this respect, the binding and catalyic sites of neutrophil elastase may be described as two tiered, one portion being hydrophobic and the other hydrophilic. How these interact with substrates is not yet known. These data are summarized in Table 3.

Table 3.   Amino Acid Residues Involved in the Activity of
           Elastolytic Enzymes

Site	Pancreatic	Neutrophil
Catalytic Triad	Intact	Intact
S-1	Glutamine	Phenylalanine
S-2	Glycine	Glycine
S-3	Glutamine	Phenylalanine
S-4	Threonine	Phenylalanine
S-5	Tyrosine	Leucine
Substrate Binding Pocket	Valine	Valine
	Threonine	Aspartic Acid
	Serine	Glycine

Sequence Homology

Comparative sequence analysis of human neutrophil elastase and cathepsin G with other serine proteinases indicates, as might be suspected, that the former enzyme is most closely aligned with porcine pancreatic elastase. In contrast, and despite the fact that cathepsin G can cleave peptide bonds after residues which would make it a member of the chymotrypsin family (16), the homology of this enzyme is far closer to that of other white blood cell proteinases. Indeed, all of these latter enzymes are characterized by the fact that they only contain three disulfide bonds, and it is perhaps this lack of at least one more disulfide bond which makes such enzymes rather poor proteinases. A summary of the comparison of each enzyme with others is given in Tables 4 and 5.

Table 4. Comparative Sequence Homology of Human Neutrophil
Elastase with Other Serine Proteinases

Enzyme	% Homology
1. Pancreatic Elastase-Porcine	43.0
2. Pancreatic Kallikrein-Porcine	39.2
3. Cathepsin G-Human	37.2
4. Collagenolytic Proteinase-Fiddler Crab	35.4
5. Preproelastase II-Rat	34.9
6. Trypsinogen-Dogfish	34.8
7. Plasminogen-Human	34.0
8. Factor D-Human	32.9
9. Chymotrypsinogen A-Bovine	31.6

Table 5. Comparative Sequence Homology of Human Neutrophil
. Cathepsin G with Other Serine Proteinases

Enzyme	% Homology
1. Cytotoxic Lymphocyte Proteinase I-Rat	57.0
2. Mast Cell Proteinase II-Rat	49.6
3. Trypsinogen I-Rat	39.4
4. Factor D-Human	38.5
5. Neutrophil Elastase-Human	37.2
6. Trypsinogen-Porcine	36.7
7. Collagenolytic Proteinase-Fiddler Crab	36.6
8. Trypsinogen II-Rat	35.8
9. Trypsinogen-Dogfish	35.1

Control of Biological Activity

Proteolytic enzymes by virtue of their substrate specificity
are usually subject to autoinactivation. In the case of the
neutrophil enzymes, however, there appears to be little of this
occurring primarily because of the adequate quantities of
protein substrates available for hydrolysis. A further method
of controlling proteinase activity is through the synthesis
of the enzyme in an inactive, zymogen form (17). While this
occurs readily in the pancreatic system there is little to
suggest that the neutrophil enzymes are subjected to control
in a similar manner. There is some evidence from cDNA analysis
of cathepsin G clones for the possible presence of a dipeptide
activation fragment in this enzyme (11); however, it is just as
likely that this dipeptide acts to process the enzyme for
localization in the azurophil granule and that the enzyme is
probably functional in its packaged form.

How, then, is control of neutrophil elastase and cathepsin
G mediated? From all studies it would appear that this occurs
through regulation by plasma and tissue proteinase inhibitors
(5). Indeed, several studies have shown that both elastase
and cathepsin G can form stable complexes with plasma inhibitors
at extremely rapid rates (6). When a comparison is made of

inhibitor specificity based on rates of reactions with a variety
of serine proteinases, target enzymes for each inhibitor can
be readily determined. Thus, in the case of the Serpins in
plasma one can easily predict the controlling inhibitors for
a variety of proteinases. This is summarized in Table 6.

Table 6. Target Enzymes for Plasma Serine Proteinase Inhibitors

Enzyme	Inhibitor
Neutrophil Elastase	α-1-PI
Cathepsin G	α-1-Achy
Trypsin	α-2-Antiplasmin
Plasmin	α-2-Antiplasmin
Chymotrypsin	α-2-Macroglobulin
Thrombin	Antithrombin III
Kallikrein	C1-Inhibitor

It is clear that the specificity of each enzyme readily
matches the structure of the reactive site of each inhibitor.
For example, elastase can cleave peptide bonds after methionine
residues and this also represents the reactive site of α-1-PI.
Similarly, thrombin prefers arginyl residues and, again,
this represents the reactive site P-1 residue of Antithrombin
III. Clearly, however, other residues in these inhibitors
must play important roles in determining inhibitor specificity
since Antithrombin III, C1-Inhibitor, and α-2-antiplasmin all
have arginyl residues at their reactive site P-1 positions.

In α-1-PI deficient individuals it is well known that
control of neutrophil elastase activity is severely impaired.
As a result such subjects are more prone to the development
of pulmonary emphysema (18). However, it is also known
that depletion of active inhibitor can also be obtained by
inhibitor inactivation through either oxidation (19) or
proteolytic inactivation (20). Thus, even in individuals
with normal circulating plasma levels of inhibitor it is
possible for connective tissue damage to occur.

The story with cathepsin G, however, is not nearly as well
developed. In the first place little is known about the true
function of this enzyme which, based on experiments with
synthetic substrates (21), is a relatively poor proteinase.
Although it can act as an angiotensin II converting enzyme
(22) and a bradykinin inactivating enzyme (23) it is likely
that these are not its normal substrates. Indeed, it has
been shown that this enzyme can act as a powerful bacteriocidal
agent, either in its active or DFP-inactivated form. Whether
this is a normal role remains to be elucidated. Nevertheless,
its control by α-1-Achy is remarkably tight and, considering
the fact that this proteinase inhibitor is a major acute
phase protein, one can only conclude that the rapid rise
in inhibitor concentration under inflammatory conditions is
to regulate cathepsin G levels in tissues invaded by neutrophils.
Alternatively, of course, the inhibitor could also be controlling
other chymotrypsin-like enzymes such as mast cell or cytotoxic
lymphocyte related proteinases. Clearly, this is an important
area for further research.

As for other inhibitors such as BMPI which are synthesized locally, it is not clear what their role might be in terms of controlling elastase and cathepsin G. It is easy to assume that BMPI, for example, limits proteolysis in the upper airways where this inhibitor is synthesized. However, neutrophils are seldom found in this part of the lung. Therefore, one must question a functional role against neutrophil enzymes. Instead, one could consider other sources of proteinases such as the mast cell which contains significant quantities of tryptase and chymase. These cells are prevalent in the upper airways and BMPI may actually be acting as an inhibitor of enzymes released from them during mast cell degranulation. Similar reasoning must also be taken into account for this inhibitor in seminal plasma, tears, and other tissues.

SUMMARY

When neutrophils invade inflamed areas of the body to remove either dead or foreign components they inadvertently release potent enzymes which can, if not properly controlled, cause severe damage to healthy tissue. This can lead to a myriad of diseases including emphysema, rheumatoid arthritis, and glomuerlopnephritis, all of which are really problems of abnormal connective tissue turnover due to uncontrolled protelysis by neutrophil elastase and cathepsin G.

An important step in elucidating the functions of both elastase and cathepsin G has been made by virtue of the fact that the amino acid sequence of each has been determined. Furthermore, the crystal structure of one, neutrophil elastase, is now understood. With this knowledge in mind and with the potential for a similar understanding of the mechanism of action of cathepsin G, it should soon be possible to produce synthetic inhibitors of each enzyme which can act as adjunct inhibitors to those naturally circulating in the blood or present in other tissues. As a result there is great hope for reducing the severity of injury produced by these enzymes and, therefore, in decreasing the risk for development of the debilitating diseases associated with abnormal proteolysis by neutrophil proteinases.

ACKNOWLEDGMENTS

This research was supported by grants from the National Lung Institute, National Institutes of Health, Bethesda, Maryland.

REFERENCES

1. U. Bretz and M. Baggiolini. Biochemical and Morphological Characterization of Azurophil and Specific Granules of Human Polymorphonuclear Leukocytes. J. Cell Biol. 63:251 (1974).
2. C. Kuhn and R.M. Senior. The Role of Elastase in the Development of Emphysema. Lung 155:185 (1975).

3. A. Janoff. Biochemical Links Between Cigarette Smoke and Pulmonary Emphysema. J. Appl. Physiol. 55:285 (1983).

4. A. Janoff and J. Blondin. Depletion of Cartilage Matrix by a Neutral Proteinase Fraction of Human Leucocyte Lysosomes. Proc. Soc. Exp. Biol. Med. 135:302 (1970).

5. J. Travis and G. Salvesen. Human Plasma Proteinase Inhibitors. Ann. Rev. Biochem. 52:655 (1983).

6. K. Beatty, J. Bieth, and J. Travis. Kinetics of Association of Serine Proteinases with Native and Oxidized Alpha-1-Proteinase Inhibitor and Alpha-1-Antichymotrypsin. J. Biol. Chem. 255:3931 (1980).

7. C.E. Smith and D.A. Johnson. Human Bronchial Leukocyte Proteinase Inhibitor. Rapid Isolation and Kinetic Analysis with Human Leukocyte Proteinases. Biochem. J. 225:463 (1985).

8. U. Seemuller, M. Arnhold, H. Fritz, K. Wiedenmann, W. Machleidt, R. Heinzel, H. Appelhaus, H. Gunter Gassen, and F. Lottspeich. The Acid-Stable Proteinase Inhibitor of Human Mucous Secretions. FEBS Letters 199:43 (1986).

9. J.C. Powers and D.L. Carroll. Reactions of Azyl Carbazates with Proteolytic Enzymes. Biochem. Biophys. Res. Commun. 67:639 (1975).

10. S. Sinha, W. Watorek, S. Karr, J. Giles, W. Bode, and J. Travis. Primary Structure of Human Neutrophil Elastase. Proc. Natl. Acad. Sci . USA 84:2228 (1987).

11. G. Salvesen, D. Farley, J. Shuman, A. Przybyla, C. Reilly, and J. Travis. Molecular Cloning of Human Cathepsin G. Structural Similarity to Mast Cell and Cytotoxic Lymphocyte Proteinases. Biochemistry 26:2289 (1987).

12. R. Senior, E. Campbell, J. Landis, R. Cox, C. Kuh, and H. Koren. Elastase of U-937 Monocyte Like Cells. Comparison with Elastases Derived from Human Monocytes and Neutrophils and Murine Macrophage Like Cells. J. Clin. Invest. 69: 384 (1982).

13. R. Senior and E. Campbell. Cathepsin G in Human Mononuclear Phagocytes. Comparison Between Monocytes and U-937 Monocyte-Like Cells. J. Immunol. 132:2547 (1984).

14. R. Baugh and J. Travis. Human Leukocyte Granule Elastase: Rapid Isolation and Characterization. Biochemistry 15: 836 (1976).

15. W. Bode, A-Z. Wei, R. Huber, E. Meyer, J. Travis, and S. Neumann. X-Ray Crystal Structure of the Complex of Human Leukocyte Elastase (PMN-Elastase) and the Third Domain of the Turkey Ovomucoid Inhibitor. EMBO J. 5:2453 (1986).

16. A. Gerber, J. Carson, and B. Hadorn. Partial Purification and Characterization of a Chymotrypsin-Like Enzyme from Human Neutrophil Leukocytes. Biochim. Biophys. Acta 364:103 (1974).

17. H. Neurath. Mechanism of Zymogen Activation. Fed. Proc. 23:1 (1964).

18. S. Eriksson. Pulmonary Emphysema and Alpha-1-Antitrypsin Deficiency. Acta Med. Scand. 175:197 (1964).

19. D. Johnson and J. Travis. The Oxidative Inactivation of Human Alpha-1-Proteinase Inhibitor. Further Evidence for Methionine at the Reactive Center. J. Biol. Chem. 254: 4022 (1979).

20. D. Johnson and J. Travis. Inactivation of Human Alpha-1-Proteinase Inhibitor by Thiol Proteinases. Biochemical J. 163:639 (1977).

21. T. Tanaka, Y. Minematsu, C. Reilly, J. Travis, and J. Powers. Human Leukocyte Cathepsin G: Subsite Mapping

with 4-Nitroanilides. Chemical Modification and Effect of Possible Cofactors. <u>Biochemistry</u> 24:2040-2047.

22. C.F. Reilly, D. Tewksbury, N. Schechter, and J. Travis. Rapid Conversion of Angiotensin I to Angiotensin II by Neutrophil and Mast Cell Proteinases. <u>J. Biol. Chem.</u> 257:8619 (1982).

23. C.F. Reilly, N. Schechter, and J. Travis. Inactivation of Bradykinin and Kallidin by Cathepsin G and Mast Cell Chymase. <u>Biochem. Biophys. Res. Commun.</u> 127:443 (1985).

THE DEGRADATION OF COLLAGEN BY A

METALLOPROTEINASE FROM HUMAN LEUCOCYTES

U. Kohnert[1], R. Oberhoff[1], J. Fedrowitz[1], U. Bergmann[1],
J. Rauterberg[2] and H. Tschesche[1]

[1]Universität Bielefeld [2]Institut f. Arteriosklerose-
Fakultät für Chemie   forschung
Postfach 86 40   Domagkstr. 3
D-4800 Bielefeld 1, F.R.G. 4400 Münster, F.R.G.

INTRODUCTION

Neutrophil granulocytes are a main component of the non-specific
defence system. In order to destroy invading microorganisms leukocytes
must be able to leave the blood vessels and to have access to the microorganisms in the surrounding connective tissue. The leukocytes must thus
have the enzymatic capacity to penetrate the basement membrane which lines
the vascular lumen and to degrade interstitial collagen. The same enzymatic capacity to degrade extracellular matrices is obviously a prerequisite
for the metastasis of malignant cells.
Human polymorphonuclear (PMN) leukocytes have been shown to have the
necessary proteinases for intra and extravasation (1) which implies the
degradation of basement membrane proteins, such as collagen type IV.
Degradation of interstitial collagen (collagen types I,II,III) is initiated by collagenase which splits the triple helical molecule at a specific site yielding a 1/4 and a 3/4 fragment (1). Although these fragments
which lose their helical conformation at body temperature are susceptible
to proteolysis by several proteases, further degradation in vivo is thought
to be mainly dependent on the action of gelatinase, which is highly active
against denatured collagen (2-4).
It has further been observed that type IV collagen is subject to degradation by human neutrophil elastase (5).
Apart from the synergic action of elastase and collagenase with gelatinase in
interstitial collagen breakdown, some authors have found that gelatinase degrades the basement membrane collagen type IV and also type V collagen (4,6).

In this paper we report on the purification and further characterization of a metalloproteinase from human PMN leukocytes which is probably identical with the gelatinase previously reported on. We focused our attention on the substrate specificity in order to find out whether this enzyme might be involved in basement membrane degradation by leukocytes.

MATERIALS AND METHODS

All chemical reagents were of analytical grade purity. Human leukocytes were obtained from the buffy coat of healthy donors (DRK, Hagen, FRG). Sephacryl S-200, Sephacryl S-300, Sepharose Cl-4B and Q-Sepharose ff were purchased from Pharmacia (Freiburg, FRG). DEAE A-52-Servacel and fluorescamin were obtained from Serva (Heidelberg, FRG). Type IV procollagen dimer was the gift of Dr. K. Kühn (Martinsried, FRG). DNP-peptide was purchased from Bachem (Bubendorf, Switzerland).

Preparation of Enzyme Substrates

Gelatin was prepared from soluble type I collagen by thermal denaturation (15 min at 100°C). Collagen types I, II, III, V and VI were prepared by methods described elswhere (7-10).
Type IV collagen from bovine anterior lens capsules was prepared by a modification of the method described in (11).

Assay of Enzymatic Activity

Proteolytic activity was measured quantitatively by a modification of the fluorescamine method previously described in (12,13).
The degradation of DNP-Peptide was analysed as described in (14).

In addition, proteolytic activity was determined by incubating the enzyme and substrate in 0.05 M Tris/HCl pH 7.0, 0.2 M NaCl, 5 mM $CaCl_2$. The degradation products were analysed by SDS-PAGE as described below.

Isolation and Purification of Gelatinase from Human PMN Leukozytes

Leukocytes were prepared in the presence of PMSF, benzamidine and catalase from buffy coat as described in (15). The final pellet (50 g) was suspended in 200 ml of buffer (0.02 M citric acid/NaOH, pH 6.0, 5 mM $CaCl_2$, 0.5 M $ZnCl_2$). After homogenisation (Potter-Elvehjem-homogenisator) for 8 min the suspension was treated in the Branson sonifier B15 for 15 min with a pulsed programme (i.e. 7.5 min sonification spread over 15 min) and centrifuged (48,000 g, 1h, 4°C).

All the following steps were carried out at 4°C. The suspension was stirred for 30 min and after centrifugation (48,000 g, 1h, 4°C) the supernatant protein was precipitated with ammonium sulphate to a saturation of 60%. The precipitate was collected by centrifugation (20,000 g, 1h, 4°C) and dissolved in 15 ml of 0.05 M Tris/HCl pH 7.0, 0.2 M NaCl, 5 mM $CaCl_2$ (buffer A). This was then extensively dialysed using the same buffer, and subjected to gelfiltration on a Sephaycryl S-200 column ((80 x 3.5 cm), using buffer A for equilibration and elution. Fractions with gelatinolytic activity were pooled and dialysed extensively against 0.05 M Tris/HCl, pH 8.0, 5 mM $CaCl_2$. After dialysis the enzyme was applied to a DEAE A-52 Servacel-column (16 x 2.8 cm) which was previously equilibrated with the dialysis buffer. Approximately 1.5 column volumes were eluted with the equilibration buffer, after which the enzyme was eluted from the column using a 0 - 0.35 M NaCl gradient. Fractions containing gelatinolytic activity were pooled, dialysed using 0.05 M Tris/HCl, pH 8.0, 5 mM $CaCl_2$ and applied to a Q-Sepharose-column (23 x 1.4 cm). Bound material was eluted using a 0 - 0.6 M NaCl gradient. The enzyme was concentrated by lyophilisation, further purified by gelfiltration on Sephacryl S-300 (70 x 2.8 cm) with the buffer A and finally purified by dye-ligand-affinity chromatography on Blue Sepharose CL-4B (24 x 1.8 cm), which was prepared according to (16) using Sepharose CL-4B instead of Sephadex G-200. The column was equilibrated with buffer A and eluted with 0.05 M Tris/HCl, pH 8.0, 1 M NaCl, 5 mM $CaCl_2$.

Molecular Weight Determination

The molecular weight of the enzyme was determined by SDS-PAGE and by gelfiltration on Sephacryl S-300. SDS-PAGE was carried out on 7.5% gel as described in (17) employing a calibration kit containing ferritin subunit, phosporylase b, bovine serum albumin, ovalbumin and carboanhydrase. The Sephacryl S-300-column was calibrated using ferritin, catalase, bovine serum albumin, ovalbumin, and myoglobin.

Gels were stained either with Amidoblack 10B, Coomassie Blue or with the silver stain method described in (18).

Determination of the pH-Optimum of the Active Enzyme

Enzyme and substrate were mixed with buffers adjusted to various pH values: for pH 5-8 Tris/maleic acid and for pH 7-10 Tris/HCl were used. The final

concentration of the buffer was 0.1 M, each containing 0.2 M NaCl and 5 mM
CaCl$_2$.

Analysis of Type V Collagen Degradation by Electron Microscopy

Analysis by electron microscopy was carried out as described in (19). Briefly the split collagen was precipitated with ATP to segment long spacing
crystallites and subjected to electron microscopic examination after staining with phosphotungstic acid and uranyl acetate.

RESULTS

Purification of the Enzyme

Purification was commenced with gelfiltration on Sephacryl S-200, followed
by ion exchange chromatography on DEAE A-52 Servacel and Q-Sepharose
(Fig. 1), and a second gelfiltration on Sephacryl S-300 (Fig. 2).
Final purification was achieved by affinity chromatography on Blue Sepharose
CL-4B (Fig. 3) which allows a rapid and nearly quantitative isolation of the
type IV/V collagen-degrading metalloproteinase. Elution was carried out with
a stepwise change to a higher concentration of NaCl and pH-value (see materials and methods). Using a gradient did not improve enzyme purity.

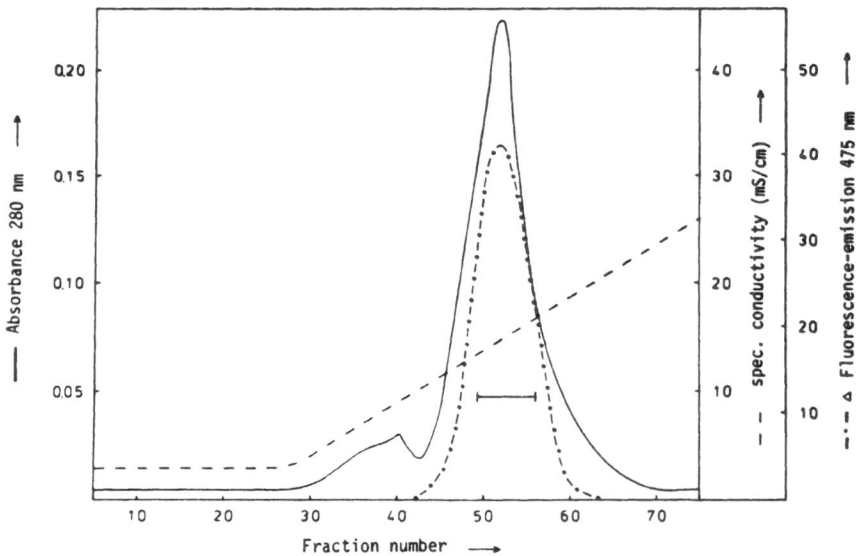

Fig. 1. Anion exchange chromatography of the partially purified enzyme
on Q-Sepharose ff, (see materials and methods for details).

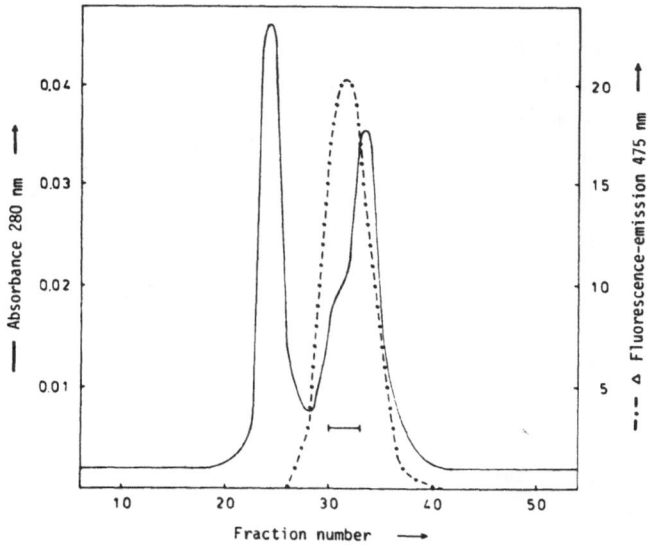

Fig. 2. Gelfiltration of the partially purified enzyme
on Sephacryl S-300, (see materials and methods
for details).

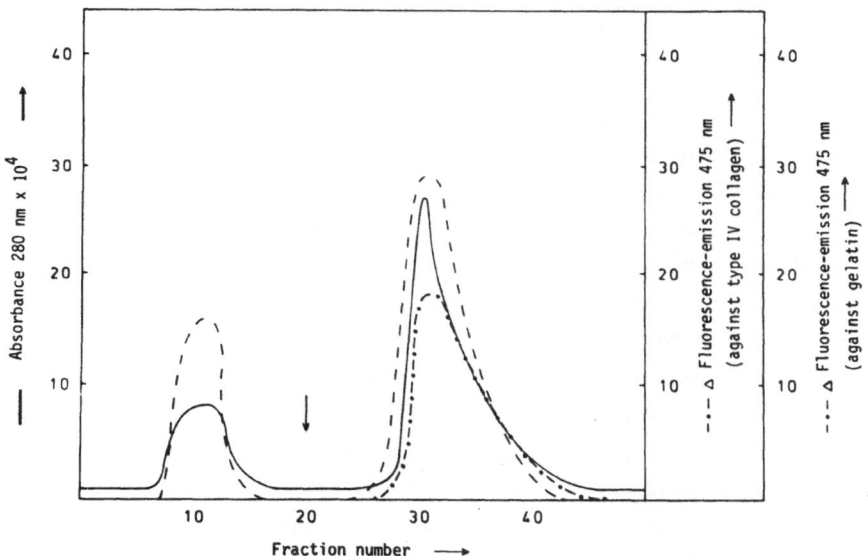

Fig. 3. Final purification of the enzyme on Blue-Sepharose CL 4B,
(see materials and methods for details).

Characterization of the Enzyme

On passing the purified enzyme through a molecular sieve column (Sephacryl S-300) this could be eluted as a single peak corresponding to a molecular weight of 240.000 Da (Fig. 4).
By contrast on SDS-PAGE a molecular weight of 98.000 Da was determined (Fig 5).

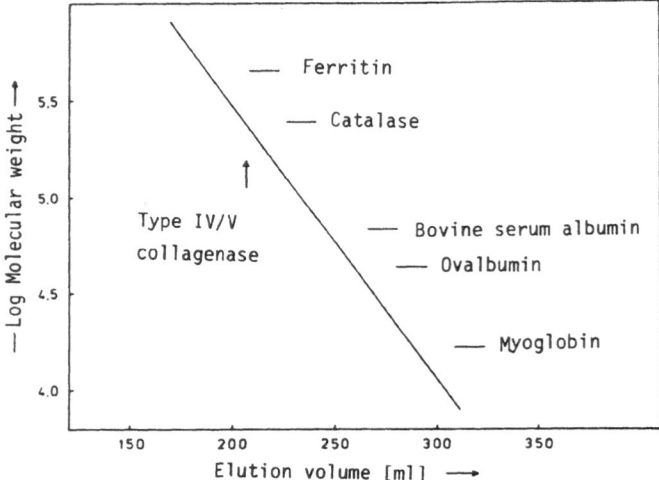

Fig. 4. Molecular weight determination of the latent enzyme by gelfiltration, (see materials and methods for details).

The purified enzyme has a pH-optimum of 7-8 (Fig. 6). It is completely inhibited by EDTA, α_2M, and tetracyclin. The inhibition by EDTA was non – reversible (not shown). The enzyme was not, or only poorly, inhibited by PMSF, NEM, aprotinin, and penicillamin.
The enzyme can be activated by mersalyl, but proteolytic activation is needed to achieve maximum activity (Tab. 1); if PMSF, benzamidine and catalase are present during the first steps of preparation, at least 90% of the enzyme was isolated in latent form.
On activation with trypsin the molecular weight of the enzyme decreases by about 17000 Da, while mersalylic acid leaves the molecular weight unchanged. Part of the enzyme is autoactivated during the purification.

When tested against different substrates, the purified enzyme showed predominant activity against gelatin (Fig. 7).
Type IV collagen from bovine anterior lens capsules was cleaved specific-

Table 1. Activation of the enzyme

Activator	Concentration	Relative Activation
Trypsin	24 μg/ml	100 %
Kallikrein	24 μg/ml	145 %
Plasmin	2 mU/ml	53 %
Cathepsin G (partially purified)	250 mU/ml	140 %
Mersalyl	38 μg/ml	66 %
Oxidized glutathion	20 μg/ml	0 %
Trypsinogen	24 μg/ml	5 %

ally after incubation with the enzyme at temperatures above 27°C (Fig. 8) whereas at temperatures above 33 $^\circ$C the substrate was completely degraded (not shown).

Preliminary experiments revealed that a procollagen type IV dimer was also completely degraded above 33°C.

Type V collagen was susceptible to proteolysis at 26°C as shown in Fig. 8.

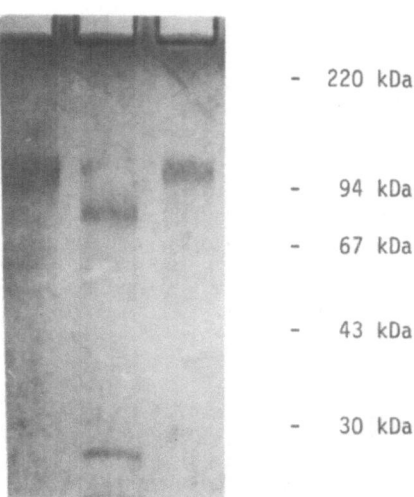

Fig. 5. SDS-Page of the purified latent enzyme;
    (I)   latent enzyme without additions
    (II) after preincubation with trypsin
    (III) after preincubation with mersalyl
    (see materials and methods for details)

The enzyme has no activity against collagens types I, II, III and VI, whereas the synthetic substrate DNP-Peptide could be cleaved (not shown, see (20)).

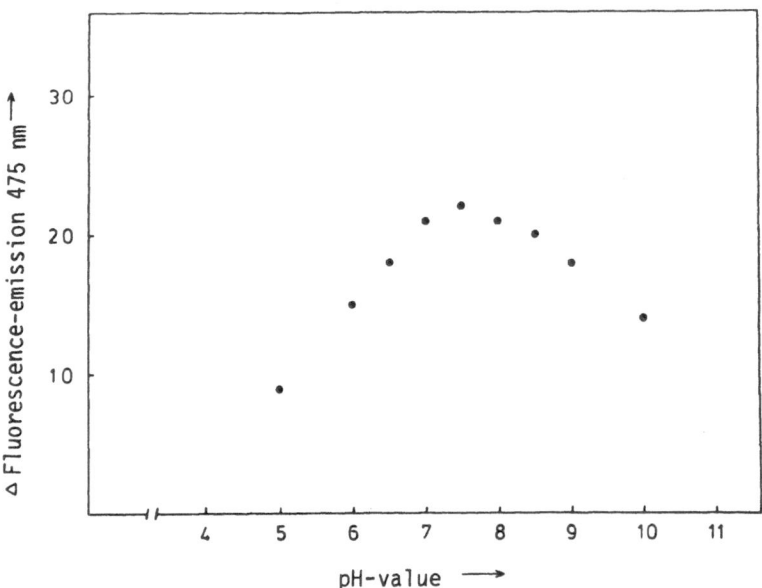

Fig. 6. pH-optimum of the enzyme
(see materials and methods for details)

Fig. 7. Degradation of gelatin (I), by the activated enzyme (VI), after 15 min (II), 30 min (III), 60 min (IV), and 120 min (V) incubation time

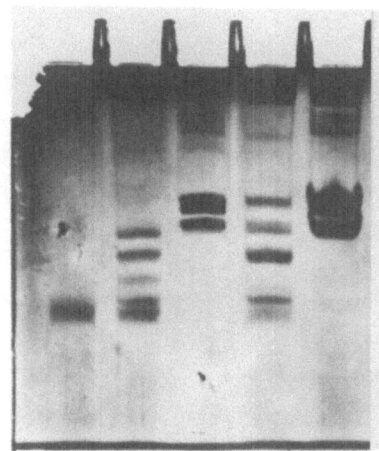

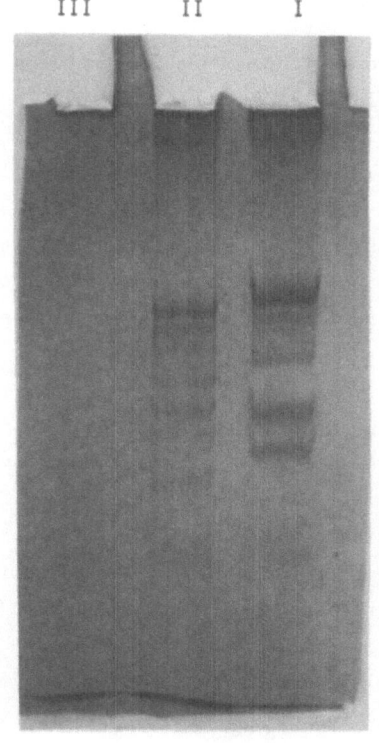

8b

without enzyme

HOOC ─────── NH$_2$

HOOC ─────── NH$_2$

8c   with enzyme

Fig. 8. Collagen degradation by the activated enzyme.

a) Degradation of type IV collagen from bovine anterior lens
   capsules. Substrate without additions (I), substrate after
   incubation with enzyme (II), and pure enzyme (III).
   Degradation of type V collagen as examined by

b) SDS-PAGE; enzyme without additions (I), enzyme with type V
   collagen (three chain form) (II), enzyme with type V collagen
   (two chain form) (IV), and pure collagen types V - three
   chain form (III) and two chain form (V).

c) electron microscopy,
   (See materials and methods for details)

DISCUSSION

In this report we describe the isolation of a metalloproteinase from human PMN leucocytes which is capable of degrading gelatin and collagen types IV and V.
This enzyme shows in respect of molecular weight, pH-optimum, and the inhibition and activation spectrum similarities to human PMN gelatinase, as reported on earlier (2-3,6,21-23).
The substrate specificity, however, shows some discrepancies.
The gelatinase found by Rantala-Ryhänen et al. (23) has no activity against native collagen types I-V, whereas Murphy et al. (6) reported activity against types IV and V. The gelatinase isolated by Hibbs et al. (4) splits type V collagen, while no information about the activity against type IV collagen was given.

When we used type IV collagen from bovine anterior lens capsules as a substrate, we found specific proteolytic activity of the purified enzyme if the incubation was carried out at temperatures between 27 and 33°C.
At higher tempratures complete degradation was observed with no distinct fragments of higher molecular weight being formed. This seems to indicate that the specific fragments formed in the first instance are denatured at higher temperatures, and that the gelatin peptides released from the tripel helical fragments are easily degraded by the gelatinase activity of the enzyme. This could imply that the proteolytic action on soluble type IV collagen, as reported earlier, is highly dependent on the incubation temperature. The temperature dependence of this cleavage is under investigation.

It could well be possible that intra and extravasation also depends on the primary action of human leukocyte elastase, followed by the action of gelatinase. While the action of leukocyte elastase is potentially counteracted by a high concentration of α_1-proteinase inhibitor in plasma, there is no comparable high inhibitory potential against leukocyte metalloproteinases present in plasma, since the concentration of β_1-anticollagenase, i.e. tissue inhibitor of metalloproteinases (TIMP), is rather low or negligible (24).
This would explain the excistence of a special collagen degrading enzyme for the initial step of matrix degradation.
Further investigations are needed to reveal the existence and possible importance of a type IV collagenase action in the process of basement membrane penetration.

Acknowledgements

Especially we would like to emphasize the engaged and highly
competent cooperation of Mr. D. Lange in the various stages
of this work.
This work was supported by the Deutsche Forschungsgemeinschaft, Bonn
Bad Godesberg, Sonderforschungsbereich 223 and the Fonds der
chemischen Industrie, Frankfurt M.
We wish to thank D. Troyer for electron microscopy.
The authors are indebted to Miss U. Saft for her continuous
preparation of bovine anterior lens capsule type IV collagen and to
Misses E. Krimpenfort and E. Pieper for their skilfull preparations
of human cell material.

REFERENCES

1. S. J. Weiss and G. J. Peppin, Collagenolytic metalloenzymes of the human
   neutrophil, Biochem. Pharmacol, 35:3189 (1986).
2. J. Fedrowitz, H. W. Macartney, and H. Tschesche, Human polymorphonuclear leukocyte gelatinase - a metalloproteinase, Hoppe-
   Seylers Z. Physiol. Chem. 364:1121 (1983).
3. I. Sopata and A. M. Dancewicz, Presence of a gelatin-specific
   proteinase and its latent form in human leucocytes, Biochim.
   Biophys. Acta 370:510 (1974).
4. M. Hibbs, K. A. Hasty, J. M. Seyer, A. H. Kang, and C. L. Mainardi,
   Biochemical and immunological characterisation of the secreted forms
   of human neutrophil gelatinase, J. Biol. Chem. 260:2493 (1985).
5. D. J. Pipoly and E. C. Crouch, Degradation of native typeIV
   procollagen by human neutrophil elastase. Implications for leukocyte
   mediated degradation of basement membranes, Biochemistry, 26:5748
   (1987).
6. G. Murphy, J. J. Reynolds, U. Bretz, and M. Baggiolini, Partial
   purification of collagenase and gelatinase from human polymorphonuclear leucocytes, Biochem. J. 203:209 (1982).
7. J. Rauterberg and K. Kühn, Acid soluble calf skin collagen, Eur. J.
   Biochem. 19:398 (1971).
8. J. Rauterberg, H. Allmann, W. Henkel, and P. P. Fietzek, Isolation and
   charakterization of CNBr derived peptides of the 1(III) chain of
   pepsin solubilized calf skin collagen, Hoppe Seyler's Z. Physiol.
   Chem. 357:1401 (1976).
9. E. J. Miller and R. K. Rhodes, Preparation and charakterization of the
   different types of collagen, Methods Enzymol. 82:33 (1982).
10. R. Jander, J. Rauterberg, and W. Glanville, Further charakterization
    of the three polypeptide chains of bovine and human short-chain
    collagen (intima collagen), Eur. J. Biochem. 133:39 (1983).
11. C. L. Mainardi, S. N. Dixit, and A. H. Kang, Degradation of type IV
    (basement membrane) collagen by a proteinase isolated from human
    polymorphonuclear leukocyte granules, J. Biol. Chem. 255:5435
    (1980).
12. H. Tschesche, H. W. Macartney, and J. Fedrowitz, Gelatinase, in:
    "Methods of enzymatic analysis, Vol. V", H. U. Bergmeyer, ed., Verlag
    Chemie, Weinheim (1984).
13. H. Tschesche and H. W. Macartney, Collagenase, in: "Methods of enzy-
    matic analysis Vol. V", H. U. Bergmeyer, ed., Verlag Chemie, Weinheim
    (1984).
14. Y. Masui, T. Takemoto, S. Sakakibara, H. Hori, and Y. Nagai, Synthetic
    substrates for vertebrate collagenase, Biochem. Med. 17:215 (1977).

15. S. Engelbrecht, E. Pieper, H. W. Macartney, W. Rautenberg, H. R. Wenzel, and H. Tschesche, Separation of the human leucocyte enzymes alanine aminopeptidase, cathepsin G, collagenase, elastase and myeloperoxidase, Hoppe-Seylers Z. Physiol. Chem. 363:305 (1982).
16. H.-J. Böhme, G. Kopperschläger, J. Schulz, and E. Hofmann, Affinity chromatography of phosphofructokinase using cibacron blue F3G-A, J. Chromatogr. 69:209 (1972).
17. F. W. Studier, Analysis of bacteriophage T7 early RNAs and proteins on slab gels, J. Mol. Biol. 79:237 (1973).
18. J. Heukeshoven and R. Dernick, Simplified method for silver staining of proteins in polyacrylamide gels and the mechanism of silver staining, Elektrophoresis 6:103 (1985).
19. J. Rauterberg and K. Kühn, Electron microscopic method of ordering the cyanogen bromide peptides of the 1 chain from acid soluble calfskin collagen, FEBS Letters 1:230 (1968).
20. J. Fedrowitz, Isolierung und Charakterisierung von Gelatinase aus menschlichen Leukocyten, PhD Thesis, Universität Bielefeld (1986).
21. G. Murphy, U. Bretz, M. Baggiolini, and J. J. Reynolds, The latent collagenase and gelatinase of human polymorphonuclear neutrophil leucocytes, Biochem. J. 192:517 (1980).
22. I. Sopata, Further purification and some properties of a gelatin-specific proteinase of human leucocytes, Biochim. Biophys. Acta 717:26 (1982).
23. S. Rantala-Ryhänen, L. Ryhänen, F. V. Nowak, and J. Uitto, Proteinases in human polymorphonuclear leucocytes, Eur. J. Biochem. 134:129 (1983).
24. H. G. Welgus and G. P. Stricklin, Human skin fibroblast collagenase inhibitor, J. Biol. Chem. 258:12259 (1983).

PLASMA MEMBRANE PROTEASES AS USEFUL TOOL IN HISTOCHEMICAL TOXICOLOGY[+]

Renate Graf and Reinhart Gossrau

Department of Anatomy, Free University

1000 Berlin (West) 33, FRG

INTRODUCTION

Previous studies in histochemical toxicology have shown that plasma (surface) membrane ectoenzymes seem to be more easily affected by exogenous noxae such as drugs and/or their metabolites than other enzymes, e.g. those in lysosomes, Golgi apparatus or the endoplasmic reticulum of rat and mouse organs (for literature see Gossrau et al., 1987 a). This might be due also to the localization of the surface membrane enzymes at the border between the extra- and intracellular space which xenobiotics have to cross on their way into or through the respective cells.

In the present paper it will be shown that the demonstration of certain plasma membrane ectoproteases (Kenny, 1977; McDonald and Schwabe, 1977) which all belong to the subgroup of ectoexopeptidases (McDonald and Barrett, 1986) may often deliver more insight into the response of cells of female laboratory rodents to salicylic acid and salicylate respectively and glucocorticoids also used in human therapy when compared with surface membrane-associated phosphatases and exoglycosidases and various other enzymes linked to lysosomes, Golgi apparatus, endoplasmic reticulum and mitochondria.

MATERIALS AND METHODS

Pregnant Wistar rats with and without Zn-deficiency were kept under conditions and treated with salicylic acid (SA) or dexamethasone (DEXA) as described by Gossrau et al. (1987 b). In addition, non-pregnant and pregnant rats of gestational day (GD) 16 received a single dosage of 750 mg SA/kg perorally. Control rats were treated with either physiological saline or olive oil. The pregnant rats were sacrificed on GD 19 or 20, the non-pregnant animals 3 or 4 days after SA administration in ether narcosis. Afterwards, the thyroid and parathyroid gland, lung, liver and kidney of the pregnant and non-pregnant rats as well as the placenta and yolk sac were processed for qualitative enzyme histochemistry with the methods of choice as given by Lojda et al. (1979), Gossrau et al. (1987 a) and Gossrau and Lojda (unpublished azo-indoxyl or indigogenic method for lysosomal α-D-galactosidase), for biochemical measurement of various plasma membrane and lysosomal proteases using continuous fluorometry according to Gossrau (1981) and Gossrau et al. (1987 c) and for semithin-section morphology after

---

[+]Supported by the German Research Foundation (Sfb 174)

Karnovsky fixation (Geyer, 1973) and embedding in Technovit[R] according to the prescriptions as given by the manufacturer (Kulzer, Wehrheim, FRG).

RESULTS

In the following only those enzyme histochemical data will be described which were not paralled by morphological lesions, i.e. lesions also found in semithin sections. This type of findings will be reported elsewhere (Graf et al., manuscript in preparation). Zn-deficiency alone does not affect any of the investigated enzymes in any of the studied organs of non-pregnant rats, maternal rats and their fetuses.

The designation of the exopeptidases bases on the nomenclature proposed by McDonald and Barrett (1986).

The parathyroid gland and lung were not affected.

Thyroid gland. After a single dose of SA glutamyl aminopeptidase (EAP; formerly designated aminopeptidase A; Fig. 1,2) activity was increased in the plasma membrane of capillary endothelial cells in numerous areas of the glandular stroma of non-pregnant rats. In contrast, the activity of dipeptidyl peptidase (DPP) IV (Fig. 3,4) linked to these structures was often decreased whereas the surface membrane-associated non-specific alkaline phosphatase (alP; Fig. 5,6) and Mg-dependent adenosine triphosphatase (ATPase) did not respond in the capillary endothelium. In the follicular epithelial cells mitochondrial succinate dehydrogenase (SDH) and NADH tetrazolium reductase (NADHTR) and lysosomal DPP I and II, non-specific acid phosphatase (aP), ß-D-glucuronidase (ßGluc), ß-N-acetyl-D-glucosaminidase (NAG), α-D-galactosidase (αGal), acid α-D-glucosidase (aαGluc), acid ß-D-galactosidase (aßGal), and non-specific estereases (nE) behaved as in controls as was the case for microsomal alanyl aminopeptidase (mAAP) in the stromal connective tissue and for all investigated enzymes in the thyroid gland of pregnant rats.

Liver. In pregnant rats, no alterations were noticed for any of the enzymes described above after treatment with a single dose of SA, and, in addition, no changes for glucose-6-phosphatase (G6Pase) in the endoplasmic reticulum and γ-glutamyl transpeptidase (γ-GTP; formerly designated γ-glutamyl transferase) occasionally linked with low activity to the plasma membrane at the biliary pole of some periportal hepatocytes and with high activity to the luminal surface membrane of the epithelial cells of the terminal ductules of the bile duct system (canals of Hering) and its interlobular ducts.

In contrast, after a single dosage of SA much more periportal (peripheral) hepatocytes with comparatively higher γ-GTP activity (Fig. 7,8) were present in non-pregnant rats. The epithelial cells of the bile ducts did not respond. As in controls no γ-GTP activity was found in the perivenous (central) hepatocytes of the traditional liver lobule. No response was observed for DPP IV (Fig. 9,10), mAAP, alP, 5'-nucleotidase (5'-Nase) and Mg-ATPase in the hepatocyte plasma membrane at its biliary pole everywhere in the liver lobule. This finding was substantiated for DPP IV and mAAP by minimal incubation, i.e. incubation until the first appearance of coloured reaction product under the view of the microscope. The activity of mitochondrial SDH and NADHTR and lysosomal DPP I and II, aP, NAG, aßGal, ßGluc, αGal and nE in any of the liver cell types was unaltered, too. The same was true for DPP IV in the luminal plasma membrane of the epithelial cells of the canals of Hering and interlobular bile ducts. - Sometimes, G6Pase activity seemed to be somewhat lowered when compared with untreated animals. SA administration combined with Zn-deficiency led to less dramatic increases of γ-GTP activity in non-pregnant rats when compared with the situation after SA administration alone.

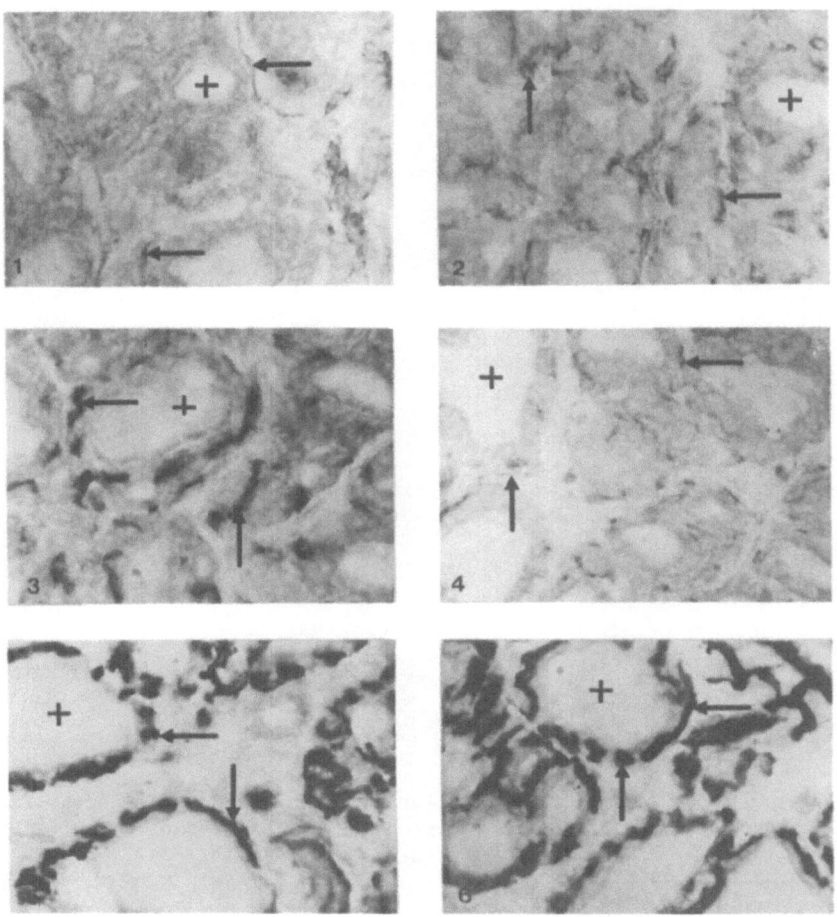

Fig. 1-6. Thyroid gland, arrows capillaries, + follicular lumen. Fig. 1,2.
EAP reaction. Fig. 1. Control. Fig. 2. SA treatment. Fig. 3,4. DPP IV
reaction. Fig. 3. Control. Fig. 4. SA administration. Fig. 5,6. alP reaction.
Fig. 5. Control. Fig. 6. SA treatment. Magnification x450

Kidney. SA treatment alone caused no clear-cut effects detectable by enzyme histochemistry or semithin section morphology in pregnant rats. However, combined Zn-deficiency and SA administration decreased glucoamylase Fig. 11,12) activity in the brush border of the proximal tubules. Microvillous mAAP, ɤ-GTP, EAP (Fig. 13,14), DPP IV, alP, 5'-Nase and Mg-dependent

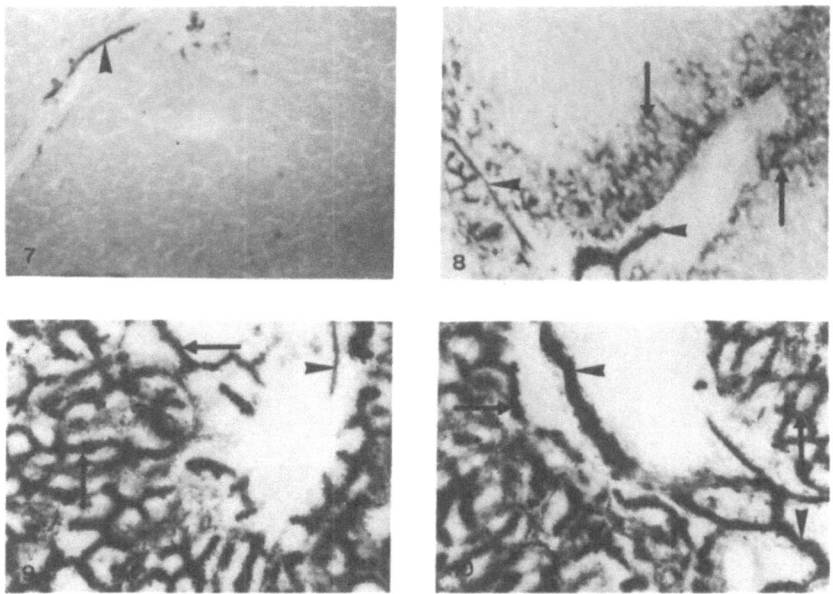

Fig. 7-10. Liver, arrows bile capillaries, arrow heads bile ducts. Fig. 7,8. ɤ-GTP reaction. Fig. 7. Control. Fig. 8. SA administration. Magnification x200. Fig. 9,10. DPP IV reaction. Fig. 9. Control. Fig. 10. SA administration. Magnification x450

ATPase appeared not to respond here. Clear-cut responses were also not found for alP, DPP IV and EAP in the plasma membrane of glomerular cells, for DPP IV and ɤ-GTP in the luminal surface membrane of distal tubules, for ɤ-GTP in the basal labyrinth of proximal tubular cells, for alP in the capillary endothelium of the medulla and for SDH, NADHTR and G6Pase in the proximal and/or distal tubules.

Placenta. Administration of DEXA considerably increased γ-GTP staining (Fig. 15,16) in the plasma membrane of spongiotrophoblast cells or/and outside them and DPP IV activity (Fig. 17,18) in the surface membrane of fetal (labyrinthic) capillary endothelial cells in the labyrinth whose peripheral regions (near to the spongiotrophoblast) appeared to be more affected than the central areas (near to the chorionic plate) and decreased DPP IV activity of decidual cells. The activities of EAP and γ-GTP were also enhanced in the fetal capillary endothelium; however, compared with DPP IV the increase in activity was lower. Other enzymes could not be detected either in the normal or in the capillary endothelium after DEXA

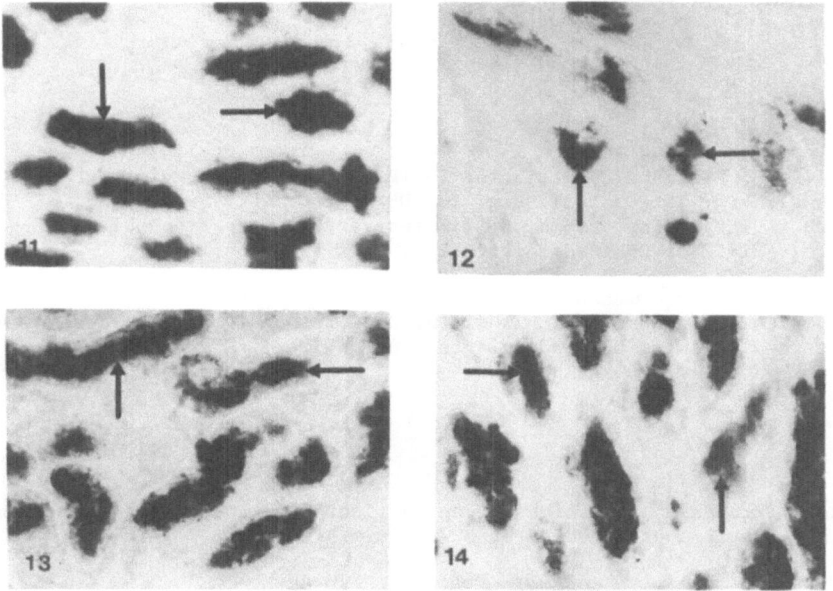

Fig. 11-14. Kidney, arrows brush border. Fig. 11,12. Glucoamylase reaction. Fig. 11. Control. Fig. 12. Combined SA treatment and Zn-deficiency. Fig. 13,14. EAP reaction. Fig. 13. Control. Fig. 14 see Fig. 12. Magnification x450

treatment. Furthermore, these proteases, plasma membrane-associated mAAP, aIP, Mg-dependent ATPase and 5'-Nase and the mitochondrial dehydrogenases and reductases (respectively) and lysosomal hydrolases did not respond in any of the cells of the placenta, i.e. those in the basal decidua, spongiotrophoblast, labyrinthic cytotrophoblast and syncytiotrophoblast, chorionic plate and intraplacental portion of the yolk sac (sinus of Duval; Graf and Gossrau, 1986 a).

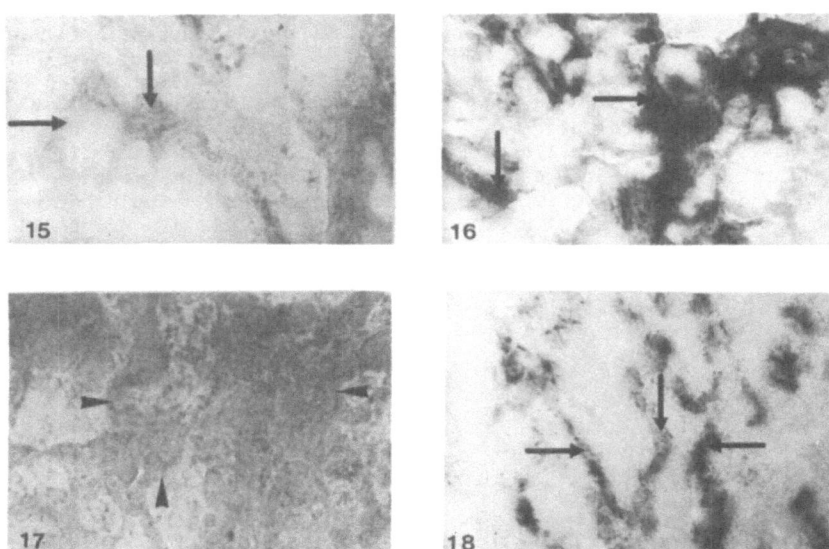

Fig. 15-18. Placenta. Fig. 15,16. ℾ-GTP reaction in spongiotrophoblast cells
(arrows). Fig. 15. Control. Fig. 16. DEXA treatment. Fig. 17,18. DPP IV reaction in labyrinth, arrows capillaries, arrow heads trophoblast. Fig. 17.
Control. Fig. 18. DEXA administration. Magnification x450

Fluorometric measurements of mAAP, EAP and DPP I, II and IV revealed
activities without significant differences between control placentas and
placentas following DEXA administration (Table 1).

Table 1. Protease activities (given in fluorescence units) in the normal
rat placenta and after DEXA administration

	control	DEXA
EAP	1.5	1.4
mAAP	11.2	12.5
DPP IV	2.1	2.2
DPP I	1.4	1.3
DPP II	11.0	12.0

Yolk sac. DEXA treatment enhanced the activity of mAAP, DPP IV and
ℾ-GTP (fig. 19,20) in the brush border of single or groups of visceral
epithelial cells in the abembryonic yolk sac region whereas others did not
respond or their activity seemed to be even lowered. This was also true for
alP, 5'-Nase and Mg-ATPase in the microvillous zone of the visceral epithelial cells. Mitochondrial SDH and NADHTR and lysosomal (and secretory granule-
associated) DPP I and II, aP, ßGluc, a∝Gluc, ∝Gal, aßGal, NAG and nE were
not affected either.

Fluorometric measurements of mAAP, ℾ-GTP and DPP I, II and IV showed
similar activities in DEXA-treated and non-treated rats (Table 2).

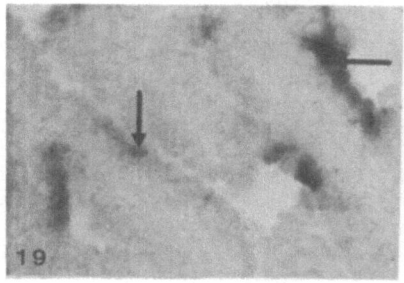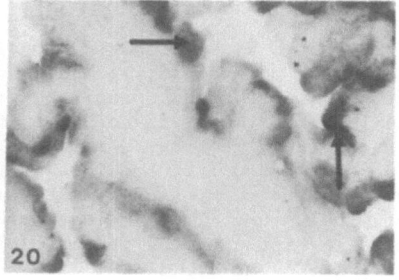

Fig. 19,20. γ-GTP reaction in brush border (arrows) of visceral yolk sac epithelial cells. Fig. 19. Control. Fig. 20. DEXA-treatment. Magnification x450

Table 2. Protease activities (given in fluorescence units) in the normal visceral yolk sac of rats and after DEXA treatment

	control	DEXA
EAP	1.1	1.2
mAAP	3.5	3.4
DPP IV	18.8	17.9
DPP I	116.2	120.1
DPP II	88.5	83.5

DISCUSSION

Traditionally, biochemical, morphological and physiological methods are employed to study possible effects and side effects of drugs (for references see Gossrau et al., 1987 a). In the present investigation enzyme histochemical means were used in combination with semithin section morphology and fluorometric analysis. Special attention was paid to the potential of these three methodological approaches in general and to the potential of different enzymes in particular for the tracing of side effects of the drugs salicylic acid and dexamethasone in the presence or absence of Zn-deficiency.

Both compounds are frequently used in human therapy. Salicylic acid has an antiphlogistic, antipyretic, and analgesic effect and represents the active metabolite of acetylsalicylic acid and dexamethasone belongs to the synthetic glucocorticoids which act as antiinflammatory, immunosuppressive, and antiallergic drugs (for references see Forth et al., 1984). Zn is essential for the normal function of many enzymes, e.g. microsomal alanyl aminopeptidase and non-specific alkaline phosphatase (Vallee and Galdes, 1984), may also act as an enzyme inhibitor, e.g. of dipeptidyl peptidase IV (Heymann and Mentlin, 1984) and is among further effects (for references see Gossrau et al., 1987 b) considered to be a membrane-stabilizing agent (Prasad, 1979; Bettger and O'Dell, 1981). Furthermore, the potentiating effect of Zn-deficiency was previously shown for the maternal and fetal rat kidney, thymus, placenta and yolk sac, when, e.g. a Zn-deficient diet was combined with the administration of the antiepileptic drug valproic acid or salicylic acid (Vormann et al., 1986; Vormann and Günther, 1986; Gossrau et al., 1987 a, b; Graf et al., manuscript in preparation). These alterations were accompanied, however, by morphological damages which are not the topic of this paper (see Graf et al., manuscript in preparation).

From the present observations evidence occurs that proteases often appear to be more sensitive indicators of plasma membrane lesions than other surface membrane-associated enzymes (except for glucoamylase in the kidney). Furthermore, it could be shown that proteases being localized in the same plasma membrane domain do respond differently to the same drug and/or its metabolite, and finally that plasma membrane lesions may exist which are not accompanied by alterations detectable by morphological and biochemical means. This indicates the usefulness of histochemical protease methods in histochemical pharmacology and toxicology (Wachsmuth, 1981; Bach et al., 1982; Marley, 1985). The most important aspect, however, is that without the histochemical procedures for surface membrane-associated proteases (and of other enzymes in other locations; Gossrau et al., 1987 a) lesions which actually exist may not be detected; they are ignored and will not be taken into account when possible side effects of drugs are considered.

The reason why plasma membrane proteases are more easily affected than other enzymes and do respond differently to the same drug is not yet known. It is also unknown at present why e.g. hepatocytes of non-pregnant rats are sensitive to salicylic acid whereas the same cells in the liver of pregnant rats are not. This is difficult to explain also for the reason since other drugs such as the glucocorticoid triamcinolone in the kidney lead to more severe lesions than in non-pregnant animals (for references see Schardein, 1976; Graf and Gossrau, 1986 b). Therefore, at least in general the assumption of a protective effect of either non-pregnancy or pregnancy against certain drugs is not valid.

The differences between protease histochemistry and protease biochemistry in the placenta and yolk sac might be of methodological origin. For the measurements in the placenta and yolk sac supernatants of homogenates from the whole organs were used containing cells with elevated plasma membrane protease activity but also other cells with reduced protease activities. In addition, the total number of placental and yolk sac cells with increased protease activity might be rather small. Thereby, the increase of activity might be either too low for a positive fluorometric signal or becomes indistinct in the fluorometer. Without histochemical protease methods the placental and yolk sac lesions would not have been detected.

The most striking observation was the nearly selective response and activation of γ-glutamyl transpeptidase at the biliary pole of the plasma membrane of hepatocytes in the periportal region after administration of salicylic acid to non-pregnant rats. Usually, this drug is absorbed by small intestinal enterocytes and transported to the liver. Here, it is also taken up by the hepatocytes where its biotransformation (conjugation and oxidation processes) and secretion as a glycin and glucuronide conjugate as well as oxidation product into the blood (and bile) takes place (for references see Forth et al., 1984; Karzel and Liedtke, 1985). Therefore, not only the free salicylic acid and salicylate respectively but also its metabolites might be responsible for the increased γ-glutamyl peptidase activity. This explanation must not be valid for the thyroid gland with the opposite activity changes of capillary endothelium-linked glutamyl aminopeptidase and dipeptidyl peptidase IV activity and the non-affected activities of non-specific and specific phosphatases after salicylic acid treatment alone in non-pregnant rats and the kidney of pregnant rats where except for microvillous glucoamylase the activities of brush border and basal labyrinth proteases and phosphatases were not clearly affected when administration of salicylic acid is combined with Zn-deficiency. Compared to the liver lower serum concentrations of salicylic acid and salicylic acid metabolites should e.g. be present in the thyroid gland, and the combination of salicylic acid and Zn-deficiency reduces serum Zn further which may contribute to the observed renal lesions.

52

Whether the enhanced hepatic γ-glutamyl transpeptidase activity in non-pregnant rats may affect the possible or established functions of this enzyme in the liver, e.g. in glutathion turnover and perhaps also DNA and protein synthesis, detoxification processes, leucotrien and prostaglandine metabolism or amino acid transport (Silbernagl et al., 1978; Hammarström, 1983; Meister and Anderson, 1983) can not yet be answered.

The activity increase of plasma membrane-associated dipeptidyl peptidase IV, glutamyl aminopeptidase, γ-glutamyl aminopeptidase and microsomal alanyl aminopeptidase in special placental and yolk sac structures after dexamethasone administration coincides with dexamethasone binding sites resembling glucocorticoid receptors in both organs (Heller et al., 1981; Carbone et al., 1986) and glucocorticoid accumulation in the yolk sac (Waddell, 1971) and suggests an induction effect of dexamethasone on these proteases. Difficulties arise again if the possible functional consequences of the enhanced activities of surface membrane-linked proteases are considered for the placenta and yolk sac. For the human and monkey placenta where dipeptidyl peptidase IV is associated with the plasma membrane of the fetal capillary endothelium its possible regulatory role in blood flow is discussed via the action of the enzyme on substance P present in the fetal (and maternal) blood (for references see Heymann and Mentlin, 1984; Gossrau et al., 1987 c). Similar possibilities might exist for dipeptidyl peptidase IV (and perhaps also for the angiotensin II degrading glutamyl aminopeptidase) linked to the plasma membrane of the labyrinthic capillary endothelial cells in the rat placenta. Still obscure is also whether the locally enhanced microvillous protease activities in the visceral yolk sac epithelium are of practical importance for the nourishment or passive immunisation of the rat fetus in which this organ is involved (for references see Beck et al., 1984).

The reported data cannot be transfered or extrapolated without further ado to the human situation. Caution is advisable due to possible species differences in morphology and the pharmacokinetic behaviour of salicylic acid and dexamethasone but also dose differences and other factors. Especially the rat and human placenta are differently organized in many respects (Gossrau et al., 1987 c) and the employed salicylic acid concentration is unusually high and not practiced in human therapy respectively; to a lesser degree the same is true for dexamethasone (for references see Forth et al., 1984). However, it has also to be considered that protease lesions might be caused in man, too by these substances in lower concentrations than those used here but are not detected because the corresponding structures might be free of the respective and responding protease provided not the special enzyme is selectively disturbed but rather the site of its location, e.g. a certain domain of the plasma membrane. The surface membrane of the capillary endothelium belongs to the structures with species-dependent proteases patterns.

However, despite all these limitations the fact remains that new insight into the side effects of drugs and also about the different biological behaviour of plasma membranes may be obtained when histochemical methods for proteases are used.

SUMMARY

Compared with mitochondrial, lysosomal and endoplasmic reticulum enzymes and plasma membrane-associated hydrolases plasma membrane-linked proteases in the rat thyroid gland, liver, kidney, placenta and yolk sac were more easily affected after the administration of the drugs salicylic acid and dexamethasone. Either a decrease or an increase of protease activity was observed in the surface membrane of capillary endothelial cells, hepatocytes, fetal placental cells and visceral yolk sac epithelial cells whereby proteases being localized in the same domain of the plasma membrane showed a different response.

ACKNOWLEDGEMENTS

The authors wish to thank Mrs. B. Bochow for technical assistance, Mrs. U. Sauerbier for the photographic work and Mrs. A. Hechel for her help with the preparation of the manuscript.

REFERENCES

Bach, P.H., Bonner, F.W., Bridges, J.W., and Lock, E.A., 1982, 'Nephrotoxicity, Assessment and pathogenesis', Wiley and Sons, Chichester, New York, Brisbane, Toronto, Singapore.
Beck, F., Huxham, J.M., and Al Alonsi, L., 1984, Transport in yolk sac epithelium. Verh. Anat. Ges. 78:99.
Bettger, W.J., and O'Dell, B.L., 1981, A critical physiological role of zinc in the structure and function of biomembranes, Life Sci. 28:1425.
Carbone, R., Baldridge, A.B., Magen, A.B., Andrew, C.L., Koszalka, T.R., and Brent, R.L., 1986, Glucocorticoid and progesterone receptors in the yolk sac placenta, Placenta 7:425.
Forth, W., Henschler, D., and Rummel, W., 1984, "Allgemeine Pharmakologie und Toxikologie", Wissenschaftsverlag, Bibliographisches Institut, Mannheim, Wien, Zürich.
Geyer, G., 1973, "Ultrahistochemie", Fischer, Jena.
Gossrau, R., 1981, Investigation of proteinases in the digestive tract using 4-methoxy-2-naphthylamine (MNA) substrates, J. Histochem. Cytochem. 29:464.
Gossrau, R., Graf, R., Günther, R., Merker, H.-J., Vormann, J., Nau, H., and Stahlmann, R., 1987 a, in press, Enzyme cytochemistry for the risk assessment of drug administration during pregnancy with special reference to proteases, Acta histochem. Suppl..
Gossrau, R., Merker, H.-J., Günther, T., Graf, R., and Vormann, J., 1987 b, Enzymatic and morphological response of the thymus to drugs in normal and zn-deficient pregnant rats and their fetuses, Histochemistry 86:321
Gossrau, R., Graf, R., Ruhnke, M., and Hanski, C., 1987 c, Proteases in the human full-term placenta, Histochemistry 86:405.
Graf, R., and Gossrau, R., 1986 a, Enzyme cytochemistry of the yolk sac diverticula in the mature murine rodent placenta, Placenta 7:491.
Graf, R., and Gossrau, R., 1986 b, Enzyme cytochemistry of pregnant and non-pregnant mouse organs after triamcinolone acetonide treatment. Histochem. J. 18:666.
Hammarström, S., 1983, Leukotrienes. Ann. Rev. Biochem. 52:355.
Heller, C., Coivine, H., de Nicola, A.F., 1981, Binding of ^{3}H dexamethasone by rat placenta, Endocrinology 108:1697.
Heymann, E., and Mentlin, R., 1984, Beeinflußt Dipeptidylpeptidase IV Blutdruck und Gerinnung?, Klin. Wschr. 62:2.
Karzel, K., and Liedtke, R.K., 1985, "Einführung in die Arzneimitteltherapie", 2. Aufl., UTB, Fischer, Stuttgart.
Kenny, A.J., 1977, Proteinases associated with cell membranes, in: 'Proteinases in mammalian cells and tissues', A.J. Barrett, ed., North Holland, Amsterdam, Oxford, New York.
Lojda, Z., Gossrau, R., and Schiebler, T.H., 1979, Enzyme histochemistry, A laboratory manual, Springer, Berlin, Heidelberg, New York.
Marley, M., 1985, The detection of drugs by histochemical procedures, in: 'Analytical methods in human toxicology', A.S. Curvy, ed., part 1, Verlag Chemie, Weinheim, Deerfield Beach, Florida, Basel.
McDonald, J.K., and Barrett, A.J., 1986, 'Mammalian proteases', A glossary and bibliography, Vol. 2, Exopeptidases, Academic Press, London,

Orlando, San Diego, New York, Montreal, Sydney, Tokyo, Toronto.

McDonald, J.K., and Schwabe, C., 1977, Intracellular exopeptidases, in: 'Proteinases in mammalian cells and tissues', Barrett, A.J., ed., North Holland, Amsterdam, New York, Oxford.

Meister, A., and Anderson, M.E., 1983, Glutathione, Ann. Rev. Biochem. 52: 711.

Prasad, A.S., 1979, Clinical, biochemical and pharmacological role of zinc, Ann. Rev. Pharmacol. Toxicol. 20:393.

Schardein, J.L., 1976, Drugs as teratogens, CRC Press, Cleveland, Ohio.

Silbernagl, S., Pfaller, W., Henle, H., and Wendel, A., 1978, Topology and function of renal γ-glutamyl transpeptidase, in: 'Functions of glutathione in liver and kidney, H. Sels and A. Wendel, eds., Springer, Berlin, Heidelberg, New York.

Vallee, B.L., and Galdes, A., 1984, The metallobiochemistry of zinc enzymes, Adv., Enzymol. 56, 283.

Vormann, J., Höllriegel, V., Merker, H.-J., and Günther, T., 1986, Effect of salicylate on zinc metabolism in fetal and maternal rats fed normal and zinc deficient diets, Biol. Trace Element Res. 9:55.

Vormann, J., and Günther, T., 1986, Development of fetal mineral and trace element metabolism in rats with normal as well as magnesium and zinc-deficient diets, Biol. Trace Element Res. 9:37.

Wachsmuth, E.D., 1981, The potential role of histochemistry in pharmacology and toxicology, in: Histochemistry: The widening horizons, P.J. Stoward and J.M. Pollak, eds., Wiley and Sons, Chichester.

Waddell, W.J., 1971, The distribution of cortisone-C in pregnant mice, Teratology 4:355.

ACTIVATION OF LEUKOCYTES DURING PROLONGED PHYSICAL EXERCISE

K. Kokot[1], R.M. Schaefer[1], M. Teschner[1], U. Gilge[1], R. Plass[2], and A. Heidland[1]

Departments of [1]Internal Medicine and [2]Physical Education, University of Wuerzburg, FRG D-8700 Würzburg

INTRODUCTION

In recent years, many forms of physical exercise have been considerably propagated in western societies. Therefore, many investigators have turned their interest towards the metabolic alterations, induced by enhanced physical training. Moorthy and Zimmermann reported an increase in the number of circulating leukocytes (1) during prolonged exercise, whereas Hedfors et al. (2) showed a phase of lymphocytosis following short-time exercise. Body core temperature was found to be increased after lengthy exercise (3). Furthermore, an increment of plasma levels of acute-phase proteins, such as C-reactive protein, ceruloplasmin, fibrinogen, and $alpha_1$-antitrypsin, has been demonstrated (4). Studies in subjects on long-term physical training showed chronic depression of plasma iron (5) and zinc (6) levels. Even frank anemia has been reported in joggers and competition runners (5). Cannon and Kluger were the first to relate this metabolic pattern, which displays a remarkable similarity to the physiological response to an infectious or inflammatory challenge, to an increased release of endogenous pyrogen, which is secreted by monocytes and macrophages (7).

To assess the functional state of polymorphonuclear (PMN) leukocytes, degranulation, as measured by the release of elastase, and generation of reactive oxygen species, determined as chemiluminescence have been recently employed. Enhanced degranulation, giving rise to elevated plasma levels of PMN elastase has been reported in severe bacterial infections (8) and during hemodialysis (9). PMN leukocytes emit light or chemiluminescence (10) in response to phago-cytosis or soluble (phorbol) stimuli. Even though the precise mechansims are not completely understood, the generation of chemiluminescence appears to depend on production of several oxygen species, including superoxide anion, hydrogen peroxide, hydroxyl radical, and singlet oxygen, which subsequently may interact with other cell constituents resulting in the final phenomenon of chemiluminescence (11).

The present study was conducted to establish the effect of physical exercise on the functional state of PMN leukocytes, as measured by the rate of degranulation and chemiluminescence.

METHODS

Ten healthy male volunteers between 22 and 32 years of age took part in the study. They all had some interest in prolonged physical exercise, ranging from occasional to daily jogging. The studies were conducted in morning hours (08^{00}-10^{00}) during cool weather. Each of the subjects had to run over a distance of both 2,000 and 10,000 meters at the maximal possible speed. Between these 2 bouts of exertion the runners were at least allowed to rest for 48 hours. Whole blood samples were obtained from a cubital vein before and immediately after exercise. For determination of PMN elastase all samples were anticoagulated with sodium citrate, centrifuged and the plasma was separated to prevent leakage of leukocyte constituents. The specimens were stored at $-20^{\circ}C$ and thawed only once for processing. Heparinized blood samples were used for the measurements of chemiluminescence.

Blood cells were counted by an electronic counter (ELT 8, Molter, Neckargmünd, FRG). Quantitative estimations of the plasma levels of elastase in complex with alpha$_1$-proteinase inhibitor were carried out with a commercially available enzym-linked immunoassay (12). For measurement of chemiluminescence whole blood samples were diluted in Hank's-solution (1:20) without calcium and magnesium. 20 µl of the diluted whole blood samples were given into 440 µl of normal Hank's-solution. Then, 20 µl of lucigenin 10 mM (Sigma, Munich, FRG) were added and the basal light emission of leukocytes was registered at $38^{\circ}C$ during 5 minutes in a Biolumat, model LB 9500 (Berthold, Wildbach, fRG).

Chemiluminescence was stimulated by adding 20 µl of phorbol-12-myristate-13-acetate (PMA) 0.5 mM (Sigma, Munich, FRG). Light emission was recorded until peak values had passed and a tendency towards baseline values was evident.

All values are reported as mean $\pm$ SE. The significances of differences between the study periods were analyzed by the Mann-Whitney rank sum test. Significance was defined as a p value of less than 0.05.

RESULTS

Before exercise the leukocyte count was within the normal range. Following short-term exercise, the number of circulating leukocytes increased by 21 %. By contrast, after long-term exercise there was a significant increment of circulating leukocytes (+ 61 %) (Table 1).

Table 1. Effect of physical exercise on white cell count
(cells/µl)

	short-term (2,000 meters)	long-term (10,000 meters)
before running	6,470 $\pm$ 499	6,020 $\pm$ 640
after running	7,820 $\pm$ 530	9,670 $\pm$ 860
p	N.S.	0.01

Mean values $\pm$ SE from 10 subjects: N.S. denotes not
significant.

   For determination of the generation of reactive oxygen
species by leukocytes, phorbol-stimulated chemiluminescence
was measured. Following short-term exercise, there was a
reduction of chemiluminescence compared to pre-exercise
values (- 12 %). However, this difference did not reach
significance. After long-term exertion the decrease in
chemiluminescence of leukocytes was more pronounced (-18%).
The difference before and after exercise was hightly
significant, suggesting a reduction in the formation of
reactive oxygen species (Table 2).

Table 2. Effect of physical exercise on chemiluminescence
(cpm x 10^{-3} cells)

	short-term (2,000 meters)	long-term (10,000 meters)
before running	1,960 $\pm$ 130	2,520 $\pm$ 190
after running	1,730 $\pm$ 150	2,060 $\pm$ 160
p	N.S.	0.01

Mean values $\pm$ SE from 10 subjects: N.S. denotes not
significant.

   As a further index of leukocyte activity, degranulation
of white cells was observed, as measured by the release of
elastase. Prior to exercise, plasma levels of PMN elastase
were in the normal range. Immediately after short-term
exercise, there was a small increase (+ 15 %) of elastase
levels. Following prolonged physical activity, a marked
and significant increment (+ 200 %) of elastase plasma
levels could be observed, suggesting enhanced degranulation
of PMN leukocytes during long-term exercise (Fig. 1).

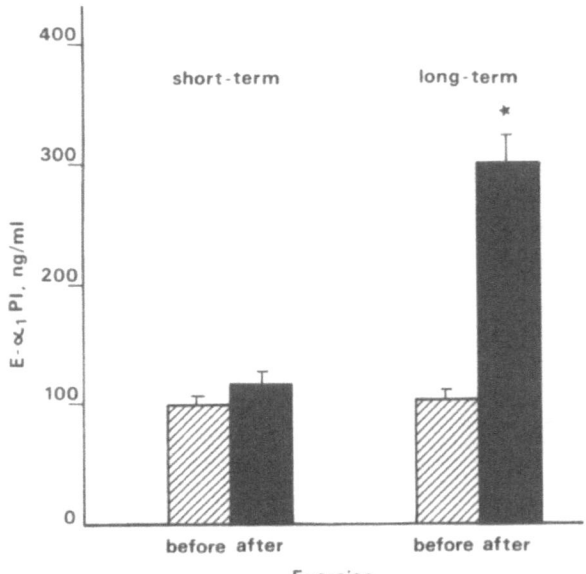

Fig. 1. Effect of short-term (running 2,000 meters) and
long-term (running 10,000 meters) physical exercise on plasma
levels of PMN elastase. Values are given as mean ± SE
obtained from 10 subjects immediately before and after the
run.
* p ∠ 0.01 before versus after physical exercise.

DISCUSSION

In accordance with previous findings (1), we observed an increase of circulating leukocytes following acute physical exercise (Table 1). The increment was 21 % following short-term activity and 61 % after a lengthy period of exertion. Being more pronounced with long-term exertion, leukocytosis seems to depend on the degree of physical exercise.

In order to evaluate, not only the number of circulating white cells, but although their functional activity, we determined both, parameters of degranulation and of production of reactive oxygen species. As an index of oxygen radical formation, chemiluminescence was measured. Following physical exercise, there was a reduction of light emission from leukocytes, which was more pronounced after running 10,000 meters (- 18 % versus - 12 %), as compared to the 2,000 meter run. This reduction of chemiluminescence in response to a phorbol stimulus suggest that, immediately after acute exercise, the formation of reactive oxygen species in leukocytes is diminished. Possibly, the decrease in light emission could be explained as the manifestation of a certain degree of exhaustion which would follow a phase of enhanced generation of oxygen radicals during physical activity (Table 2).

As a parameter of degranulation, the release of PMN elastase following physical exertion was evaluated. PMN leukocytes hold two distinct populations of granules in human neutrophils, namly azurophilic and specific granules (13) and elastase is located exclusively in the azurophilic granules. Following short-term exercise, there was only a slight increment in plasma levels of elastase. By contrast, long-term exertion induced high levels of this compound, suggesting marked degranulation of azurophilic granules of PMN leukocytes (Fig. 1).

It should be mentioned that Petrova et al. (14) found reduced serum levels of lysozyme, a constituent of both azurophilic and specific granules (13) in 91 sportsmen. In contrast to these findings, Haralambie (15) reported an increment in serum lysozyme levels immediately after completion of a marathon run. This discrepancy between those two studies is most likely due to a difference in the time period elapsing between the end of the physical exertion and the collection of blood samples. Thus, it seems that when blood is obtained immediately after exercise, evidence for enhanced degranulation of PMN leukocytes is found, which was the case in our study and in the investigations of Haralambie (15).

Taken together our findings support the view that PMN leukocytes undergo some activation during prolonged physical exercise, as demonstrated by enhanced plasma levels of PMN elastase and a decrease of phorbol-stimulated chemi-luminescence immediately after lengthy exercise. To determine whether these alteration of leukocyte functions give rise to a reduced capability to fight bacterial infections, further studies are warranted.

SUMMARY

Previous studies have demonstrated an increment of
circulating leukocytes and enhanced secretion of interleukin-
1 by monocytes and macrophages during physical exercise. In
the present study the effect of physical exertion on the
activity of polymorphonuclear (PMN) leukocytes was
investigated. Following both short-term (running 2,000 meters)
and long-term (running 10,000 meters) exertion, phorbol-
stimulated chemiluminescence, as an indicator of leukocytic
oxygen radical formation and release of leukocytic elastase,
as a parameter of degranulation, were determined immediately
after running. The number of circulating leukocytes increased
both after short-term (+ 21 %) and long-term (+ 61 %) exercise.
There was a minor release of PMN elastase following short-
term activity causing plasma levels of this compound to rise
from 100 ± 4.0 ng/ml to 116 ± 12.3 ng/ml. Long-term exercise,
on the other hand, induced a significant increase of elastase
plasma levels from 107 ± 9.1 ng/ml to 300 ± 23.4 ng/ml,
suggesting a remarkable release of this proteinase from
neutrophils. Based on these findings we conclude that during
physical exercise degranulation of PMN leukocytes occurs.
Moreover, the fact that phorbol-stimulated chemiluminescence
is decreased after running demonstrates an impaired
capability of white cells to generate oxygen radicals.

ACKNOWLEDGEMENTS

The technical assistance of M. Roeder and the
secretarial help of E. Hammer are gratefully acknowledged.

REFERENCES

1. A.V. Moorthy and S.W. Zimmerman, Human leukocyte responses
    to an endurance run, Eur. J. Appl. Physiol. 38: 271
    (1978)
2. E. Hedfors, G. Holm and B. Öhnell, Variations of blood
    lymphocytes during work studies by cell surface markers,
    DNA synthesis and cytotoxicity, Clin. Exp. Immol. 24:
    328 (1976)
3. J.S.J. Haight and W.R. Keatinge, Elevation in the set
    point for body temperature regulation after prolonged
    exercise, J. Physiol. 229: 77 (1973)
4. G. Haralambie, J. Keul and F. Theumert, Changes in serum
    proteins, -iron and -copper in swimmers before and after
    altitude training, Eur. J. Appl. Physiol. 35: 21 (1976)
5. A. Hunding, R. Jordal and P.E. Paulev, Runner's anemia
    and iron deficiency, Acta Med. Scand. 209: 315 (1981)
6. G. Haralambie, Serum zinc in atheletes in training, Int.
    J. Sports Medicine 2: 135 (1981)
7. J.G. Cannon and M.J. Kluger, Endogenous pyrogen activity
    in human plasma after exercise, Science 220: 617 (1983)
8. M. Jochum, K.H. Duswald, S. Neumann, J. Witte and H.
    Fritz, Proteinases and their inhibitors in septicema,
    Adv. Exp. Med. Biol. 167: 391 (1984)
9. W.H. Hörl, M. Jochum, A. Heidland and H. Fritz, Release
    of granulocyte proteinases during hemodialyse, Am. J.
    Nephrol. 3: 213 (1983)

10. R.C. Allen, R. Stjernholm and R. Steele, Evidence for the generation of an electronic excitation state(s) in human polymorphonuclear leukocytes and its participation in bacterial activity, <u>Biochem. Biophys. Res. Commun.</u> 47: 679 (1972)

11. B.O. Cheson, R.L. Christensen, R. Sperling, B.E. Kohler and B.M. Babior, The origin of the chemiluminescence of phagocytosing granulocytes, <u>J. Clin. Invest.</u> 58: 789 (1976)

12. S. Neuman, G. Gunzer, N. Hennrich and H. Lang, "PMN-Elastase assay": enzyme immunoassay for human poly-morphonuclear eleastase complexed with $alpha_1$-proteinase inhibitor, <u>J. Clin. Chem. Clin. Biochem.</u> 22: 693 (1984)

13. K. Havemann and M. Gramse, Physiology and pathophysiology of neutral proteinases of human granulocytes, <u>Adv. Exp. Med. Biol.</u> 167: 1 (1984)

14. I.V. Petrova, S.N. Kuzmin, T.s. Kurshakova, R.S. Suzdalnitskii and B.B. Persin, Fagotsitarnai aktivnost neitrofilov i gumoralnye factory obshchego i mestogno immuniteta pri intensivnykh fizischeskikh magruzkakh, <u>Zh. Mikrobiol. Epidemiol. Immobiol.</u> 12: 53 (1983)

15. G. Haralambie, Stoffwechselveränderungen nach lang-dauernder körperlicher Belastung beim Menschen, <u>Thesis,</u> Freiburg, FRG (1975)

INHIBITION OF HUMAN NEUTROPHIL ELASTASE BY POLYGUANYLIC ACID AND OTHER SYNTHETIC POLYNUCLEOTIDES

Sanford Simon[1], Micha Vered[2], Alex Rinehart[1], John Cheronis[3], and Aaron Janoff[2]

Department of Biochemistry[1] and Pathology[2]
SUNY at Stony Brook
Stony Brook, New York 11794 USA
[3]Cortech, Inc.
Boulder, Colorado 80301

INTRODUCTION

We have recently described the inhibition of human neutrophil elastase (HNE) by RNA purified from Streptococcus pneumoniae (1). Inhibition of this enzyme by a wide variety of polyanions has also been reported by others. Such agents include ribonucleic acids from yeast (1,2), E. coli (2), polyribosylribitol phosphate from H. influenzae capsules (1), and several sulfated polysaccharides (3-5). In addition to these naturally-occurring polyanions, synthetic RNA homopolymers were also tested previously (2).

In the present study, we compared the inhibition of HNE given by different synthetic polyribonucleotides as well as the relationship between inhibition and polymer size. In addition, the ability of several polyanions to inhibit ongoing elastinolysis was tested. Also, the effects of ionic strength, of added amphiphilic agents and of prior HNE-inactivation upon complex formation were investigated. Lastly, HNE and two other highly cationic serine proteases were compared with respect to complex-formation with polyguanylic acid. The results we obtained show that HNE-inhibition by polyanions is highly specific and suggest that hydrophobic as well as ionic interactions may regulate the inhibition of HNE by polyribonucleotides.

MATERIALS AND METHODS

Polycytidylic acid (poly C), polyadenylic acid (poly A), polyuridylic acid (poly U), polyguanylic acid (poly G), and polyinosinic acid (poly I) were purchased from Pharmacia Fine Chemicals, NJ. High MW polyribosylribitol phosphate, purified from H. influenzae capsular polysaccharide, was from Praxis Biologics, Inc., NY. This material was separated from low MW contaminants and lactose (which had been added as a stabilizer by the manufacturer) by gel filtration through Sephadex G-25. Human neutrophil elastase (HNE) was from Elastin Products Co., MO. and was purified from cystic fibrosis sputum. Porcine pancreatic elastase was from Sigma

Chemical Co., MO. Cathepsin G was a generous gift from Dr. James C. Travis. Succinyl-alanyl-alanyl-alanyl-p-nitroanilide (SLAPN) and methoxy-succinyl-alanyl-alanyl-prolyl-phenylalanyl-p-nitroanilide were from Sigma Chemical Co. ^{14}C-elastin from bovine ligament was a gift from Dr. Shie-Ye Yu. ^{14}C-methoxysuccinyl-alanyl-alanyl-prolyl-valine chloromethylketone (MeSAAPVCMK) was a gift from Dr. James C. Powers. ^{3}H-diisopropylfluoro-phosphate (DFP) was from New England Nuclear, MA. Bile salts (Na glyco-cholate and Na taurocholate) were also obtained from Sigma Chemical Co.

Kinetic assays with the synthetic chromogenic substrates, succinyl-ala-ala-ala-p-nitroanilide (SLAPN) and methoxysuccinyl-ala-ala-pro-val-p-nitroanilide (MeSAAPVPN), were carried out with an LKB 4050 spectro-photometer using a modification of the procedure of Bieth et al. (6), in which rate of formation of p-nitroaniline is measured.

Elastinolysis was assayed by release of tricholoroacetic acid soluble radioactivity from ^{14}C-labelled insoluble elastin, prepared according to the procedure of Bielefeld et al. (7), after two hours of digestion with neutrophil elastase at 37°C. Assays for inhibitory activity of the synthetic RNA homopolymers or of polyribosylribitol phosphate were per-formed as above, except that the enzyme was preincubated with the desired concentration of inhibitor in 0.01 M Tris-HCl, 0.15 M NaCl, pH 8.0 for 10 minutes and then added to the suspended elastin substrate. Alternatively, in other experiments, enzyme and ^{14}C-elastin were preincubated 10 min at 37°C and then inhibitor was added.

Fractionation of poly G polymers according to their MW was performed using an FPLC system (Pharmacia Fine Chemicals) and a Superose-12 HR column (1x30 cm). The column buffer was 0.01 M Tris-HCl, 0.15 M NaCl, pH 8.0.

Resolution of free, labelled enzymes vs complexes of poly G and labelled enzymes was also carried out using the above FPLC system. Briefly, enzymes and poly G were allowed to interact in the above buffer for 15 min at 37°C. Ratios of enzyme to poly G (w:w) are given in the legends to Figures 2-4. NaCl concentration in the buffer was varied from zero to 1.0M during the preincubation of enzymes with poly G as well as in the FPLC buffer, during chromatographic separation. After pre-incubation, the enzyme-poly G mixtures were chromatographed as above.

RESULTS

Inhibition of HNE-Mediated Amidolysis by Various Synthetic Ribopoly-nucleotides

Figure 1 shows the effect of five different RNA homopolymers on SLAPN-hydrolysis by HNE. It can be seen that poly C was ineffective, while poly G and poly I were the most effective inhibitors of amidolysis by the neutrophil enzyme. In poly C, the molecular structure is that of a helix in which the phospho-ribosyl moieties face inward; this is not the case for poly G or Poly I (8). Thus, ionic interactions between the negatively charged phosphate groups of poly C and the positive charges on HNE would be prevented, and this may be the explanation for the failure of poly C to inhibit the enzyme. Although not shown in Figure 1, we also observed no inhibition of HNE by poly I:poly C duplexes, whereas poly I homopolymers, as well as the single stranded random copolymers, poly [G,U] and poly [I,U], were highly inhibitory. This suggests that the 3-dimen-sional structure of polyribonucleotides is also an important determinant of interaction with HNE, in addition to charge associations. Of the two most effective RNA homopolymers, poly G and poly I, the former was se-lected for further study.

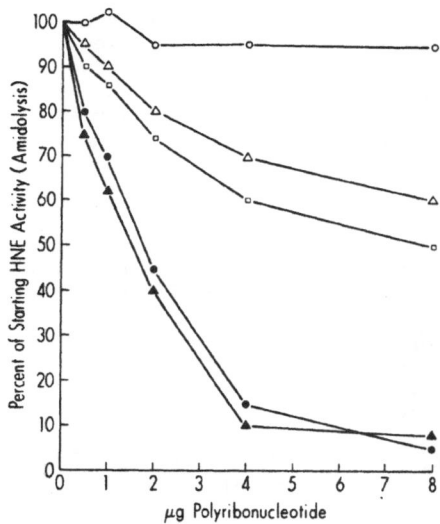

Figure 1. Effect of synthetic homopolyribonucleotides on SLAPN-hydrolysis by HNE (see Methods for description of assay). (O – O), Poly C; (△ - △), poly A; (□ - □), poly U; (● - ●), poly G; (▲ - ▲), poly I. HNE was 2μg (0.066nmole) in all cases. SLAPN was 1mM.

When the commercial preparation of poly G was fractionated according to polymer size by chromatography on Superose-12 as described in Methods, inhibitory activity vs HNE was directly related to Mr (app) of the homopolymer. Highest inhibitory activity was found in the void volume fractions, containing polymers of $Mr \geq 10^6$. Still, even polymers of poly G with Mr (app) as low as 6×10^4 displayed some HNE inhibition.

Inhibition of HNE-Mediated Elastinolysis by Poly G and by H. influenzae Capsular Polysaccharide

In Table I we show that fractions of poly G of Mr 5000 inhibited elastinolysis by 2 ug HNE either when preincubated with the enzyme 10 minutes before addition of substrate, or when added 10 minutes after the addition of enzyme to substrate to start an ongoing elastinolysis reaction. The dependence of inhibition on poly G concentration was fit to a model of mixed partial inhibition and the apparent K_I and residual catalytic activity were estimated by graphical analysis of double reciprocal plots (9). When poly G was preincubated with the enzyme 10 minutes before addition of substrate, the apparent K_I was approximately one tenth the apparent K_I observed when the inhibitor was added 10 minutes after addition of enzyme to substrate. The catalytic activity was reduced to an extrapolated value of 5% of the uninhibited enzyme when HNE was preincubated with different concentrations of poly G, and was reduced to an extrapolated value of 37% of the uninhibited enzyme when different concentrations of poly G were added to an ongoing elastinolysis reaction. By comparison, polyribosylribitol phosphate (PRP) from H. influenzae was completely ineffective in inhibiting ongoing elastinolysis by HNE, and actually appeared to activate ongoing elastinolysis weakly at high concentrations to approximately 108% of the activity in the absence of PRP. When different concentrations of PRP were preincubated with HNE before addition of elastin, catalytic activity was reduced to an extrapolated value of less than 5% of the activity of the uninhibited enzyme. The apparent K_I for PRP was approximately 300 times as great as that for poly G as a mixed partial inhibitor when preincubated with HNE.

Table I

Inhibition of HNE-Mediated Elastinolysis by Polyanions Preincubated with HNE or Added to an Ongoing Elastinolysis Reaction

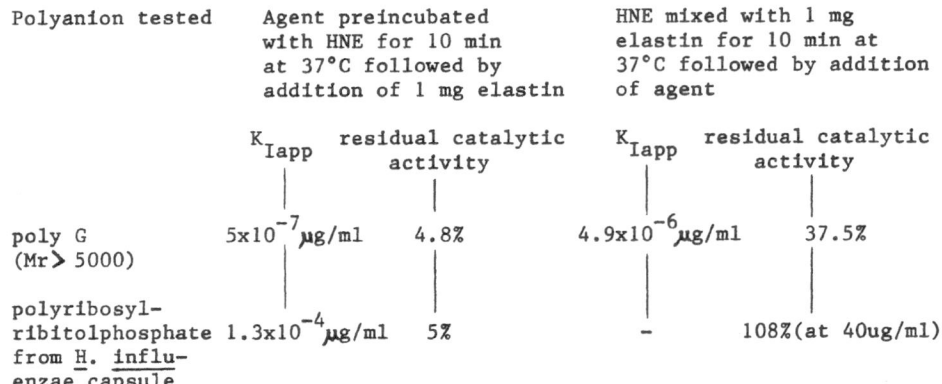

Polyanion tested	Agent preincubated with HNE for 10 min at 37°C followed by addition of 1 mg elastin		HNE mixed with 1 mg elastin for 10 min at 37°C followed by addition of agent	
	K_{Iapp}	residual catalytic activity	K_{Iapp}	residual catalytic activity
poly G (Mr > 5000)	5×10^{-7} µg/ml	4.8%	4.9×10^{-6} µg/ml	37.5%
polyribosyl-ribitolphosphate from H. influenzae capsule	1.3×10^{-4} µg/ml	5%	–	108%(at 40ug/ml)

Mechanism of HNE + Poly G Complex Formation - Effects of Catalytic Inactivation and Ionic Strength

In Figure 2, we show that catalytic inactivation of HNE with diisopropylfluorophosphate (DFP) does not prevent subsequent complex-formation between the enzyme and poly G (unfractionated polymers). This is evident from the fact that HNE treated with ^{3}H-DFP (^{3}H-DIP-HNE) and then reacted with poly G in buffer containing 0.15M NaCl, elutes immediately after the void volume in the same fractions as does high Mr poly G itself. Thus, a free active-site serine hydroxyl group is not required for association of HNE with poly G. This same observation had been made previously, by us (9), when studying the mechanism of complex-formation between HNE and the pneumococcal elastase inhibitor (now identified as RNA (1))

Whereas inactivation of the catalytic-site serine residue does not interfere with complex-formation between HNE and poly G, high ionic strength does interfere. This is also evident from Figure 2, if one compares the amount of ^{3}H-DIP-HNE appearing immediately after the void volume (i.e., poly G-complexed ^{3}H-DIP HNE) with the amount of uncomplexed ^{3}H-DIP-HNE (eluting after fraction 40) at ionic strengths of 0.15, 0.3, and 0.6 (M NaCl). Clearly, ^{3}H-DIP-HNE is almost completely complexed by poly G at physiologic ionic strength (0.15M NaCl), but remains almost entirely free at four-fold greater ionic strength (0.6M NaCl). Complex-formation is still significant in the presence of 0.3M NaCl. Since it probably can be assumed that active HNE would bind at least as well as ^{3}HDIP-HNE, this last observation is somewhat discordant with the report by Lestienne and Bieth, who found that complex-formation between active HNE and yeast transfer RNA could be reversed by addition of 0.3M NaCl to the reaction (2). However, their finding may be explained by the relatively higher background ionic strength of their reaction buffer. Lestienne and Bieth used 0.05M Na phosphate buffer, whereas we employed a weakly-ionizing Tris buffer (pH 8.0) at 0.01M concentration.

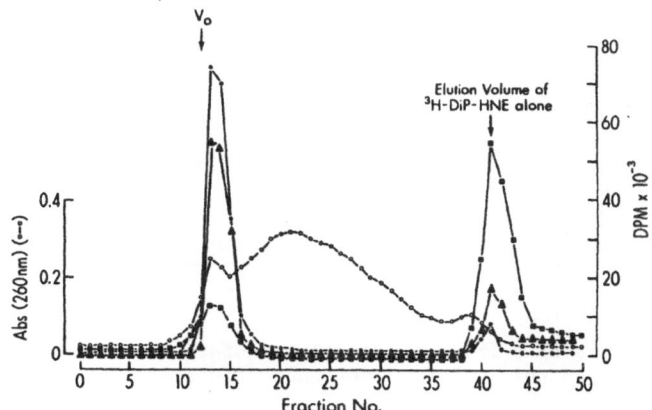

Figure 2. Effect of increasing ionic strength on complex formation between HNE and poly G. HNE was treated with ^{3}H-labelled DFP and 10µg of the labelled, inactivated enzyme were incubated for 10 min at 37°C with 20µg of unfractionated polymers of poly G. Complexed and free enzyme were then separated by Superose-12 column chromatography as described in Methods. Absorbance at 260 nm (o-o) was used to monitor the distribution of poly G among the effluent fractions, while radioactivity of aliquots of these same fractions identified ^{3}H-DiP-HNE. Incubation and chromatography were carried out in the presence of 0.15M NaCl (●-●), 0.3M NaCl (▲-▲), and 0.6M NaCl (■-■). The buffer was 0.1M Tris-HCl, pH 8.0.

Specificity of HNE Complex-Formation with Poly G

Lestienne and Bieth showed that the interaction with polyribonucleotides was specific to HNE, since neither porcine pancreatic elastase nor bovine or human chymotrypsins could be inhibited by yeast transfer RNA (2). They suggested that this specific feature of the interaction between HNE and polyanions was still consonant with an electrostatic binding-mechanism, because HNE contains a greater number of arginine residues (twenty-two) than the other enzymes they tested (twelve arginines or fewer). In Figure 3, we show that ^{3}H-DiP-porcine pancreatic elastase does not form complexes with poly G even at low ionic strength (confirming the findings of Lestienne and Bieth (2)). However, we also show that complex-formation does not take place with ^{3}H-DiP-cathepsin G either. This last finding is consonant with our previous observation that the activity of cathepsin G is not blocked by pneumococcal HNE-inhibitor (10). These observations do not fit Lestienne and Bieth's hypothesis, since there are at least twenty-nine arginine residues in cathepsin G (11) and possibly as many as thirty-one (S. Sinha, personal communication) and its isoelectric pH is even more basic than that of HNE (> 13.0 vs 10-11, respectively). Therefore, high basic charge may not be the only factor involved in the specific interaction between HNE and polyribonucleotides. The density of charges and/or the spatial relationship between charged regions and the enzyme's active-site may also influence the specific character of the reaction between HNE and these polyanions.

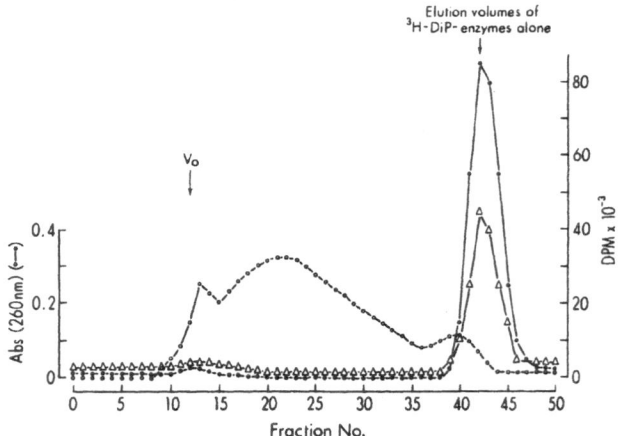

Figure 3. Failure of porcine pancreatic elastase and cathepsin G to form complexes with poly G. Porcine pancreatic elastase and cathepsin G were treated with ^{3}H-labelled DFP and 10μg of the labelled, inactivated enzymes were incubated for 10 min at 37°C with 20μg of unfractionated polymers of poly G. Complexed and free enzymes were then separated as in Figure 2, using A 260nm (o-o) and radioactivity of aliquots to monitor the column effluent for distribution of poly G and enzymes, respectively. Buffer was 0.01M Tris-HCl, pH 8.0. Ionic strength was 0.15M NaCl. (●-●), ^{3}H-DiP-pancreatic elastase; (▲-▲), ^{3}H-DiP-cathepsin G.

Peptide Chloromethylketone-Inactivated HNE Does not Form Complexes with Poly G

In Figure 4 it can be seen that, in contrast to ^{3}H-DiP-HNE, neutrophil elastase that has been inactivated by ^{14}C-MeSAAPVCMK no longer forms complexes with poly G. In the case of ^{3}H-DiP-HNE, catalysis by HNE is blocked, but the more distal subsites of the substrate-binding pocket of the enzyme are not fully occupied. In the case of CMK-inactivated HNE, however, catalysis is blocked and four to five subsites of the substrate-binding pocket of the enzyme are occupied by the tetrapeptide moiety (-alanyl-alanyl-prolyl-valine). These results suggest that complex-formation between poly G and HNE requires an unoccupied extended substrate-binding pocket. Alternatively, occupation of the binding-pocket by the tetrapeptide and/or interaction of the chloromethylketone with the active-site histidine and serine residues (12) may trigger allosteric conformational changes in the enzyme which prevent subsequent associations with poly G at sites which are distant from the binding-pocket. Presumably, interaction of DFP with the active-site serine residue fails to induce such a conformational change. Further studies (e.g. using fluorescence spectra or circular dichroism analysis) should help to clarify this interpretation.

Similar observations were made by us previously, in our studies on complex-formation between HNE and its inhibitor in pneumococcal extracts. Again, DiP-HNE formed complexes with the pneumococcal agent (10), while HNE inactivated with a peptide-chloromethylketone did not (13). As described by us recently (1), the pneumococcal inhibitor of HNE is also a polyribonucleotide.

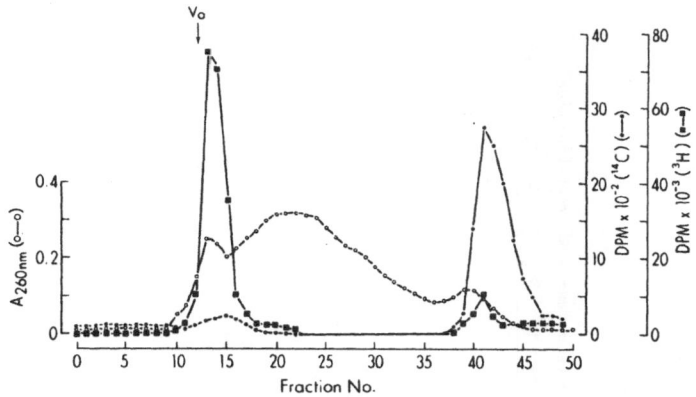

Figure 4. Effect of inactivation of HNE with a peptide-chloromethyl-
    ketone upon complex-formation with poly G. HNE was either treated
    with ^{3}H-DFP or with ^{14}C-MeSAAPVCMK to inactivate the enzyme. Then
    10μg of either one or the other of the inactivated enzymes were
    reacted with 20μg poly G (unfractionated polymers) as in Figure 2,
    and the reaction mixture chromatographed on Superose-12 to resolve
    labelled complexes from free enzyme. Buffer was 0.01M Tris-HCl, pH
    8.0, containing 0.15M NaCl$_3$ (o-o), A260 nm of the effluent fractions
    (poly G); (■-■), dpm of ^{3}H recovered in the effluent fractions;
    (●-●), dpm of ^{14}C recovered in the effluent fractions. Chromato-
    graphy of ^{3}H-DiP-HNE + poly G was carried out separately from chroma-
    tography of ^{14}C-MeSAAPVCMK-HNE + poly G, although the results have
    been plotted together in the same figure. The ^{3}H-DiP-HNE + poly G
    data, in fact, are re-plotted from Figure 2 (0.15M NaCl curve).

Bile Salts Suppress HNE-Inhibition by Poly G

     Figure 5 shows the results of an experiment designed to test the
effect of bile salts upon the interactions between HNE and poly G. It can
be seen that, in the presence of 0.01M taurocholic acid or 0.01M glycocholic acid, there was significant suppression of the HNE-inhibition
normally caused by poly G. These results indicate that steroidal amphiphiles can reduce the inhibition of HNE by poly G. Since steroidal
amphiphiles, by themselves (even at the relatively high concentration
employed in Figure 5), stimulate the amidolytic activity of HNE (14), the
relief of poly G inhibition shown in Figure 5 cannot be due to inactivation of the enzyme by bile salts. Rather it may be that their interference with poly G inhibition reflects a common mechanism of HNE-interaction
by polyribonucleotides and bile salts. These results suggest that in
addition to ionic interactions, hydrophobic associations may be yet
another mechanism involved in the specific binding of HNE to poly G.

DISCUSSION

     In this paper, we have shown that HNE can be inhibited by synthetic
RNA homopolymers, a finding which is consistent with our observations of
HNE-inhibition by polyribonucleotides released from pneumococci (1).
These results confirm and extend the report by Lestienne and Bieth (2),
who also observed inhibition of HNE by yeast RNA, transfer RNA from E.
coli, and synthetic homopolyribonucleotides. The mechanism of inhibition

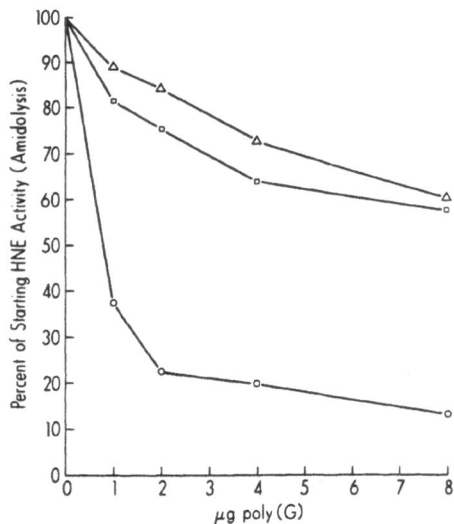

Figure 5. HNE-inhibition by poly G in the presence of bile salts. HNE
amidolytic activity vs SLAPN (expressed as percent of the activity
observed when enzyme alone or enzyme plus bile salt was tested in the
absence of poly G) is here plotted against ug of added poly G (unfractionated polymers). HNE = 2μg (0.066nmoles). Buffer = Tris-HCl
(0.01M), pH 8.0, + 0.15M NaCl. Final substrate concentration = 1mM.
(o-o), no bile salt added; (△-△), taurocholic acid (0.01M);
(□-□), glycocholic acid (0.01M).

does not appear to simply involve obstruction of binding to insoluble
elastin by the polyribonucleotides, since amidolysis of two small peptide
substrates is also inhibited. The kinetic results reported here, and
previously by us (13) for the pneumococcal inhibitor (now shown to be RNA
as well) are consistent with a reduction in the catalytic capacity of the
enzyme in the presence of polyribonucleotides. It has been recently
pointed out that amidolysis of SLAPN may involve participation of only the
active site serine and histidine, unlike the amidolysis of MeSAAPVPN,
which appears to involve the full catalytic triad (15). Our findings
suggest, therefore, that polyribonucleotides may inhibit HNE by interfering with the simplest catalytic mechanism employed by the enzyme for
amide cleavage.

Specifically, we found that polyguanylic and polyinosinic acids were
especially effective inhibitors, while polycytidylic acid was inactive,
presumably because its phospho-ribosyl moieties face inward (7) and are
unavailable for binding to positively-charged groups on HNE. In the
double stranded polymer poly I:poly C, which is inactive as an inhibitor
the bases are tied up in Watson-Crick pairs, but the single stranded
random copolymers, poly [G,U] and poly [I,U], are still strongly inhibitory, confirming the apparent requirement for access both to the polyanionic backbone and to the more hydrophobic bases. The bases in the
naturally occuring RNA's which we have shown to be inhibitory in our
recent report (1) are also largely not tied up in Watson-Crick pairs,
although some loops and hairpins are probably present.

We showed that inhibition of HNE by poly G was most effective when
the highest Mr polymers of poly G were employed. A similar association

between Mr and inhibitory activity was found by Lentini et al. for inhibition of HNE by sulfated polysaccharides (5).

A new observation, not previously reported, concerns the ability of poly G to inhibit ongoing elastinolysis. This observation is consistent with the limited kinetic data which indicate that catalysis, rather than substrate binding, is most markedly affected by interaction with polynucleotides.

The dependency of complex-formation upon ionic-associations between the polyribonucleotide and HNE was also demonstrated, as was the HNE-specific character of this reaction. Such observations were also reported previously (2). However, a novel feature of the binding mechanism, discovered in the course of the present study, is that it may involve both ionic and hydrophobic interactions between poly G and HNE. Thus, steroidal agents (bile salts) can interfere with the inhibition of HNE by poly G (Figure 5). Purines and pyrimidines might serve as the hydrophobic moieties interacting with HNE in the case of polyribonucleotides. Positive identification of the site at which polynucleotides bind to HNE will await the completion of enzyme structural determinations currently underway at a number of laboratories.

Agents which are capable of both hydrophobic and ionic interactions with HNE may be more effective inhibitors of this enzyme than are agents which bind HNE through electrostatic associations alone. Thus, we noted that a polyanionic polysaccharide lacking hydrophobic groups, such as the capsular polyribosylribitol phosphate from H. influenzae, was ineffective at inhibiting elastinolysis by HNE, if digestion was already in progress when the polyanion was added (Table I). Similar results were obtained with heparin (data not shown). By contrast, poly G retained significant inhibitory activity under these same conditions (see Table I). Indeed, the failure of polyribosylribitol phosphate to inhibit ongoing elastinolysis may help explain the severe lung injury that often accompanies H. influenzae pneumonia (1).

Some of the findings described above may prove useful in designing novel HNE-inhibitors for application in human inflammatory diseases, where HNE-mediated proteolysis of connective tissue substrates (16) or of components of the plasma cascades (17) is likely to be of pathogenetic significance. Additional studies designed to investigate this possibility are currently underway.

REFERENCES

1. Vered M, Simon S, Dearing R, Janoff A: Inhibition of human neutrophil elastase by bacterial polyanions. Exp. Lung Res.:00-00, 1987.
2. Lestienne, P, Bieth JG: Inhibition of human leukocyte elastase by polynucleotides. Biochimie 65: 49-52, 1983.
3. Baici A, Salgam P, Fehr K, Boni A: Inhibition of human elastase from polymorphonuclear leukocytes by a glycosaminoglycan polysulfate (Arteparon). Biochem Pharmacol 29: 1723-1727, 1980.
4. Baici A, Bradamante P: Interaction between human leukocyte elastase and chondroitin sulfate. Chem Biol Interact 51: 1-11, 1984.
5. Lentini A, Ternai B, Ghosh P: Synthetic inhibitors of human leukocyte elastase: Part 1- Sulfated polysaccharides. Biochem Internat 10:221-232, 1985.
6. Bieth J, Spiess B, Wermuth C: The synthesis and analytical use of a highly sensitive and convenient substrate of elastase. Biochem Med 11: 350-357, 1974.

7. Bielefeld, DR, Senior, RM, and Yu, SY: A new method for determination of elastolytic activity using ^{14}C labeled elastin and its application to leukocyte elastase. Biochim. Biophys. Res. Comm. 67:1553-1559, 1975.

8. Broido MS, Kearns DR: ^{1}H NMR evidence for a left-handed helical structure of poly (ribocytidylic acid) in neutral solution. J Am Chem Soc 104:5207-5216, 1982.

9. Segel IH: Enzyme Kinetics. Behavior and Analysis of Rapid Equilibrium and Steady-State Enzyme Systems. J. Wiley and Sons, New York. 1975, pp.178-187.

10. Vered M, Simon SR, Janoff A: Subcellular localization and further characterization of a new elastase inhibitor from pneumococci. Infect Immun 49: 52-60, 1985.

11. Twumasi DY, Liener IE: Proteases from purulent sputum. Purification and properties of the elastase and chymotrypsin-like enzymes. J Biol Chem 252: 1917-1926, 1977.

12. James MNG, Brayer GD, Delbaere LTJ, Sielckei AR, Gertler A: Crystal structure studies and inhibition kinetics of tripeptide chloromethyl ketone inhibitors with Streptomyces griseus Protease B. J Mol Biol 139: 432-438, 1980.

13. Vered M, Schutzbank T, Janoff A: Inhibitors of human neutrophil elastase in extracts of Streptococcus pneumoniae. Am Rev Respir Dis 130: 1118-1124, 1984.

14. Vered M, Janoff A, Simon S: Stimulation of human leukocyte elastase activity by bile salts and detergents. J. Cell. Biochem. Suppl. 10A: 255, 1986.

15. Stein, R: Catalysis by human leukocyte elastase. 4. Role of secondary subsite interactions. J. Am. Chem. Soc. 107:5767-5775, 1985.

16. Janoff A: Elastase in tissue injury. Ann Rev Med 36: 207-216, 1985.

17. Havemann K, Gramse M: Physiology and pathophysiology of neutral proteinases of human granulocytes. Adv Exp Med 167: 1-20, 1984.

INHIBITION OF HUMAN NEUTROPHIL ELASTASE
BY ACID-SOLUBLE INTER-α-TRYPSIN INHIBITOR

Alain Gast and Joseph G. Bieth

Laboratoire d'Enzymologie, INSERM Unité 237, Université Louis Pasteur, B.P. 10, 67048 Strasbourg Cédex (France)

In 1961 Steinbuch and Loeb (1961) described a new plasma protein named protein π. A few years later, Heide et al. (1965) purified a new plasma proteinase inhibitor, inter-a-trypsin inhibitor (ITI). These two proteins proved to be identical. ITI is a single-chain, very labile, Zn-containing glycoprotein (Steinbuch, 1976). It is a reversible inhibitor of serine proteinases which is much more potent on bovine trypsin (K_i = 2.1 x 10^{-11} M) than on bovine chymotrypsin (K_i = 1.08 x 10^{-9} M) (Aubry and Bieth, 1976). ITI inhibits human pancreatic trypsin and chymotrypsin with much less efficiency (Aubry and Bieth, 1976) and is inactive against porcine pancreatic elastase (Meyer et al., 1975). The physiological function of ITI is not well understood.

A variety of body fluids including serum, urine, bronchial and cervical mucus secretions contain low molecular mass derivatives of ITI. These fragments are generated by limited proteolysis of ITI and are soluble and stable in perchloric acid (Bretzel and Hochstrasser, 1976, Hochstrasser et al., 1976). Hochstrasser and coworkers have sequenced HI-14, the 14-kDa fragment derived from ITI by limited proteolysis. This fragment appears to be a Kunitz-type two-domain inhibitor having P_1 = Arg at the C-terminal domain and P_1 = Met at the N-terminal one (Wachter and Hochstrasser, 1979, Albrecht et al., 1983). The C-terminal domain is responsible for the inhibition of trypsin and chymotrypsin whereas the N-terminal domain is able to react with human neutrophil elastase. The latter is also inhibited by intact ITI (Albrecht et al., 1983, Jochum and Bittner, 1983).

Neutrophil elastase is involved in a variety of inflammatory diseases including lung damage (Bieth, 1986). Upper airways secretions contain a number of neutrophil elastase inhibitors including α1-proteinase inhibitor, bronchial inhibitor and perchloric acid soluble ITI derivatives (Kueppers and Bromke, 1983). In vitro inhibition of a proteinase by a proteinase inhibitor is not a sufficient criterion to claim that the inhibitor plays a physiological antiproteinase function. Knowledge of the in vivo concentration of the inhibitor and the kinetic constants describing the enzyme/inhibitor interaction to some extent helps delineating this physiological function (Bieth, 1980, Bieth, 1984). We therefore decided to measure the K_i for the interaction between human neutrophil elastase and the perchloric acid soluble ITI fragments in order to decide whether the fragmented inhibitor contributes to the anti-elastase screen of the upper respiratory tract. The inhibition of bovine pancreatic chymotrypsin was included for the sake of comparison. An abstract of the present paper has been published previously (Gast and Bieth, 1985).

Materials and methods

ITI was a generous gift from Dr. M. Steinbuch, Centre National de Transfusion Sanguine, Orsay. The inhibitor was shiped as a 30 g/l solution and kept frozen until used. This solution was reacted with perchloric acid (final w/v concentration = 5 %) during 16 h at 4°C, centrifuged at 10,000 x g for 15 min, neutralised with 5 N potassium hydroxide and centrifuged again. The inhibitor resulting from this treatment will be referred to as acid-soluble inhibitor or acid-soluble ITI.

The operational normality of native or acid-soluble ITI solutions was determined with porcine trypsin (Choay). The latter was itself titrated with p-nitrophenol-p'-guanidinobenzoate (Cyclo Chemicals) according to the procedure of Chase and Shaw (1967). Increasing amounts of ITI were incubated for 2 min at 25°C with constant amounts of trypsin (final conc. = 10^{-7} M) in 0.2 M Tris-HCl, 20 mM $CaCl_2$, pH 8.0. Residual trypsin activities were then measured at 25° C with 1 mM benzoyl-L-arginine-p-nitroanilide (Sigma). Linear inhibition curves were obtained for both ITI preparations (data not shown). Since one mole of ITI inhibits one mole of trypsin (Aubry and Bieth, 1976), these titration experiments could be used to calculate the molarities of active inhibitor present in solutions of native and acid-soluble ITI.

Human neutrophil elastase and cathepsin G, isolated from purulent sputum (Martodam et al., 1979), were active-site titrated with acetyl-(Ala)$_2$-azaAla-p-nitrophenylester (Enzyme System Products) (Powers et al., 1984), and N-trans-cinnamoylimidazole (Sigma) (Schonbaum et al. 1961), respectively. Increasing amounts of native or acid-soluble ITI were reacted for 15 min at 25°C with constant amounts of neutrophil elastase (final conc. = 1.9 x 10^{-7} M) in 0.2 M Tris-HCl pH 8.0. Residual elastase activities were then measured at 25°C with 1 mM succinyl-(Ala)$_3$-p-nitroanilide (Choay) (Bieth et al., 1974). A similar procedure was used to test the inhibitory effect of ITI on cathepsin G (final conc. = 4 x 10^{-7} M). The preincubation time was 30 min at 25°C and enzymatic activities were measured with 0.2 mM succinyl-(Ala)$_2$-Pro-Phe-p-nitroanilide (Bachem) (Nakajima et al., 1979).

Bovine pancreatic α-chymotrypsin (Worthington) was active-site titrated with N-trans-cinnamoylimidazole (see above) and used at a final concentration of 5.4 x 10^{-8} M. It was reacted with variable concentrations of ITI during 10 min at 25°C in 0.2 M Tris-HCl, 20 mM CaCl$_2$ pH 8.0. The residual enzymatic activities were measured with 0.1mM succinyl-(Ala)$_2$-Pro-Phe-p-nitroanilide.

Results

Native and acid-soluble ITI were able to inhibit trypsin, chymotrypsin and neutrophil elastase but not neutrophil cathepsin G. The figure shows the inhibition of human neutrophil elastase by native and acid-soluble ITI. Both inhibition curves are concaved, indicating moderately potent inhibition.

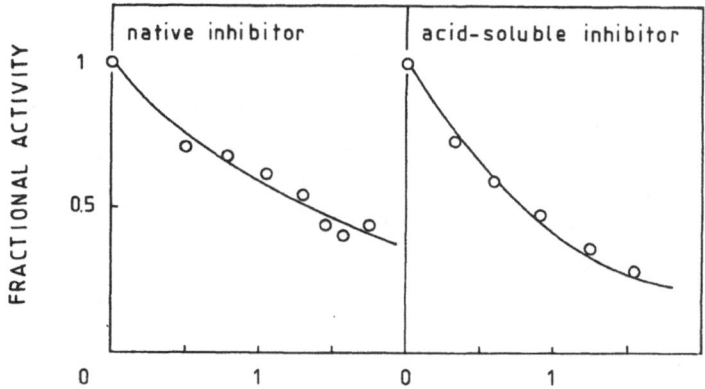

MOLE INTERα TRYPSIN INHIBITOR / MOLE ELASTASE

Figure. Inhibition of human neutrophil elastase by native and acid-soluble inter-α-trypsin inhibitor at pH8.0 and 25°C.

It can also be seen that acid-soluble ITI is a somewhat better inhibitor than native ITI. The inhibition curves obtained with bovine α-chymotrypsin were steeper but still concaved near the equivalence points (data not shown). The data were analysed assuming that ITI and proteinases form a reversible inactive complex. The reversibility of the chymotrypsin-ITI interaction has been demonstrated previously (Aubry and Bieth, 1976). The reversibility of elastase - ITI association may be inferred from the concaved shape of the inhibition curves (for comments, see Bieth, 1974). The equilibrium dissociation constants K_i of the complexes and their confidence intervals were calculated by non-linear regression analysis using a program fitting the data to the following equation (see for instance Bieth, 1974) :

$$a = 1 - \frac{(Eo) + (Io) + K_i\,(app) - \left\{ \left[(Eo) + (Io) + K_i\,(app) \right]^2 - 4\,(Eo)\,(Io) \right\}^{1/2}}{2\,(Eo)}$$

where a is the fractional enzymatic activity (see ordonate of the figure), (Eo) and (Io) are the total concentration of enzyme and inhibitor and K_i (app) is the apparent equilibrium dissociation constant of the complex related to K_i through the following relationship :

$$K_i\,(app) = K_i\,(1 + (So)/Km)$$

where (So) and Km have their usual meanings. Km values from the literature (Bieth et al., 1974, Nakajima et al., 1979) were used to convert K_i (app) into K_i.

The various K_i values are reported in the table. It can be seen that acid-soluble ITI exhibits a higher affinity for elastase and chymotrypsin than native ITI. Inspection of the confidence intervals shows that these differences are significant. The K_i (native ITI) /K_i (acid-soluble ITI) ratios are 2.3 for elastase and 3.3 for chymotrypsin.

Discussion

Low molecular mass acid-soluble ITI derivatives are usually prepared by reacting ITI or perchloric acid precipitated plasma proteins with a proteinase (Bretzel and Hochstrasser, 1976, Albrecht et al., 1983). In preliminary experiments we found that perchloric acid treatment of our ITI preparation yielded a solution that was able to inhibit trypsin, chymotrypsin and neutrophil elastase. Moreover, the acid-soluble inhibitor was able to react with anti-ITI antibodies as evidenced by radial immuno-diffusion. We therefore did not treat our ITIsolution with a proteinase prior to acidification and we compensated the lack of quantitative recovery of acid-soluble ITI by active-site titrating the recovered inhibitor with active-site titrated trypsin. From the above it follows that our ITI solution must have been partially contaminated by

Table   Inhibition of human neutrophil elastase and bovine pancreatic
        α-chymotrypsin by inter α trypsin inhibitor (ITI)

Enzymes	K_i (M)	
	Native ITI	Acid-soluble ITI
Elastase	$(5.3 \pm 0.3) \times 10^{-8}$	$(2.3 \pm 0.1) \times 10^{-8}$
Chymotrypsin	$(5.0 \pm 0.5) \times 10^{-10}$	$(1.5 \pm 0.5) \times 10^{-10}$

fragmented inhibitor so that the term "native ITI" used in this paper must be considered with some caution. ITI is a very labile protein (Steinbuch, 1976) which may be cleaved by contaminating plasma kallikrein (Bretzel and Hochstrasser, 1976). It is thus not surprising that part of our native ITI may in fact have been proteolyzed. As a consequence, the K_i values reported for native ITI must be considered as tentative the so more that the native and the acid-soluble inhibitor have significantly different dissociation constants for the two proteinases.

The K_i for the bovine chymotrypsin-native ITI interaction reported here (5×10^{-10} M) is close to that we have found in 1976 (1.08×10^{-9} M) (Aubry and Bieth, 1976). On the other hand, the K_i for the chymotrypsin-acid-soluble

ITI complex (1.5×10^{-10} M) is of the same order of magnitude as that of the chymotrypsin - 14 kDa ITI derivative (7.2×10^{-10} M) (Albrecht et al., 1983). The K_i for the neutrophil elastase - ITI complex has not been measured previously. Albrecht et al. (1983) simply stated that it is higher than 10^{-9} M, a finding that is consistent with our data. It is worth noticing that both chymotrypsin and elastase are more potently inhibited by the acid-soluble inhibitor than by native ITI. The two proteinases bind at two different domains of the inhibito r (Albrecht et al., 1983). Both domains are thus "activated" by the removal of the bulk of the carrier protein and this might suggest that ITI expresses its physiological antiproteinase function, if any, once converted into low molecular mass derivatives.

To prevent efficiently proteolysis in vivo, a reversible proteinase inhibitor like ITI must fulfil at least two conditions : (i) it must act fast enough to minimize substrate breakdown during enzyme-inhibitor complex formation (ii) it must have a pseudo-irreversible behavior with its target proteinase (Bieth, 1980, Bieth, 1984). The latter condition is fulfilled if $(I_0)/K_i \gg 10^3$ where (I_0) is the in vivo concentration of the inhibitor. Using this assumption and the K_i values for the native and fragmented ITI neutrophil elastase complexes, we get $(I_0) \geqslant 5 \times 10^{-5}$ M for the former and $(I_0) \geqslant 2 \times 10^{-5}$ M for the latter. The plasma ITI concentration is about 2×10^{-6} M (Steinbuch, 1976). ITI can therefore not play a significant antielastase function in plasma even in individuals with genetic deficiency of α_1-proteinase inhibitor, the fast-acting irreversible elastase inhibitor (Bieth, 1986). The same is true for upper airways secretions where the ITI concentration is not known precisely but does certainly not exceed the plasma concentration. To participate in the lung antielastase screen, ITI should have a K_i in the same order of magnitude as that reported for the bronchial inhibitor. The latter has a K_i of 3×10^{-10} M for neutrophil elastase and behaves as a pseudo-irreversible inhibitor of this enzyme in bronchial secretions since $(Io)/K_i$ is estimated to be greater than 1.6×10^4 (Boudier et al., this volume).

Acknowledgment

We wish to thank Dr. M. Steinbuch for her gift of inter-α-trypsin inhibitor.

References

Albrecht G. J., Hochstrasser K. and Salier J. P., 1983, Elastase inhibition by the inter-α-trypsin inhibitor and derived inhibitors of man and cattle, Hoppe-Seyler's Z. Physiol. Chem., 364 : 1703-1708.

Aubry M. and Bieth J. G., 1976, A kinetic study of the inhibition of human and bovine trypsins and chymotrypsins by the inter-α-inhibitor from human plasma, Biochem. Biophys., Acta 438 : 221-230.

Bieth J. G., Spiess B. and Wermuth C.G., 1974, The synthesis and analytical use of a highly sensitive and convenient substrate of elastase, Biochem. Med., 11 : 350-357.

Bieth J. G., 1980, Pathophysiological interpretation of kinetic constants of protease inhibitors, Bull. Europ. Physiopath. Resp., 16, (suppl.) : 183-195.

Bieth J. G., 1984, In vivo significance of kinetic constants of protein proteinase inhibitors, Biochem. Med., 32 : 387-397.

Bieth J. G., 1986, Elastases : catalytic and biological properties, in " Regulation of matrix accumulation", R.P. Mecham ed., Academic Press, New York.

Bretzel G. and Hochstrasser K., 1976, Liberation of an acid-stable proteinase inhibitor from the human inter-α-trypsin inhibitor by the action of kallikrein, Hoppe-Seyler's Z. Physiol. Chem., 357 : 487-489.

Chase T. and Shaw E., 1967, p-nitrophenyl-p'-guanidino benzoate HC1 : a new active site titrant for trypsin, Biochem. Biophys. Res. Commun., 29 : 508-514.

Gast A. and Bieth J. G., 1985, Inhibition of human leucocyte elastase and bovine pancreatic a-chymotrypsin by native and acid-treatment inter-α- inhibitor, Biol. Chem. Hoppe Seyler., 366 : 790.

Heide K., Heimburger N. and Haupt H., 1965, An inter-α-trypsin inhibitor of human serum, Clin. Chim. Acta., 11 : 82-85.

Hochstrasser K., Bretzel G., Feuth H., Hilla W. and Lempart K., 1976, The inter-α-trypsin inhibitor as precursor of the acid-stable proteinase inhibitors in human serum and urine, Hoppe Seyler's Z. Physiol. Chem., 357 : 153-162.

Jochum M. and Bittner A., 1983, Inter-α-trypsin inhibitor of human serum : an inhibitor of polymorphonuclear granulocyte elastase, Hoppe-Seyler's Z. Physiol. Chem., 364 : 1709-1715

Kueppers F. and Bromke B. J., 1983, Protease inhibitors in tracheobronchial secretions, J. Lab. Clin. Med., 101 : 747-757.

Martodam R. P., Baugh R. J., Twumasi D. and Liener I. E., 1979, A rapid procedure for the large scale purification of elastase and cathepsin G from human sputum, Prep. Biochem., 9 : 15-31.

Meyer J. F., Bieth J. G. and Metais P., 1975, On the inhibition of elastase by serum. Some distinguishing properties of α_1-antitrypsin and α_2-macroglobulin, Clin. Chim. Acta., 62 : 43-53.

Nakajima K., Powers J. C., Ashe B.M. and Zimmerman M., 1979, Mapping the extended substrate binding site of cathepsin G and human leukocyte elastase. Studies with peptide substrates related to the α_1-protease inhibitor reactive site, J. Biol. Chem., 254 : 4027-4032.

Powers J. C., Boone R., Carrol D. L., Gupton B. F., Kam C. M., Nishino N., Sakomato M. and Tuhy P. M., 1984, Reaction of azapeptides with human leukocyte elastase and porcine pancreatic elastase. New inhibitors and active site titrants, J. Biol. Chem., 259 : 4288-4294.

Schonbaum G. R., Zerner B. and Bender M. L., 1961, The spectrophotometric determination of the operational normality of an α-chymotrypsin solution, J. Biol. Chem., 236 : 2930-2935.

Steinbuch M. and Loeb J., 1961, Isolation of an α_2-globulin from human plasma, Nature., 192 : 1196.

Steinbuch M., 1976, The inter-α-trypsin inhibitor, Methods Enzymol., 45 : 760-772.

Wachter E. and Hochstrasser K., 1979, Kunitz-Type Proteinase Inhibitors Derived by Limited Proteolysis of the Inter-α-Trypsin Inhibitor, III. Sequence of the two kunitz-type domains inside the native inter-α-trypsin inhibitor, its biological aspects and also of its cleavage products, Hoppe Seyler's Z. Physiol. Chem., 360 : 1305-1311.

DEVELOPMENT OF EGLIN c AS A DRUG: PHARMACOKINETICS

H.P. Nick, A. Probst, and H.P. Schnebli

Ciba Geigy Ltd.
CH-4002 Basel

Leukocytic elastase, (especially in concert with cathepsin G) is a very powerful, nonspecific tissue degrading enzyme, capable of attacking virtually every protein including the structural proteins elastin and collagen. The enzyme has been clearly incriminated in the development of pulmonary emphysema and probably in the pulmonary damage observed in chronic bronchitis, septicaemia, shock and ARDS and is conspicuously present in the lungs of cystic fibrosis patients. Based on this it has been proposed that highly specific and potent inhibitors of elastase should retard progressive tissue destruction in some of these lung diseases.

Eglin c, from the leech *Hirudo medicinalis*, is a very potent and specific inhibitor of human leukocyte elastase and cathepsin G. It was first described by Ursula Seemüller in 1977[1]. Eglin c consists of a single polypeptide chain of 70 amino acids[2,3]. Striking features of this molecule are the lack of sulfur containing amino acids and its outstanding resistance against denaturation by heat, acid and proteolysis.

In order to develop Eglin c as a therapeutic agent, the gene for this polypeptide was cloned and expressed in *E. coli*[4]. Recombinant Eglin c has the same inhibitor properties as the material isolated from the leech, and differs only by the acetyl-blocked N-terminus. Eglin c interacts very rapidly[5] with its target enzymes (human leukocyte elastase and cathepsin G) to form stable one to one complexes. It does not significantly inhibit any of a large number of other mammalian proteinases tested, except for the digestive enzyme chymotrypsin and the mast cell chymase[6]. In particular it does not inhibit the enzymes of the vital clotting-, fibrinolysis- and complement cascades. Due to its peptidic nature and size, application of Eglin c to animals or patients is only possible by the i.v. route, intratracheally or topically.

Pharmacological effects of Eglin c

Eglin c was found to be highly effective in preventing the elastase induced experimental emphysema in hamsters[7]. In animals given Eglin c intratracheally one to eight hours prior to elastase treatement a dose-dependent protection against the changes in lung function caused by the elastase was observed at the examination 56 days after induction[8]. Complete protection against the effects of elastase could be seen with doses as low as 0.5 mg/animal (1 hour pretreatment). Eglin c itself did not cause any changes in lung function and produced no toxic effects to liver or kidneys. Histological examination and morphometry of the left lungs confirmed that Eglin c was highly effective in preventing development of emphysema due to the instilled elastase. In addition, Eglin c at 0.5 mg/animal effectively protected the hamsters against the elastase induced secretory cell metaplasia in the large intrapulmonary airways.

Eglin c was active in several separate experimental shock models: In the pig septicaemia model[9] it had a highly significant effect on survival and lung edema and partially spared plasma AT III, factor XIII and α-2-M. In a traumatic shock model, Eglin c attenuated the decrease in AT III and factor XIII[10].

All experiments in general pharmacology demonstrated that Eglin c is extremely well tolerated: No significant effects on the cardiovascular, the central nervous system or the blood clotting, the fibrinolysis and complement system were found.

Eglin c did not cause any serious adverse or toxic effects in rabbits and baboons after single intravenous bolus injections of 400 mg/kg (maximal injectable dose) or daily intravenous administration for 14 days (baboons) or 30 days (rabbits) of up to 100 mg/kg. No changes were noted in body and organ weights, blood and urine parameters, and no gross or microscopic changes were attributable to the compound (W. Classen, unpublished).

The main safety problem is the potential of Eglin c to induce allergic reactions. Eglin c does not induce histamine release from basophils by itself, and also does not interfere with the physiologic release due to immunologic or non-immunologic stimuli (N. Subramanian, unpublished). All animals in the subchronic toxicity studies were specifically investigated for the possible development of antibodies. At the end of the studies (2 weeks baboon i.v.; 4 weeks rabbit i.v.) none of the animals, even in the recovery groups had developed specific anti-Eglin c antibodies (Ch. Rordorf, unpublished). In addition, no evidence of immediate or delayed hypersensitivity was detected in the baboons during and after the two-week administration period.

Pharmacokinetics of Eglin c

Pharmacokinetics and disposition of Eglin c have been investigated in rats, rabbits and baboons after intravenous bolus administration of 10 mg/ml ^{3}H-labeled or cold substance.

Radiometry and an ELISA were employed to determine the total of labeled substances and unchanged Eglin c, respectively, in the biological materials.

The ^{3}H-label of Eglin c was biologically unstable. In rats and rabbits which received ^{3}H-labeled Eglin c the levels of unchanged substance and of total radioactivity correlated well up to 120 minutes post dose only (Fig. 1 A). Afterwards the ^{3}H radioactivity is mostly due to tritiated water, which is eliminated from the body very slowly.

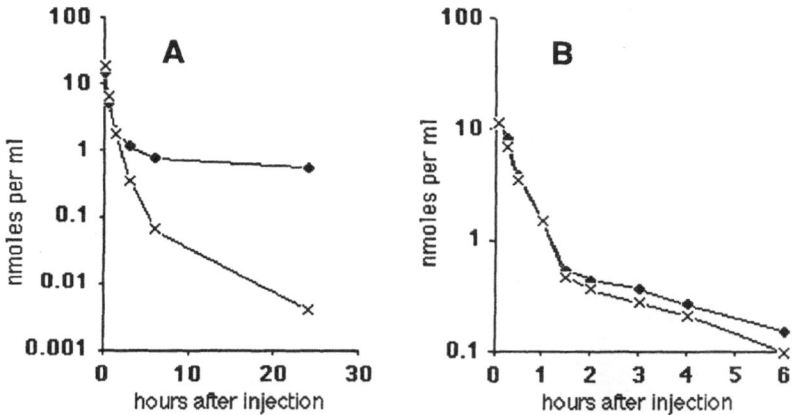

Fig. 1 **A:** Concentrations of total radioactive substances (^{3}H radioactivity, diamonds) and of unchanged Eglin c (ELISA, crosses) in plasma of rabbits (mean of two rabbits). Data from both (radiometry and ELISA) correlate well up to 120 minutes. Thereafter radioactivity decreases slower, due to the tritiation of water in the organism. **B:** Concentrations of unchanged Eglin c (ELISA) in plasma of two baboons after intravenous bolus injection. The two characteristic log-linear segments indicate a classical two compartment open-model system behaviour of which the first serves for the calculation of the half live for the distribution phase and the second for the calculation of the more important half live of the elimination phase.

The kinetics of unchanged Eglin c in plasma of rats, rabbits and baboons (Fig 1 B), as determined by ELISA, were qualitatively similar and can be described by a two-compartement open-model system. The plasma elimination half lives estimated using this kinetic model differ among the species, and amount to about 35 minutes in rats, 35-50 minutes in rabbits and 2-2.5 hours in baboons (Table 1).

Table 1 Plasma elimination half lives of Eglin c in various animal species; the respective half lives of the distribution phase and the elimination phase were calculated from the slopes of the two log-linear segments (exemplified in Fig. 1 B for baboon) of the plasma concentration time curves.

Species	Distribution Phase (min)	Elimination Phase (min)
Rat	10-15	35
Rabbit	10	35-50
Baboon	15	120-150

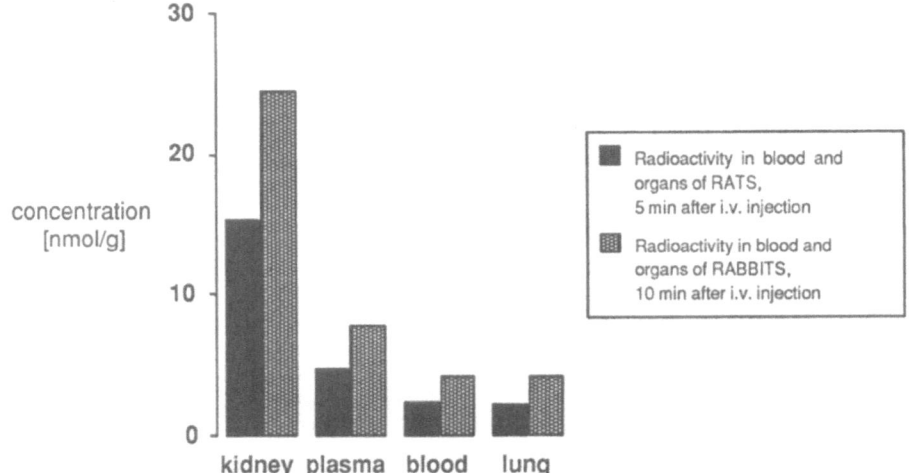

Fig. 2 Concentrations of total radioactive substances in blood and selected organs of rats and rabbits 5 and 10 minutes, respectively, after intravenous bolus injection of 10 mg/kg ^{3}H labeled Eglin c.

In rats and rabbits the patterns of ^{3}H distribution, shortly after administration of ^{3}H-labeled Eglin c (5 and 10 minutes, respectively), were similar and characterized by very high ^{3}H concentrations in the kidneys. High radioactivity concentrations were also observed in blood and lung (Fig. 2). In the remaining tissues and organs (e.g. heart, aorta, spleen muscle, sciatic nerve, adrenal, thyroid, sallivary, pancreas, stomach, small intestine) ^{3}H levels were lower than in blood. Since during the first period after administration radiometry and ELISA data correlated well in plasma, it is assumed that in these distribution experiments ^{3}H radioactivity represents mainly unchanged substance.

In all species tested (rat, rabbit, baboon) excretion of Eglin c, as assessed by ELISA, is rapid and occurs mainly through the kidneys. In the urine of rats, administered Eglin c is recovered quantitatively in unchanged form within 24 hours. In rabbits, the amount of unchanged Eglin c recovered in urine ranges between 60 and 70% of the dose, whereas in baboons urinary recovery amounts to 40-45% of the dose.

It can be concluded that in rats, rabbits and baboons Eglin c displays comparable dispositions, characterized by high total and renal clearances.

In other experiments, rats were infused intravenously with 10 mg/kg/h of Eglin c which resulted in levels of up to 2×10^{-7} mol per kg of wet lung weight (approximately 1/10 of the plasma concentration) after three hours (R. Stalder, unpublished). This indicates that therapeutically relevant amounts of Eglin c can reach the lung from the circulation.

Four hours after an intratracheal instillation of 0.4 mg ^{3}H-labelled Eglin c in hamsters, 33% of the tritium was still found in the lung tissue and bronchoalveolar lavage fluid. In the lavage fluid 83% of the radioactivity was associated with functionally active Eglin c (P. Stone and G.L. Snider, unpublished).

In pigs the tissue penetration of Eglin c was measured by monitoring the appearance of the molecule in the lymphe (ductus thoracticus) after a 4 hour i.v. infusion of about 4 mg/kg/h. Concentrations of Eglin c in lymphe were higher by a factor of about 2 compared to plasma. This was also found, although with some delay, in pigs with severe (experimental) sepsis and shock (M. Siebeck, H. Hoffmann, M. Jochum and H. Fritz, unpublished).

Conclusions

Half lives for the elimination of Eglin c from the plasma of the species tested are short, in the range of 30-50 minutes. However, given by continuous i.v. infusion over several hours, pharmacologicaly relevant Eglin c concentrations can be reached in lung lavage fluid and the interstitium. From the lung (its main target organ), Eglin c disappears more slowly than from plasma..

REFERENCES

1 U. Seemüller, M. Meier, K. Ohlsson, H.P. Müller, and H. Fritz, Isolation and characterisation of a low molecular weight in-hibitior (of chymotrypsin and human granulocytic elastase and cathepsin G) from leeches. Hoppe Seyler's Z. Physiol. Chem. 358: 1105 (1977).
2 U. Seemüller, M. Eulitz, H. Fritz, and A. Strobl, Structure of the elastase-cathepsin G inhibitor of the leech Hirudo medici-nalis. Hoppe Seyler's Z. Physiol. Chem. 361: 1841 (1980).
3 R. Knecht, U. Seemüller, M. Liersch, H. Fritz, D.G. Braun, and Y.J. Chang, Sequence determination of eglin c using combined microtechniques of amino acid analysis, peptide isolation, and automatic Edman degradation. Anal. Biochem. 130: 65 (1983).
4 H. Rink, M. Liersch, P. Sieber, and F. Meyer, A large fragment approach to DNA synthesis: total synthesis of a gene for the

protease inhibitor eglin c from the leech Hirudo medicinalis and its expression in E. coli. <u>Nucl. Acids Res.</u> 12:
6369 (1984).

5   N.J. Braun, J.L. Bodmer, G.D. Virca, G. Metz-Virca, R.
Maschler, J.G. Bieth, and H.P. Schnebli, Kinetic studies on
the interaction of eglin C with human leukocyte elastase
and cathepsin G. <u>Biol. Chem. Hoppe-Seyler</u> 368: 299 (1987).

6   E. Fink, R. Nettelbeck, and H. Fritz, Inhibition of mast
cell chymase by eglin c and antileukoprotease (HSUI - I).
Indications for potential biological functions of these inhibitors. <u>Biol. Chem. Hoppe-Seyler</u> 367: 567 (1986).

7   G.L. Snider, P.J. Stone, E.C. Lucey, R. Breuer, J.D. Calore,
T. Seshadri, A. Catanese, R. Maschler, and H.P. Schnebli,
Eglin-c, a polypeptide derived from medicinal leech, pre-
vents human neutrophil elastase-induced emphysema and
bronchial secretory cell metaplasia in the hamster. <u>Am. Rev.
Resp. Dis.</u> 132: 1155 (1985).

8   E.C. Lucey, P.J. Stone, T.G. Christensen, R. Breuer, J.D.
Calore, and G.L. Snider, Effect of varying the time interval between intratracheal administration of eglin-c and
human neutrophil elastase on prevention of emphysema and
secretory cell metaplasia in hamsters. With observation on
the fate of eglin-c and the effect of repeated instillations.
<u>Am. Rev. Resp. Dis.</u> 134: 471 (1986).

9   M. Jochum, H.F. Welter, M. Siebeck, and H. Fritz, Proteinase
inhibitor therapy of severe inflammation in pigs: first
results with eglin, a potent inhibitor of granulocyte
elastase and cathepsin G. <u>in:</u> Proteinases in Inflammation
and Tumor Invasion (H.Tschesche, ed.), Walter de Gruyter,
Berlin, New York, pp. 53 (1986).

10 C.E. Hock and A.M. Lefer, Beneficial effects of a neutral
protease inhibitor in traumatic shock. <u>Pharmacol. Res.
Commun.</u> 17: 217 (1985).

MONOCLONAL ANTIBODIES RECOGNIZING INTER-ALPHA-TRYPSIN-INHIBITOR AND ITS RELATED FRAGMENTS - EVIDENCE FOR THE INVOLVEMENT OF THE PROTEINASE INHIBITOR IN CUTANEOUS (PATHO-)PHYSIOLOGY

C. Justus, K. Hochstrasser* and M. D. Kramer

Ernst-Rodenwaldt-Institut, Fachbereich IV
Geschwister-Scholl-Str. 3, 65 Mainz, FRG
and *Klinikum Großhadern, 8000 München, FRG

INTRODUCTION

Proteolytic enzymes and their regulatory counterparts - the proteinase inhibitors - are likely to play an important role in the physiology and pathophysiology of human epidermis and dermis[1]. Numerous proteinases have been identified in the human skin [1,2,3] and have subsequently been characterized by chemical and immunological methods.

In the context of our studies on the role of proteolytic and anti-proteolytic compounds in human skin we asked whether inhibitory molecules with known activity against serine proteinases can be localized to human skin. As part of our work we raised monoclonal antibodies against Inter-alpha-Trypsin-Inhibitor (IaTI) and applied them for immunohistochemical studies on normal human skin.

IaTI is an inhibitor for serine proteinases first described and isolated from plasma in 1962; its function, however, was established not before 1965[4]. It is a single chain glycoprotein (Mr 180 000) containing 8.4 % carbohydrate. The purified inhibitor is unstable; however this does not affect its function. Molecular weight species of less than 50 000 - generated by proteolytic and/or as yet unknown mechanisms - can be obtained with no loss of antiproteolytic activity. IaTI inhibits human trypsin and chymotrypsin, but can also inhibit human acrosin, plasminogen activator and plasmin[4].

In the present publication we summarize the preliminar characterization of monoclonal antibodies recognizing either exclusively the prototype OaTI molecule (Mr 180 kDa) or the high molecular weight, total IaTI molecule and a lower molecular weight, trypsin-inhibitory fragment (HI 14) thereof. Here we describe the application of the moAbs (i) for construction of an enzyme immuno-assay suitable for quantification of IaTI and (ii) for immuno-histochemical studies on frozen sections of normal human skin.

MATERIALS AND METHODS

Human IaTI (the prototype molecule of 180 kDa) was purified by conventional methods from human plasma and was kindly provided by Dr. Heimburger (Behringwerke AG, Marburg, FRG). A low molecular weight, IaTI related inhibitory fragment (Mr of 14 kDa; termed "HI 14") was purified from human serum by affinity chromatography via immobilized pancreatic trypsin[5]..

Hybridoma derivation and selection and stabilization of moAb producing hybridoma lines was performed as described[6].

Performance of the enzyme linked immuno-sorbent assays (Elisa) for quantification of moAbs or immuno-reactive IaTI and the immuno-fluorescence methods are described in the legends to figures.

RESULTS

The presented moAbs were derived from two fusions of mice immunized with the prototype IaTI molecule (180 kDa). They were recognized by screening in Elisa assays for murine IgG antibodies. A representative Elisa experiment is shown in fig. 1. The moAbs I_1 (antibody isotype IgG1), I_2 (IgG1), I_4 (IgG2a) and I_{12} (IgG1) were tested for reactivity with the prototype IaTI molecule (180 kDa) and with the low molecular weight fragment of IaTI (HI 14; Mr 14 kDa), which contains the trypsin-inhibitory domain of the prototype inhibitor. From the presented data (fig. 1) it appears that moAb I_1 and I_2 react with both compounds (lower panel in fig. 1) whereas the moAbs I_4 and I_{12} react with the high molecular weight IaTI molecule only. These findings are confirmed by immuno-dot-blotting experiments, the results of which are given as inserts to fig. 1. Taken together we conclude from these experiments that moAbs I_1 and I_2 recognize antigenic epitopes that are located in or nearby the trypsin inhibitory domain of IaTI, whereas moAb I_4 and I_{12} recognize epitopes located moree distantly from this functional domain within the remaining non-inhibitory part of IaTI.

Fig. 2 illustrates reactivity of moAb I_1 and I_2 in immuno-fluorescence studies on frozen sections of normal human epidermis. Reactivity is prominent in the infrabasal dermal part of normal human epidermis (see fig. 3.A). However, by using moAb I_1 we observed - in addition to reactivity in the dermis - a significant net-shaped reaction within the lower epidermal compartment (see fig. 2.B and 2.C). This reflects most likely decoration of the intercellular space inbetween keratinocytes of the lower epidermis. Under pathological conditions (such as psoriatic alteration) this intra-epidermal reactivity seems to be more pronounced (data not shown); however, final conclusions as to this interesting matter have to await further systematic experimentation.

Future experimental approaches will particularly need sensitive assay systems for quantification of soluble IaTI in specimen of diverse origin. With respect to dermatological studies these specimen will include for example blister fluid and extracts of skin biopsies. For this reason we utilized our moAbs to establish a specific Elisa for quantification of immuno-reactive IaTI. Either moAb I_4 or I_{12} were coated to flat bottom microtiter plates and served as specific "catching" antibodies. Fig. 3 illustrates the result of a standardization experiment in which a dose-response curve was established with two-fold dilutions of purified IaTI between 50 and 0.7 ng/ml. The hydrolysis of the peroxidase substrate ortho-phenylene-diamine (as detected by absorbance at 492 nm) increased with IaTI concentration. This indicates that the assay - either based on moAb I_4 or I_{12} - allows detection of as low amounts as 1.5 ng/ml, which is in the picomolar range.

Flat bottom microtiter plates were coated at a concentration of 100 ng/well with either IaTI (●) or HI 14 (■). After incubation with hybridoma culture supernatants and copious washing, bound mono-clonal antibodies (identification given in the upper left corner of each quarter) were quantified via peroxidase labeled mouse specific anti-immunoglobulin antibodies.

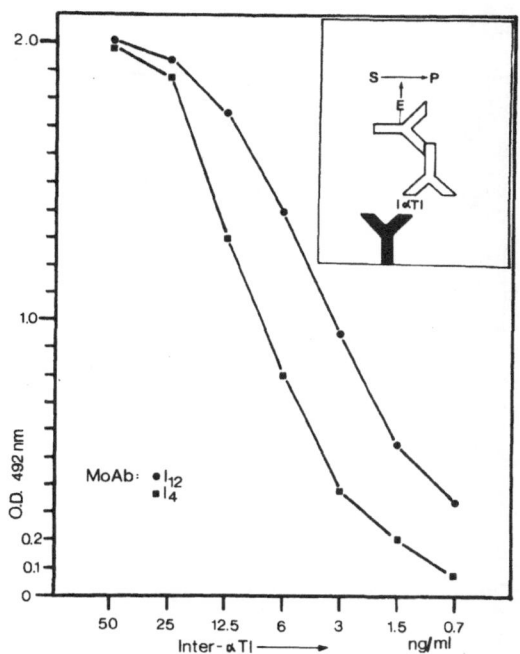

Fig. 1. Reactivity of moAbs I_1, I_2, I_4, and I_{12} with the prototype IaTI molecule (Mr - 180 kDa) and the related low molecular weight, trypsin-inhibitory fragment HI 14 (Mr - 14 kDa) in Elisa and immuno-dot-blotting.

In separate experiments graded amounts (100, 50, 25, 0 ng) of either IaTI (I), HI 14 (H), or Plasminogen (P; as control antigen) were immobilized onto nitrocellulose paper by means of a microtiter sized dot blotting apparatus. Reactivity with individual moAbs was visualized by appropriate peroxidase labeled anti-immunoglobulin reagents and diaminobenzidine as substrate.

Frozen sections of normal human skin (kindly provided by Dr. Tilgen, University Clinics, Dept. of Dermatology, Heidelberg, FRG) were analyzed by means of moAbs I_2 (2.A) or I_1 (2.B, 2.C). fig. 2.A: Hatched line indicates stratum corneum. Arrows are located within the dermal papillae below the epidermis and point at the epidermo-dermal junction. Note the strong reactivity within the infraepidermal dermis. Scale bar - 40 um. Fig. 2.B: Closed arrow is located intradermally and points at the epidermal junction. Open arrow indicates intraepidermal reactivity. Scale bar - 17 um. Fig. 2.C: Illustrates the net-shaped, intra-epidermal reactivity of moAb I_1.

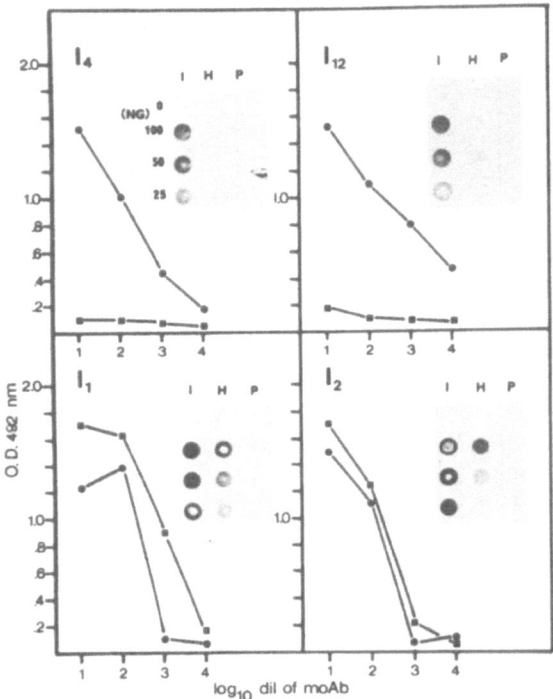

Fig. 2. Immunofluorescence staining with moAbs I_1 and I_2 of frozen sections of normal human skin.

Immunofluorescence studies were performed by a two-step procedure using fluoresceine-isothiocyanate labeled specific anti mouse immunoglobulin antibodies as second antibody. Specimen were examined using a Zeiss fluorescence microscope and photographed with a 3M 640 T color slide film (29 Din). The presented illustrations are black and white copies of these color slides.

The principle of this assay is demonstrated as insert in the upper right corner. The closed symbol represents either moAb I_4 or I_{12}.

In short: Affinity purified monoclonal antibody I_4 or I_{12} was coated as "catching antibody" onto flat bottom microtiter plates. After blocking of non-specific binding-sites by 0.2 % gelatine in phosphate buffered saline, graded amounts of IaTI (180 kDa; concentration given in the

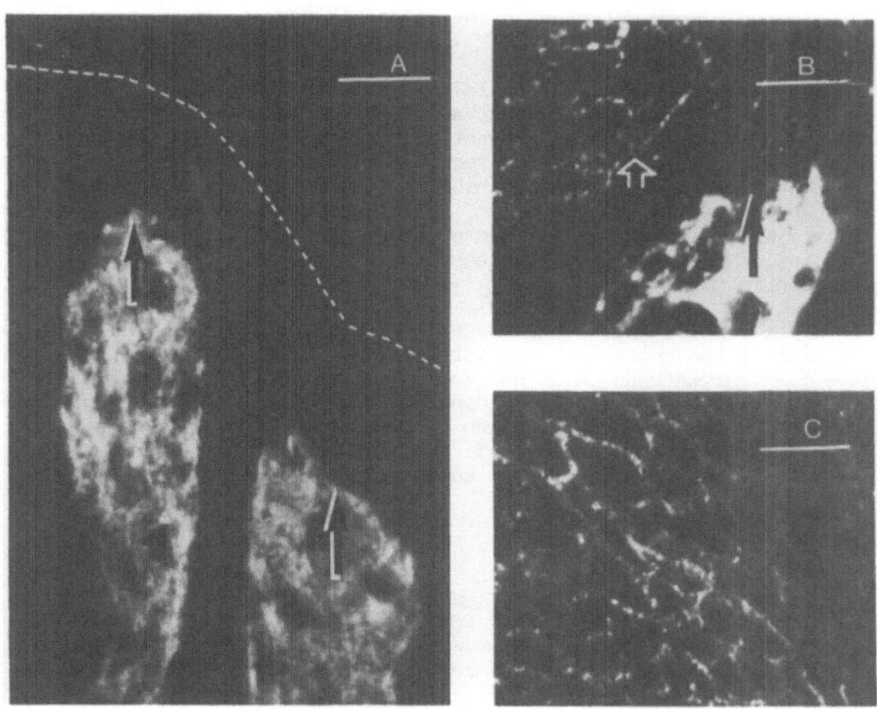

Fig. 3. Solid-phase immuno-assay for quantification of IaTI

abscissa) were given into the microwells. After binding and copious washing a polyclonal rabbit anti IaTI antiserum (kindly provided by Dr. N. Heimburger, Behringwerke AG, Marburg, FRG) followed by a peroxidase labelled rabbit specific anti-immunoglobulin antibody preparation was used as detection system. The color change produced by turnover ot the peroxidase substrate OPD was quantified by absorbance measurement at 492 nm using an automated Elisa reading apparatus.

DISCUSSION

By means of our novel IaTI specific moAbs we found evidence for the presence of this proteinase inhibitor in normal human skin: IaTI specific immuno-reactivity can be localized to the intercellular space inbetween cells of the lower epidermis and is diffusely distributed in the dermis. Moreover, we found in separate experiments that IaTI is contained in the fluid of suction blisters experimentally induced on normal epidermis (data not shown).

Whether the observed IaTI is synthesized in loco or - more likely - whether it originates from plasma via the interstitial fluid cannot be decided on the basis of our data. Notably, other soluble plasma components have previously been detected in the epidermis and are thought to fulfill important functions there[7].

What may be the function of IaTI in skin? By counteracting trypsin- and chymotrypsin-like proteinases, dermal and epidermal IaTI may play an important role in the proteinase-antiproteinase equilibrium in skin. Most interestingly, it appears from a recent publication[8] that compounds molecu- larly identical to the trypsin-inhibitory domain of IaTI act as growth factor on certain eucaryotic cells. This finding, together with the data on the localisation of IaTI presented here, opens an interesting perspective to studies on proliferative and regenerative phenomena in normal and diseased skin.

Future studies will be addressed to the following topics: (i) charac- terization of the epidermis associated IaTI molecule(s) (by means of immuno-chemical methods using our moAbs). (ii) identification of these enzymes that interact with IaTI in normal and/or pathological skin (this may be achieved by co-precipitation of the unknown enzymes in complex with the inhibitor by means of IaTI specific moAbs) and (iii) investiga- tions on the possible interaction between IaTI (or its split products) and keratinocytes in vitro.

REFERENCES

1. V. K. Hopsu-Havu, J. E. Fräki, and N. Järvinen, Proteolytic enzymes in the skin, in "Proteinases in mammalian cells and tissues," Barrett, ed., Biomedical Press, Elsevier/North-Holland (1977).

2. C. Justus, S. Müller, and M. D. Kramer, A monoclonal antibody recogni- zing plasminogen/plasmin - Altered reactivity in psoriatic lesions, Br. J. Dermatol. (in press).

3. J. E. Fraki, G. S. Lazarus, and R. S. Gilgor, Correlation of epidermal plasminogen activator activity with disease activity in psoriasis, Br. J. Dermatol. 108:39 (1983)

4. W. Gebhard and K. Hochstrasser, Inter-alpha-trypsin inhibitor and its close relatives, in "Proteinase Inhibitors," Barrett and Salvesen, eds., Biomedical Division, Elsevier Science Publishers BV (1986)

5. K. Hochstrasser and E. Wachter, Kunitz-type proteinase inhibitors de- rived by limited proteolysis of the Inter-alpha-Trypsin Inhibitor, Hoppe Seyler's Z. Physiolog. Chem., 364:1679 (1983).

6. J. H. Peters, H. Baumgarten, and M. Schulze, Monoklonale Antikörper. Herstellung und Charakterisierung. Springer-Verlag, Berlin, Heidel- berg, New York, Tokyo (1985).

7. G. Coruh and D. Y. Mason, Serum proteins in human squamous epithelium, Br. J. Dermatol. 102:497 (1980)

8. E. L. McKeehan, Y. Sakagami, H. Hiroyoshi, and K. McKeehan, Two apparent human endothelial cell growth factors from human hepatoma cells are tumor-associated proteinase inhibitors, J. Biol. Chem. 261:5378 (1986)

INHIBITION OF HUMAN CHYMOTRYPSIN-LIKE PROTEASES BY α_1-PROTEINASE INHIBITOR AND α_1-ANTICHYMOTRYPSIN

Annette Hayem, Dominique Marko, Anne Laîne, and Monique Davril

Unité INSERM N°16

Place de Verdun 59045 Lille Cedex France

Among human chymotrypsin-like proteases, cathepsin G originates from leukocytes and up to now, its pathological role has not yet been elucidated even if it might be implicated in lung destruction during emphysema (Boudier et al., 1984) ; however, it is involved in all inflammatory processes. In an other hand, chymotrypsin A and pancreatic elastase 2 are secreted by pancreas and released in the circulation during acute pancreatic diseases. The proteolytic activity of these enzymes is regulated by the inhibitory capacity of some proteins. In the serum, α_1-proteinase inhibitor (α_1PI) and α_1-antichymotrypsin (α_1Achy) are the main serine-protease inhibitors, both of them being able to inhibit chymotrypsin-like proteases.

In this paper, we determined the kinetic constants of inhibition of these enzymes in order to infer from that "in vitro" data some "in vivo" conclusions.

MATERIAL AND METHODS

Cathepsin G was obtained from purulent sputum using the procedure of Martodam et al., (1979). Chymotrypsin A was purified from human pancreatic juice (Vercaigne-Marko et al., 1987) ; human pancreatic elastase 2 (HPE 2) was prepared from human pancreas according to LARGMAN et al. (1976) and was a gift from Dr M. Rabaud (U-8 INSERM, Pessac, France).

α_1PI and α_1Achy were purified in the laboratory using immunoaffinity chromatography (Davril et al., 1987). Their purity was controlled by crossed immunoelectrophoresis and polyacrylamide gel electrophoresis under denaturing conditions. Their inhibitory capacity was assessed against bovine trypsin and bovine chymotrypsin respectively.

The association rate constants (k_{ass}) were determined at 25°C under the conditions summarized in Table I ; the substrates were from Bachem ; enzymes and inhibitors were allowed to react in molar ratios near the equivalence point.

All reagents were of analytical grade.

Table I
Conditions for the determination of association rate constants

Enzymes	Inhibitors	Substrates		Buffers
Cathepsin G	α_1PI	Suc-(Ala)$_2$-Pro-Phe-Nan	1 mM	0.1M Hepes pH 7.4, 0.5M NaCl 9% DMSO
	α_1Achy		4 mM*	0.05M TRIS/HCl pH 8.0 0.5M NaCl
Chymotrypsin A	α_1PI	Suc-(Ala)$_2$-Pro-Phe-Nan	0.2 mM	NaCl/Pi pH 7.4
	α_1Achy			
HPE 2	α_1PI	Glt-(Ala)$_2$-Pro-Phe-Nan	1 mM	0.2 M TRIS/HCl pH 8.0
	α_1Achy	Suc-(Ala)$_2$-Pro-Phe-Nan	5 mM	

*Beatty et al., 1980

Table II
Association rate constants at 25°C ($M^{-1}xs^{-1}$)

Enzymes	α_1PI	α_1Achy
Cathepsin G	1.75×10^5	5.1×10^7*
Chymotrypsin A	3.3×10^6	1.0×10^5
HPE 2	5.6×10^5	8.9×10^5

*given by Beatty et al., 1980

Table III
Delay time of inhibition in normal serum

Enzymes	α_1PI	α_1Achy
Cathepsin G	0.28 s	10 ms
Chymotrypsin A	12 ms	1.4 s
HPE 2	0.18 s	0.62 s

RESULTS AND DISCUSSION

The k_{ass} values for the association of each enzyme with both inhibitors are indicated in Table II. The k_{ass} value for cathepsin G and α_1PI is of the same order of magnitude as that determined by Beatty et al. (1980). These results confirm that α_1Achy is the specific inhibitor of cathepsin G. Concerning the inhibition of chymotrypsin A, α_1PI is more efficient than α_1Achy, since it reacts 30 times faster. An equivalent efficiency of both inhibitors reacting "in vitro" is noticed in the inhibition of HPE 2 since the two k_{ass} values are of the same order of magnitude.

As can be seen from these results, both inhibitors are able to inhibit these three chymotrypsin-like proteases which present the same specificity. Moreover, both inhibitors belong to the serpin family and are very close to each other in the phylogenetic tree.

The values of the inhibition delay times calculated assuming that α_1PI and α_1Achy serum concentrations are 50 µM and 9 µM respectively in normal serum, are indicated in Table III. As can be seen from these results, α_1PI is the physiological inhibitor of chymotrypsin A while α_1Achy preferentially inhibits cathepsin G. As far as HPE 2 is concerned, the respective roles of the two inhibitors would depend on the enzyme concentration in plasma (Davril et al., 1987).

These conclusions have to be reexamined in α_1PI deficiencies and inflammatory processes.

REFERENCES

Beatty, K., Bieth, J., Travis, J. (1980). Kinetics of association of serine proteinases with native and oxidized α_1-proteinase inhibitor and α_1-antichymotrypsin. J. Biol. Chem., 255:3931-3934.

Boudier, C., Laurent, P., Bieth, J. (1984). Leukoproteinases and pulmonary emphysema : cathepsin G and other chymotrypsin-like proteinases enhance the elastolytic activity of elastase on lung elastin. in: Advances in Experimental Medicine and Biology, 167:313-317.

Davril, M., Laine, A., Hayem, A. (1987). Studies on the interactions of human pancreatic elastase 2 with human α_1-proteinase inhibitor and α_1-antichymotrypsin. Biochem. J. (in press)

Largman, G., Brodrick, J.W., Geokas, M.C. (1976). Purification and characterization of two human pancreatic elastases. Biochemistry, 15:2491-2500.

Martodam, R.R., Baugh, R.J., Twumasi, D.Y., Liener, I.E. (1979). A rapid procedure for the large scale purification of elastase and cathepsin G from sputum. Prep. Biochem., 9:15-31.

Vercaigne-Marko, D., Carrère, J., Ducourouble, M.P., Davril, M., Laine, A., Amouric, M., Figarella, C., Hayem, A. (1987). "In vivo" and "in vitro" inhibitions of human pancreatic chymotrypsin A by serum inhibitors. Biol. Chem. Hoppe-Seyler, 368:37-45.

IMMUNOREACTIVE PANCREATIC SECRETORY TRYPSIN INHIBITOR IN GASTROINTESTINAL MUCOSA

M.Bohe, C.Lindström, and K.Ohlsson

Departments of Surgery, Pathology, and Surgical
Pathophysiology
University of Lund, Malmö General Hospital
214 01 Malmö, Sweden

INTRODUCTION

Pancreatic secretory trypsin inhibitor (PSTI), a specific trypsin inhibitor, has been isolated from pancreatic tissue and pancreatic juice[1,2].Increased serum levels of immunoreactive PSTI (irPSTI) have been reported in patients with pancreatic diseases[3,4] but also in patients with extrapancreatic carcinomas, inflammatory gastrointestinal diseases and after major abdominal surgery[5,6,7,8,9]. In patients operated on with total pancreatectomy normal serum levels of irPSTI have been found[9]. These results indicate an additional extrapancreatic source of PSTI. The aim of the present study was to analyse different types of gastrointestinal mucosa for the content and localization of irPSTI.

MATERIAL

Mucosal specimens were obtained at resectional surgery from various cases.,The following types of mucosa were represented: gastric mucosa of normal and intestinal metaplastic type, normal duodenal, jejunal, ileal and colonic mucosa. The mucosal specimens were fixed in 10% buffered formalin and embedded in paraffin for immunohistochemical analyses, or frozen at -20 $^{\circ}$C.

METHODS

The peroxidase-antiperoxidase method described by Sternberger et al[10] was used for the immunohistochemical analyses. Rabbit antiserum against human PSTI was used in dilution 1:2,000. Further sections from each block were used as control using non-immune rabbit serum and antiserum previously absorbed with PSTI. Rabbit antiserum against PSTI was available at the laboratory[2]. PSTI was purified as described earlier[2].

The ileal and colonic mucosal specimens were washed in icecold saline and frozen at -20°C. The mucosa was then homogenized in four parts 0.05 mol acetic acid containing 500 KIE aprotinin (Trasylol[R])/ml. The homogenates were then freeze-thawed five times and centrifuged at 30,000xG for 30 minutes at +4°C. The supernatants were used for further analyses.

Characterization of PSTI from mucosal homogenates. Bovine trypsin was coupled to CNBr-activated Sepharose 4 B according to the manufacturer's instructions. The homogenates were applied on the Sepharose-linked trypsin   column (0.9x10cm) equilibrated with 0.1 mol tris HCl buffer containing 0.15 M NaCl pH 7.0. After thorough washing with the starting buffer until absorbance at 280 nm reached zero, PSTI and Trasylol were eluted with 0.1 mol formic acid containing 0.5 M NaCl pH 3.0. The PSTI and Trasylol containing fractions were dialyzed with 0.05 M NaAc pH 4.3 and applied on a CP Sephadex C-50 (1.6x20cm)   equilibrated with 0.05M NaAc pH4.3 and eluted with a NaCl gradient (0-0.6 mol).

The inhibiting activity of PSTI against trypsin was tested using BZ-Ile-Glu-Gly-Arg-4NA (Serva, New York) as a substrate[11].

RESULTS

Normal gastric mucosa: In gastric mucosa of fundus type a marked PSTI immunoreactivity was demonstrated in the upper parts of the mucosa corresponding to the foveolar cells of the gastric pits. The lower two thirds of the mucosa containing the acid and pepsinogen producing cells mainly lacked irPSTI (fig.1).

In gastric mucosa of antrum type a marked PSTI immunoreactivity was found in all parts of the mucosa especially in the middle and upper thirds corresponding to the foveolar cells (fig.2).

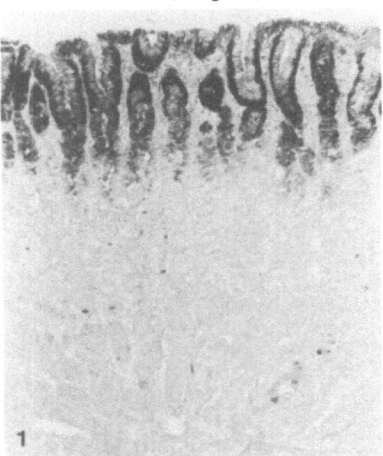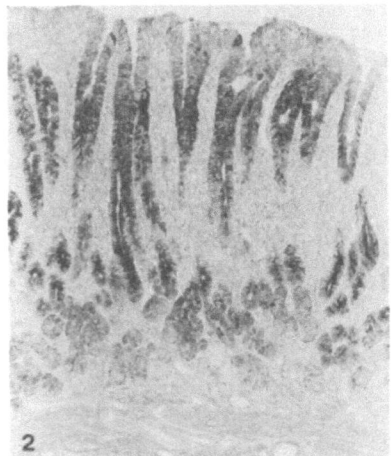

Fig 1. Gastric mucosa of fundus type. There is a high level of PSTI in the upper (foveolar) part of the mucosa.The lower 2/3 of the mucosa contains the oxyntic glands, mainly lacking in PSTI. x 60

Fig 2. Antral mucosa. There is a high level of PSTI in all parts of the mucosa, especially in the middle and upper 1/3, corresponding to the foveolar cells.x50

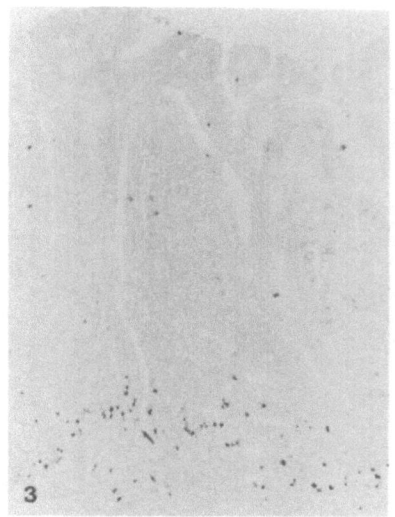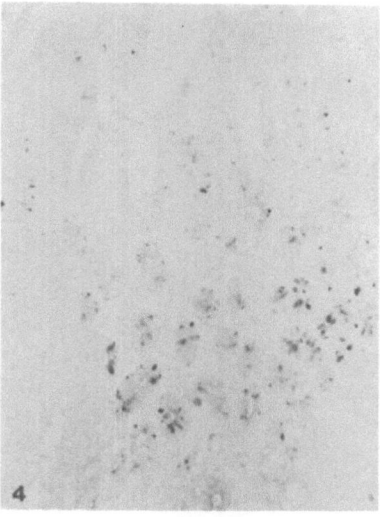

Fig 3. Antral mucosa with chronic gastritis and focal
       intestinal metaplasia. There is a marked
       deficiency of PSTI in the metaplastic mucosa. x 60

Fig 4. Duodenal mucosa. Occurence of PSTI mainly in
       Paneth cells and basal goblet cells. x 60

Gastric mucosa of intestinal metaplastic type: In gastric
mucosa with intestinal metaplasia there was a marked
deficiency of irPSTI in the upper parts of the mucosa and
only a slight PSTI immunoreactivity in the basal parts of the
mucosa mainly corresponding to cells of Paneth type and of
goblet type (fig. 3).
Duodenum: In duodenal mucosa a marked PSTI
immunoreactivity was found in the Paneth cell region and in
some goblet cells in the basal parts of the crypts. The
Brunner's glands exhibit a varying degree of immunoreactivity
(fig.4).
Ileal and jejunal mucosa: PSTI immunoreactivity was found
in the Paneth cells and the goblet cells in the basal parts of
the glandular crypts (fig.5).
Colon: In normal colonic mucosa a marked positive PSTI
immunoreactivity was found in the goblet cells in the basal
parts of the crypts (fig 6). In cases with metaplastic areas
PSTI immunoreactivity was also found in the Paneth cells.
Controls: In sections using non-immune rabbit antiserum
or antiserum previously absorbed with PSTI (1 microg/ml) no
staining reaction was found.
Ileal mucosa contained 1.3 +- 0.2 microg/g irPSTI (m +-
SD). Colonic mucosa contained 1.2+-0.2 microg/g wet weight of
irPSTI(m +- SD). After separation on a SP Sephadex the
immunoreactive PSTI eluted in three peaks similar to PSTI from
pancreatic juice[2].
PSTI from the gastrointestinal mucosa inhibited trypsin
in a 1-1 molar way.

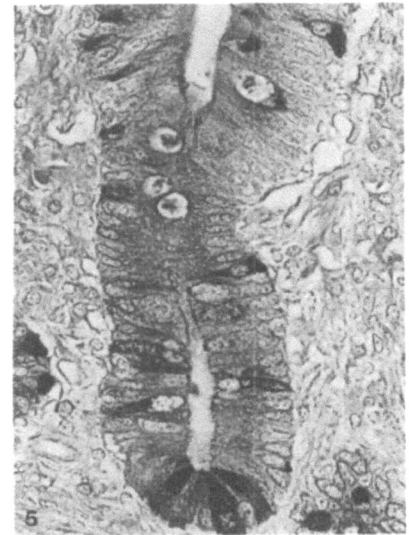 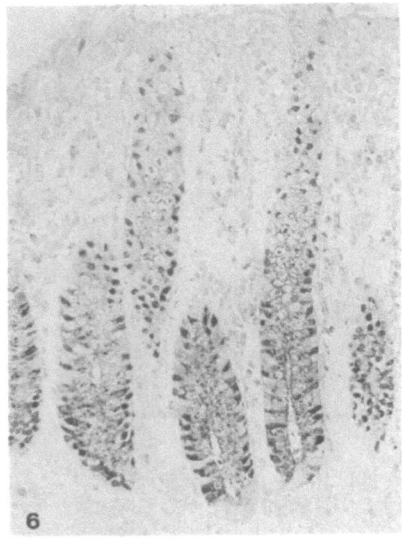

Fig 5. Normal jejunal mucosa. PSTI demonstrated in the Paneth cells and goblet cells. x 600

Fig.6 Colonic mucosa. PSTI demonstrated in the goblet cells. x 160

SUMMARY

This study has shown that ileal and colonic mucosa contained roughly 1.2 microg/g of irPSTI. irPSTI from gastrointestinal mucosa eluted in a smilar way to that of native PSTI after chromatographic separation and inhibited trypsin in a 1-1 molar way. The PSTI immunoreactive material was localized in the Paneth cells and the goblet cells in the small and large intestine. In normal gastric mucosa it was found in the foveolar cells while the acid and pepsinogen producing cells lacked PSTI immunoreactive material. In gastric mucosa with intestinal metaplasia a marked deficiency of PSTI was found. These findings indicate that gastrointestinal mucosa could be an additional source of irPSTI. Further studies are needed to elucidate if PSTI is involved in the defence of the gastrointestinal mucosa.

REFERENCES

1.   Kazal,L.A., Spicer, D.S., and Brahinsky, R.A., Isolation of a crystalline trypsine inhibitor-anticoagulant protein from pancreas. J.Am.Chem.Soc.70: 3034-3040 (1948).
2.   Eddeland, A., Ohlsson, K., Purification and immunochemical quantitation of human pancreatic secretory trypsin inhibitor. Scand.J.Clin. Lab. Invest. 38:261-267 (1978).

3. Eddeland, A., Ohlsson, K., A radioimmunoassay for measurement of human pancreatic secretory trypsin inhibitor in different body fluids. Hoppe-Seyler's Z. Physiol. Chem. 359: 671-675 (1978).

4. Otsuki, M., Oka, T., Suehiro, I., Okabayashi, Y., Ohki, A., Yuu, H., and Baba, S., Serum pancreatic secretory trypsin inhibitor in pancreatic disease. Clin. Chim. Acta 142:231-240 (1984).

5. Huhtala, M.-L., Kahanpää, K., Seppälä, M., Halila, H., Stenman, U.-H., Excretion of a tumor-associated trypsin inhibitor (TATI) in urine of patients with gynecological malignancy. Int.J.Cancer 31: 711-714 (1983).

6. Murata, A., Ogawa, M., Matsuda, K., Matsuura, N., Kosaki, G., Inoue, M., Ueda, G., Kurachi, K., Immunoreactive pancreatic secretory trypson inhibitor in gynecological diseases. Res.Commun. Chem. Path. Pharm. 41: 493-499 (1983).

7. Matsuda, K., Ogawa, M., Murata, A., Kitahara, O., Kosaki, G., Elevation of serum immunoreactive pancreatic secretory trypsin inhibitor contents in various malignant diseases. Res. Commun. Chem. Path Pharm. 40: 301-305 (1983).

8. Lasson, Å., Borgström, A., Ohlsson, K., Elevated PSTI-levels during severe inflammatory disease, renal insufficiency and after various surgical procedures. Scand.J.Gastroenterol. 21: 1275-1280 (1986).

9. Matsuda, K., Ogawa, M., Shibata, T., Nishibe, S., Miyauchi, K., Matsuda, Y., Mori, T., Postoperative elevation of serum pancreatic secretory trypsin inhibitor. Am. J. Gastroenterol. 80: 694-698 (1985).

10. Sternberger, L.A., Hardy Jr., P.H., Cuculis, J.J., Meyer, H.G., The unlabelled antibody enzyme method of immunochemistry. Preparations and properties of soluble antigen-antibody complex (horseradish peroxidase-antihorseradish peroxidase) and its use in identification of spirochetes. J. Histochem. Cytochem. 18: 315-333 (1970).

11. Claeson, G., Aurell, L., Karlsson, G, Gustavsson, S, Friberger, P., Arielly, S., Simonssen, R. In: Scully, M.F., Kakkar, V.V. (eds.), Chromogenic Peptide Substrates, Churchill Livingstone, London, pp. 20-31 (1979).

SEMISYNTHETIC INHIBITORS OF HUMAN LEUKOCYTE ELASTASE AND THEIR PROTECTIVE EFFECT ON LUNG ELASTIN DEGRADATION IN VITRO

J. Beckmann, A. Mehlich, H. R. Wenzel and H. Tschesche

Universität Bielefeld
Fakultät für Chemie
Postfach 86 40
D-4800 Bielefeld 1, F. R. G.

Proteinases and the control of their activity by inhibitors are involved in many biological processes (1). A proteinase/proteinase inhibitor imbalance hypothesis has been proposed to account for several pathological situations such as rheumatoid arthritis (2,3) or pulmonary emphysema (4). The latter is a reduction of lung function due to a greater proteolytic degradation of elastin either by an increase of proteinase release, deactivation of local proteinase inhibitors or a hereditary deficiency of α_1-proteinase inhibitor (α_1-PI). A disturbance of the elastin cross linking also appears to be involved (5,6).

The elastase of human polymorphonuclear leukocytes (HLE) (7), which was shown to produce emphysema in several animal models possibly plays a prominent role (8). The appearance of a HLE-induced emphysema could be prevented by administering a corresponding proteinase inhibitor (9). Inhibitors of this elastase and their interaction with the enzyme, especially in the presence of the physiological substrate elastin, are thus of particular interest.

Low molecular weight protein inhibitors of serine proteinases react with their target enzymes in a substrate like manner (10). The peptide bond P_1-P_1' in the Schechter and Berger notation (11) corresponds to the bond which is split in a substrate. Its hydrolysis by an inhibited proteinase is impaired by the unusual kinetics of the proteolytic reaction. In many cases the proteinase-catalyzed hydrolysis resynthesis equilibrium of this particular reactive site bond is altered by the disulphide cross-links of the inhibitor, so that after an infinite reaction time a great amount of protein with an intact P_1-P_1'-bond is found. The specificity of the inhibitor is mainly determined by the amino acid residue in the P_1-position.

Recently we described a semisynthetic method which allows for the exchange of the P_1-lysine of the bovine pancreatic trypsin inhibitor (BPTI) (12) for nearly any coded or non-coded amino acid yielding inhibitor homologues with new target enzymes (13). The native inhibitor shows an extremely high affinity to bovine trypsin but the inhibition of HLE is hardly detectable. The equilibrium constants of the complexes differ by a magnitude of about eight orders (14-16). On the other hand Val^{15}-BPTI inhibits HLE strongly with a complex dissociation constant of 1.1×10^{-10} M whereas no affinity of this derivative for bovine trypsin could be detected.

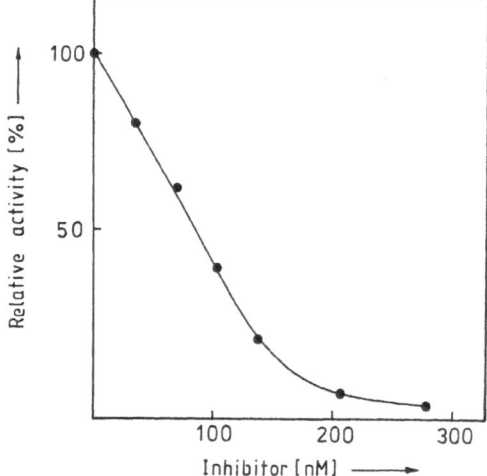

Fig. 1. Val^{15}-BPTI inhibition of the breakdown of insoluble human lung elastin by HLE

0.18 μM HLE, 2 mg/ml suspended elastin (17), 0.15 M NaCl, 0.01 % NaN_3, 0.2 M Tris/HCl pH 8.0, Val^{15}-BPTI concentrations as indicated in the diagram. The elastin solubilization was monitored by taking samples from the suspension, which was continuously stirred at 37°C, diluting with an equal volume of 0.5 M Na-acetate pH 4.5, centrifuging and reading the absorption at 280 nm.

Several of our semisynthetic BPTI-derivatives were able to block the hydrolysis of the chromogenic substrate MeO-Suc-Ala$_2$-Pro-Val-Nan (13,18). The best inhibitor of this series, Val[15]-BPTI was used for similar studies using suspended human lung elastin. Incubation of HLE, human lung elastin and various amounts of inhibitor at the same time gave a normal inhibition plot (Fig. 1), with 10 % of the relative activity for equimolar amounts of enzyme and the BPTI-derivative. As the inability of the physiological HLE inhibitor α_1-PI to totally inactivate pre-bound enzyme (19), this reflects the high affinity between elastin and elastase.

In contrast to the physiological HLE inhibitor α_1-PI, Val[15]-BPTI entirely inhibited the elastin breakdown by the pre-bound enzyme (Fig. 2). Nevertheless, 3 h were needed to stop the proteolytic solubilization of the substrate, if two equivalents of inhibitor were added.

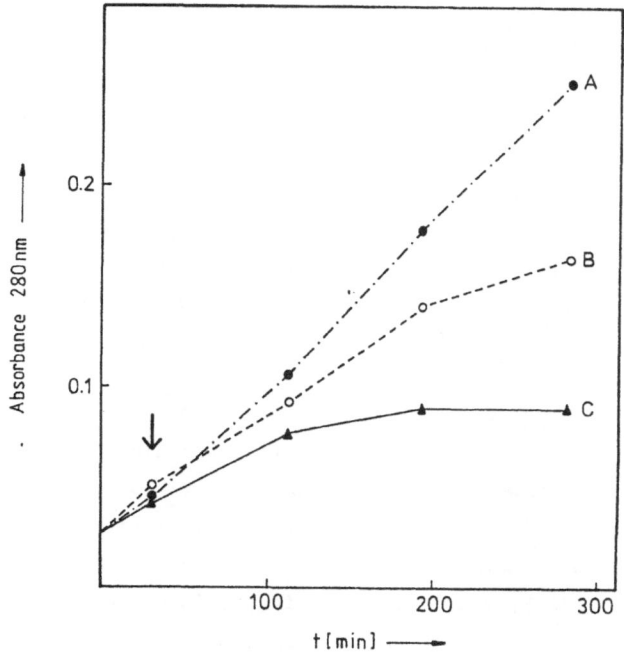

Fig. 2. Kinetics of the Val[15]-BPTI inhibition of the breakdown of human lung elastin

After 30 min pre-incubation of the enzyme and substrate 0.00 (A), 0.15 (B) and 0.36 μM Val[15]-BPTI (C) were added (arrow). Other conditions: see legend to Fig. 1.

The association reaction of a proteinase (E) and a proteinase inhibitor (I) yields the complex (EI) following a second order kinetics. If the complex formation is reversible, the minimal reaction equation is as follows:

$$E \; + \; I \; \underset{k_{off}}{\overset{k_{on}}{\rightleftharpoons}} \; EI$$

Here k_{on} and k_{off} are the association and dissociation constants respectively. For the change of the enzyme concentration one finds:

$$- \; d(E)/dt = k_{on}(E)(I) - k_{off}(EI) \hspace{3cm} \underline{1}$$

If the reaction is started by incubating the free enzyme and the inhibitor, and the dissociation velocity constant is small enough, then equation $\underline{1}$ is reduced to:

$$d(E)/dt = - \; k_{on}(E)(I) \hspace{3cm} \underline{2}$$

For $E_0 = I_0$ integration of equation $\underline{2}$ yields:

$$1/(E) \; = \; 1/(E_0) + k_{on} \; t \hspace{3cm} \underline{3}$$

Incubation of equal amounts of HLE and Val[15]-BPTI for a given time and measuring the concentration of free enzyme by means of a chromogenic substrate allowed k_{on} to be determined, according to equation $\underline{3}$ (Fig. 3).

The value of k_{on} was determined as 1.7×10^5 $M^{-1}s^{-1}$. Although this is rather low compared with the corresponding values of α_1-PI (2.4×10^7 $M^{-1}s^{-1}$), the Bowman-Birk-inhibitor ($\geq 10^7$ $M^{-1}s^{-1}$) or eglin c (1.3×10^7 $m^{-1}s^{-1}$) (20), it is higher than the lower value given in the literature for antileucoproteinase (1.06×10^5 M^-s^{-1} [21] and 1.1×10^7 $M^{-1}s^{-1}$ [22]), which is the main physiological inhibitor of this enzyme in the upper respiratory tract (23). Moreover the k_{on}-value of Val[15]-BPTI indicates, that under the conditions used for the solubilization kinetics as shown in Fig. 2 the reaction of enzyme and inhibitor to form the non-active complex should have been completed in the absence of substrate within less than 2 min.

In summary, these results indicate a possible use of Val[15]-BPTI as an agent in the therapy of pulmonary emphysema which is induced by excess elastase such as discussed for α_1-PI or eglin c (24).

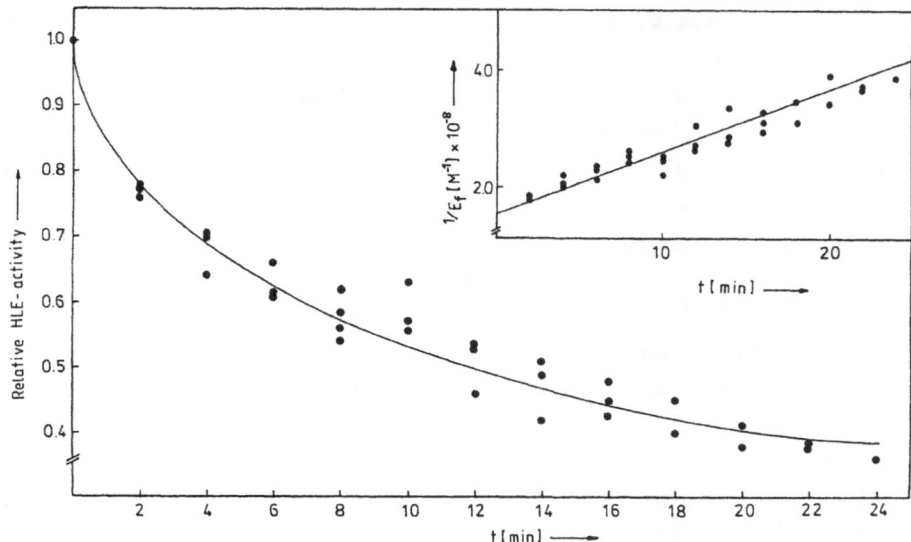

Fig. 3. Determination of k_{on} for the association of Val[15]-BPTI and HLE

6.9 nM HLE and Val[15]-BPTI, 1 M NaCl, 0.1 % Polyethylenglycol 6000, 0.02 % NaN_3, 0.2 M Tra/HCl pH 8.0 were incubated for a length of time as shown at 25 °C and 10 μl 0.59 mg MeO-Suc-Ala$_2$-Pro-Val-Nan/ml DMSO were added. The change of absorption at 406 nm was taken to be proportional to the free HLE concentrations.

SYNTHESIS OF BPTI-DERIVATIVES WITH REACTIVE-SITE AMINO ACID SEQUENCES SIMILAR TO α_1-PI

As mentioned above α_1-PI is a physiological inhibitor of HLE and a main inhibitor of this enzyme in the lower respiratory tract (6). The affinity of this protein to HLE is extremely high (25). Furthermore oxidation of the P$_1$ Met[358] (26,27) of α_1-PI yields an inhibitor derivative with a much lower affinity to the proteinase (25). This reaction may be reversible in vivo (28). It is thought to be important for the regulation of the inhibitor activity (27) and for the inactivation of the inhibitor by cigarette smoke (5).

We synthesized Met[15]-, Val[15]Ser[16]Ile[17]- and MetO[15,52]-BPTI using slightly modified methods, as described in (13), and by selective oxidation of the methionines of Met[15]-BPTI by DMSO/HCl respectively (29).

It turned out that the affinity to HLE of the Met[15]-derivative, which contains the same P_1-residue as α_1-PI is lower than the affinity of Val[15]-BPTI to the proteinase (Tab. 1). The P_1' and P_2' amino acid residues of Val[15]Ser[16]Ile[17]-BPTI are identical to the amino acid residues in the corresponding positions of α_1-PI. Moreover, all known inhibitors of the Bowman-Birk-family contain a P_1'-serine residue (10). The semisynthetic exchange of this amino acid yielded a derivative with a lower affinity to target enzymes (30). The Val[15]Ser[16]Ile[17]-BPTI also has a lower affinity to HLE than the Val[15]-derivative (Tab. 1). Obviously it is not possible to deduce optimal reactive site amino acid sequences of BPTI-derivatives simply by comparison with other inhibitors.

On the other hand oxidized α_1-PI and MetO[15,52]-BPTI behave in an analogous manner, as no affinity of the latter to HLE could be detected (Tab. 1). This shows again the prominent role of the P_1 amino acid residue for inhibitor activity.

Tab. 1. Dissociation constants K of the complexes between the BPTI-derivatives and HLE. The exchanges at the reactive site are indicated.

BPTI-derivative	K (nM)
Val[15]	0.11
Val[15]Ser[16]Ile[17]	4.6
Met[15]	2.9
MetO[15,52]	> 100
native	3500 (16)

The enzyme activities were measured spectrophotometrically with MeO-Suc-Ala$_2$-Pro-Val-Nan at pH 8.0 in the presence of 1 M NaCl. The dissociation constants were estimated according to standard procedures (31,32) using different enzyme concentrations and incubation times for each inhibitor derivative.

REFERENCES

1. H. Holzer, H. Tschesche, eds., Biological Functions of Proteinases, Springer-Verlag, Berlin-Heidelberg-New York (1979).
2. A. J. Barrett, The Possible Role of Neutrophil Proteinases in Damage to Articular Cartilage, Agents Actions, 8:11 (1978).
3. H. Menninger, W. Mohr, Neutrophile Granulozyten und ihre Enzyme bei der entzündlichen-rheumatischen Knorpeldestruktion, Therapiewoche, 31:2134 (1981).
4. W. W. West, A. Nagai, J. E. Hodgkin, W. M. Thurlbeck, The National Institutes of Health Intermittent Positive Pressure Breathing Trial-Pathology Studies. 3. The Diagnosis of Emphysema, Am. Rev. Respir. Dis., 135:123 (1987).
5. A. Janoff, Elastases and Emphysema, Am. Rev. Respir. Dis., 132:417 (1985).
6. H. M. Morrison, The Proteinase - Antiproteinase Theory of Emphysema: Time for a Reappraisal, Clin. Sci., 72:151 (1987).
7. J. G. Bieth, Elastases: Structure, Function and Pathological Role, Front. Matrix Biol., 6:1 (1978).
8. G. L. Snider, E. C. Lucey, P. J. Stone, Animal Models of Emphysema, Am. Rev. Respir. Dis., 133:149 (1986).
9. G. L. Snider, P. J. Stone, E. C. Lucey, R. Breuer, J. D. Calore, T. Seshadri, A. Catanese, R. Maschler, H.-P. Schnebli, Eglin c, a Polypeptide Derived from the Medicinal Leech, Prevents Human Neutrophil Elastase-Induced Emphysema and Bronchial Secretory Cell Metaplasia in the Hamster, Am. Rev. Respir. Dis., 132:1155 (1985).
10. M. Laskowski, Jr., I. Kato, Protein Inhibitors of Proteinases, Ann. Rev. Biochem., 49:593 (1980).
11. I. Schechter, A. Berger, On the Size of the Active Site in Proteases. I. Papain, Biochem. Biophys. Res. Commun., 27:157 (1967).
12. H. Fritz, G. Wunderer, Biochemistry and Applications of Aprotinin, the Kallikrein Inhibitor from Bovine Organs, Arzneim.-Forsch./Drug Res., 33:479 (1983).
13. H. Tschesche, J. Beckmann, A. Mehlich, E. Schnabel, E. Truscheit, H. R. Wenzel, Semisynthetic Engineering of Proteinase Inhibitor Homologues, Biochim. Biophys. Acta, 913:97 (1987).
14. J.-P. Vincent, M. Lazdunski, Trypsin-Pancreatic Trypsin Inhibitor Association. Dynamics of the Interaction and Role of Disulfide Bridges, Biochemistry, 11:2967 (1972).
15. E. Schnabel, W. Schröder, G. Reinhardt, [Ala$_2^{14,38}$]Aprotinin: Preparation by Partial Desulphurization of Aprotinin by Means of Raney Nickel and Comparison with Other Aprotinin Derivatives, Biol. Chem. Hoppe-Seyler, 367:1167 (1986).
16. P. Lestienne, J. G. Bieth, The Inhibition of Human Leukocyte Elastase by Basic Pancreatic Trypsin Inhibitor, Arch. Biochem. Biophys., 190:358 (1978).
17. B. C. Starcher, M. J. Galione, Purification and Comparison of Elastins from Different Animal Species, Anal. Biochem., 74:441 (1976).18.K. Nakajima, J. C. Powers, B. M. Ashe, M. Zimmerman, Mapping the Extended Substrate Binding Site of Cathepsin G and Human Leukocyte Elastase, J. Biol. Chem., 254:4027 (1979).
19. C. F. Reilly, J. Travis, The Degradation of Human Lung Elastin by Neutrophil Proteinases, Biochim. Biophys. Acta, 621:147 (1980).
20. N. J. Braun, J. L. Bodmer, G. D. Virca, G. Metz-Virca, R. Maschler, J. G. Bieth, H. P. Schnebli, Kinetic Studies on the Interaction of Eglin c with Human Leukocyte Elastase and Cathepsin G, Biol. Chem. Hoppe-Seyler, 368:299 (1987).
21. C. E. Smith, D. A. Johnson, Human Bronchial Leucocyte Proteinase Inhibitor, Biochem. J., 225:463 (1985).

22. F. Gauthier, U. Fryksmark, K. Ohlsson, J. G. Bieth, Kinetics of the Inhibition of Leukocyte Elastase by the Bronchial Inhibitor, Biochim. Biophys. Acta, 700:178 (1982).

23. K. Ohlsson, H. Tegner, U. Akesson, Isolation and Partial Characterization of a Low Molecular Weight Acid Stable Protease Inhibitor from Human Bronchial Secretion, Hoppe-Seyler's Z. Physiol. Chem., 358:583 (1977).

24. D. R. Boswell, R. Carrell, α_1-Antitrypsin: Molecules and Medicine, Trends Biochem. Sci., 11:102 (1986).

25. K. Beatty, N. Matheson, J. Travis, Kinetic and Chemical Evidence for the Inability of Oxidized α_1-Proteinase Inhibitor to Protect Lung Elastin from Elastolytic Degradation, Hoppe-Seyler's Z. Physiol. Chem., 365:731 (1984).

26. D. Johnson, J. Travis, Structural Evidence for Methionine at the Reactive Site of Human α_1-Proteinase Inhibitor, J. Biol. Chem., 253:7142 (1978).

27. R. W. Carrell, J.-O. Jeppsson, C.-B. Laurell, S. O. Brennan, M. C. Owen, L. Vaughan, D. R. Boswell, Structure and Variation of Human α_1-Antitrypsin, Nature 298:329 (1982).

28. H. Carp, A. Janoff, W. Abrams, G. Weinbaum, R. T. Drew, H. Weissbach, N. Brot, Human Methionine Sulfoxide-Peptide Reductase, an Enzyme Capable of Reactivating Oxidized Alpha-1-Proteinase Inhibitor In Vitro, Am. Rev. Respir. Dis., 127:301 (1983).

29. Y. Shechter, Selective Oxidation and Reduction of Methionine Residues in Peptides and Proteins by Oxygen Exchange between Sulfoxide and Sulfide. J. Biol. Chem., 261:66 (1986).

30. S. Odani, T. Ikenaka, Studies on Soybean Trypsin Inhibitors. XIV. Change of the Inhibitory Activity of Bowman-Birk Inhibitor upon Replacements of the α-Chymotrypsin Reactive Site Serine Residue by Other Amino Acids, J. Biochem., 84:1 (1978).

31. N. M. Green, E. Work, Pancreatic Trypsin Inhibitor. 2. Reaction with Trypsin, Biochem. J., 54:347 (1953).

32. J. G. Bieth, In Vivo Significance of Kinetic Constants of Protein Proteinase Inhibitors, Biochem. Med., 32:387 (1984).

HUMAN BRONCHIAL PROTEINASE INHIBITOR: RAPID PURIFICATION PROCEDURE AND INHIBITION OF LEUCOCYTE ELASTASE IN PRESENCE AND IN ABSENCE OF HUMAN LUNG ELASTIN

C. Boudier*, D. Carvallo**, M. Bruch*, C. Roitsch**, M. Courtney** and J.G. Bieth*

(*) INSERM Unité 237 Université Louis Pasteur B.P. 10, 67048 Strasbourg and (**) Transgène S.A. 11, rue de Molsheim, 67000 Strasbourg, France

INTRODUCTION

Bronchial proteinase inhibitor (brI) is the most important inhibitor of human leucocyte elastase (HLE) present in upper airways secretions[1]. Its physiological function is probably to protect the bronchoalveolar tissues from HLE-induced proteolytic damage[2,3]. Published purification procedures of brI from bronchial sputum are based on Sepharose-bound chymotrypsin or trypsin, and yield an inhibitor preparation with a low specific activity and a multi-band pattern in SDS polyacrylamide gel electrophoresis[4-6]. In this paper we describe a purification procedure involving conventional chromatographic steps only and yielding a highly active inhibitor. The latter was used to study the kinetics of inhibition of HLE in a buffer system that mimics the physiological medium. Inhibition of HLE by brI was also investigated in presence of human lung elastin.

PURIFICATION AND N-TERMINAL SEQUENCING OF brI

BrI was purified from non-purulent sputum collected from chronic bronchitis patients. The material soluble in trichloroacetic acid was successively chromatographed on CM trisacryl and Sephadex G75 columns. The resulting preparation was found to be 90 % active on HLE and showed a single band on SDS-polyacrylamide gel electrophoresis (figure 1).

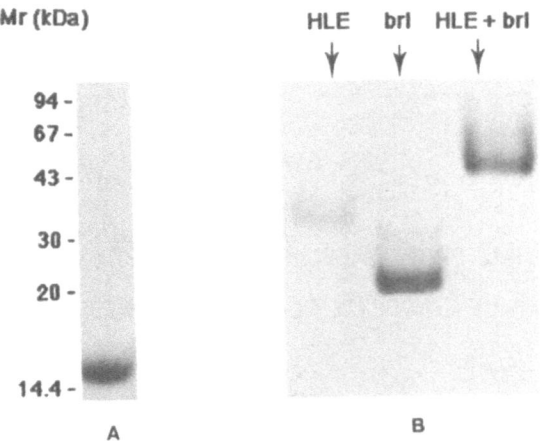

Figure 1. **A** SDS polyacrylamide gel electrophoresis of pure brI. **B**
demonstration by electrophoresis performed at pH 7.4 in
non-denaturing conditions of the complete association of
brI (7µg) with an equimolar quantity of HLE (14µg)

The amino acid composition was found to be similar to that published by
others[5-7] and particularly close to that deduced from the complete sequence
of the human seminal plasma inhibitor I[8] and of the human secretory
leucocyte protease inhibitor[9]. The N-terminal sequence was determined up to
step 29 (see Table 1) and was in full agreement with that found for the two
inhibitors cited above. This strongly suggests that the three proteinase
inhibitors are identical entities. Further technical data are given in ref 10.

Table 1. N-terminal sequence of brI

5   10
Ser-Gly-Lys-Ser-Phe-Lys-Ala-Gly-Val-Cys

15   20
Pro-Pro-Lys-Lys-Ser-Ala-Gln-Cys-Leu-Arg

25
Tyr-Lys-Lys-Pro-Glu-Cys-Gln-Ser-Asp

The reversible association of brl with HLE may be described by sheme 1

$$HLE + brl \underset{k_{diss}}{\overset{k_{ass}}{\rightleftharpoons}} HLE : brl \qquad (1)$$

where k_{ass} and k_{diss} are the association and dissociation rate constants, respectively. All the kinetic experiments were performed at 37°C in a buffer containing 0.1 M Hepes pH 7.5, 0.1 M NaCl, 10 g/l albumin. The association kinetics were measured under second order conditions. Reaction mixtures containing HLE (15 nM) and brl (24 nM) were incubated for different times before addition of 0.17 mM MeO-Suc-Ala$_2$-Pro-Val-p-nitroanilide. k_{ass} was graphically determined from a classical second order plot of the data. The value of k_{ass} given in Table 2 is the mean of four determinations.

Table 2. Kinetic constants describing the
interaction of brl with HLE at pH 7.4
and 37°C

k_{ass} $(M^{-1}s^{-1})$	k_{diss} (s^{-1})	K_i (M)
7×10^6	3×10^{-3} (a)	3×10^{-10}
	2.1×10^{-3} (b)	

a   Experimentally determined
b   Calculated from $k_{diss} = K_i \times k_{ass}$

The apparent equilibrium dissociation constant K_i (app) was determined by reacting constant amounts of HLE (24 nM) with increasing amounts of brl for 30 min at 37°C. Residual enzymatic activities were measured with MeO Suc-Ala$_2$-Pro-Val-p-nitroanilide. K_i(app) was calculated using a non-linear regression analysis program based on equation 2.

$$a = 1 - \frac{(E_o) + (I_o) + K_i\,(app) - \{[(E_o) + (I_o) + K_i\,(app)]^2 - 4\,(E_o)\,(I_o)\}^{1/2}}{2\,(E°)} \qquad (2)$$

where a is the fractional enzymatic activity (i.e. the ratio of the rate in presence of brl to the rate in absence of brl and (E_o) and (I_o) are the total concentrations of HLE and brI, respectively. K_i is related to K_i(app) through the following relationship:

$$K_i = \frac{K_i \, (app)}{1 + (S°)/Km} \qquad (3)$$

where (S_o) and K_m are the total substrate concentration and the Michaelis constant, respectively. k_{diss} calculated from $k_{diss} = K_i \times k_{ass}$ was found to be equal to $2.1 \times 10^{-3} \, s^{-1}$. This constant was also determined as described below. The dissociation of a preformed HLE-brI complex (total concentration = 8nM) was followed in a spectrophotometer cuvette containing 2.54 mM MeO-Suc-Ala$_2$-Pro-Val-p-nitroanilide, a substrate concentration equal to about 20 K_m. The substrate-induced dissociation of the enzyme-inhibitor complex is a reversible first order dissociation reaction equilibrated by a second order reaction. After a certain time an equilibrium between enzyme, inhibitor, substrate and their respective complexes is established (i.e. the rate of substrate breadown becomes constant). In our experimental conditions more than 40% dissociation of the HLE-brI complex was achieved when this equilibrium was reached. Equation 4 described this kinetic system [11].

$$\ln \frac{Ax_e + x(A - x_e)}{A(x_e - x)} = \frac{2(A - x_e)}{x_e} k_{diss} \, t \qquad (4)$$

where A is the concentration of complex at time zero, x the enzyme or inhibitor concentration at any time t and x_e the enzyme or inhibitor concentration at equilibrium. k_{diss} was calculated from the slope of a plot in accordance with equation 4 and was found to well agree with the calculated value (see Table 2).

The K_i determined in the present investigation fairly well agrees with that published recently by Thompson and Ohlsson for the interaction between neutrophil elastase and the human secretory leucocyte protease inhibitor, an inhibitor identical to brI ($K_i = 2 \pm 1.10^{-10}$M). These authors did not measure the rate constants for the enzyme-inhibitor interaction. These constants were measured previously in our laboratory[12] ($k_{ass} = 10^7 \, M^{-1}.s^{-1}$, $k_{diss} = 1.3 \times 10^{-4} \, s^{-1}$) and by Smith and Johnson[6] ($k_{ass} = 10^5 \, M^{-1}.s^{-1}$, $k_{diss} = 2 \times 10^{-5} \, s^{-1}$). Except for our previously published k_{ass} value, all constants dramatically differ from those found in the present paper. These discrepancies might arise from the use of protease affinity chromatography-prepared brI in the previous papers[6,12] or from differing buffers and temperature.

The kinetic constants depicted in Table 2 may be used to delineate the in vivo efficiency of brI. The in vivo potency of an inhibitor depends mainly upon two factors (i) the stability of the enzyme-inhibitor complex and (ii) the rate at which the inhibition takes place[13]. A reversible inhibitor may have a pseudo irreversible behaviour if the in vivo concentration (I_o) of this inhibitor is much larger than K_i. For instance if $(I_o)/K_i \geq 10^3$, there is ≤ 0.1 percent free enzyme present at equilibrium and $k_{ass}(I_o) \gg k_{diss}$ so that biological substrates can virtually not dissociate EI[14]. The concentration of brI in bronchial mucus is about 5 μM[15]. Hence, $(I_o)/Ki > 10^4$ so that brI may be considered as an irreversible leucocyte elastase inhibitor in bronchial secretions. As a further consequence of this behaviour, the delay time of inhibition concept[13] may be applied to this inhibitor. The delay time of inhibition d(t), the time required for almost full inhibition in vivo is given by $d(t) = 5 k_{ass} (I_o)$ and is equal to about 140 ms for the inhibition of HLE by brI in bronchial secretions. It has been suggested that if d(t) < 1s, an irreversible or pseudo irreversible inhibitor may play a physiological antienzyme function[13]. brI may therefore be considered as a potent antielastase in the upper respiratory tract. The same is probably not true for the lower respiratory tract where the brI concentration is very low[15]. brI might nevertheless contribute to the antielastase screen of the lower respiratory tract for reasons given below.

INHIBITION OF HLE BY brI IN PRESENCE OF ELASTIN

The kinetic results reported above have demonstrated the potency of brI in inhibiting elastase in absence of elastin, the physiological substrate of this enzyme. We have therefore tested the behaviour of the HLE:brI complex in presence of elastin, the competition between brI and elastin for the binding of elastase, and the ability of brI to bind elastase preadsorbed onto insoluble elastin which mimic physiological situations[17].

The stability of the HLE-brI complex was investigated in presence of elastin in the following conditions : the residual activities of mixtures containing constant amounts of HLE (1 μM) and various amounts of brI were measured with lung elastin. The inhibition curves were linear up to 100 % inhibition, confirming the lack of EI dissociation. The results of this experiment are in full agreement with the pseudo irreversible behaviour predicted above.

When leukocyte elastase (1 μM) was added to elastin supensions containing various concentrations of brl and the rate of elastolysis measured, a straight inhibition curve showing some deviation near the equivalence point was observed. The extrapolated inhibition curve crossed the abcissa at a brl:HLE molar ratio near to 1 indicating that brl favourably competes with elastin for the binding of HLE. With a 1.5 molar excess of brl over HLE we observed full inhibition of elastolytic activity.

When HLE was preadsorbed onto elastin prior addition of brl and the rate of elastolysis measured, an inhibition curve similar to that described above was obtained. By contrast, a similar experiment with α_1-proteinase inhibitor, yielded an inhibition curve that levelled off at 75% inhibition in agreement with previous findings[18,19]. The data are summarized in Table 3 where it can be seen that a 1.8 molar excess of brl over HLE gives 100% inhibition of elastolysis whereas with the same molar excess, α_1-proteinase inhibitor yields only 75% inhibition. Even more interesting is the observation that the fraction of elastase that resists to inhibition by α_1- proteinase inhibitor can easily be inhibited by brl (bottom of table 3).

Table 3. Inhibition of elastin-bound HLE by α_1-proteinase inhibitor, brl, and the two inhibitors added sequentially

HLE (µM)	α_1PI (µM)	br(µM)	%inhibition
1	0.5		48
1	1		71.5
1	1.2		75
1	1.8		75
1		0.5	50
1		1	91.5
1		1.2	96
1		1.8	100
1	1.2		75
1	1.2	0.3	90
1	1.2	0.6	100

CONCLUSION

Our isolation procedure yields pure and active brl whose sequence is probably identical to those of the human seminal plasma proteinase inhibitor I[8] and the human secretory leucocyte protease inhibitor[9]. The kinetic analysis of the HLE-brl interaction reveals that in the upper respiratory tract brI behaves as a fast-acting pseudoirreversible inhibitor which may efficiently prevent HLE-induced proteolysis. Our data show that brI may also play a role in the lower respiratory tract by efficiently inhibiting elastin-bound HLE, a property not shared by α_1-proteinase inhibitor, the major antielastase of the lower respiratory tract. This property is also of therapeutic relevance since bioengeneering of brI is in active progress.

REFERENCES

1. H. Tegner, Quantitation of human granulocyte protease inhibitors in non-purulent bronchial lavage fluids, Acta Otolaryngol., 85: 282 (1978).
2. K. Ohlsson and H. Tegner, Inhibition of elastase from granulocytes by the low molecular weight bronchial protease inhibitor, Scand. J. Clin. Lab. Invest., 36: 437 (1976).
3. H. Schiessler, K. Hochstrasser and K. Ohlsson, Human mucous secretions biochemistry and possible biological function.In Neutral Proteases of Human Poly-morphonuclear Leukocytes (Haveman, K. and Janoff, A., Eds) , pp. 195 - 207, Urban & Schwarzenberg, Baltimore/Munich (1978).
4. K. Hochstrasser, K. Reichert, S. Schwarz, and E. Werle, Isolierung und Charakterisierung eines Proteaseninhibitors aus menschlichem Bronchialsekret, Hoppe Seyler's Z. Physiol. Chem., 353: 221 (1972).
5. K. Ohlsson, H. Tegner and U. Akesson, isolation and partial characterization of a low molecular weight acid stable protease inhibitor from human bronchial secretions. Hoppe-Seyler's Z. Physiol. Chem , 358: 583 (1977).
6. C.E. Smith and D.A. Johnson, Human bronchial leucocyte proteinase inhibitor. Rapid isolation and kinetic analysis with human leucocyte proteinases, Biochem. J., 225: 463 (1985).
7. E.C. Klasen and J.A. Kramps, The N-terminal sequence of antileukoprotease isolated from bronchial secretion, Biochem. Biophys. Res. Commun. , 128: 285 (1985).
8. U. Seemuler, M. Arnhold, H. Fritz, K. Wiedenmann, W. Machleidt, R. Heinzel, H. Appelhans, H.-G. Gassen and F. Lottspeich, The acid-stable proteinase inhibitor of human mucous secretions (HUSI-I, anti-leukoprotease) Complete aminoacid sequence as revealed by protein and cDNA sequencing and structural homology to whey protein and red sea turtle proteinase inhibitor, FEBS Lett. , 199: 43 (1986).
9. R.C. Thompson and K. Ohlsson, Isolation, properties, and complete aminoacid sequence of human secretory leukocyte protease inhibitor, a potent inhibitor of leukocyte elastase, Proc. Natl. Acad. Sci. USA , 83: 6692 (1986).

10. C. Boudier, D. Carvallo C. Roitsch, J. G. Bieth and M. Courtney, Purification and characterization of human bronchial proteinase inhibitor, <u>Arch. Biochem. Biophys.</u>, 253: 439 (1987).

11. C. Capellos and B. H. J. Bielski, Kinetic Systems. Mathematical description of chemical kinetics in solution, Krieger Pu. Co., Huntington, N.Y., USA (1980).

12. F. Gauthier, U. Fryksmark, K. Ohlsson, and J.G. Bieth, Kinetics of the inhibition of leucocyte elastase by the bronchial inhibitor, <u>Biochim. Biophys. Acta</u>, 700: 178 (1982).

13. J.G. Bieth, Pathophysiological interpretation of kinetic constants of protease inhibitors, <u>Bull. Europ. Physiopath. resp.</u>, 16:(suppl.) 183 (1980).

14. J. G. Bieth, In vivo signifiance of kinetic constants of protein proteinase inhibitors, <u>Biochem. Med.</u>, 32: 387, (1984).

15. J.M. Tournier, J. Jacquot, P. Sadoul and J.G. Bieth, Non-competitive enzyme immunoassay for the measurement of bronchial inhibitor in biological fluids, <u>Anal. Biochem.</u>, 131: 345 (1983).

16. C. Boudier, A.Pelletier, A. Gast, J.M. Tournier, G. Pauli and J.G. Bieth, The elastase inhibitory capacity and the α_1-proteinase inhibitor and bronchial inhibitor content of bronchoalveolar lavage fluids from healthy subjects, <u>Biol. Chem. Hoppe-Seyler</u>, in press (1987).

17. M.Bruch and J.G. Bieth, Influence of elastin on the inhibition of leucocyte elastase by α_1-proteinase inhibitor and bronchial inhibitor, <u>Biochem. J.</u>, 238: 269 (1986).

18. C.F. Reilly and J. Travis, The degradation of human lung elastin by neutrophil proteinases, <u>Biochim. Biophys. Acta</u>, 621:147 (1980).

19. W. Hornebeck and H. P. Schnebli, Effects of different elastase inhibitors on leukocyte elastase pre-adsorbed to elastin. <u>Hoppe-Seyler's Z. Physiol. Chem.</u>, 363: 455 (1982).

FUNCTIONAL STUDIES OF HUMAN SECRETORY LEUKOCYTE

PROTEASE INHIBITOR

Kjell Ohlsson, Magnus Bergenfeldt and
Peter Björk

Department of Surgical Pathophysiology
University of Lund
Malmö General Hospital
214 01 Malmö, Sweden

The balance between the concentrations of proteases and their inhibitors is of vital importance to the physiology of a number of tissues including the lung. The proteases in the lung mainly come from infiltrating neutrophils and macrophages and when released from these cells can cause extensive tissue degradation unless they are inactivated by protease inhibitors. One of the major protease inhibitors found in the lung tissue is alpha-1-protease inhibitor (alpha-1-PI) which blocks the activity of neutrophil elastase (1) and cathepsin G (2).

Mucous secretions of the respiratory tract - and the genital tract - have long been known to contain acid-stabile, low molecular weight inhibitors of granulocyte and pancreatic proteases (3). Two forms of inhibitors have been purified. One form is the native, single-chain inhibitor from parotid saliva (4). The other form is a cleaved variety of bronchial mucus and seminal plasma (5-7). We first identified the elastase inhibitor in parotid secretions by virtue of its ability to react with antibodies raised against the elastase inhibitor from bronchial mucus (8,9). Since the inhibitor appeared as a homogenous protein not showing degradation products or protease inhibitor complexes we set out to purify the protein from this source even though it was present at a low concentration. A simple purification procedure was developed based on ion-exchange chromatograpy on SP-sephadex C50 at pH 6.0. The column (2.5 x 40 cm) was equilibrated with 0.05 M acetat buffer pH 6.0 containing 0.005 M NaCl and loaded with 6 l of parotid secretions brought to pH 6.0 with acetate buffer and acetic acid and clarified by centrifugation. The column was washed with equilibration buffer until the absorbance of the effluent returned to baseline. It was then eluted with a linear gradient of NaCl (400 ml of the acetate buffer with 0.005 M NaCl and 400 ml with 1.0 M NaCl). Fraction volumes of 7.5 ml were collected at the flow rate of 30 ml/hour and

Table 1

Steps	Total volume ml	Total protein mg	Total amount of SLPI mg	Recovery %
Parotid secretion	6000	5280	5.7	100
SP-Sephadex C-50	75	736	4.6	81
Sephadex G-50	18	4.9	4.0	70

The amount of SLPI was determined by radioimmunoassay.
Total protein was determined according to Lowry using
albumin as a standard. Titration of granulocyte
elastase with the isolated SLPI gave a final amount of
purified inhibitor of 4.2 mg (4).

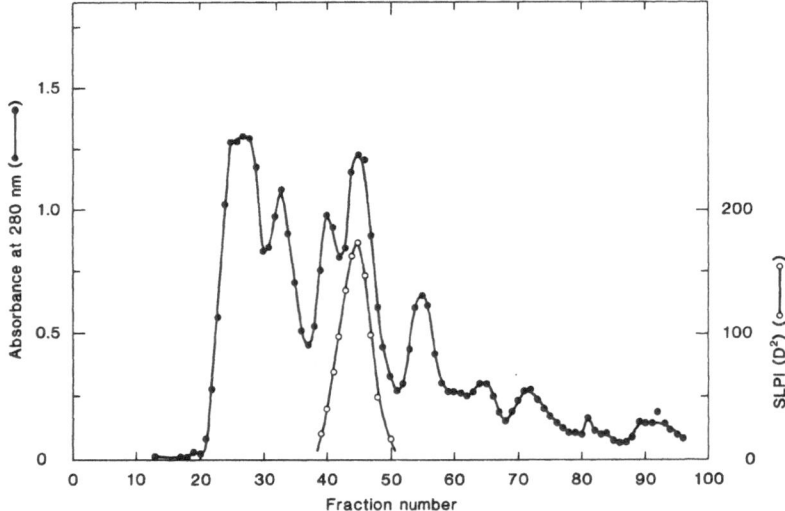

Fig. 1   Ion-exchange chromatography of 6 liters of
         parotid secretion on SP-Sephadex C50. (●—●)
         Absorbance at 280 nm.(o—o) Immunoreactive
         SLPI. (For details: see text)

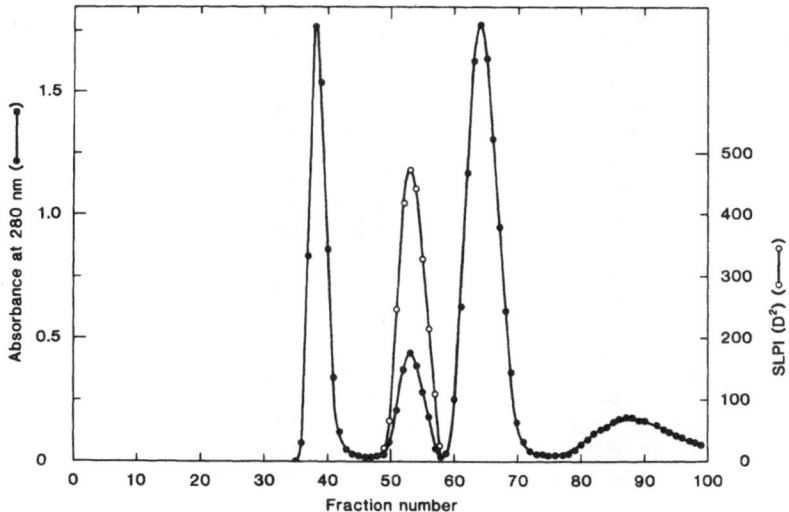

Fig. 2   Elution of SLPI immunoreactive pool of fig. 1
from Sephadex G-50 gel filtration column.
(●——●) Absorbance at 280 nm.
(o——o) Immunoreactive SLPI. (For details: see
text).

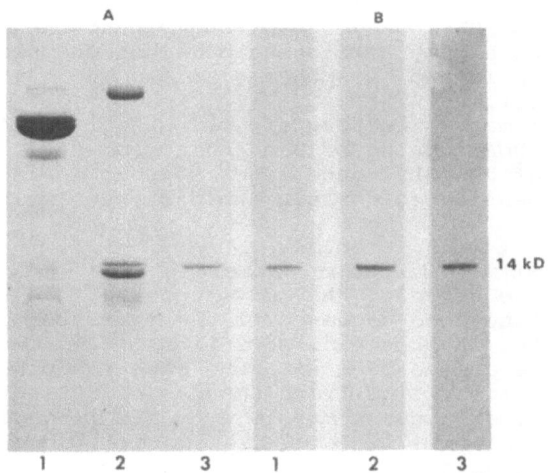

Fig. 3   SDS-polyacrylamide gel electrophoresis of 1.
parotid secretion, 2. SLPI after SP-Sephadex
C-50 chromatography and 3. after gel
filtration on Sephadex G-50. All samples were
reduced with 10 mM DTT before electrophoresis.
(a) The proteins were stained with Coomassie R
250 after electrophoresis.   (b) Western
blotting of the gel using rabbit antiserum to
bronchial mucus inhibitor and
peroxidase-coupled goat anti-rabbit IgG (Dako
Immunoglobulin, Copenhagen) and
3-amino-9-ethyl carbazole. (4)

measured for immunoreactive SLPI by single radial immunodiffusion. SLPI eluted as a single peak (Fig.1) and these fractions were combined, concentrated with ultra filtration and chromatographed on sephadex G50 (1.6 x 100 cm) equilibrated with 0.05 M tris HCl buffer, pH 7.4 containing 0.6 M NaCl. Fractions of 2.0 ml were collected at a flow rate of 20 ml/hour and measured for immunoreactive SLPI. SLPI eluted as a single peak between 100-118 ml of this chromatographic step (Fig.2). With this procedure about 4 mg of inhibitor could be obtained from 6 l of parotid secretions. (table I) The inhibitor purified in this way is essentially 100 % active against leukocyte elastase, cathepsin G and trypsin. It runs as a single band on SDS-PAGE (Fig.3) and in acid urea gels, and it shows a single peak on reverse phase HPLC.

The protein was reduced and carboxymethylated and was repurified by reverse phase HPLC before being sequenced by standard protein chemical methods. A long run of sequence was obtained by N-terminal Edman degradation, but we were surprised to find that this sequence bore no resemblence to the reported amino terminal sequence of the protease inhibitor from human seminal plasma, which had been reported to be identical to that from bronchial mucus. The inhibitor from parotid secretions was then tentatively named secretory leukocyte protease inhibitor (SLPI) (4). The rest of the sequence was determined by Edman degradation of a series of peptides obtained from reduced, carboxymethylated SLPI by digestion with different specific proteases. The complete structure of the protein could thus be elucidated. The amino acid sequence of SLPI suggests strongly that the protein consists of two highly homologous domains. One probably inhibiting trypsin and the other elastase and chymotrypsin like enzymes (4).
Solving the primary structure of SLPI greatly facilitated genomic and c-DNA cloning of DNA coding for SLPI. The organization of the SLPI gene lends support to the idea that the protein does indeed contain two distinct inhibitory domains (10).

Comparisons of the amino-acid sequence of the SLPI and the nucleotid sequence of the cervix uteri inhibitor (7) and the partial amino-acid sequence of the inhibitor in seminal plasma (3,7) and in bronchial secretions (6,11) made it evident that we are dealing with the same inhibitor in all types of secretions. The trace amounts of SLPI present in plasma probably originates mainly from the lung (12).
A synthetic gene for SLPI has now been expressed in E-coli ensuring us a good supply of this protein for investigation of its physiological role and its potential therapeutical applications. (Dr. R.C. Thompson, Synergen, personal communication)

Initial measurements indicated that SLPI inhibits leukocyte elastase and catepsin G through forming a 1:1 complex with these enzymes. To determine whether the protein could be an important inhibitor of leukocyte proteases in vivo, we measured its affinity for these enzymes in vitro (4). Detailed titration showed a Kd of $2+/-1x10^{-10}$M and $5+/-2x10^{-9}$M for the SLPI-elastase and the SLPI cathepsin G

complexes respectively. In a similar experiment the Kd of
the complex of SLPI with human anionic trypsin was found to
be 4×10^{-9}M.

Influence of plasma protease inhibitors and the SLPI   on
leukocyte elastase-induced consumption of selected plasma
proteins in vitro in man

     The effects of SLPI on the partition of leukocyte
elastase between alpha-1-PI and alpha-2-macroglobulin
(alpha-2-M) and on the residual elastase activity in
mixtures of plasma and elastase were studied by adding
various amounts of SLPI (final concentration 0-12umol/l) to
100 ul of fresh normal plasma adjusted to a final volume of
400 ul using 0.05M tris-HCl buffer, pH 7.4, 0.15 M NaCl.
Thereafter, leukocyte elastase was added to a final
concentration of 12 umol/l while stirring vigorously. This
amount of leukocyte elastase had been shown initially to
give complete saturation of alpha-1-PI and free elastolytic
and fibrinolytic activity of the reaction mixtures. Without
SLPI C_3, fibronectin, alpha-2-antiplasmin (alpha-2-AP) and
antithrombin III (AT III) were completely inactivated. After
60 minutes' incubation at 37^oC the reaction mixtures were
analyzed. No C_3d or fibronectin split products were present
at an SLPI concentration of 6 umol/l. In addition a major
part of the functional alpha-2-AP and AT III activity was
retained at a 2.4 umolar level of SLPI. At the 6.0 umolar
level of SLPI functional alpha-2-AP and AT III was
completely retained. The elastase-like activity on low
molecular weight substrates caused by the reaction mixtures
was blocked to more than 90% at the level of 6 umol/l of
SLPI. The elastolytic and fibrinolytic activity of the
reaction mixtures was blocked at the level of 2.4 umol/l.
     Furthermore, the added SLPI had an effect on the
partitioning of leukocyte elastase between alpha-1-PI and
alpha-2-M. Up to a level of about 3 umol/l of SLPI the
inhibitor caused increased binding of elastase to
alpha-1-PI. With further increasing SLPI concentrations the
binding of elastase to alpha-1-PI decreased succcessively.
The elastase binding to alpha-2-M showed a slow successive
decrease with increasing SLPI concentrations.Native and
recombinant SLPI gave identical results.
The plasma components analysed in this study have all been
shown to be consumed or degraded by proteases in peritonitis
and sepsis in man. Based on experimental and clinical data,
the leukocyte proteases, elastase, and cathepsin G, have
been proposed to play an important role in these consumption
fenomena (13-15). In critically ill patients with purulent
peritonitis total saturation of alpha-1-PI in the peritoneal
exudate with concomittant complete C_3 cleavage has been
observed, largely caused by leukocyte proteases (16).
Increased levels of alpha-1-PI-bound elastase are also seen
in plasma in peritonitis but this finding is more pronounced
in sepsis. The involvement of elastase in the systemic
pathobiochemistry of the disease can thus be assumed. The
plasma concentration of inhibitors such as alpha-1-PI and

alpha-2-M is, however, high and capable of fast inactivation of elastase. The enzyme may, however, be effective locally when released in massive amounts as is the case in purulent peritonitis. The decreased plasma level of important factors such as AT III, factor XIII, plasminogen, C3, C4 and alpha-2-M seen in sepsis may also be the result of local consumption in areas with massive leukocyte infiltration such as the lung.

The consumption of plasma proteins by proteases during disease can be mimicked by adding excess human leukocyte elastase to fresh plasma as shown in the present study. The cleavage of C3, fibronectin, AT III, alpha-2-AP can be completely blocked by adding SLPI to a concentration of 3-6umol/l in the present model. The increased binding of elastase to alpha-1-PI on the addition of lower amounts of SLPI might be explained by a protective function of SLPI against degradation of alpha-1-PI on the addition of the very large amount of leukocyte elastase used.

The plasma concentration of SLPI is normally too low for this experimental finding to be physiological significant in blood considering the huge alpha-1-PI concentration. The tenfold increased plasma level of SLPI seen in pneumonia indicates, however, a substantial local release of the inhibitor into the lung tissues in this disease. It is thus concievable that SLPI may fullfil a physiologic function protecting not only the mucous membranes of the lung but also the lung tissue in concert with alpha-1-PI. Especially as SLPI in contrast to alpha-1-PI has the capacity to inhibit elastin-bound leukocyte elastase (17).

SLPI blocks leukocyte elastase-induced destruction of trachial mucosa and its mucociliary activity

Trachea from healthy rabbits was placed in an experimental chamber and the mucociliary activity was recorded. The trachial preparation was incubated at 37°C with leukocyte elastase dissolved in 0.1M tris HCl buffer, pH 7.4 with 0.15M NaCL. The concentration of elastase varied between 0 and 500 ug/ml. In some experiments SLPI was added to a final concentration between 50 and 200 ug/ml before or 5 minutes after the addition of leukocyte elastase giving a final concentration of 250ug/ml.

The leukocyte elastase inhibited mucociliary activity in a dose-dependent way. After about 2 hours' incubation of the trachial preparation with elastase in a concentration of 250 ug/ml without SLPI all mucociliary activity ceased. SLPI in a final concentration of 100 ug/ml blocked this elastase effect completely (Fig.4). These results also indicate a physiologic role for SLPI in the protection against leukocyte elastase-induced destruction of the respiratory mucosa, especially as we have earlier shown very high concentrations of leukocyte elastase in purulent sputa in bronchitis (18).

Influence of SLPI on the activity of proteases released from PMN leukocytes during phagocytosis

Standard incubation mixtures contained 10^7 leukocytes, 40×10^7 opsonized yeast cells, 2-10% human serum and medium

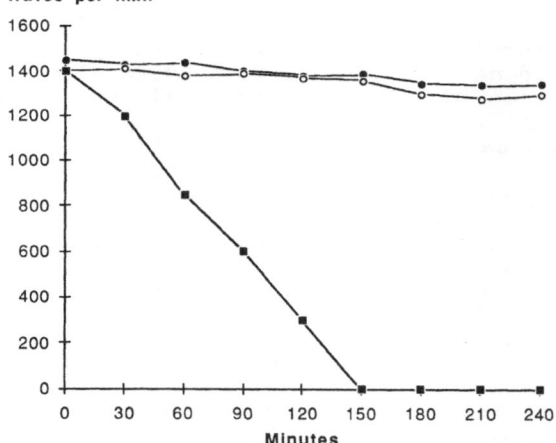

Fig. 4   Influence of human leukocyte elastase and/or
         SLPI on the mucociliary activity of rabbit
         tracheal mucosa (for details: see text).
         (●——●) SLPI (0.1 mg/ml), (o——o) leukocyte
         elastase (0.25 mg/ml) and SLPI (0.1 mg/ml),
         (■——■) leukocyte elastase; 0.25 mg/ml.

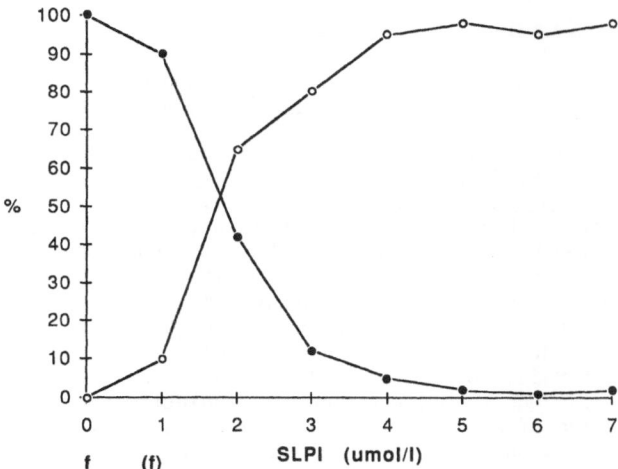

Fig. 5   Phagocytosis experiment with human PMN
         leukocytes and opsonized yeast cells (for
         details: see text).Influence of SLPI on the
         binding of elastase, released from the
         leukocytes, to alpha-1-PI (●——●) and on
         C3-degradation (o——o).f = fibrinolysis.

(Parker 199) to a final volume of 350 ul. SLPI concentration
0-7 umol/l. After 10 minutes' of pre-incubation at 37°C the
reaction is started by the addition of the yeast cells.
Incubations are carried out for 60 minutes with continuous
shaking. The reaction mixtures were analysed for elastase
alpha-1-PI complexes and for C3 and C3d. Reaction mixtures
without SLPI showed complete C3 degradation and free
proteolytic activity as measured on fibrin agarose plates.
At an SLPI concentration of about 5 umol/l the C3 cleavage
was completely blocked and all leukocyte elastase was found
in complex with SLPI while the major part of alpha-1-PI was
found in complex with other leukocyte proteases (Fig. 5).
The results show that SLPI influences the proteases released
from leukocytes on phagocytosis and their partition between
the plasma protease inhibitors in such a way that at least
fibrinolytic and C3 cleaving activity is completely blocked.

Taken together the data reviewed and those presented
indicate that the SLPI type inhibitor found in human mucous
secretions are encoded by a single gene expressed in all
mucous epitelial cells. It evidently offers effective
protection of the ciliated mucous membranes against those
leukocyte proteases, perhaps preferentially elastase, able
to destruct this type of mucosa. Furthermore, the findings
presented may have therapeutic significance. The inhibitor
may offer effective protection against leukocyte
elastase-induced effects on components of the various
cascade systems in diseases like sepsis and purulent
peritonitis. Effective SLPI concentrations are probably
attainable by intravenous or intraperitoneal administration
of the recombinant protein. Preliminary results indicate
that SLPI may prolong the life of experimental animals in
septic and endotoxic shock.

REFERENCES

1   Ohlsson, K. Neutral leukocyte proteases and elastase inhibited
     by plasma α_1-antitrypsin. Scand. J. Clin. Lab. Invest. (1971)
     28: 251-253.
2   Venge, P., Olsson, I., Odeberg, H. Cationic proteins of human
     granulocytes. V. Interaction with plasma protease inhibitors.
     Scand. J. Clin. Lab. Invest (1978) 35: 737-744.
3   Schiessler, H., Hochstrasser, K., Ohlsson, K. in Neutral Protea-
     ses and Human Polymorphonuclear Leukocytes, eds. Havemann, K &
     Janoff, A. (Urban & Schwarzenberg, Baltimore),(1978) pp. 195-207.
4   Thompson, R.C., Ohlsson, K. Isolation, properties, and complete
     amino acid sequence of human secretory leukocyte protease inhi-
     bitor, a potent inhibitor of leukocyte elastase. Proc. Natl.
     Acad. Sci.(1986), vol 83, pp. 6692-6696.
5   Ohlsson, K., Tegner, H., Åkesson, U. Isolation and partial cha-
     racterization of a low molecular weight acid stable protease
     inhibitor from human bronchial secretion. Hoppe-Seyler's Z. Phy-
     siol. Chem. (1977) 358, 583-589.
6   Klassen , E.C., Kramps, J.A. The N-terminal sequence of antileu-
     koprotease isolated from bronchial secretion. Biochem. Biophys.
     Res. Commun. (1985) 128, 285-289.
7   Seemüller, U., Arnhold, M., Fritz, H., Wiedenmann, K., Machleidt,
     W., Heinzel, R., Appelhans, H., Gassen, H.-G., Lottspeich, F. The
     acid-stable proteinase inhibitor of human mucous secretions (HU-
     SI-I, antileukoprotease). FEBS, (1986) Vol. 199, No.1; 43-48.

8   Ohlsson, M., Fryksmark, U., Polling, Å., Tegner, H., Ohlsson, K. Localization of antileukoprotease in the parotid and submandibular salivary glands. <u>Acta Otolaryngol (Stockh)</u> (1984) 98, 147-151.

9   Ohlsson, M., Rosengren, M., Tegner, H., Ohlsson, K. Quantification of granulocyte elastase inhibitors in human mixed saliva and in pure parotid secretion. <u>Hoppe-Seyler's Z. Physiol. Chem</u> (1983) 364, 1323-1328.

10 Stetler, G., Brewer, M.T., Thompson, R.C. Isolation and sequence of a human gene encoding a potent inhibitor of leukocyte proteases. <u>Nucleic Acids Research,</u> (1986) vol 14, No. 20; 7883-7896.

11 Smith, C.E., Johnson, P.A. Human bronchial leucocyte proteinase inhibitor. Rapid isolation and kinetic analysis with human leucocyte proteinases. <u>Biochem. J.</u> (1985) 225, 463-472.

12 Fryksmark, U., Prellner, T., Tegner, H., Ohlsson, K. Studies on the role of antileukoprotease in respiratory tract diseases. <u>Eur J Respir Dis,</u> (1984) 65, 201-20).

13 Egbring, R., Schmidt, W., Fuchs, G., Havemann, K. Demonstration of granulocytic proteases in plasma of patients with acute leukemia and septicemia with coagulation defects. <u>Blood,</u> (1977) 49: 219-231.

14 Aasen, A.O., Ohlsson, K. Release of granulocyte elastase in lethal canine endotoxin shock. <u>Hoppe Seyler's Z. Physiol. Chem.</u> (1978) 359: 683-690.

15 Jochum, M., Witte, J., Schiessler, H., Selbmann, H.K., Ruck-deschl, G., Fritz, H. Clotting and other plasma factors in experimental endotoxemia: Inhibition of degradation by exogenous proteinase inhibitors. <u>Eur. Surg. Res.</u> (1981) 13: 152-168.

16 Ohlsson, K. Collagenase and elastase released during peritonitis are complexed by plasma protease inhibitors. <u>Surgery,</u>(1976) Vol. 79, No. 6: 652-657.

17 Bruch, M., Bieth, J.G. Influence of elastin on the inhibition of leucocyte elastase by α_1-proteinase inhibitor and bronchial inhibitor. <u>Biochem. J.</u> (1986) 238: 269-273.

18 Ohlsson, K., Tegner, H. Granulocyte collagenase, elastase and plasma protease inhibitors in purulent sputum. <u>Europ. J. Clin. Invest.</u> (1975) 5: 221-227.

THE ROLE OF CHYMASE IN IONOPHORE-INDUCED HISTAMINE RELEASE FROM HUMAN PULMONARY MAST CELLS

T. Hultsch, Madeleine Ennis(1), and H.H. Heidtmann

Department of Internal Medicine and (1) Institute for
Theoretical Surgery, University of Marburg, D-3550 Marburg
F.R.Germany

ABSTRACT

Human pulmonary mast cells contain the serine proteases tryptase and chymase. Chymase is present in much smaller quantities than tryptase. The definite physiological role of both enzymes remains to be elucidated, angiotensin processing has been proposed as one possible function of chymase.

A dose-dependent inhibition of A 23187-induced histamine release from dispersed human lung mast cells was observed after pretreatment with diisopropylfluorophosphate (DFP) or 1-1-tosyamide-2-phenylethyl chloromethyl ketone (TPCK) but not with N-𝓛-p-tosyl-1-lysine chloromethyl ketone (TLCK). In contrast, no inhibition was observed under the same conditions with isolated rat peritoneal mast cells. These results indicate that a chymase is probably an important factor in a late phase of human lung mast cell activation. Current work focuses on the isolation of human lung chymase to further investigate this topic.

INTRODUCTION

Mast cells contain serine proteases. Much knowledge has been accumulated about the enzymes in rat mast cells, however current interest has focused on those in human mast cells. In contrast to the results obtained with rat peritoneal mast cells, the dominant enzyme in human mast cells is of a trypsin-like substrate specificity (tryptase), although recent work has shown the presence of chymotrypsin-like activity (chymase) in human lung and skin mast cells.

Several physiological roles for human mast cell chymase have been proposed e.g. extracellular activation of angiotensin 1 to angiotensin 2. However, a function of tryptase and/or chymase in human mast cell activation remains to be elucidated. We have therefore begun investigating this topic by examining the action of trypsin- and chymotrypsin-inhibitors on histamine release from human lung mast cells. The calcium ionophore A 23187 was chosen as the stimulus in order to bypass the initial membrane events preceeding calcium influx and to directly activate the final common pathway of stimulus-secretion-coupling.

MATERIALS AND METHODS

Human lung mast cell isolation and the experimental proceedure were performed essentially as described by Ennis[1]. Mast cells (human or rat peritoneal) were preincubated for 20 min with the following inhibitors or their solvents:

an inhibitor of both chymotrypsin and trypsin:
diisophropylfluorophosphate
DFP = 3.6 - 360 µM

a chymotrypsin inhibitor
1-1-tosylamide-2-phenylethyl chloromethyl ketone
TPCK = 1.0 - 100 µM

a trypsin inhibitor
N-⌼-p-tosyl-1-lysine chloromethyl ketone
TLCK = 1.0 - 100 µM.

Following preincubation, the mast cells were stimulated with the calcium ionophore A 23187 (human: 10 µM, rat 0.5 µM) for a further period (human: 5 min; rat: 90 sec, 5 min). The reactions were stopped by the addition of ice-cold Tyrode's buffer and the cells and supernatants separated by centrifugation.

Histamine was determined in both portions using a Technicon Auto-analyser. Histamine release is expressed as a percentage of the total histamine measured in the cells and supernatants. All values have been corrected for the spontaneous release occurring in the absence releasing agents or inhibitors. The inhibitory effect of the drugs is expressed as percentage inhibition:

$$\% \text{ inhibit.} = 100 \left(1 - \frac{\text{histamine release in the presence of inhibitor}}{\text{histamine release in the absence of inhibitor}}\right)$$

All values are reported as means and S. E. M. for 4 experiments.

RESULTS

The solvents used (DMSO, IPA) had no effect on either the ionophore induced or the spontaneous histamine release. DFP, a potent inhibitor of serin proteases and TPCK, a specific inhibitor for proteases with chymotrypsin-like substrate specificity, produced a dose-dependent inhibitor of the A 23187 induced histamine release (Fig. 1). TLCK, a specific inhibitor for proteases with trypsin-like substrate specificity, did not inhibit (A 23187 alone: 18.8 ± 1.0%; A 23187 + TLCK: 18.0 ± 2.0%; n = 4). In contrast, none of the serine protease inhibitors tested was able to inhibit the A 23187 induced histamine release from rat peritoneal mast cells (Table 1).

DISCUSSION

DFP is a potent inhibitor for serine proteases and inhibits histamine release from e.g. chopped human lung and rat peritoneal mast cells after immunological stimulation[2]. However, it has no effect on 48/80-induced histamine release[3]. In this study, none of the serine protease inhibitors tested was able to inhibit ionophore-induced histamine release from the rat mast cells. In contrast, both DFP and TPCK elcitied a dose-dependent inhibition of A 23187-induced histamine release from human pulmonary mast cells but TLCK was without effect.

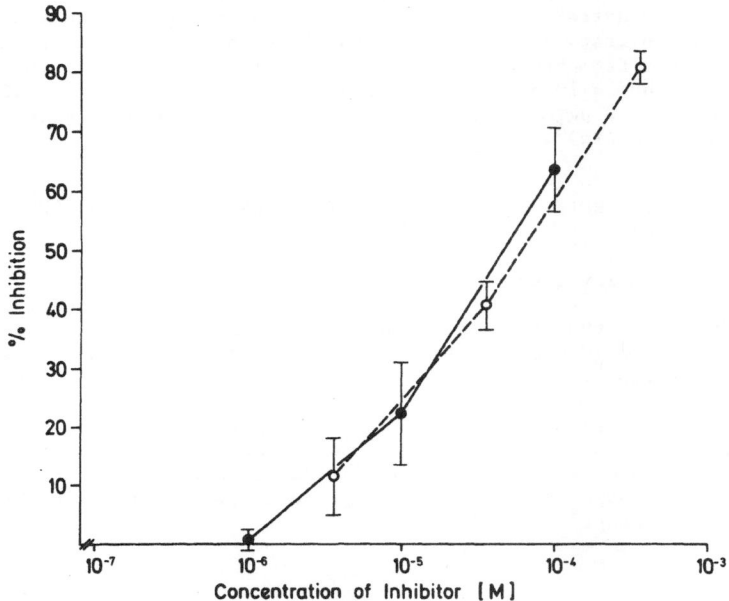

Fig. 1: Inhibition of A 23187 induced histamine release from human lung mast cells by DFP (open circles) and TPCK (filled circles). The unblocked release was 18.8 ± 1.0%.

Table 1

Histamine release (%) from rat peritoneal mast cells.

| | Incubation Time | |
	90 sec	5 sec
A 23187 (0.5 µM)	17.9 ± 1.8	25.4 ± 2.5
A + DFP (360 µM)	21.4 ± 3.2	26.3 ± 2.7
A + TLCK (100 µM)	16.3 ± 4.6	21.4 ± 4.5
A + TPCK (100 µM)	17.3 ± 3.8	22.3 ± 4.9

Our results provide further evidence for the functional heterogeneity of mast cells. Ishizaka and coworkers have shown that a serine protease is involved in an early stage of stimulus-secretion-coupling, as demonstrated with rat peritoneal mast cells[2]. The use of the calcium ionophore enabled us to selectively investigate the later phases of mast cell activation, since this agent bypasses the membrane events leading to calcium influx. Our results demonstrate the a protease with chymotrypsin-like substrate specificity (chymase) is involved in the late phases of the histamine release process from human pulmonary mast cells. This is an interesting result, since human lung mast cells contain only a small amount of chymase (0.1 - 1.0 µg/million cells) but large amounts of tryptase (12 µg/million cells)[4].

WHAT COULD BE THE ROLE OF CHYMASE IN IONOPHORE-INDUCED HUMAN LUNG MAST CELL ACTIVATION?

Chymase might be involved in:

1. Dissolution of the granular storage matrix. Electron microscopical studies provide evidence for changes in the granular matrix morphology prior to secretion[5].

2. Alteration of the granular membrane leading to e.g. the opening of ion channels and thus contributing to major changes in granular content composition/morphology preceeding histamine release. Electron microscopical studies have shown that granular swelling is the first morphological change[5].

REFERENCES

1. M. Ennis, Histamine release from human pulmonary mast cells, Agents Actions 12: 60 (1982).
2. T. Ishizaka and K. Ishizaka, Activation of mast cells for mediator release through IgE receptors, Prog. Allergy 34: 188 (1984).
3. H. Kido, N. Fukusen and N. Katunuma, Antibody and inhibitor of chymase inhibit histamine release in immunglobulin E-activated mast cells, Biochem. Internat. 10: 863 (1985).
4. H.R. Katz, R.L. Stevens and K.F. Austen, Heterogeneity of mammilian mast cells differentiated in vivo and in vitro, J. Allergy Clin. Immunol. 76: 250 (1985).
5. J.P. Caufield, R.A. Lewis, A. Hein and K.F. Austen, Secretion in dissociated human pulmonary mast cells, J. Cell Biol. 85: 299 (1980).

PROTEOLYTIC ACTIVITIES IN BRONCHOALVEOLAR LAVAGE FLUID CORRELATE TO STAGE AND COURSE OF INTERSTITIAL LUNG DISEASE

M.Schmidt, and E.Brugger

Medizinische Universitäsklinik Würzburg, Pneumologie

Josef- Schneider-Straße 2, D-8700 Würzburg

INTRODUCTION

In the clinican's view interstitial lung diseases are a very inhomo-geneous group of disorders, and in the last few years cell biologists showed us the reasons why. In exogen allergic alveolitis (EAA) or sarcoidosis stage I (SA-I) there is a predominantly benign course of disease, and one may find a lmphocytemacrophage alveolitis in bronchoalveolar lavage fluid (BAL-F). Advanced stages of EAA or sarcoidosis stage II & III (SA-II&III) are charac-terized by an increase of polymorphonuclear neutrophils (PMN). More often idiopathic pulmonary fibrosis (IPF) shows a progressive course, and in BAL-F we find a neutrophil alveolitis with PMN, eosinophils and mastcells as a sign of bad prognosis[1]. But if only differential cell counts in BAL-F are considered, lavage data show no good correlation to stage or course of diseases. This is due to the different kind of cell populations and cell interactions. Many investigators typified lymphocyte (LY) or alveolar macro-phage (AM) subpopulations, or identified chemotactic and activating factors. So we have a better impression of interactions between LY, PMN, AM, and connective tissue cells (e.g. fibroblasts) today.

Our hypothesis is, that the activation of AM and the following influx of PMN into bronchoalveolar space should be reflected in the emzyme release. During inflammation and connective tissue turn over many proteinases are released, and probably in progressive interstitial lung disease the inhi-biton capacities are impaired. So we described the protease burden in BAL-F

in some lung disorders before[2]. The consideration of proteinases beside
cell counts should result in a better correlation between lavage data and
state of disorder. The aim of our study was to evaluate the use of proteinases measurement in BAL-F in order to get more reliable lavage data for
stage and course of interstitial lung diseases.

PATIENTS

We investigated nine patients with SA-I (five smokers), nine patients
with SA-II&III (three smokers), ten patients with IPF (without vascular or
collagen disease) (three smokers), six patients with EAA (four smokers), and
compaired them to nine control persons (one smoker). All patients were
untreated with drugs, and in an active state of disease using clinical
criteria and histological findings (each diagnosis was proven histologically).

METHODS

We performed bronchoalveolar lavage as described before[2]. In native
BAL-F (+4°C) we performed total cell counts with a Fuchs-Rosenthal chamber.
After cytocentrifugation and staining with haematoxilin-eosin we counted 400
cells for differential cell counts. In BAL-F supernatant we measured total
protein content[3], protein split products[3], and albumin[4].

In order to seize a broad spectrum of proteinase activities we used
azocasein[3] at pH 7.0 as a suitable non-specific substrate for rough BAL-F
material (azoasein degrading activities, "ADA"). Leukocyte elastase was
measured as immunoreactive PMN-elastase (Eα1Pi)[6].

One day before lavage we measured the following lung function parameters: vital capacity (VC), diffusion capacity for carbon monoxide (D-CO),
and maximum oxigen uptake on a treadmill (V˙O2max) using pneumotachographs,
analyzers for carbon monoxide, carbon dioxide, and oxigen (E.Jaeger, Würzburg, FRG). After one year we made a lung function follow up with VC and
D-CO.

RESULTS

1) In SA-I we found elevated cell counts (predominantly LY) as known from many previous investigators (TAB.1). ADA/albumin was increased (p<0.05), but compared with IPF or SA-II&III in a lower range. Eα1Pi/albumin was not elevated (TAB.2). We found no difference between smokers and non-smokers. Increased total cell counts, i.e. high inflammatory states, correlated with low V˙O2max (p<0.05). Beside this, only ADA correlated with lung function: VC and D-CO were low in case of elevated ADA (p<0.05). One year later the increased ADA correlated with low D-CO (p<0.05), i.e., without corticosteroid therapy patients with elevated ADA deteriorated lung function during follow up (TAB.3).

2) Compared with SA-I in SA-II&III ADA/albumin was elevated strikingly (p<0.05), the increase in Eα1Pi/albumin was not significant (TAB.2). We found no difference between smokers and nonsmokers. Increased ADA correlated with low VC (p<0.01) and low D-CO (not significant). After corticosteroid therapy we found no correlation to lung function follow up (TAB.3).

3) In IPF we found the highest ADA/albumin (p<0.05), Eα1Pi/albumin (not significant), and split products (p<0.05) (TAB.2). Smokers had elevated values (not significant). There always were positive correlations between ADA and lung function. Early stages of IPF showed still good VC (p<0.01), D-CO (p<0.01), V˙O2max (p<0.001) together with elevated ADA. Advanced stages of disease were correlated with lower values. After one year D-CO improved (p<0.05), if ADA had been elevated before (TAB.3).

4) EAA showed similar cell distribution than SA-I (TAB.1) and similar low ADA/albumin and Eα1Pi/albumin (TAB.2). Smokers had higher values (not significant). High total cell counts correlated with low D-CO (p<0.05), other lavage data showed no correlation. But one year later the patients, who had had increased ADA before (smokers !), did not improve D-CO under corticosteroids (p<0.05) (TAB.3).

TABLE 1

Total cell counts (TCC, cells/ml*10^3), differential cell counts (percent of inflammatory cells), AM: alveolar macrophages, PMN: polymorponuclear leukocytes, LY: lymphocytes (arithm.mean ± SEM; *:p<0.05; **:p<0.01; ***:p<0.001; U-test vs. controls)

	TCC	AM	PMN	LY
SA-I	7.09±4.42**	50.0±5.4***	9.7±4.9	40.3±5.4***
SA-II&III	4.62±0.43**	58.8±5.3***	6.3±2.4	34.8±6.4**
IPF	5.58±0.59**	78.5±3.6	12.8±3.0*	8.7±2.9
EAA	7.15±0.33**	47.8±11.3*	11.3±5.9	40.8±10.3**
Controls	2.07±0.29	85.0±1.9	5.8±1.4	9.2±1.2

TABLE 2

Azocasein degrading activities (ADA/albumin; pH 7.0; PU/mg), Neutrophil-Elastase-α1-Antitrypsin Complex (Eα1Pi/albumin; µg/mg), and protein split products (after acid precipitation, percent of total protein) (arithm.mean ± SEM; *:p<0.05; U-test vs. controls)

	ADA/albumin	Eα1Pi/albumin	split products
SA-I	520.9±133.4*	0.506±0.133	2.35±0.84
SA II&III	756.9±336.4*	0.963±0.037	1.46±0.29
IPF	791.1±156.1*	2.26±1.93	6.01±0.75*
EAA	490.5±214.5*	0.594±0.48	4.25±0.92
Controls	150.5±65.2	0.486±0.248	1.25±0.03

TABLE 3

Correlation between azocasein degrading activities (pH 7.0) and lung function parameters. Vital capacity (VC), diffusion capacity (DCO), maximum oxigen uptake (V'O2max), and VC and DCO twelve months after bronchoalveolar lavage (Spearman's "R", *: p<0.05; **:p<0.01; ***:p<0.001)

	SPEARMANS "R" BETWEEN AZOCASEIN DEGRADING ACTIVITIES AND ...			... VC ... DCO after 12 months	
	... VC	... DCO	... V'O2max	... VC	... DCO
SA-I	-0.667*	-0.685*	-0.486	+0.048	-0.800*
SA-II&III	-0.800**	-0.573	-0.357	+0.335	+0.571
IPF	+0.790**	+0.867**	+0.955***	-0.139	+0.619*
EAA	-0.143	0.000	+0.089	-0.486	-0.673*

DISCUSSION

Proteolytic activities in BAL-F

In contrast to sarcoidosis and controls elevated collagenase activities are described in IPF[7]. Early stages of sarcoidosis show no detectable elastase activity[8], but advanced cases may produce some collagenase activity too[9].

In our study SA-I showed very low ADA and Eα1Pi. There may be two explanations for this phenomenon: 1) there could be a sufficient inhibitor shield or, 2) since neutrophils play a minor part in alveolitis of sarcoidosis, a special subpopulation of macrophages could be suggested[10], which does not secrete proteinases or neutrophil chemotactic factors. In spite of increased inflammatory cells ADA was lowered in our patients.

Patients with SA-II&III showed higher ADA/albumin and elevated Eα1Pi/albumin (p<0.05) than patients in stage I. Progressive disease seemed to be characterized by increasing proteolysis.

In EAA ADA/albumin and Eα1Pi/albumin were comparable with patients with sarcoidosis stage I. We found no connection between inflammatory cells and proteolytic activity.

In contrast to other diseases, IPF showed marked elevated ADA/albumin and Eα1Pi/albumin, and increased protein split products, too. Since IPF was the only disease with numerous neutrophils in BALF, neutrophil sequestration into alveolar space seems to be the reason for protease imbalance.

Lung function

In patients with lymphocyte-macrophage alveolitis (i.e. EAA and SA-I) the low, but detectable proteolytic acitivities correlated to lung function at the date of lavage or at the date of follow up one year later. In sarcoidosis high ADA was connected with low values for VC and D-CO. In stage I these patients showed a further decrease in D-CO after one year. Patients with SA-I were untreated; elevated ADA indicated a progressive course without corticosteroids. In EAA elevated ADA seemed to prevent an improvement of D-CO with corticosteroids within the first year of follow up. Evidentially this finding of lung function deterioration in case of elevated ADA points to activity of disease and activation of inflammatory cells in bronchoalveolar space. So ADA seems to be a simple parameter for disease activity in lymphocyte-macrophage alveolitis.

In IPF high ADA was connected with almost normal lung function parameters. The lower ADA was, the worse were VC, D-CO, and V'O2max at the date of lavage. Our hypothesis to this point is, that early stages of IPF show acute inflammatory infiltration and the proteolytic activities from inflammatory cells are increased in neutrophil alveolitis. So elevated ADA indicated an early stage of IPF, and patients with elevated ADA improved D-CO within one year during corticosteroid therapy.

CONCLUSION

In our patients with sarcoidosis, exogen allergic alveolitis and idiopathic pulmonary fibrosis we found a better connection between lung function (as a parameter of stage and course) and proteolytic activities than between lung function and differential cell counts. These data are to confirm our hypothesis, that measurement of proteolytic activities in BAL-F may give some interesting hints to stage and course in interstitial lung diseases.

REFERENCES

1) P.L. Haslam, Bronchoalveolar Lavage, Seminars in Respiratory Medicine 6:55 (1984).

2) M. Schmidt, E. Brugger, L. Maiwald, and H. Schweisfurth, Free proteolytic activities in bronchoalveolar lavage fluid, Eur.J.Respir.Dis. 69(Suppl 146):183 (1986).

3) O.H. Lowry, N.Y. Rosebrough , A.L. Farr, and R.J. Randall, Protein measurement with the Folin phenol reagent, J.Biol.Chem. 226:497 (1951).

4) M.M. Bradford, A rapid and sensitive method for the quantitation of microgram quantities of protein utilizing the principle of protein-dye binding, Analyt.Biochem. 72:248 (1972)

5) J. Langner, A. Wakil, M. Zimmermann, S. Ansorge, P. Bohley, H. Kirschke, and B. Wiederanders, Aktivitätsbestimmung proteolytischer Enzyme mit Azokasein als Substrat, Acta biol.med.germ. 31:1 (1973)

6) S. Neumann, G. Gunzer, N. Hennrich, and H. Lang, "PMN-Elastase Assay": Enzyme immunoassay for human polymorphonuclear elastase complexed with α-1 proteinase inhibitor, J.Clin.Chem.Clin.Biochem. 22:693 (1984)

7) J.E. Gadek, J.E. Kelman, G. Fells, S.E. Weinberger, A.L. Horowitz, and H.Y. Reynolds, Collagenase in the lower respiratory tract of patients with idiopathic pulmonary fibrosis. New Engl.J.Med. 301:737 (1979)

8) R.G. Crystal, G.W. Hunninghake, J.E. Gadek, B.A. Keogh, S.I. Rennard, and P.B. Bitterman, State of the art: the pathogenesis of sarcoidosis; in: "Sarcoidosis (1981)," J. Chretien, J. Marsac, J.C. Satiel JC, eds., Pergamon Press, Paris-Oxford-New York (1983)

9) J.F. Cordier, Y. Lasne, J.A. Grimaud, and R. Touraine, Collagenolytic activity of alveolar fluid of patients with pulmonary sarcoidosis, in: "Sarcoidosis (1981)," J. Chretien, J. Marsac, J.C. Satiel JC, eds., Pergamon Press, Paris-Oxford-New York (1983)

10) L. Lenzini, C.J. Heather, P. Rottoli, M.G. Perari, P. Sestini, and G. Carriero, Pulmonary macrophages in sarcoidosis, in: "Sarcoidosis (1981)," J. Chretien, J. Marsac, J.C. Satiel JC, eds., Pergamon Press, Paris-Oxford-New York (1983)

BEHAVIOUR OF ANGIOTENSIN CONVERTING ENZYME, HYDROXYPROLINE AND SOME PROTEASE INHIBITORS IN PULMONARY SARCOIDOSIS

Michal Masiak, Bogdan Podwysocki, and Alicja Gajewska

Department of Medical Analytics, Szewska 38 Street
50-139 Wroclaw, Poland
Clinic of Pulmonary Diseases, Division B
Grabiszyńska 105 Street, 53-439 Wroclaw, Poland

INTRODUCTION

Angiotensin converting enzyme (ACE) also called kininase II is widely used in the diagnosis and monitoring therapy of sarcoidosis. This work was undertaken in order to find out whether there is any relationship between the behaviour of serum ACE and such biological indicators as α_1-antitrypsin (α_1-AT), α_2-macroglobulin (α_2-M) and hydroxyproline (HYP) in patients with sarcoidosis and interstitial lung fibrosis.

MATERIALS AND METHODS

On the basis of radiological changes patients were identified as having sarcoidosis stadium I (n=39) or II (n=24). In serum samples activity of ACE was determined by the method of Cushman and Cheung[1] modified by Lieberman[2]. Additionally ACE activity in serum samples from patients with tuberculosis (n=16), lung cancer (n=26) and interstitial lung fibrosis (n=13) were tested. The method of single radial immunodiffusion on ready plates manufactured by Behringwerke AG was used to determine the serum concentration of α_1-AT and α_2-M in patients with sarcoidosis and interstitial lung fibrosis. Total, protein bound and free HYP was measured by the method of Prockop and Kivirikko[3]. For statisitical analyses Student's t-test and test of rang correlation by Spearman was applied.

RESULTS

We confirmed a significantly increased activity of ACE both in stadium I and II of sarcoidosis. An elevation of ACE activity was more distinct in stadium II of sarcoidosis as compared to the group of healthy subjects. A diminished activity of ACE in lung cancer (18.7+4.3; p<0.001) and tuberculosis (20.1+5.9; p<0.05) was found, whereas the same activity in interstitial lung fibrosis remained within the normal range (Table 1.) The serum concentration of α_1-AT did not deviate from normal values either in stadium I or II of sarcoidosis. But in the group with interstitial lung fibrosis α_1-AT concentration was substantially higher than in healthy subjects. In stadium I α_2-M concentration was slightly increased but normal in stadium II. A moderate elevation of α_2-M concentration in the group with interstitial lung fibrosis was also noticed as compared to healthy subjects.

Table 1. Serum levels of ACE, α_1-AT, α_2-M and HYP in tested groups. Mean values and significance level (p)

	ACE U/ml	α_1-AT g/l	α_2-M g/l	HYP μg/ml Total	HYP μg/ml Prot.b.	HYP μg/ml Free
a	34.7+10.0 < 0.001	2.2+0.7 NS	2.3+0.7 < 0.05	17.3+3.0 < 0.001	12.1+3.7 NS	5.2+2.0 < 0.001
b	44.0+16.5 < 0.001	2.4+0.9 NS	2.1+0.4 NS	18.8+2.1 < 0.001	14.0+3.2 NS	4.8+1.7 < 0.001
c	24.5+7.4 NS	2.7+0.7 < 0.001	2.6+1.1 < 0.05	17.7+2.6 < 0.001	11.6+3.7 NS	6.1+1.8 < 0.001
d	23.1+4.0	2.2+0.4	1.9+0.4	12.9+1.9	10.6+1.7	2.3+1.1

NS-not significant, Prot.b.-protein bound, a-stadium I, b-stadium II of sarcoidosis, c-interstitial lung fibrosis, d-healthy subjects.

Average serum levels of total and free HYP were significantly increased both in sarcoidosis and interstitial lung fibrosis, while the level of protein bound HYP was held within the normal range. It has been proved that there is no statistical correlation among the biochemical parameters tested in this work. Only a small positive correlation on the significance level $p < 0.05$ between total and α_1-AT was noticed.

DISCUSSION

It has been demonstrated[4,5] that the protease inhibitors α_1-AT and α_2-M are synthesized by monocytes. The results of our work may suggest that a diffuse infiltration of the lung by large numbers of monocytes/macrophages does not markedly influence the serum concentration of these proteins in sarcoidosis. Presumably more advanced and progressive inflammatory stages may be responsible for the rise of α_1-AT and α_2-M concentrations in interstitial lung fibrosis. In the course of interstitial lung diseases there appears to be a development of fibrosis which is caused by the deposition of excess collagen in the lung. An increased serum concentration of total HYP with considerable part of free form of this amino acid may be associated with intensified degradation of newly synthesized collagen in the places of lung injury. However, no difference in serum level of HYP between sarcoidosis and interstitial lung fibrosis was found. Only ACE activity in sarcoidosis was symptomatically elevated in contrast to normal values in interstitial lung fibrosis.

REFERENCES

1. D. W. Cushman and H. S. Cheung, Spectophotometric assay and properties of the angiotensin converting enzyme of rabbit lung, Biochem Pharmacol 20:1637 (1971)
2. J. Lieberman, Elevation of serum angiotensin converting enzyme (ACE) level in sarcoidosis, Am J Med 59:365 (1975)
3. D. J. Prockop and K. I. Kivirikko, Relationship of hydroxyproline excretion in urine to collagen metabolism, Ann Intern Med 66:1243 (1967)
4. R. Van Furth, Synthesis of α_1-antitrypsin by human monocytes, Clin Exp Immunol 51:551 (1983)
5. T. Hovi and D. E. Nosher, Cultured human monocytes synthesize and secrete α_2-macroglobulin, J Exp Med 145:1580 (1977)

6. H. Jmbermann, F. Oppenheim and C. Franzblau, The appearance of free
   hydroxyproline as the major product of degradation of newly synthesized
   collagen in cell culture, Biochem Biophys Acta 719:480 (1982)

EXPERIMENTAL STUDIES ON THE ADULT RESPIRATORY DISTRESS SYNDROME: ELASTASE INFUSION IN NORMAL AND AGRANULOCYTIC MINIPIGS

Hilmar Burchardi and Trond Stokke

Department of Anaesthesiology, University of
Goettingen, FRG and Department of Anaesthesiology
University of Oslo, Ullevål Hospital, Norway

INTRODUCTION

The initial phase of the adult respiratory distress syndrome (ARDS) is clinically characterized by an increase of the pulmonary vascular resistance, an impairment of the gas exchange function, and an increase of the pulmonary capillary permeability. Polymorphonuclear neutrophils, extensively accumulated in the lung ("pulmonary granulocytosis"), seem to play an important role in the early development. Unspecific proteases massively released from these stimulated granulocytes may participate in this complex pathomechanism. Clinical and experimental results indicate now the possible importance of proteases for the development of ARDS.

In this connection we wanted to investigate the possible effect of elastase upon lung function. In order to differentiate elastase actions from other additional effects of stimulated granulocytes, we infused elastase continuously in normal minipigs as well as in granulocyte depleted animals and studied the pathophysiological and morphological effects. The effects upon the blood coagulation system and the results of proteolytic activity determinations from these experiments are already published elsewhere (Stokke et al., 1985; Stokke et al., 1986).

MATERIALS AND METHODS

Animals

The experiments were carried out in 16 female and male minipigs (Institute of "Tierzucht und Haustiergenetik", University of Goettingen, FRG), with a mean body weight of 42 kg (28 - 76 kg).

Experimental protocol

The experiments were performed in the central animal experimental department of the University of Goettingen (Head: Dr. med. vet. K. Nebendahl). This department and the ethical standards of the experiments were controlled by a veterinary of the district government in Braunschweig.

Prior the experiments 4 animals were depleted from granulocytes by administration of 5 mg/kg dimethylmyleran (Fa. Drug Synthesis and Chemistry Branch, Silver Spring, Maryland, USA). Cefuroxim (Fa. Hoechst, Frankfurt, FRG) was given to prevent infection. After 7 days, at the beginning of the investigation no other white blood cells but lymphocytes were found (except 3 % eosinophils in one animal).

All experiments were performed in general anaesthesia (1 mg·kg^{-1}·h^{-1} azaperon, Fa. Janssen, Düsseldorf, FRG and 2.5 mg·kg^{-1}·h^{-1} metomidate, Fa. Janssen, Düsseldorf, FRG) and myorelaxation (0.2 mg·kg^{-1}·h^{-1} hexacarbacholine bromide, Fa. Hormonchemie, Munich, FRG) during artificial ventilation (Engström-Respirator) at F_IO_2 = 0.4.

Three series were investigated:
a) A control series (C, n = 3) without further treatment
b) An elastase series (E, n = 9) with continuous intravenous infusion of elastase (330 U·kg^{-1}·h^{-1} porcine pancreatic elastase, Fa. Serva, Heidelberg, FRC, ordering No. 20930). 1 U of elastase digests 1 /umol min^{-1} N-acetyl-(L-alanin)$_3$-methylester at 25°C and pH 8.5.
c) An agranulocytic-elastase series (EA, n = 4) with the same administration of elastase (described above) in granulocytes depleted animals.

Measurements of hemodynamics and gas exchange

For blood sampling and continuous pressure measurements catheters were inserted by common techniques into the aorta, the pulmonary artery, v. femoralis and v. jugularis.

The following cardiovascular parameters were measured in short intervals: systemic arterial pressure (P_{art}), pulmonary artery pressure (P_{AP}), cardiac output by Fick's principle ($\dot{Q}_{Fick}$). Pulmonary vascular resistance (PVR) was calculated using the following formula:
$$PVR = (P_{\overline{AP}} - P_{\overline{PW}}) / \dot{Q}_{Fick}$$
where:
$P_{\overline{AP}}$ = mean pressure in the pulmonary artery
$P_{\overline{PW}}$ = mean pulmonary wedge pressure.

For investigation of gas exchange and pulmonary function the following parameters were measured periodically in short intervals: Effective compliance (C) and resistance (R, data not shown) of the total respiratory system using pneumotachography (Fa. Dr. Fenyves & Gut, Basel, Switzerland, previously described (Burchardi and Hensel, 1979). Gas concentrations for determination of $AaDO_2$, $\dot{Q}_S/\dot{Q}_T$ and V_D/V_T were measured by mass spectrometry (MGA 1100A, Fa. Perkin-Elmer, Pomona, California, USA). Dead-space to tidal volume ratio (V_D/V_T) was calculated with the formula by Enghoff (1938), venous admixture ($\dot{Q}_S/\dot{Q}_T$) with the well known shunt formula (Finley et al., 1960).

Arterial and mixed venous blood gas analysis were performed by routine laboratory methods (ABL-2, Fa. Radiometer, Copenhagen, Denmark; for O_2 content: Lex-O_2-Con-TL, Fa. Lexington Instruments, Waltham, Massachusetts, USA).

Control measurements started after hemodynamic parameters were stabilized. Elastase treatment lasts for 3 - 5 hrs; then the animals were sacrified. The total lungs were fixed with formalin under endotracheal pressure of 20 cm H_2O for histological studies.

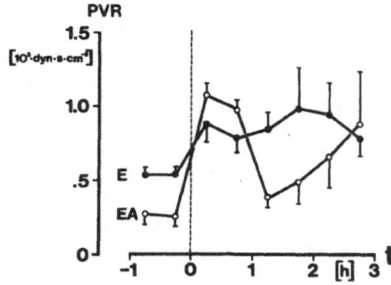

Fig. 1 Effects of elastase infusion (onset = 0 h) in normal (E) and agranulocytic (EA) minipigs:
Pulmonary vascular resistance (PVR)

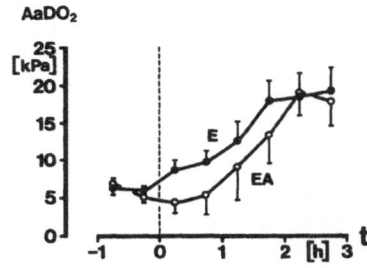

Fig. 2 Effects of elastase infusion (onset = 0 h) in normal (E) and agranulocytic (EA) minipigs:
Alveolar-arterial oxygen tension difference ($AaDO_2$)

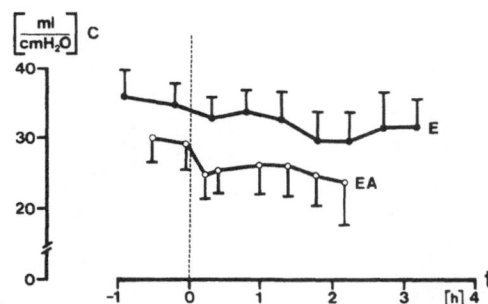

Fig. 3 Effects of elastase infusion (onset = 0 h) in normal (E) and agranulocytic (EA) minipigs:
Effective respiratory compliance (C)

RESULTS

Pulmonary vascular resistance (PVR)

Pulmonary vascular resistance increased immediately after
onset of the elastase infusion (Fig. 1). In the serie E this resistance remained increased during the whole experiment. But, in
the agranulocytic animals (serie EA), this early increase was
only temporary; a second increase was observed later-on.

Alveolar-arterial oxgyen tension difference (AaDO2)

Oxygenation deteriorated progressively soon after onset of
elastase infusion: Alveolar-arterial oxygen tension difference
(AaDO$_2$) and venous admixture ($\dot{Q}_S/\dot{Q}_T$) (data not shown) increased
markedly. There is no significant difference between the two
elastase series (E and EA) (Fig. 2).

Compliance (C)

The effective compliance of the total respiratory system
decreased significantly after elastase infusion in both series
(E and EA) (Fig. 3).

Morphology

In the non-pretreated animals (series E) elastase induced a
marked pulmonary granulocytosis and a significant interstitial
edema (Fig. 4 and 5). Whereas, in the agranulocytic animals (series EA) neither granulocytes nor edema were seen; the lung interstitium looked apparently normal.

Control series

In the control series (C) without elastase infusion all
functional parameters remained unchanged during the whole experiment lasting 5 hours (data not shown). Histological studies
revealed normal lung morphology.

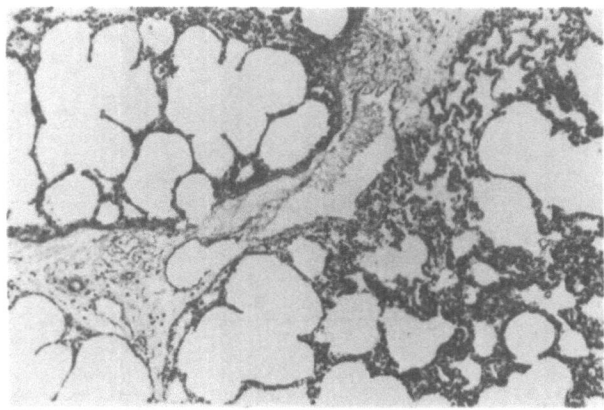

Fig. 4   Lung morphology after elastase infusion in a normal
         minipig: Pulmonary granulocytosis and interstitial edema

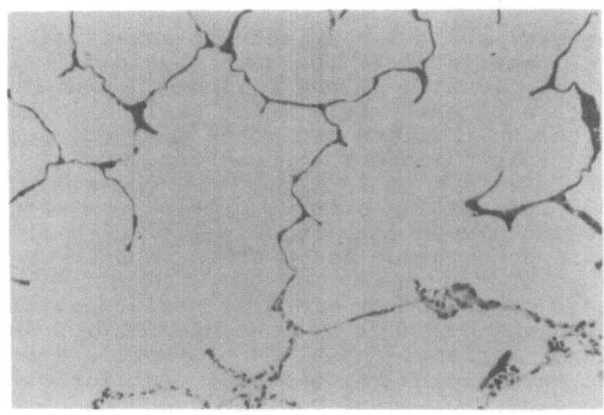

Fig. 5 Lung morphology after elastase infusion in an agranulocytic animal: Apparently normal lung tissue.

DISCUSSION

Our experiments demonstrate that continuous infusion of
pancreatic elastase in pigs impairs pulmonary functions progressively. These pathophysiological changes in lung mechanics,
pulmonary hemodynamics and gas exchange are comparable to those
in early ARDS. The functional impairment does (more or less) not
depend on the presence of polymorphonuclear neutrophils. It
seems to be induced by the elastase itself.

There is no evidence for a direct acting effect of elastase
upon pulmonary perfusion. It is much more probable that elastase
may promote these effects indirectly by stimulating other mediator systems, e. g. the complement system, the arachidonate metabolism and the kallikrein-kinin system. It is well known that
proteases activate the complement system (Johnson et al., 1976).
The resulting complement products may increase pulmonary vascular resistance, either by histamin release or stimulating the
arachidonate metabolism (Cooper et al., 1980). Especially, the
cyclooxygenase products, TXA_2 and $PGF_{2\alpha}$, are effective pulmonary vasoconstrictors. They are not only released by neutrophils but also by peripheral monocytes, alveolar or tissue macrophages and other cells which may react also in neutrophil-
depleted animals (Kennedy et al., 1980). So, the presence of
granulocytes or other blood cells seems not to be necessary for
complement-induced stimulation of the arachidonate metabolism.
On the other hand, macrophages and monocytes seem to release
their arachidonate mediators only after a certain time delay
(Neuhof, 1986). Therefore, the significant bi-phasic increase of
the pulmonary vascular resistance in our agranulocytic animals
may be caused both by histamin and cyclooxygenase products as
well.

Also, elastase-caused proteolytic fibrinolysis may stimulate the arachidonate metabolism (e. g. by fibrin degrading
products).

Further, elastase may affect the kallikrein-kinin system, as both are inactivated by the same physiological inhibitors. Restricted inactivation of the kallikrein-kinin system may result in release of bradykinin which again stimulates the arachidonate metabolism (Mullane and Moncada, 1980, Seeger et al., 1982) and provoke pulmonary vasoconstriction.

So, the stimulating effects of elastase upon the arachidonate metabolism always seem to be mediated by other systems; yet, there is no evidence for a direct action upon this process.

Increase of pulmonary vascular resistance allways influence the pulmonary gas exchange by inhomogeniety of the $\dot{V}_A/\dot{Q}$ ratio (mismatching). Regional reduction of capillary perfusion results directly in an increase of deadspace ventilation (V_D/V_T), but also in a redistribution of capillary perfusion which leads to an increase in venous admixture (Q_S/Q_T). In our data mismatching (with increase of V_D/V_T and Q_S/Q_T, as well) progressively occurs in both elastase series (also without granulocytes and interstitial edema).

So, this pulmonary function deterioration not necessarily depends on the presence of granulocytes. Other pathways, not directly linked to neutrophils, may be involved. This supports recent clinical and experimental observations which demonstrate that neutrophils not exclusively may be required for the development of ARDS (Braude et al., 1985; Flick et al., 1983; Laufe et al., 1986; Maunder et al., 1986; Ognibene et al., 1986; Schwartz et al., 1983).

On the other hand, in our experiments pulmonary endothelial permeability, indeed, seems to depend on the presence of leukocytes, since interstitial edema not occurs in the agranulocytic animals.

Although, clinically the key role of neutrophils in ARDS may be still questioned, the importance of PMNs in experimental lung damage has often been demonstrated (Taylor et al., 1985, Repine et al., 1982). In sheep microembolisation increases lung permeability only when leucocytes are present (Flick et al., 1981) . After disseminated intravascular clotting capillary permeability remained normal in leukopenic animals (Malik et al., 1982). Stimulation of polymorphonuclear neutrophils by phorbolmyristate-acetate (PMA) caused acute lung edema only in normal but not in neutropenic rabbits (Shasby et al., 1982).

Elastase is known as a potent stimulus for granulocytes; its chemotactic properties are described by Löffler (1978); they are again confirmed by our experiments in the course of which elastase infusion results in a marked pulmonary granulocytosis.

There are many mechanisms which principally might alter lung capillary permeability, like oxygen radicals, arachidonic acid metabolites, platelet-activating agents, proteases and others. Especially, oxygen radicals released from stimulated neutrophils ("burst reaction") may play a key role in capillary membrane injury (Sacks et al., 1978). In isolated perfused rabbit lungs human neutrophils, stimulated by PMA, caused acute interstitial

edema; but with granulocytes from patients with chronic granulo-
matous disease, which are deficient in oxygen radical production, PMA stimulation was ineffective (Shasby et al., 1982).

It is not certain whether elastase directly stimulates
oxygen radical release. But, complement activation products
(C_{5a}) as well as arachidonate metabolites (LTB_4) are known as
potent triggers for superoxide release from leukocytes (Heideman, 1986), and both mechanisms are stimulated by elastase.

Oxygen radicals are also known to inactivate the physiological protease inhibitors (Cocchrane et al., 1983, Travis et
al., 1984) so that high protease activity may cause local tissue damage. McDonald et al. (1979) demonstrated that granulocytic proteases are able to destroy basic membran structures.

Besides that, stimulated neutrophils may release arachidonate metabolites (especially leukotrienes) which increase vascular permeability (Dahlén et al., 1981).

CONCLUSION

Under the special conditions of our short-term experiments
in minipigs the following conclusions may be allowed:
1. Elastase enhance a marked pulmonary granulocytosis by inducing chemotactic activity.
2. Elastase impair pulmonary hemodynamics and gas exchange functions, even without polymorphonuclear neutrophils. These effects may be mediated by activation of other cascade systems
   not bound to granulocytes, like e. g. the complement system
   and the arachidonate metabolism.
3. On the other hand, capillary permeability seems to depend on
   the additional presence of neutrophils in the lung; membrane
   damage may essentially be caused by release of toxic oxygen
   radicals from granulocytes.

Our results support the hypothesis that proteases may play
a role in early ARDS; but obviously, this is only one factor
in a very complex mechanism of interactive mediator-releasing
cascade systems.

REFERENCES

Burchardi, H., and Hensel, I., 1979, Measurement of respiratory
        resistance and ventilation distribution in artificially
        ventilated patients, in: "Cardiac lung", G. Guintini and
        P. Panuccio, eds., Piccin Medical Books, Padova.
Braude, S., Apperley, J., Krausz, T., Goldman, J., and Royston,
        D, 1985, Adult respiratory distress syndrome after al-
        logenic bone-marrow transplantation: evidence for a
        neutrophil-independent mechanism,
        Lancet, I:1239.
Cocchrane, C. G., Spragg, R. G., Revak, S. D., Cohen, A. B., and
        McGuire, W. W., 1983, The presence of neutrophil elasta-
        se and evidence of oxidant activity in bronchoalveolar
        lavage fluid of patients with the adult respiratory
        distress syndrome,
        Am. Rev. Respir. Dis. (Suppl.), 127: 25.

Cooper, J. D., McDonald, J. W. D., ALI, M., MENKES, E., MASTER-
    SON, J., and KLEMENT, P., 1980, Prostaglandin production
    associated with the pulmonary vascular response to com-
    plement activation,
    Surgery, 88: 215.
Dahlén, S. E., Björk, J., Hedqvist, P., Arfors, K.-E., Hammar-
    ström, S., Lindgren, J.-A., and Samuelsson, B., 1981,
    Leukotrienes promote plasma leakage and leukocyte ad-
    hesion in postcapillary venules: In vivo effects with
    relevance to the acute inflammatory response,
    Proc. Nat. Acad. Sci. USA, 78:3887.
Enghoff, H., 1938, Volumen inefficax. Bemerkungen zur Frage des
    schädlichen Raumes,
    Upsala Läkarefören. Förhandl., 44: 191.
Finley, T. N., Lenfant, C., Haap, P., Piiper, J., and Rahn, H.,
    1960, Venous admixture in the pulmonary circulation of
    anesthetized dogs,
    J. Appl. Physiol., 15: 418.
Flick, M. R., Julien, M., and Heuffel, J. M., 1983, Leukocytes
    are not required for oleic acid induced lung injury in
    sheep (abstract),
    Physiologist 26: A55.
Flick, M. R., Peel, A., and Staub, N. C., 1981, Leucocytes are
    required for increased lung microvascular permeability
    after microembolisation in sheep,
    Circulation Res., 48: 344.
Heideman, M., 1986, Plasma mediators of oxygen free radicals in
    shock, in: "Oxygen free radicals in shock", G. P. Novel-
    li and F. Ursini, eds., Karger, Basel. Johnson, K.,
    Ohlsson, K., and Olsson, I., 1976, Effects of granulo-
    cyte neutral proteases on complement components, Scand.
    J. Immunol., 5: 421.
Kennedy, M. S., Stobo, J. D., and Goldyne, M. E, 1980, In vitro
    synthesis of prostaglandins and related lipids by popu-
    lations of human peripheral mononuclear cells,
    Prostaglandins, 20: 135.
Laufe, M. D., Simon, R. H., Flint, A., and Keller, J. B., 1986,
    Adult respiratory distress syndrome in neutropenic pa-
    tients,
    Am. J. Med., 80: 1022.
Löffler, C., 1978, Human granulocyte elastase and chymotrypsin:
    Generation of chemotactic activity in presence of se-
    rumantiproteases, in:" Neutral proteases of human poly-
    morphonuclear leukocytes", K.-A. Havemann and A. Janoff,
    eds., Urban & Schwarzenberg, Baltimore-Munich.
Malik, A. B., Johnson, A., and Tahamont, M. V., 1982, Mechanisms
    of lung vascular injury after intravascular coagulation,
    in: "Mechanisms of lung microvascular injury", A. B.
    Malik, and N. C. Staub, eds., Academic Science Press,
    New York.
Maunder, R. J., Hackman, R. C., Riff, E., Albert, R. K., and
    Springmeyer, S. C., 1986, Occurence of the adult respi-
    ratory distess syndrome in neutropenic patients,
    Am. Rev. Respir. Dis., 133: 313.
McDonald, J. A., Baum, B. J., Rosenberg, D. M., Kelman, J. A.,
    Brin, S. C., and Crystal, R. G., 1979, Destruction of a
    major extracellular adhesive glycoprotein (fibronectin)
    of human fibroblasts by neutral proteases from polymor-
    phonuclear leukocytes granules,
    Lab. Invest., 40: 350.

Mullane, K. M., and Moncada, S., 1980, Prostacyclin release and
    the modulation of some vasoactive hormones,
    Prostaglandins, 20: 25.
Neuhof, H., 1986, Mediatoren in der Pathogenese des akuten Atem-
    notsyndroms (ARDS), in: "Aktuelle Aspekte und Trends der
    respiratorischen Therapie", P. Lawin, ed., Springer,
    Berlin
Ognibene, F. P., Martin, S. E., Parker, M. M., Schlesinger, T.,
    Roach, P., Burch, C., Shelhamer, J H., and Parillo, J.
    E., 1986, Adult respiratory distress syndrome in pa-
    tients with severe neutropenia,
    N. Engl. J. Med., 315: 547.
Repine, J. E., and Bowman, C. M., and Tate, R. M., 1982, Neutro-
    phils and lung edema: state of the art,
    Chest, 81S: 47S
Sacks, T., Moldow, C. F., Craddock, P. R., Bowers, T. K., and
    Jacob, H.S., 1978, Oxygen radicals mediate endothelial
    cell damage by complement-stimulated granulocytes,
    J. Clin. Invest., 61: 1161.
Schwartz, B. A., Niewohner, D. E., and Hoidal, J. R., 1983,
    Neutrophil depletion does not protect rats from hyper-
    oxic lung injury (abstract),
    Clin. Res. 31: 746A.
Seeger, W., Neuhof, H., Graubert, E., Wolf, H., and Roka, L.,
    1982, Comparative influence of the Ca-ionophore A 23187,
    bradykinin, kallidin and eledoisin on the rabbit pulmo-
    nary vasculature with special reference to arachidonate
    metabolism,
    Adv. Exp. Med. Biol., 156: 533.
Shasby, D. M., Vanbenthuysen, K. M., Tate, R. M., Shasby, S. S.,
    McMurtry, I., and Repine, J. E., 1982, Granulocytes me-
    diate acute edematous lung injury in rabbits and in iso-
    lated rabbit lungs perfused with phorbol myristate
    acetate: Role of oxygen radicals,
    Am. Rev. Respir. Dis., 125: 443.
Stokke, T., Burchardi, H., Hensel, I., and Hörl, W. H., 1985,
    Experimental studies on the adult respiratory distress
    syndrome: effects of induced DIC; granulocytes and ela-
    stase in mini pigs,
    Eur. J. Clin. Invest. 15: 415.
Stokke, T., Burchardi, H., Hensel, I., Köstering, H., Käthner,
    T., and Rahlf, G., 1986, Continuous intravenous infusion
    of elastase in normal and agranulocytic minipigs -
    effects on lungs and the blood coagulation system,
    Resuscitation 14: 61.
Taylor, R. G., McCall, C. E., Thrall, R. S., Woodruff, R. D.,
    and O'Flaherty, J. T., 1985, Histopathologic features of
    phorbol myristate acetate-induced lung injury,
    Lab. Invest., 52: 61.
Travis, J., Beatty, K., and Matheson, N., 1984, Oxidation of al-
    pha-1-proteinase inhibitor: significance for pathobiolo-
    gy, in: "Proteases: Potential role in health and dis-
    ease", H. Hörl, and A. Heidland, eds., Plenum, New York.
Wedmore, C. V., and Williams, T. J., 1981, Control of vascular
    permeability by polymorphonuclear leukocytes in inflam-
    mation,
    Nature, 289: 646.

ARGINYLATION, SURFACE HYDROPHOBICITY AND DEGRADATION OF CYTOSOL PROTEINS

FROM RAT HEPATOCYTES

Peter Bohley, Jürgen Kopitz and Gabriele Adam

Physiologisch-chemisches Institut der Universität
D 74 Tübingen, Hoppe-Seyler-Str. 4 (West Germany)

INTRODUCTION

Intracellular proteolysis is necessary to adapt the protein composition of cells to environmental changes, to correct errors in transcription or in translation, to degrade inactive pre- and pro-peptides and to liberate amino acids as an energy source from superfluous protein molecules in case of need[1].

There are remarkable differences in the half-lives of proteins in hepatocytes which may vary from a few minutes to many weeks, whereas the mean lifetime of hepatocytes is in the order of months or even years. On the other hand, extracellular proteins, in particular collagen and elastin, are relatively inert and therefore show very long half-lives. In spite of the enormous diversity of half-lives of intracellular proteins, the total protein content in hepatocytes is remarkably constant and this implies that synthesis and degradation are tightly controlled.

Dietary changes may produce alterations in the enzyme concentrations and these again are subject to a fairly tight control, which is one of the most amazing phenomena of protein metabolism[1].

Furthermore, the limited degradation (proteolytic processing leading to mature and active protein molecules) is the most general of all covalent modifications of proteins in the cell[1,2,3,4]. Most proteins contain information within their sequences that directs them to the correct subcellular compartment or to the extracellular environment and many of them are proteolytically modified after synthesis from the precursor sequence encoded by the messenger ribonucleic acid[4].

Processing enzymes convert inactive proproteins to the mature, active polypeptides and both endo- and exopeptidases may be involved in such processing of proteins e.g. albumin, peptide hormones, growth factors and lysosomal enzymes.

Signal peptidases are responsible for the removal of signal sequences from newly synthesized proteins. These signal sequences are presequences which are necessary for the correct targeting of proteins to or through membranes or to extracellular spaces[5].

The proteases involved in such intracellular processings are in general not well characterized yet, because they have been difficult to purify and to characterize due to their low concentration, high substrate specifities, and their remarkable instability in vitro. Altogether, the degradation of inactive pro- and pre-peptides after the limited proteolysis during the intracellular processing of proteins accounts to less than 20% of total proteolysis in hepatocytes. Furthermore, it has been shown, that up to 40% of all newly synthesized proteins can be degraded in the first hour after a very short time labelling[6], but only a very few individual proteins show actually half-lives of less than one hour. This means that a considerable part of the very fast turnover of newly synthesized proteins may relate to their "metastable" state[6] in which a risk exists of nascent proteins being rapidly degraded before they become stabilized by covalent modifications, by association with their substrates or cosubstrates, by the selfassembly of monomers to form an oligomeric protein, by correct integration into an organelle or by posttranslational protection after correct disulfide bond formation or other protecting reactions[6].

We have some experimental justification for defining[1] three main groups with respect to protein turnover in hepatocytes (and many other cells):

1. very-fast-turnover proteins with half-lives of less than one hour (consisting mainly of metastable proteins and inactive pre- and pro-peptides, but also of enzymes as ornithine decarboxylase)
2. fast-turnover proteins with half-lives between 1 and 24 hours and
3. slow-turnover proteins with half-lives of more than 24 hours[1].

Enzymes with rapid turnover of their protein part are sited at important metabolic control points[7]. This corresponds well with the hypothesis that turnover of cellular proteins will only be rapid when the great expenditure of energy involved is balanced by some metabolic advantage, such as the means to regulate metabolic pathways. It is necessary for hepatocytes to degrade proteins that have no current value in the cell because a change in metabolic state has made them superfluous. The effect of such adaptation is that one group of enzymes is replaced by another more appropriate group without any change in the total amount of protein in the hepatocytes[1].

The most heterogeneous turnover rates are found in cytosol, where glycolytic enzymes belong to the slow-turnover group and where many enzymes of the amino acid metabolism belong to the fast-turnover or even to the very-fast-turnover group[1]. Compared to this, protein turnover in nuclei and especially in mitochondria is slow, although also in these organelles different turnover rates for different proteins have been found[1]. Presumably, peroxisomes and also ribosomes are degraded as an entity in hepatocytes.

Although lysosomes constitute less than 1% of total cell protein in hepatocytes, they contribute to more than 60% of total protein turnover in these cells. A remarkable part of fast-turnover proteins is degraded extralysosomally, for reviews of extralysosomal proteolysis see [1-3,7-9].

The lysosomes of hepatocytes contain endopeptidase as well as exopeptidase activities in very high concentration which may reach the range of even millimolar in some cases[10]. Whereas some cathepsins are now recognized as exopeptidases (as cathepsin A, which is a carboxypeptidase and cathepsin C, which is a dipeptidylaminopeptidase), some others act as endopeptidase and as exopeptidase (as cathepsin B, which shows weak endopeptidase activity and also acts as peptidyldipeptidase and cathepsin

H, which is an endoaminopeptidase showing aminopeptidase as well as
endopeptidase activity).

The most active proteinase in hepatocytes is cathepsin L, which is
located in the lysosomes of perivenous hepatocytes and degrades proteins
at least 10 times faster than the other cellular cysteine proteinases
e.g. the cathepsins A,B,C,H,I,J,K,M,N,S and T can do[11-15].

Substrate proteins for cathepsin L (E.C. 2.4.22.15) are cytosolic
enzymes (except lactate dehydrogenase), myosin, actin, troponins, calmodulin, parvalbumins, tubulin, vimentin, collagen[15], elastin, hemoglobin, histones, insulin, glucagon and many others. A well suited peptide
substrate is benzyl-oxycarbonyl-phenylalanyl-phenylalanyl-diazomethyl-
ketone[16] (Z-Phe-PheCHN$_2$) which is in good agreement with the substrate
specificity of cathepsin L: hydrophobic amino acids in the position P$_3$
and P$_2$ are necessary and a further apolar amino acid in P$_1$ may be helpful for a more rapid splitting of the respective peptide bond.

Cathepsin L has been found in many species[17] and organs[17,18] and
its amino acid sequence shows a remarkable homology to the plant cysteine
proteinase papain[19]. The cloned cDNA of cathepsin L encodes a 334-residue
protein containing both a 17-amino acid preregion and a 96-amino acid pro-
region[19]. A 36 kDa precursor was converted intracellularly into a 28 kDa
protein and subsequently into a 21 kDa protein which is in agreement
with earlier findings which showed a molecular weight of 28 kDa for the
unreduced proteinase and the existence of two chains with 22 kDa and
6 kDa, respectively, after reduction of cathepsin L[12,13]. The great number
of different proteinases in hepatocytes[1] made it necessary to compare the
activity against natural substrates as cytosol proteins for calculation
of the share of each of these proteinases in the total proteolytic
activity and the result of such comparison was that cathepsin L is by
far the most active proteinase in hepatocytes and contributes to more
than 40% of the total activity in these cells[14]. This property and the
specificity for apolar regions in the substrate proteins are the causes
for our special interest for this enzyme in hepatocyte proteolysis.

SUBSTRATE PROPERTIES DETERMINING PROTEOLYSIS IN HEPATOCYTES

However, not only the specificity of proteinases can be responsible
for the selectivity of intracellular proteolysis. Also properties of the
substrate proteins themselves must be determinants of this process[1,20,21
22].

For instance, a positive correlation between high molecular weights
had been found and it seemed very plausible to argue that proteins with
higher molecular weights implying more susceptible bonds, more different
conformational stages, less stabile native conformations and more possible
errors in amino acid sequence might be degraded faster[7]. There are,
however, exceptions, and for many proteins this correlation has not been
found in later investigations (for references see[1]). Other covalent and
noncovalent modifications of the substrate are proteins which have been
suggested to mark them for degradation are: acetylation, deamidation,
glycosylation, methylation,oxidation, phosphorylation and ubiquitination
1,3,6,7,14,10-26. Furthermore, the amino acid sequences of proteins with
short half-lives contain one more regions rich in proline (P), glutamic
acid (E), serine (S), and threonine (T). These PEST-regions are generally,
but not always, flanked by clusters containing several positively charged
amino acids and hydrophobicity plots for a subset of PEST regions from
rapidly degraded proteins showing the common feature that a deep hydrophilic valley is always followed by a rather sharp hydrophobic peak

(Fig. 2 in[27]). Actually, a positive correlation between surface hydrophobicity and short half-lives of proteins was found [1,14,20,21,22].

This might be of special importance for control of intracellular
proteolysis because this surface hydrophobicity of the substrate proteins
can be changed reversibly by dissociation of monomers from an oligometric
protein, by dissociation of the proteins from an organelle (protecting
membrane), by dissociation of stabilizing ligands (substrates, cosubstrates, cofactors) or by the splitting of stabilizing disulfide bonds [1,
20,21,22]. Such increase in surface hydrophobicity could be a common factor
in the more rapid degradation of many kinds of proteins, not only because
of peptide bonds of adjacent or nearby apolar residues, but also because
it facilitates the aggregation of proteins and presumably the subsequent
entry into a degradative environment [1,14,20,21,22].

More recent work has shown an interesting correlation between the
in-vivo-half-lives of proteins and their amino-terminal residue[26]:
according to the N-end rule[26], proteins with amino-terminal arginine,
lysine, leucine, phenylalanine, aspartic acid, glutamine, tyrosine,
glutamic acid and isoleucin and perhaps also asparagine, histidine and
tryptophane should belong to the very-fast-turnover or to the fast-turnover group of cellular proteins[26].

The function of the previously described posttranslational addition
of single amino acids to protein amino-terminal may also be accounted for
by the N-end rule[26]. Namely, the cytosolic enzyme arginyltransferase (E.C.
2.3. 2.8.) or arginyl-tRNA: protein transferase[28,29] was well characterized
long ago[28], but the physiological function of this interesting enzyme
which is able to add amino acids to the free N-ends of proteins posttranslationally was unkown[26]. On the other hand, it was well known that transfer
RNA is an essential component of the ubiquitin- and ATP-dependent proteolytic system[30]. It was, however, totally unknown yet to which extent the
cytosol proteins can be arginylated posttranslationally and which properties of these substrate proteins may be important for such additions of
further amino acids to the free N-ends of mature cytosol proteins.

Because we know from many investigations that surface hydrophobicity
is one of the properties of substrate proteins which can be very important
for their preferential intracellular degradation [1,3,7,14,20,21,22], see
also above, we investigated whether correlations exist between surface
hydrophobicity of cytosol proteins from rat hepatocytes and their arginylation with H-arginyl-tRNA catalyzed by arginyltransferase (E.C. 2.3.
2.8.).

PREPARATION OF ^{3}H-ARGINYL-^{14}C-LEUCINE-LABELLED CYTOSOL PROTEINS

Rat liver parenchymal cells were isolated by a collagenase perfusion
method as described in[31], cultured as described in[32], and labelled with
^{14}C-leucine for 16 hours[32]. After washing steps, the labelled cells were
harvested, homogenized and centrifuged for 15 minutes at 400 000 x g as
described in[32], the ^{14}C-leucine-labelled cytosol proteins were purified
as in[32] and than used for the posttranslational ^{3}H-arginyl-tRNA[28] and
incubated this highly labelled product (>2 MBq/mg) for 15 minutes at 37° C
with arginyltransferase (E.C. 2.3.2.8) and with the ^{14}C-labelled cytosol
proteins (see above) in DTT-KCl-ATP-MgCl$_2$-medium[28]. Afterwards, we degraded the remaining H-arginyl-tRNA by incubation with ribonuclease A.
Special precautions were necessary to avoid a secondary rapid degradation
of the arginylated part of proteins by the high amount of aminopeptidases
in cytosol[32]. In a following filtration step, we separated the ^{3}H-argi-
nylated-^{14}C-leucine-labelled cytosol proteins from the tRNA-degradation-

products, from free ^{3}H-arginine and from the aminopeptidase inhibitors. A control with leucine aminopeptidase from bovine eye lenses (which is totally free of endopeptidase activity[33]) showed that actually the amino ends of the cytosol proteins had been arginylated[32].

HYDROPHOBIC CHROMATOGRAPHY OF THESE DOUBLE-LABELLED CYTOSOL PROTEINS

The purified ^{3}H-arginylated-^{14}C-leucine-labelled cytosol proteins were separated on a phenylsuperose column (PHARMACIA, HR 5/5) by FPLC with a flow rate of 5 ml/hour and the following elution buffers: 10 ml 20 mM Tris/150 mM KCl pH 7.2; 5 ml 20 mM Tris pH 7.2; 5 ml 10 mM Tris/ 50% Ethylene glycol; 5 ml 10 mM Tris/2 % Tween 80/50% Ethylene glycol and finally 5 ml 0.1 NaOH. Probe and fraction volume was always 0.5 ml. The separation of the double-labelled cytosol proteins on octyl- and phenyl-sepharose occured with the same elution buffers, but in a batch-procedure with stepwise elution of the proteins and subsequent precipi-tation with more than the 10 fold volume of acetone at -30° C, 30 minutes centrifugation at 5000 x g and solubilization of the dried pellets in 0.2 NaOH. The radioactivity of the solubilized proteins was determined after mixing with 4 ml Aqualuma plus in a LKB 1219 Rackbeta liquid scintillation counter with automatic dpm calculation for ^{3}H and for ^{14}C.

As is shown in Figure 1 in[32], there is a preferential binding of ^{3}H-arginylated proteins to phenylsuperose: all ^{3}H / ^{14}C-ratios in the column fractions eluted with 20 mM Tris/150 mM KCl are much lower than the ^{3}H / ^{14}C-ratios of the protein mixture before the separation.

Also the reduction of the ionic strength alone is not sufficient to elute proteins with higher ^{3}H / ^{14}C-ratios and these ratios only start to rise after addition of ethylene glycol and especially of Tween 80 to the elution buffers. A small part of the total amount of cytosol proteins can be eluted with NaOH only and may mainly represent irreversibly denatured cytosol proteins, which in total account to more than 6% of all arginylated proteins and to less than 3% of the ^{14}C-leucine-labelled cytosol proteins.

Separations of these double-labelled cytosol proteins on octyl-sepharose and on phenylsepharose confirmed this result[32]. It is, however, evident from these results that in general separations of cytosol on phenylsuperose with FLPC protect the proteins better against denaturation than separation on octylsepharose or on phenylsepharose with the batch-procedure.

AGGREGATION OF THESE CYTOSOL PROTEINS IN VITRO

The purified ^{3}H-arginylated-^{14}C-leucine-labelled cytosol proteins were incubated with 0.2 mg proteinase K/ml in 20 mM Tris/150 mM KCl pH 7.8 for 5, 10, 15, 30 and 60 minutes at 37° C and subsequently the undegraded part of proteins was precipitated in 5% trichloracetic acid and centri-fuged for 10 minutes at 13 000 x g. The radioactivities in supernatants and in pellets were determined as described above. Although this enzyme is an endopeptidase, it degrades the arginylated proteins much more rapidly than the bulk of proteins, e.g. we found 75% + 0.8% degradation of ^{3}H-labelled proteins after 5 minutes and only 42% + 0.9% degradation of ^{14}C-labelled proteins after 5 minutes and even after 60 minutes remain more than 20% of all ^{14}C-labelled proteins undegraded, whereas less than 2% of all ^{3}H-labelled proteins are not degraded after these 60 minutes.

We conclude from these results that not only surface hydrophobicity and aggregation but also proteolytic susceptibility of the arginylated proteins from rat hepatocyte cytosol is increased. Quite clear, we plan now to investigate the ability of cathepsins and proteinases from hepatocytes to degrade these arginylated proteins.

The arginylation itself, which introduces an additional very hydrophilic group to the modified substrate proteins, seems not to be responsible for a secondary increase in surface hydrophobicity of these cytosol proteins although alone from our investigation this possibility as yet cannot be excluded totally.

On the contrary, we assume that arginylation occurs selectively at such proteins from hepatocyte cytosol which show an increased surface hydrophobicity compared to the bulk of cytosol proteins and as was shown earlier, these proteins belong mainly to the very-fast-turnover group or at least to the fast-turnover group[1,14,20,21,22].

We conclude that arginyltransferase (E.C. 2.3.2.8) preferentially arginylates cytosol proteins with an increased surface hydrophobicity.

DISCUSSION

Intracellular proteolysis is a first-order process and the selectivity of degradation appears to be determined by features of substrate proteins structure rather than by the specificity of cellular proteases (see above). There is a remarkable dependence of ATP for the degradation of many cellular proteins which might be due to a specific requirement of ATP for ubiquitination, phosphorylation, glycosylation, repression of an inhibitor, increase in proteolytic susceptibility of some substrate proteins, maintenance of the low intralysosomal pH, uptake of substrate proteins by lysosomes or even for rapid resynthesis of short-lived activators like glutathione, or some cytoskeletal elements, of intracellular receptors or, finally, for as yet unkown short-lived proteinases (for references see[1]).

Additionally, it is still not clear to what extent an intracellular circulation or cellular autofiltration might be energy-dependent and important for regulation of intracellular proteolysis[1].

In spite of many efforts, we are still far away from a comprehensive understanding of the molecular mechanisms of regulation of proteolysis in cells. We can, however, list some possible factors that may be involved in such regulation (for references see[1]): compartmentation, amount and activity of proteinases and of their inhibitors, posttranslational modifications of substrate proteins (see below), levels of free amino acids, insulin (the main inhibiting hormone), glucagon (the main stimulating hormone), glucocorticoids, thyroxine, growth factors, prostaglandins, interleukin I, calcium ions, aminoacylation of t-RNAs and passive stretch muscle.

The level of a protein can be controlled both by its rate of synthesis and by its rate of degradation. Thus, protein degradation may also be involved in the control on intracellular protein levels in general but also in the altering of levels of specific proteins.

The molecular basis for metabolic control of the intracellular proteolysis involves primary steps which mark the respective proteins for degradation. Some of the many possible modifications suggested to mark proteins, as e.g. acetylation, carbamylation, deamination and formation of mixed disulfides.

However, oxidation of histidine or methionine residues, ubiquitination and perhaps even phosphorylation could occur at the majority of cellular proteins and thus mark them for rapid degradation.

In particular, surface hydrophobicity is a property which belongs in some respect to nearly all cellular proteins. Quite clear, the best characterized conformation of proteins is the unique, globular conformation that most proteins form spontaneously during and after their synthesis under physiological conditions. In this confirmation the main part of apolar residues are buried, but some apolar residues can always be found at the exterior of the molecule exposed to the solvent. It is clear why the properties of a protein change substantially when the protein unfolds. Unfolded proteins are less soluble and more hydrophobic because many more nonpolar residues are in contact with the solvent. Because of the specificity of cell proteinases against the apolar part at the surface of the substrate proteins, it is evident that proteins with more hydrophobic amino acids at their surface might be degraded faster. Furthermore, such proteins tend to aggregate more rapidly with each other or with any apolar surface in the cell. Consequently, surface hydrophobicity may be important for selectivity and also for control of intracellular proteolysis, because it can be changed reversibly by the dissociation of monomers from an oligomeric protein, by the dissociation of stabilizing ligands as substrates or cosubstrates (that means, substrates protect their respective enzymes against proteolysis), by splitting of disulfide bonds or of only one single peptide bond followed by an unfolding of the polypeptide chain and by any conformational change of these substrate proteins in the cell.

All such increases in surface hydrophobic areas could therefore be common factors in the accelerated degradation of many kinds of proteins, as already described above. Especially, it seems to be important to know in which way combined effects of altered substrate properties (e.g. oxidation and surface hydrophobicity or ubiquitination and surface hydrophobicity) can act on degradation of cell proteins. Recently, we found that ubiquitinated cytosol proteins from HeLa-cell cultures tend to aggregate more rapidly and bind preferentially to phenylsuperose (unpublished results with Heribert Wegner 1987).

It is still totally unknown, why an additional arginylation in hepatocytes should be helpfull for intracellular proteolysis. Surface hydrophobicity alone could be sufficient to explain the selectivity and control of this process. However, the preferential arginylation of such cytosol proteins which show more apolar amino acids at their surfaces might be one of the preconditions for a subsequent ubiquitination[26] which could lead to a further conformational change of the modified protein with an additional increase in surface hydrophobicity. This altogether could facilitate the final degradation of this severalfold modified protein even more.

However, it seems to be a superfluous complication of the whole process when in an already conformationally changed protein further changes are postulated which are thought to be necessary for an even better degradation.

On the other hand, arginylation might be reversible, because the aminopeptidase activity in cytosol is very high[33] and this could be a further way to control intracellular proteolysis. Precondition for a possibly arginylation is a free N-end and the main part of cytosol proteins is normally acetylated (for references and discussion see[1]). Recently a further type of N-terminal acylation has been reviewed in detail but

A POSSIBLE ROLE OF ARGINYLATION IN CELLULAR PROTEOLYSIS

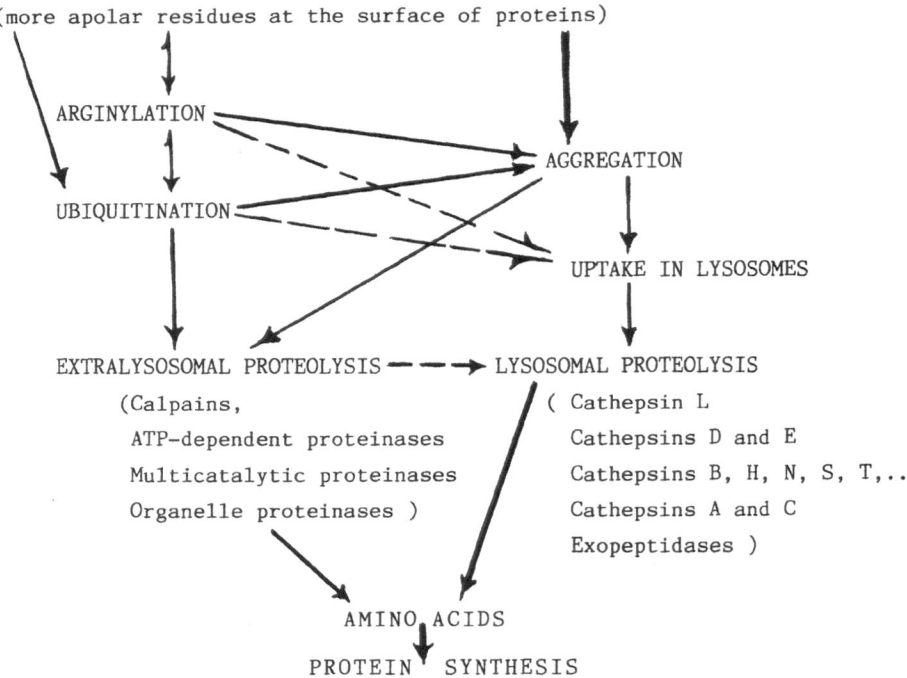

MORE STABLE CONFORMATIONS OF SUBSTRATE PROTEINS

(few apolar residues at the surface of proteins)

Dissociation into monomers
Dissociation of substrates
Dissociation of cosubstrates
Dissociation from organelles
Splitting of disulfide bonds
Splitting of single peptide bonds
Splitting by oxidation (Met, His,..)
Other conformational changes

LESS STABLE CONFORMATIONS OF SUBSTRATE PROTEINS

(more apolar residues at the surface of proteins)

ARGINYLATION

AGGREGATION

UBIQUITINATION

UPTAKE IN LYSOSOMES

EXTRALYSOSOMAL PROTEOLYSIS — — → LYSOSOMAL PROTEOLYSIS

(Calpains,
ATP–dependent proteinases
Multicatalytic proteinases
Organelle proteinases)

(Cathepsin L
Cathepsins D and E
Cathepsins B, H, N, S, T,..
Cathepsins A and C
Exopeptidases)

AMINO ACIDS

PROTEIN SYNTHESIS

In the scheme given above, we try to put together some main events in intracellular proteolysis with special emphasis on arginylation and surface hydrophobicity which we investigated here in more detail. The whole process is naturally a cyclic one, because the proteins newly synthesized from the amino acids which had been liberated before by cellular proteolysis are now again the substrate proteins which undergo the same processes as their predecessors. In general, it is still totally unknown to which extent the different cellular pathways for proteolysis participate in this important process and how they are regulated in detail. There is only one pathway for protein synthesis (as only one way for our birth) but there are very many different pathways for degradation of cellular proteins (as also so many different possibilities for our death). We are still far away from an elucidation of the molecular mechanisms of intracellular proteolysis.

without any relation to cellular proteolysis[34]. Thus we still wait for experimental work on the relationship between N-terminal methylation of proteins[34] and the role of this posttranslational in vivo modification for proteolysis in cells. There was no difference in the degradation of in vitro generally methylated and unmethylated proteins in hepatome tissue culture cells after microinjection[35].

Because only the correlation of such general physicochemical properties of the substrates as molecular weight or even surface hydrophobicity with specific individual degradation rates seems to some authors inherently unlikely[36], an additional modification of the substrate proteins as a signal for degradation has been postulated[36]. Perhaps is arginylation really one of such additional signals, although the molecular mechanisms of its role in the whole process are unknown yet. For instance, it is still unknown whether arginylation may play a role in the energy-dependent process[37] of autophagic degradation of endogenous liver cell protein and it seems to be worthful therefore to investigate possible relationships between arginylation and autophagy although this process itself seems to be not selective. The selectivity of cellular proteolysis, however, could be cause by steps before autophagy as for instance arginylation and subsequent aggregation of the short-lived proteins.

On the other hand, arginylation cannot play an important role in the degradation of many other cellular proteins. Such proteins are not only those which are acylated and cannot be arginylated for this reason (see above) but also the many proteins which are located in the other cell organelles. For instance, a selective degradation of proteins of the inner mitochondrial membrane by a proteinase associated with the intermembrane space fraction of rat liver mitochondria has been described recently [38] and ATP has been described in this special case as molecule which protects the protein substrate against the selective degradation[38]. This means that intracellular localization of proteinases and their respective substrate proteins is the most important precondition for intracellular proteolysis.

Arginylation should occur only in the cytosol, because arginyltransferase seems to be located exclusively in this compartment and can therefore not be important for the degradation of mitochondrial proteins.

ACKNOWLEDGEMENTS

We thank for kind help in this work Prof.Dr.D.Mecke and Dr.R.Gebhardt and their group working with hepatocyte cultures, in particular H.Burger, H.Heini, M.Locher, C.Mayer, and B.Ugele.

The work has been supported by the "Deutsche Forschungsgemeinschaft" by a grant "Intrazelluläre Proteolyse" (Bo 831 / 1 -1) which is also greatly acknowledged.

REFERENCES

1. P.Bohley, Intracellular proteolysis , in: New Comprehensive Biochemistry, A.Neuberger, and K.Brocklehurst, eds., Amsterdam, 1987, Elsevier, 16, p. 307

2. J.C.Waterlow, P.Garlick, and D.J.Millward, Protein turnover in mammalian tissues and in the whole body, Amsterdam, 1978, North Holland Publ. Co.

3. J.S.Bond, and R.J.Beynon, Proteolysis and physiological regulation, Mol. Aspects Med. 9: 173 (1987).

4. F.Wold, In vivo chemical modification of proteins, Annu.Rev.Biochem. 50: 783 (1981).

5. P.Walter, R.Gilmore, and G.Blobel, Protein translocation across the endoplasmatic reticulum, Cell 38: 5 (1984).

6. D.N.Wheatley, S.Grisolia, and J.Hernandez-Yago, Significance of the rapid degradation of newly synthesized proteins in mammalian cells : a working hypothesis, J.Theoret.Biol. 98: 283 (1982).

7. A.L.Goldberg, and A.C.St.John, Intracellular protein degradation in bacterial and mammalian cells. Part 2, Annu.Rev.Biochem. 45: 747 (1976).

8. J.S.Bond, and P.E.Butler, Intracellular proteases, Annu.Rev.Biochem. 56: 333 (1987).

9. S.Pontremoli, and E.Melloni, Extralysosomal protein degradation, Annu. Rev.Biochem. 55: 455 (1986).

10. R.T.Dean, and A.J.Barrett, Lysosomes, Essays Biochem. 12: 1 (1976).

11. H.Kirschke, J.Langner, B.Wiederanders, S.Ansorge, and P.Bohley, Cathepsin L. A new proteinase from rat liver lysosomes, Eur.J.Biochem. 74: 293 (1977).

12. A.J.Barrett, and H.Kirschke, Cathepsin B, Cathepsin H, and Cathepsin L, Methods Enzymol. 80: 535 (1981).

13. H.Kirschke, and A.J.Barrett, Cathepsin L - a lysosomal cysteine proteinase, in: Intracellular protein catabolism V, E.Khairallah, J.S.Bond, and J.W.C.Bird, eds., New York, 1985, Alan R. Liss Inc., Progr.Clinical and Biol.Res. 180, p. 61

14. P.Bohley, C.Hieke, H.Kirschke, and S.Schaper, Protein degradation in rat liver cells, in: Intracellular protein catabolism V, E.Khairallah, J.S. Bond, and J.W.C.Bird, eds., New York, 1985, Alan R.Liss Inc., Progr. Clinical and Biol.Res. 180, p. 447

15. H.Kirschke, A.A.Khembhavi, P.Bohley, and A.J.Barrett, Action of rat liver cathepsin L on collagen and other substrates, Biochem.J. 201: 367 (1982).

16. H.Kirschke, and E.Shaw, Rapid inactivation of cathepsin L by Z-Phe-Phe-CHN2 & Z-Phe-Ala-CHN2, Biochem.Biophys.Res.Commun. 101: 454 (1981).

17. S.Riemann, H.Kirschke, B.Wiederanders, A.Brouwer, E.Shaw, and P.Bohley, Inhibition of cysteine proteinase activity by Z-Phe-Phe-diazomethane and of aspartic proteinase activity by pepstatin in different organs from some animals and isolated cells from rat liver, Acta Biol.Med.Germ. 41: 83 (1982).

18. Y.Bando, E.Kominami, and N.Katunuma, Purification and tissue distribution of rat cathepsin L, J.Biochem. 100: 35 (1986).

19. D.A.Portnoy, A.H.Erickson, J.Kochan, J.V.Ravetch, and J.C.Unkeless, Cloning and characterization of a mouse cysteine proteinase, J.Biol.Chem. 261: 14697 (1986).

20. P.Bohley, Intrazelluläre Proteolyse, Naturwissenschaften 55: 211 (1968).

21. P.Bohley, H.Kirschke, J.Langner, M.Miehe, S.Riemann, Z.Salama, E.Schön, B.Wiederanders, and S.Ansorge, Intracellular protein turnover, in: Biological functions of proteinases, 30. Mosbach-Colloquium, H.Holzer, and H.Tschesche, eds., Berlin, Heidelberg, New York, 1979, Springer-Verlag, p. 17

22. P.Bohley, and S.Riemann, Intracellular protein catabolism IX: Hydrophobicity of substrate proteins is a molecular basis of selectivity, Acta Biol.Med.Germ. 36: 1823 (1977).

23. D.Finley, and A.Varshavsky, The ubiquitin system: functions and mechanisms, Trends Biochem.Sci. 10: 343 (1985).

24. E.R.Stadtmann, Oxidation of proteins by mixed-function oxidation systems: implication in protein turnover, ageing and neutrophil function, Trends Biochem.Sci. 11: 11 (1986).
25. A.J.Rivett, Regulation of intracellular proteinturnover: Covalent modification as a mechanism of marking proteins for degradation, Curr.Top. Cell.Regul. 28: 291 (1986).
26. A.Bachmair, D.Finley, and A.Varshavsky, In vivo half-life of a protein is a function of its amino-terminal residue, Science 234: 179 (1986).
27. S.Rogers, R.Williams, and M.Rechsteiner, Amino acid sequences common to rapidly degraded proteins: The PEST hypothesis, Science 234: 364 (1986).
28. R.L.Soffer, Enzymatic Modification of Proteins II. Purification and properties of the arginyl transfer ribonucleic acid - protein transferase from rabbit liver cytosol, J.Biol.Chem. 245: 731 (1970).
29. R.L.Soffer, Aminoacyl-tRNA transferases, Enzymes 40: 91 (1974).
30. A.Ciechanover, S.L.Wolin, J.A.Steitz, and H.F.Lodish, Transfer RNA is an essential component of the ubiquitin- and ATP-dependent proteolytic system, Proc.Natl.Acad.Sci.USA 82: 1341 (1985).
31. R.Gebhardt, and W.Jung, Biliary secretion of sodium fluorescein in primary monolayer cultures of adult rat hepatocytes and its stimulation by nicotinamide, J.Cell Sci. 56: 233 (1982).
32. P.Bohley, J.Kopitz, and G.Adam, Surface hydrophobicity, arginylation and degradation of cytosol proteins from rat hepatocytes, Biol.Chem. Hoppe-Seyler (submitted for publication).
33. H.Hanson, Peptidasen (Exopeptidasen), in: Handbuch der physiologisch- und pathologisch-chemischen Analyse, 10.Aufl., Hoppe-Seyler/Thierfelder, Berlin, Heidelberg, New York, 1966, Springer-Verlag, Band VI/C p.1
34. A.Stock, S.Clarke, C.Clarke, and J.Stock, Review Letter: N-terminal methylation of proteins: structure, function and specificity, FEBS Letters 220: 8 (1987).
35. R.Katznelson, and R.G.Kulka, Degradation of microinjected methylated and unmethylated proteins in hepatoma tissue culture cells, J.Biol.Chem. 258: 9597 (1983).
36. R.J.Mayer, and F.Doherty, Review Letter: Intracellular protein catabolism: State of the art, FEBS Letters 198: 181 (1986).
37. P.J.A.M. Plomp, E.J.Wolvetang, A.K.Groen, A.J.Meijer, P.B.Gordon, and P.O.Seglen, Energy dependence of autophagic protein degradation in isolated rat hepatocytes, Eur.J.Biochem. 164: 197 (1987).
38. M.C.Duque-Magalhães, and J.M.Gualberto, Regulation of mitochondrial proteolysis. Selective degradation of inner membrane polypeptides, FEBS Letters 210: 142 (1987).

PROTEINASE INHIBITORS AS ACUTE PHASE REACTANTS: REGULATION OF SYNTHESIS
AND TURNOVER

Aleksander Koj, Danuta Magielska-Żero
Anna Kurdowska and Joanna Bereta

Institute of Molecular Biology
Jagiellonian University
Al. Mickiewicza 3, 31-120 Kraków, Poland

INTRODUCTION

A dynamic equilibrium between the blood or tissue proteinases and
their inhibitors is drastically disturbed during acute inflammation
elicited by various noxious stimuli /Fritz, 1980; Koj 1985b; Fritz et al.,
1986/. A massive release of proteolytic enzymes from injured cells and
from infiltrating leucocytes and macrophages must be promptly neutralized
by a range of antiproteases present in body fluids /Fig.1/. Since the
reaction between proteinases and inhibitors is in most cases irreversible
and resulting complexes are quickly removed the enhanced proteolytic
activity could seriously deplete the body reserves of these antiproteases.
As demonstrated by the elegant studies of Ohlsson /1974/ and Fritz and
co-workers /1980, 1986/ such situations occur in certain cases of acute
pancreatitis and septic shock or endotoxaemia. Severe depletion of plasma
level of proteinase inhibitors, and especially of α_2-macroglobulin /α_2M/,
has a bad prognostic significance. However, the acute phase response
usually facilitates replenishment of proteinase inhibitors due to their
enhanced liver synthesis.

In man, the level of α_1-antichymotrypsin /α_1ACh/ begins to rise in
plasma soon after injury and is followed by rise of α_1-proteinase inhibitor
/α_1PI/ /Aronsen et al.,1972; Kueppers and Black, 1974/ but other human
plasma proteinase inhibitors poorly respond to inflammation. Rather
surprisingly a different pattern is observed in each mammalian species and
this bears a direct relationship to molecular mechanisms underlying the
acute phase response. Although macrophages and leucocytes were shown to
produce small amounts of α_1PI /Rogers et al.,1983; Takemura et al.,1986;
Lamontagne et al.,1985/ the bulk of plasma proteinase inhibitors is of
liver origin. Nevertheless, production of some proteinase inhibitors by infiltrating inflammatory cells may contribute to formation of a protective
microenvironment.

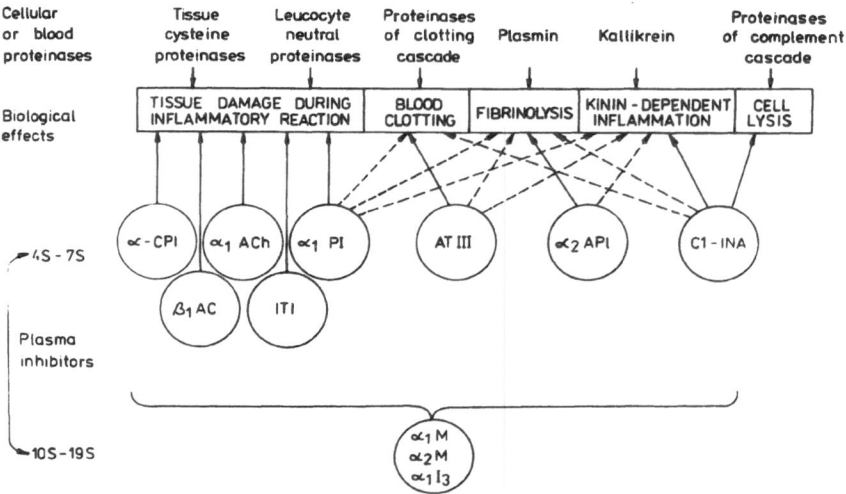

Fig.1. Homeostatic functions of some plasma antiproteinases /modified
from Koj, 1985/. Solid arrows indicate main targets of a given
proteinase inhibitor and dashed arrows - secondary targets /less
effective inhibition/.

ACUTE PHASE REACTANTS AMONG THREE FAMILIES OF PLASMA PROTEINASE
INHIBITORS

        The macroglobulin family is characterized by broad inhibitory
specificity and the presence of a labile thiol ester involved in protei-
inase binding /Travis and Salvesen, 1983/. It comprises at least 3
distinct high mol.wt. inhibitors. The best known example is human $\alpha_2 M$
built of four identical subunits consisting of 1451 amino acid residues
and including the bait region, reactive thiol ester and receptor binding
site /Sottrup-Jensen et al.,1984 ; van Leuven et al.,1986/. A homologous
protein has been described in rat, rabbit, guinea pig, dog, pig, horse
and other mammalian species /Schaeufele and Koo, 1982; Nelles and
Schnebli, 1982; Westrüm et al.,1983; Dubin et al.,1984; Suzuki and
Sinohara, 1986/. Rodents have in plasma another macroglobulin migrating
usually /but not always/ in the α_1 region and consisting of four pairs
of 2 subunits: heavy chain of approximately 160 kDa and light chain of
approx. 40 kDa /Schaeufele and Koo, 1982; Nelles and Schnebli, 1982;
Hudson and Koo, 1982/. As recently demonstrated by Heinrich and coworkers /Geiger et al.,1987/ synthesis of this macroglobulin in rat
hepatocytes occurs via a single chain precursor but the secreted glycosylated form is a heterologous octamer.

        Another high mol.wt. proteinase inhibitor containing a thiol ester
was described in rat by Esnard and Gauthier /1980/ and Esnard et al./1985/
under the name of $\alpha_1 I_3$, and in mouse by Saito and Sinohara /1985/ under
the name of murinoglobulin. The protein consisting of a single polypeptide
chain of approx. 180 kDa is abundant in rat and mouse plasma but absent
in human blood /Table 1/.

172

Table 1. Acute phase reactants among principal families of plasma proteinase inhibitors in man, rat and mouse /for references see text/

Species	MAN			RAT			MOUSE		
Protein	M_r kDa	mg/ml	AP-response	M_r kDa	mg/ml	AP-response	M_r kDa	mg/ml	AP-response
Macroglobulins									
α_1M		Not known		740^a160 $_b40$	2-4	0	740^a160 $_b40$	5-7	-
α_2M	725 /180/	2-4	0	720 /180/	<0.05	+++		Not known	
α_1I$_3$		Not known		180	5-7	--	180	2	--
Serpins									
α_1PI	53	2-4	+	55	1-2	+	53	3^x	0
α_1ACh	65	0.3	++		Not known			Not known	
Contrapsin	Not known			66	?	+	65	2	+
AT III	58	3	0	60	0.4	0	60	0.5	0
Kininogens									
HMW	114	0.2	0	100	0.1	0		Not known	
LMW	64	0.4	0	60	?	0		Not known	
T-kininogen		Not known		66	0.9^{xx}	+++		Not known	

x/ Higher level in males; xx/ Higher level in females; +, ++ or +++ indicate relative increase in plasma level; - or -- indicate relative decrease in plasma level; 0, no change during the acute phase response.

As demonstrated by many authors /for references see Koj, 1985a/ and shown in Table 1 rat α_2M is the only positive AP-protein in the macroglobulin family and its contents in plasma may rise from 0.03 to 3 mg/ml within 48 h of the acute inflammation. Its gene has been partly sequenced and mechanism of its induced synthesis is discussed in this volume by Heinrich /see pp.183/. On the other hand, α_1-macroglobulin concentration in rat plasma changes only negligibly during the acute phase response of the rat /Koj et al.,1982; Lonberg-Holm et al.,1987/, whereas $\alpha_1 I_3$ level decreases considerably /Esnard and Gauthier, 1980; Lonberg-Holm et al., 1987/. It is interesting that mouse, a species closely related to rat, responds to inflammatory stimuli by reduced synthesis of both macroglobulin species /Fig.2/.

The members of the large family of serpins show considerable variations in the specificity toward target enzymes and inhibit not only trypsin and chymotrypsin but also leucocyte elastase and cathepsin G as well as several enzymes of the clotting and complement cascades /Travis and Salvesen, 1983; Carrell and Travis, 1985; Travis, 1986/. In man α_1ACh is one of the major acute phase reactants and its level in plasma may be quadrupled within the first day after injury /Aronsen et al.,1972; Lindmark and Eriksson, 1985/ with the accompanying appearance of the modified protein /Morii and Travis, 1983/. On the other hand, α_1PI increases slowly in man and its level may be doubled after 3 days /Aronsen et al.,1972; Kueppers and Black, 1974/.

A different pattern was observed in the rat where known serpins are poor acute phase reactants and α_1PI is only slightly elevated after injury /Koj et al.,1978/. In the mouse α_1PI was classified as a neutral protein /Baumann et al.,1986b/ but showing considerable strain and sex differences /Minnich et al.,1984/. Antithrombin III does not respond to inflammation

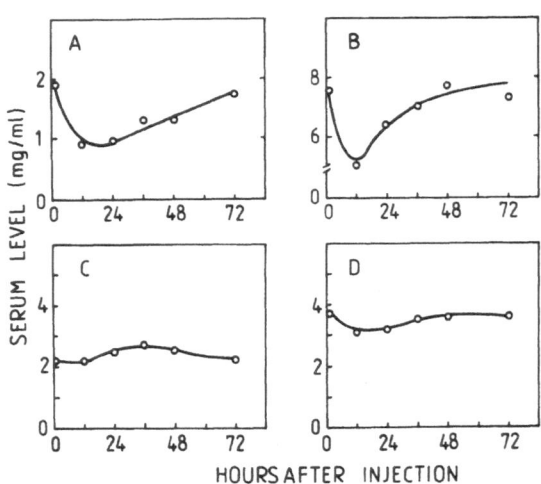

Fig.2. Changes in the serum levels of $\alpha_1 I_3$ = murinoglobulin /A/, α-macroglobulin /B/, contrapsin /C/ and α_1PI /D/ after i.p. injection of endotoxin into ICR female mice /after Yamamoto et.al., 1985/

in all so far tested mammalian species /Koj, 1985a/. It is interesting
that α_1ACh was not found in mouse or rat but these two rodent species have
a special trypsin inhibitor known as contrapsin. This protein shows
a considerable amino acid homology to human α_1ACh but due to mutated
reactive centre inhibits only trypsin /Lamontagne et al.,1981; Takahara
and Sinohara, 1982; Hill et al.,1984; Kuehn et al.,1984/. Increased concentration of contrapsin during the acute phase response /Yamamoto et al.,
1985; Baumann et al.,1986b/ suggests that its genomic regulatory sequence
might be similar to human α_1ACh.

Rats appear to be unique in their ability to produce considerable
amounts of an inhibitor of cysteine proteinases /Esnard and Gauthier,
1983/. The protein, known formerly as α_1-acute phase globulin /Koj, 1974/,
or rat major acute phase protein /Cole et al.,1985/, represents a special
type of kininogen /T-kininogen/ which can release kinin upon treatment
with trypsin /Furuto-Kato et al.,1984; Kageyama et al.,1984; Moreau et al.,
1986/. Other high and low molecular weight kininogens /HMW and LMW
kininogens/ found in the blood of several mammalian species /Hayashi et
al.,1985; Kellerman et al.,1986; Salvesen et al.,1986/ do not change
their concentration during the acute phase response. Although human LMW
kininogen and rat T-kininogen contain three cystatin-like domains only
two of them are able to inhibit cysteine proteinases and each with different specificity /Salvesen et al.,1986/: domain 2 was found to be a good
inhibitor of calpain, and to a lesser extent papain and cathepsin L,
whereas the latter two enzymes were preferentially inactivated by domain 3.

METHODS OF STUDYING OF THE REGULATORY MECHANISMS OF PROTEINASE INHIBITORS
SYNTHESIS

Some authors postulated that release of lysosomal hydrolases and
products of tissue degradation may stimulate synthesis of proteinase
inhibitors /cf. Koj, 1974/. The suspected mediators are peptides released
from the inhibitors during interaction with proteinases, or proteinase-
-inhibitor complexes. Such complexes of α_1PI and α_2M with elastase or other
proteinases increase in vivo during the early stages of inflammatory
reaction /Brower and Harpel, 1983; Fritz, 1980; Fritz et al.,1986/. The
complexes are recognized as foreign molecules and quickly cleared from
the circulation by hepatocytes, macrophages and fibroblasts /Fuchs et al.,
1984; Feldmann et al.,1983; Gliemann et al.,1985; Van Leuven et al.,1986/.
Also immature glycoproteins /both not glycosylated polypeptides or mannose-
-rich forms/ are quickly taken up by the perfused rat liver /Steube et al.,
1986/. However, there is no proof that proteinase-inhibitor complexes or
protein degradation products can directly stimulate synthesis of acute
phase proteins. As demonstrated by Ritchie and co-workers /1982/ plasmin-
-derived fragments of fibrinogen or fibrin stimulate macrophages to
production of hepatocyte stimulating factors /cf. also Fig.4/.

Although studies in vivo or in the perfused liver provided basic
information on the mechanism of induced synthesis of acute phase proteins
the primary cultures of rat and mouse hepatocytes or established lines
of hepatoma are now currently used enabling a more direct and effective
approach /Koj et al.,1984; Baumann et al.,1986a/. In our laboratory
hepatocytes or hepatoma cells are cultured for 2 days in Parker or Eagle
media enriched in glutamine, insulin, dexamethasone and preparations of
cytokines isolated from peritoneal macrophages or blood leucocytes. Plasma
proteins produced by the cells and secreted to the media are determined
by electroimmunoassay using monospecific antisera /Koj et al.,1987;
Magielska-Żero et al.,1987/ /cf. Fig.3/.

RESULTS AND DISCUSSION OF TISSUE CULTURE STUDIES

Rat hepatocytes cultured with dialysed supernatants from LPS-
-stimulated rat peritoneal macrophages show reduced synthesis of albumin
and $\alpha_1 I_3$, and enhanced synthesis of $\alpha_2 M$ and CPI /Table 2/. Similar results
were obtained with human blood-derived normal monocytes or leukemic cells
in agreement with the observations of other authors /for references see
Koj et al.,1987/. As demonstrated in many laboratories synthesis of acute
phase proteins is always preceded by accumulation in the cell of appropriate mRNAs /cf. Baumann et al.,1986a/. However, as shown by Birch and
Schreiber /1986/ this may be either due to increased transcription rates
of appropriate genes /e.g. rat CPI/, or increased mRNA processing and
stability /e.g. rat $\alpha_2 M$/.

In our experiments $\alpha_1 PI$ synthesis was only slightly elevated while
$\alpha_1 M$ and AT III production remained steady. Exposure of cultured hepatocytes to human recombinant interleukin 1 alpha or beta /HrIL-1, Biogen,
Geneve, Switzerland; 1 - 1000 ng/ml medium/ or human recombinant tumour
necrosis factor /HrTNFα, Bioghent, Ghent, Belgium, 10 - 1000 units/ml
medium/ lead to a significant reduction of albumin synthesis although the
response did not reach the level observed with crude cytokines, i.e.
dialysed macrophage supernatants. None of the recombinant cytokines, alone
or in combination with others, had significant effects on proteinase
inhibitors. When, however, recombinant preparations of human IL-1 or TNF
were added to hepatocyte cultures simultaneously with crude rat or human
cytokines the induced synthesis of CPI was selectively blocked /cf. Fig.3;
Koj et al.,1987/. This indicates a highly complex mechanism of action of
the tested cytokines on the expression of genes coding for acute phase
proteins and possible independent regulation of synthesis of individual
proteins. Recent reports presented during the Cold Spring Harbor Laboratory
Conference /April 29 - May 3, 1987/ suggest the involvement of interferon
β_2 in the stimulation of liver cells to the acute phase response
/J. Gauldie, personal communication/.

Table 2. Synthesis of albumin and six proteinase inhibitors by rat
hepatocytes cultured for 2 days with 0.1% BSA in PBS or with
natural or recombinant cytokines. The results are given in µg
protein produced by 10^6 cells during 24 h.

Protein secreted	Added to hepatocyte medium					
	BSA PBS	Rat cytokines	Human cytokines	HrIL-1α	HrIL-1β	TNFα
Albumin	16.2	8.7	9.6	10.1	10.8	10.2
$\alpha_1 M$	1.0	0.9	1.0	1.1	0.8	0.9
$\alpha_2 M$	2.1	10.8	9.9	1.9	2.0	1.8
$\alpha_1 I_3$	2.5	1.2	1.3	N.D.	N.D.	N.D.
$\alpha_1 PI$	8.1	12.2	9.0	8.2	8.4	8.0
AT III	0.9	0.8	0.9	0.8	0.7	0.8
$\alpha_1 CPI$	2.4	6.6	6.0	2.2	2.1	2.0

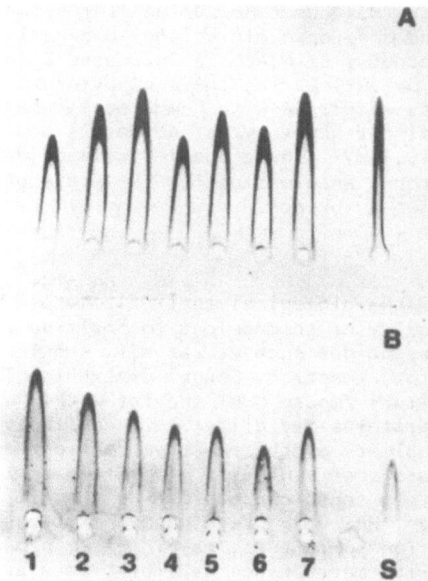

Fig. 3. Synthesis of albumin /A/ and α_1-cysteine proteinase inhibitor /B/
by rat hepatocytes cultured for 2 days with the supernatants of
LPS-stimulated human blood leukemic cells, HrTNFα and mixtures
of cytokines and TNF. Wells 1-3: media from hepatocytes cultured
with consecutive 5-fold dilutions of leukemic cytokines, wells 4
and 5 - from hepatocytes cultured with a mixture of the same
cytokines and TNF, well 6 - hepatocytes cultured with with
1000 units/ml of TNF, and well 7 - control culture /BSA/PBS/.
S - protein standards.

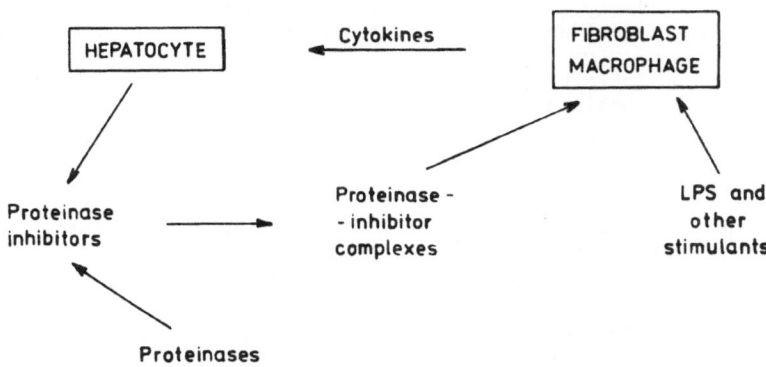

Fig. 4. Possible involvement of proteinase inhibitor complexes in the
induction of acute phase protein synthesis.

It is interesting that human Hep G2 cells respond differently to cytokines: albumin and α-fetoprotein belong to negative acute phase reactants whereas synthesis of α_1ACh is increased by natural cytokines and recombinant IL-1 or TNF. During these experiments it was noted that rat cytokines are less effective with human cells whereas human cytokines are active with normal rat hepatocytes and Morris hepatoma 7777 cells /Magielska-Żero et al.,1987/. These species-related differences should be always taken into account when evaluating the acute phase response of target cells to various cytokines.

CONCLUSIONS

As expected from the biological antiinflammatory functions of proteinase inhibitors majority of them belong to positive acute phase proteins. However, there are exceptions such as rat $\alpha_1 I_3$ showing decreased production during inflammatory reaction. Sexual dimorphism in the plasma level of some of the inhibitors /mouse α_1PI and rat α_1CPI/ complicates the picture. Homologous proteins may differ in specificity of proteinase inhibition due to variability of the reactive centre /human α_1ACh vs mouse contrapsin/. It appears that synthesis of proteinase inhibitors is regulated similarly to other acute phase proteins through specific mediators, inflammatory cytokines. However, proteinase-inhibitor complexes, or proteolytically modified inhibitors, may increase production of acute phase proteins indirectly with macrophages or fibroblasts as the regulatory cells /Fig.4/. Such a possibility should be verified experimentally in tissue cultures.

ACKNOWLEDGEMENTS

The studies on proteinase inhibitors were supported by grant 04.01.2.10 from the Polish Academy of Sciences. The authors are indebted to Dr. J.M. Dayer /Geneve/ for providing recombinant cytokines and to Dr J. Gauldie /Hamilton/ for generous supply of other materials.

REFERENCES

Aronsen, K.F., Ekelund, G., Kindmark, C.O. and Laurell, C.B., 1972, Sequencial changes of plasma proteins after surgical trauma, Scand. J.Clin.Lab.Invest.,29 /Suppl.124/:127.

Baumann, H., Hill, R.E., Sauder D.N. and Jahreis, G.P., 1986a, Regulation of major acute phase plasma proteins by hepatocyte stimulating factors of human squamous carcinoma cells, J.Cell Biol.,102:370.

Baumann, H., Latimer, J.J. and Glibetic, M.D., 1986b, Mouse α_1-protease inhibitor is not an acute phase reactant, Arch.Biochem.Biophys.,246: 488.

Birch, H.E. and Schreiber, G., 1986, Transcriptional regulation of plasma protein synthesis during inflammation, J.Biol.Chem.,261:8077.

Brower, M.S. and Harpel, P.C., 1983, Alpha-1-antitrypsin-human leucocyte elastase complexes in blood: quantification by an enzyme-linked immunosorbent assay and comparison with alpha-2-plasmin inhibitor-plasmin complexes, Blood,61:842.

Carrell, R. and Travis, J., 1985, α_1-antitrypsin and the serpins: variation and countervariation, Trends in Biochem.Sci.,10:20.

Cole, T., Inglis, A., Nagashima, M. and Schreiber, G., 1985, Major acute phase alpha/1/protein in the rat. Structure, molecular cloning and regulation of mRNA levels. Biochem.Biophys.Res.Commun.,126:719.

Dubin, A., Potempa, J. and Silberring, J., 1984, α_2-macroglobulin from horse plasma. Purification, properties and interaction with certain serine proteinases, Biochem.Int.,8:589.

Esnard, F. and Gauthier, F., 1980, Purification and physicochemical characterization of a new rat plasma proteinase inhibitor, α_1-inhibitor III, Biochim.Biophys.Acta,614:553.

Esnard, F. and Gauthier, F., 1983, Rat α_1-cysteine proteinase inhibitor. An acute phase reactant identical with α_1-acute phase globulin, J.Biol.Chem.,258:12443.

Esnard, F., Gutman, M., El Moujahed, A. and Gauthier, F., 1985, Rat plasma α_1-inhibitor$_3$: a member of the α-macroglobulin family, FEBS Lett., 182:125.

Feldmann, S.F., Ney, K.A., Gonias, S.L. and Pizzo, S.V., 1983, In vitro binding and in vivo clearance of human α_2-macroglobulin after reaction with endoproteases from four different classes, Biochem. Biophys.Res.Commun.,114:757.

Fritz, H., 1980, Proteinase inhibitors in severe inflammatory processes /septic shock and experimental endotoxaemia/: biochemical, pathophysiological and therapeutic aspects, Ciba Foundation Symposium, 75:351.

Fritz, M., Jochum, M., Geiger, R., Duswald, K.H., Dittmer, H., Kortmann, H., Neumann, S. and Lang, H., 1986, Granulocyte proteinases as mediators of unspecific proteolysis in inflammation, Folia Histochem.Cytobiol.,24:99.

Fuchs, H.E., Michalopoulos, G.K. and Pizzo, S.V., 1984, Hepatocyte uptake of α_1-proteinase inhibitor-trypsin complexes in vitro: evidence for a shared uptake mechanism for proteinase complexes of α_1-proteinase inhibitor and antithrombin III, J.Cell.Biochem.25:231.

Furuto-Kato, S., Matsumoto, A., Kitamura, N. and Nakanishi, S., 1985, Primary structures of the mRNAs encoding the rat precursors for bradykinin and T-kinin. Structural relationship of kininogens with major acute phase protein and α_1-cysteine proteinase inhibitor, J.Biol.Chem.,260:12054.

Geiger, T., Tran-Thi, T.A., Decker, K. and Heinrich, P.C., 1987, Biosynthesis of rat α_1-macroglobulin. Identification of an intracellular precursor, J.Biol.Chem.,262:

Gliemann, J., Davidsen, O., Sottrup-Jensen, L. and Sonne, O., 1985, Uptake of rat and human α_2-macroglobulin-trypsin complexes into rat and human cells, FEBS Lett.,188:352.

Hayashi, I., Kato, H., Iwanaga, S. and Oh-Ishi, S., 1985, Rat plasma high-molecular-weight kininogen. A simple method for its purification and its characterization, J.Biol.Chem.,260:6115.

Hill, R.E., Shaw, P.H., Boyd, P.A. and Hastie, N.D., 1984, Plasma protease inhibitors in mouse and man: divergence within the reactive centre regions, Nature, 311:175.

Hudson, N.W. and Koo, P.H., 1982, Subunit and primary structure of a mouse alpha macroglobulin, a human α_2-macroglobulin homologue, Biochim. Biophys.Acta, 704:290.

Kageyama, R., Kitamura, N., Ohkubo, H. and Nakanishi, S., 1985, Differential expression of the multiple forms of rat prekininogen mRNAs after acute inflammation, J.Biol.Chem.,260:12060.

Kellerman, J., Lottspeich, F., Henschen, A. and Müller-Esterl, W., 1986, Completion of the primary structure of human high molecular mass kininogen. The amino acid sequence of the entire heavy chain and evidence for its evolution by gene triplication, Eur.J.Biochem., 154:471.

Koj, A., 1974, Acute-phase reactants, in: Structure and Function of Plasma Proteins, vol.1, pp.73-131, A.C. Allison, ed., Plenum Press, London and New York.

Koj, A., 1985a, Definition and classification of acute phase proteins, in: The Acute Phase Response to Injury and Infection, pp.139-144, A.H. Gordon and A. Koj, eds., Elsevier, Amsterdam, New York, Oxford.

Koj, A., 1985b, Biological functions of acute phase proteins, in: The Acute Phase Response to Injury and Infection, pp.145-160, A.H. Gordon and A. Koj, eds., Elsevier, Amsterdam, New York, Oxford.

Koj, A., Regoeczi, E., Toews, C.J., Leveille, R. and Gauldie, J., 1978, Synthesis of antithrombin III and alpha-1-antitrypsin by the perfused rat liver, Biochim.Biophys.Acta,539:496.

Koj, A., Dubin, A., Kasperczyk, H., Bereta, J. and Gordon, A.H., 1982, Changes in the blood level and affinity to concanavalin A of rat plasma glycoproteins during acute inflammation and hepatoma growth, Biochem.J.,206:545.

Koj, A., Gauldie, J., Regoeczi, E., Sauder, D.N. and Sweeney, G.D., 1984, The acute phase response of cultured rat hepatocytes, Biochem.J., 224:505.

Koj, A., Kurdowska, A., Magielska-Żero, D., Rokita, H., Sipe, J.D., Dayer, J.M., Demczuk, S. and Gauldie, J., 1987, Limited effects of recombinant human and murine interleukin 1 and tumour necrosis factor on production of acute phase proteins by cultured rat hepatocytes, Biochem.Int.,14:553.

Kueppers, F. and Black, L.F., 1974, Alpha-1-antitrypsin and its deficiency, Am.Rev.Resp.Dis.,110:176.

Kuehn, L., Rutschmann, M., Dahlmann, B. and Reinauer, H., 1984, Proteinase inhibitors of rat serum. Purification and partial characterization of three functionally distinct trypsin inhibitors, Biochem.J.,218:953.

Lamontagne, L.R., Stadnyk, A.W. and Gauldie, J., 1985, Synthesis of alpha--1-protease inhibitor by resident and activated mouse alveolar macrophages, Am.Rev.Resp.Dis.,131:321.

Lamontagne, L., Gauldie, J. and Koj, A., 1981, Ontogeny and tissue distribution of alpha-1-antitrypsin of the mouse, Biochim.Biophys.Acta,662:15.

Lindmark, B.E. and Eriksson, S.G., 1985, Plasma α_1-antichymotrypsin in liver disease, Clin.Chim.Acta,152:261.

Lonberg-Holm, K., Reed, D.L., Roberts, R.C., Hebert, R.R., Hillman, M.C. and Kutney, R.M., 1987, Three high molecular weight protease inhibitors of rat plasma. Isolation, characterization and acute phase changes, J.Biol.Chem.,262:438.

Magielska-Żero, D., Rokita, H., Cięszka, K., Kurdowska A., Koj, A., Sipe, J.D. and Gauldie, J., 1987, Comparison of the acute phase response of cultured Morris hepatoma 7777 cells and of rat hepatocytes, Br.J.Exp.Path.,68:

Minnich, M., Kueppers, F. and James, H., 1984, Alpha-1-antitrypsin from mouse serum: isolation and characterization, Comp.Biochem.Physiol., 78B:413.

Morii, M. and Travis, J., 1983, Structural alterations in α_1-antichymotrypsin from normal and acute phase plasma, Biochem.Biophys.Res. Commun.,111:438.

Moreau, T., Gutman, N., El Moujahed, A., Esnard, F. and Gauthier, F., 1986, Relationship between the cysteine-proteinase-inhibitory function of rat T-kininogen and the release of immunoreactive kinin upon trypsin treatment.,Eur.J.Biochem.,159:341.

Müller-Esterl, W., Fritz, H., Kellerman, J., Lottspeich, F., Machleidt, W. and Turk, V., 1985, Genealogy of mammalian cysteine proteinase inhibitors. Common evolutionary origin of stefins, cystatins and kininogens, FEBS Lett.,191:221.

Nelles, L.P. and Schnebli, H.P., 1982, Subunit structure of the rat α--macroglobulin proteinase inhibitors, Hoppe Seyler's Z.Physiol.Chem., 363:677.

Ritchie, D.G., Levy, B.A., Adams, M.A. and Fuller, G.M., 1982, Regulation of fibrinogen synthesis by plasmin-derived fragments of fibrinogen and fibrin: an indirect feedback pathway, Proc.Natl.Acad.Sci.USA, 79:1530.

Rogers, J., Kalsheker, N., Wallis, S., Speer, A., Coutelle, C.H.,
Woods, D. and Humphries, S.E., 1983, The isolation of a clone for
human α_1-antitrypsin and the detection of α_1-antitrypsin in mRNA
from liver and leukocytes, Biochem.Biophys.Res.Commun.,116:375.

Saito, A. and Sinohara, H., 1985, Murinoglobulin, a novel protease inhibitor from murine plasma. Isolation, characterization and comparison
with murine α-macroglobulin and human α_2-macroglobulin, J.Biol.Chem.,
260:775.

Salvesen, G., Parkes, C., Abrahamson, M., Grubb, A. and Barrett, A.J.,
1986, Human low M_r kininogen contains three copies of a cystatin
sequence that are divergent in structure and in inhibitory activity
for cysteine proteinases, Biochem.J.,234:429.

Schaeufele, J.T. and Koo, P.H., 1982, Structural comparison of rat α_1-
and α_2-macroglobulins, Biochem.Biophys.Res.Commun.,108:1.

Sottrup-Jensen, L., Stepanik, T.M., Kristensen, T., Wierzbicki, D.M.,
Jones, C.M., Lonblad, P.B., Magnusson, S. and Petersen, T.E., 1984,
Primary structure of human α_2-macroglobulin. V. The complete structu-
re. J.Biol.Chem.,259:8318.

Steube, K., Gross, V., Häuslinger, D., Tran-Thi, T.A., Decker, K.,
Gerok, W. and Heinrich, P.C., 1986, Clearance of acute phase plasma
proteins with no, high mannose-, hybrid-, or complex-type oligo-
saccharide side chains by the isolated perfused rat liver, Biochem.
Biophys.Res.Commun.,141:949.

Suzuki, Y. and Sinohara, H., 1986, Guinea pig murinoglobulin. Purification
and some properties, Biol.Chem.Hoppe-Seyler,367:579.

Takahara, H. and Sinohara, H., 1982, Mouse plasma trypsin inhibitors.
Isolation and characterization of α_1-antitrypsin and contrapsin,
a novel trypsin inhibitor, J.Biol.Chem.,257:2438.

Takemura, S., Rossing, T.H. and Perlmutter, D.H., 1986, A lymphokine
regulates expression of alpha-1-proteinase inhibitor in human
monocytes and macrophages, J.Clin.Invest.,77:1207.

Travis, J., 1986, Target enzymes for plasma proteinase inhibitors, Folia
Histochem.Cytobiol.,24:117.

Travis, J. and Salvesen, G., 1983, Human plasma proteinase inhibitors,
Annu.Rev.Biochem.,52:655.

Weström, B.R., Karlsson, B.W. and Ohlsson, K., 1983, Immunocrossreactivity
between macroglobulins from pig, dog, rat and man including human
pregnancy-associated α_2-glycoprotein, Hoppe Seyler's Z.Physiol.Chem.,
364:375.

Van Leuven, F., Marynen, P., Sottrup-Jensen, L., Cassiman, J.J. and
van den Berghe H., 1986, The receptor-binding domain of human α_2-
-macroglobulin, J.Biol.Chem.,261:11369.

Yamamoto, K., Tsujino, Y., Saito, A. and Sinohara, H., 1985, Concentrations
of murinoglobulin and α-macroglobulin in the mouse serum: variations
with age, sex, strain and experimental inflammation, Biochem.Int.,
10:463.

REGULATION OF PROTEINASE ACTIVITY BY HIGH MOLECULAR WEIGHT INHIBITORS:

BIOSYNTHESIS OF RAT α-MACROGLOBULINS

T. Geiger, T. Andus, D. Kunz, M. Heisig, J. Bauer[+], N. Northoff[‡], F. Gauthier[*], T.-A. Tran-Thi, K. Decker, and P.C. Heinrich

Biochemisches Institut and [+]Medizinische Klinik der Universität Freiburg, D-7800 Freiburg, FRG, [‡]Blutspendezentrale des DRK, D-7900 Ulm, FRG, [*]Laboratoire de Biochimie, Faculté de Medécine, F-37032 Tours, France

Many proteolytic reactions occur in blood plasma during coagulation, fibrinolysis and complement activation. Particularly during inflammatory processes proteinases such as elastase, collagenase and cathepsin G are released by granulocytes and macrophages[1-3]. Due to their capability to destroy connective tissues in many organs, these proteinases represent a severe hazard for the organism. It is therefore essential that activated proteinases in plasma are controlled. One way to regulate proteinase activities is by specific inhibitors[4-6]. Proteinase inhibitors account for about 20% of total proteins in normal rat plasma. Among them there are three proteinase inhibitors of high molecular weight and a wide inhibitory spectrum: α_1-macroglobulin, α_2-macroglobulin and α_1-inhibitor3[7-13]. They belong to the same superfamily and are characterized by their particular mechanism of interaction with proteinases resulting in enzymatically active complexes[7,12].

We have isolated all three high molecular weight proteinase inhibitors from rat plasma. Fig. 1 shows the sodium dodecyl sulfate polyacrylamide gel electrophoretic separation of α_1-macroglobulin, α_2-macroglobulin and α_1-inhibitor3. Whereas α_2-macroglobulin and α_1-inhibitor3 consist of only one high molecular weight subunit, α_1-macroglobulin very likely contains a heavy chain and a light chain. From the work of other investigators it is well established that α_2-macroglobulin consists of four identical subunits[8-10], while α_1-inhibitor3 is composed of of only one polypeptide chain[14]. In contrast to α_2-macroglobulin, little is known about the quaternary structure of α_1-macroglobulin. An octame-

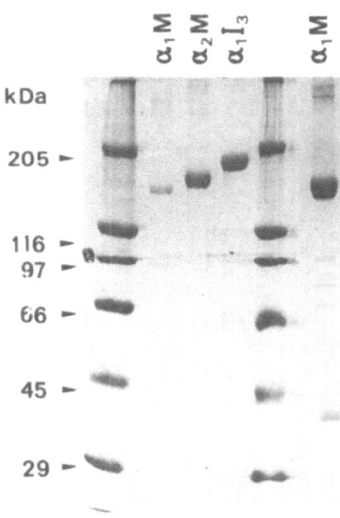

Fig. 1. SDS-polyacrylamide gel electrophoresis of rat α_1-macroglobulin, α_2-macroglobulin and α_1-inhibitor3

Table 1

THE α-MACROGLOBULIN FAMILY

Species	Protein	Concentration (mg/ml)		Molecular weight of subunit(s)	Number of subunits	Carbohydrate content %
		Normal plasma	AP-plasma			
Rat	$\alpha_1 M$	3.8	3.9	168 000 38 000	8 (?)	15
	$\alpha_2 M$	<0.02	2.0	182 000	4	15.9
	$\alpha_1 I_3$	6.0	2.1	186 000	1	15
Human	$\alpha_2 M$	2-4	2-4	179 000	4	10.2
	PZP	<0.01	1-1.4	180 000	4	10-12

ric structure of four heavy (160 kDa) and four light chains (39 kDa) has been proposed[8, 10, 11] (Table 1). All three α-macroglobulins contain internal thioester bonds and are characterized by their wide inhibitory spectrum to proteinases[7, 14]. Although α_1-macroglobulin, α_2-macroglobulin and α_1-inhibitor3 are such abundant proteinase inhibitors in the normal and inflamed rat, respectively, little is known about their biosynthesis.

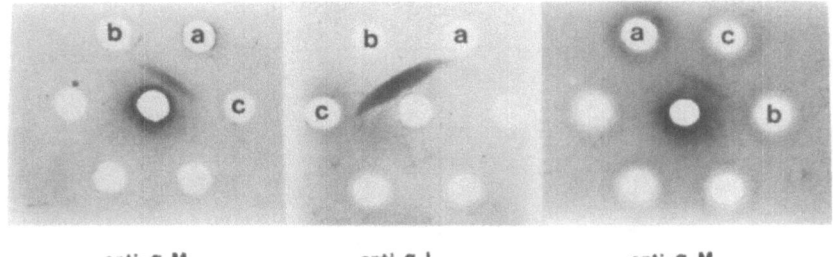

anti-α_1M   anti-α_1I$_3$ anti-α_2M

Fig. 2. Ouchterlony double diffusion assay

a = α_1-macroglobulin, b = α_1-inhibitor3, c = α_2-macroglobulin

(A) Biosynthesis of α_1-macroglobulin, α_2-macroglobulin and α_1-inhibitor3 in rat hepatocyte primary cultures

Hepatocyte primary cultures were incubated with $|^{35}S|$methionine, cells and medium were separated and α_1-macroglobulin, α_2-macrogloblin and α_1-inhibitor3 were immunoprecipitated with specific antisera from cells and their respective media. The antisera do not show cross-reactivity (Fig. 2). Fig. 3 shows that the intracellular forms of α_2-macrogloblin and α_1-inhibitor3 (C) exhibit a higher electrophoretic mobility than those immunoprecipitated from the media (M). We have demonstrated by pulse-chase experiments, by incubation with either endoglucosaminidase H or sialidase, by labeling with $|^3H|$fucose, $|^3H|$mannose or $|^3H|$galactose, and by binding to and elution from concanavalin A-Sepharose that the intracellular forms of α_2-macrogloblin and α_1-inhibitor3 contain oligosaccharide side chains of the high-mannose type, whereas the secreted α_2-macroglobulins are characterized by complex type carbohydrate side chains (data not shown). Unlike in the case of α_2-macrogloblin and α_1-inhibitor3 we found for α_1-macroglobulin an intracellar form of a higher apparent molecular weight than the medium form. It is interesting that we always detected a second rather faint polypeptide of low molecular weight. Recently we have demonstrated that the intracellular high molecular weight form of α_1-macroglobulin must be N-glycosylated and proteolytically processed to a heavy and light chain before secretion can occur[15].

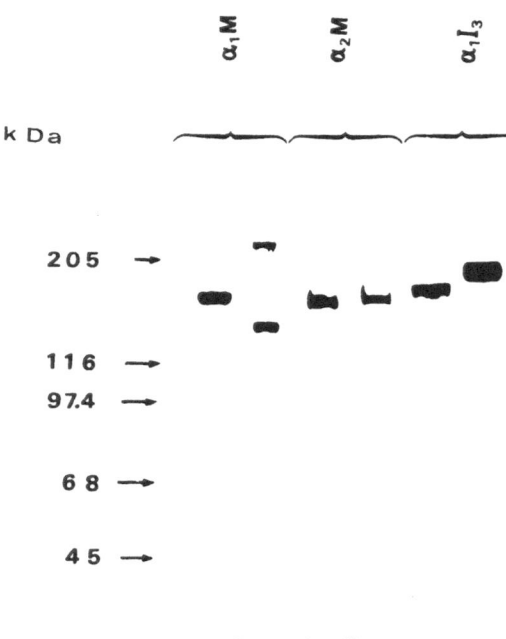

Fig. 3. SDS-polyacrylamide gel electrophoresis of α_1-macroglobulin, α_2-macroglobulin and α_1-inhibitor3 immunoprecipitated from hepatocytes and hepatocyte media

(B) α-Macroglobulin gene expression during experimental inflammation

An interesting aspect of the biosynthesis of the α-macroglobulins is their regulation during acute inflammation. As summarized in Table 1 and Fig. 4, plasma levels of α_2-macroglobulin increase about 100-fold[9, 17-19], α_1-inhibitor3 levels decrease to about 30% of controls[12], whereas α_1-macroglobulin concentrations remain essentially unchanged during acute inflammation[12].

We have shown by in vitro translation of poly(A)RNA[19] and by dot-blot hybridization with specific cDNAs[20,21] that in the case of α_2-macroglobulin the increase in plasma is preceded by an increase in α_2-macroglobulin mRNA levels in the liver (Fig. 5, upper part) and in the case of α_1-inhibitor3 the decrease in plasma is due to a decrease in α_1-inhibitor3 mRNA levels in the liver (Fig. 5, lower part)[22]. When serum

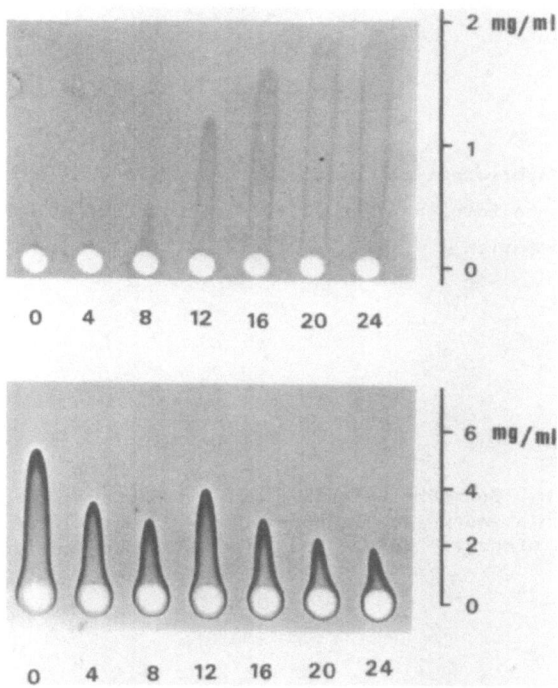

Fig. 4. Rocket immunoelectrophoresis of rat sera at different times after turpentine treatment

Upper part: α_2-macroglobulin; lower part: α_1-inhibitor3

and mRNA levels of α_2-macroglobulin were determined in turpentine-treated hypophysectomized rats, which have very low glucocorticoid levels, no corresponding increases were observed. Only the combined administration of turpentine and the synthetic glucocorticoid dexamethasone led to a stimulation of α_2-macroglobulin translation, indicating a permissive role of glucocorticoid hormones in the acute-phase response[23].

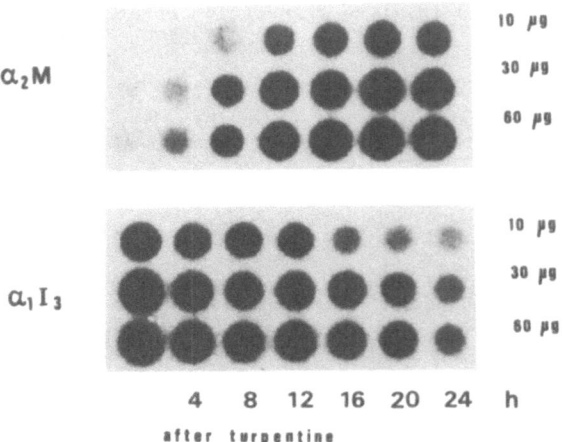

Fig. 5. Dot hybridization of α_2-macroglobulin cDNA and α_1-inhibitor3 cDNA to total RNA from livers of rats suffering from an acute inflammation

Acknowledgements

We thank H. Gottschalk for her help with the preparation of this manuscript. This work was supported by grants from the Deutsche Forschungsgemeinschaft, Bonn, and the Fonds der Chemischen Industrie, Frankfurt.

References

1. A. Koj, Acute-phase reactants. Their biosynthesis, turnover and biological significane, in: Structure and Function of Plasma Proteins, Allison, A.C., ed. Vol. 1, pp. 73-125, Plenum Press, London (1974).

2. H. Fritz, and M. Jochum, Granulocyte proteinases as mediators of unspecific proteolysis in inflammation: a review, in: Selected Topics in Clinical Enzymology, D.M. Goldberg, and M. Werner, eds., Vol. 2, pp. 305-328, Walter de Gruyter, Berlin-New York (1984).

3. A. Koj, Definition and classification of acute-phase proteins, in: Research Monographs in Cell and Tissue Physiology, A.H. Gordon, and A. Koj, eds., Vol. 10, p. 139, Elsevier Science Publishers (1985).

4. M. Laskowski, and I. Katc, 1980, Protein inhibitors cf proteinases, Annu. Rev. Biochem. 49:593.

5. H. Holzer, and P.C. Heinrich, 1980, Control of Proteolysis, Annu. Rev. Biochem. 49:63.

6. J. Travis, and G.S. Salvesen, 1983, Human plasma proteinase inhibitors, Annu. Rev. Biochem. 52:655.

7. P. M. Starkey, and A.J. Barret, α2-Macroglobulin, a physiological regulator of proteinase activity, in: Proteinases in Mammalian Cells and Tissues, A.J. Barret, ed., pp. 663-696, North-Holland Publishing Company, Amsterdam, New York (1977).

8. J. T. Schaeufele, and P.H. Koo, 1982, Structural comparison of rat α1- and α2-macroglobulins, Biochem. Biophys. Res. Commun. 108:1.

9. A. H. Gordon, 1976, The α-macroglobulins of rat serum, Biochem. J. 159:643.

10. L. P. Nelles, and H.P. Schnebli, 1982, Subunit structure of the rat α-macroglobulin proteinase inhibitors, Hoppe Seyler's Z. Physiol. Chem. 363:677.

11. J. Bergsma, M.K. Boelen, A. Duursma, W.G. Schutter, M.W. Bouma, and M. Gruber, 1985, Complexes of rat α1-macroglobulin and subtilisin are endocytosed by parenchymal liver cells, Biochem. J. 226:75.

12. F. Gauthier, and K. Ohlsson, 1978, Isolation and some properties of a new enzyme binding protein in rat plasma, Hoppe-Seyler's Z. Physiol. Chem. 359:987.

13. F. Esnard, and F. Gauthier, 1980, Purification and physicochemical characterization of a new rat plasma proteinase inhibitor, α1-inhibitor III, Biochim. Biophys. Acta 614:553.

14. F. Esnard, N. Gutman, A. El Moujahed, and F. Gauthier, 1985, Rat plasma α1-inhibitor3: a member of the α-macroglobulin family, FEBS Lett. 182:125.

15. T. Geiger, T.A. Tran-Thi, K. Decker, and P.C. Heinrich, 1987, Biosynthesis of rat α1-macroglobulin. Identification of an intracellular precursor. J. Biol. Chem. 262:4973.

16. F. van Leuven, 1982, Human α2-macroglobulin: structure and function, Trends Biochem. Sci. 7:185.

17. H. Okubo, O. Miynage, M. Nagano, H. Ishibashi, J. Kudo, R. Ikuta, and K. Shibata, 1981, Purification and immunological determination of α2-macroglobulin in serum from injured rats. Biochim. Biophys. Acta 668:257.

18. G. Schreiber, and G. Howlett, Synthesis and secretion of acute-phase proteins, in: Plasma Protein Secretion by the Liver, H. Glaumann, T. Peters, Jr., C. Redman, eds, pp. 423-449, Academic Press, London-New York (1983).

19. W. Northemann, T. Andus, V. Gross, and P.C. Heinrich, 1983, Cell-free synthesis of rat α2-macroglobulin and induction of its mRNA during experimental inflammation. Eur. J. Biochem. 137:257.

20. W. Northemann, M. Heisig, D. Kunz, and P.C. Heinrich, 1985, Molecular Cloning of cDNA sequences for rat α2-macroglobulin and measurements of its transcription during experimental inflammation. J. Biol. Chem. 260:6200.

21. G. Schreiber, A.R. Aldred, T. Thomas, H.E. Birch, P.W. Dickson, T. Guo-Fen, P.C. Heinrich, W. Northemann, G.J. Howlett, F.A. De Jong, and A. Mitchell, 1986, Levels of messenger ribonucleic acids for plasma proteins in rat liver during acute experimental inflammation, Inflammation 10:59.

22. M. Schweizer, K. Takabayashi, T. Geiger, T. Laux, G. Biermann, J.-M. Buhler, F. Gauthier, L.M. Roberts, and P.C. Heinrich, 1987, Identification and sequencing of cDNA clones for the rodent negative acute-phase protein α1-inhibitor 3, Eur. J. Biochem. 164:375.

23. J. Bauer, M. Birmelin, G.-H. Northoff, W. Northemann, T.-A. Tran-Thi, H. Ueberberg, K. Decker, and P.C. Heinrich, 1984, Induction of rat α2-macroglobulin in vivo and in hepatocyte primary cultures: synergistic action of glucocorticoids and a Kupffer cell derived factor. FEBS Lett. 177:89.

INDUCTION OF THE PROTEINASE INHIBITOR α_2-MACROGLOBULIN IN RAT HEPATOCYTES BY A MONOCYTE-DERIVED FACTOR

T. Andus[1], H. Northoff[2], J. Bauer[3], U. Ganter[3], D. Männel[4], T.-A. Tran-Thi[1], K. Decker[1], and P. C. Heinrich[1]

[1]Biochemisches Institut, Univ. Freiburg, D-7800 Freiburg, [2]DRK-Blutspendezentrale, D-7900 Ulm, [3]Medizinische Univ.-Klinik, D-7800 Freiburg, [4]DKFZ, Institut für Immunologie, D-6900 Heidelberg, FRG

Disturbances of the physiologic homeostasis such as infections, tissue injury, tumor growth and immunologic disorders lead to a highly complex reaction of the organism, the so called acute-phase response[1-3]. The acute-phase response is characterized by fever, leukocytosis, a negative nitrogen balance, depression of serum iron and zinc levels, elevation of serum copper and dramatic changes in the synthesis of hepatic acute-phase proteins[4,5]. The plasma concentration of the proteinase-inhibitor α_2-macroglobulin (α_2M) for example increases 100-500-fold in the rat during acute inflammation. The concentration of the proteinase-inhibitor α_1-inhibitor3 ($\alpha_1 I_3$) belonging to the same macroglobulin family decreases to about 30% simultaneously. These changes are generated by mediators secreted from mononuclear phagocytes. Hepatocyte-stimulating factor (HSF), interleukin 1 (IL-1), tumor necrosis factor α (TNFα) and interferon ß (IFN ß) are the most important of theses mediators.

We examined the regulation of the synthesis of HSF and IL-1 in human blood monocytes and discriminated HSF from IL-1, TNFα , IFNα , IFN ß$_1$ and IFN γ.

1.　α_2-Macroglobulin and α_1-inhibitor3 synthesis in rat hepatocytes

We used rat hepatocyte primary cultures to measure the ability of human monocytes to regulate the synthesis of acute-phase proteins in the liver. After preparation rat hepatocytes were cultured for 5 days and then incubated at 37° C for 12 h with conditioned media from human monocytes. Thereafter $|^{35}S|$methionine was added to methionine-free culture medium for the labeling of proteins. After incubation at 37°C for 3 h, the media were separated from the cells and used for immunoprecipitation. By use of monospecific antisera the acute-phase proteins were immunoprecipitated from the hepatocyte medium and subjected to sodium dodecyl sulfate polyacrylamide gel electrophoresis (SDS-PAGE) and fluorography. For the quantification of the radioactivity incorporated into the different acute-phase proteins, the corresponding bands identified by fluorography were excised from the gels, solubilized and counted in a liquid scintillation spectrometer.

The effect of increasing amounts of HSF containing human monocyte supernatants on the synthesis of the positive acute phase protein α_2M

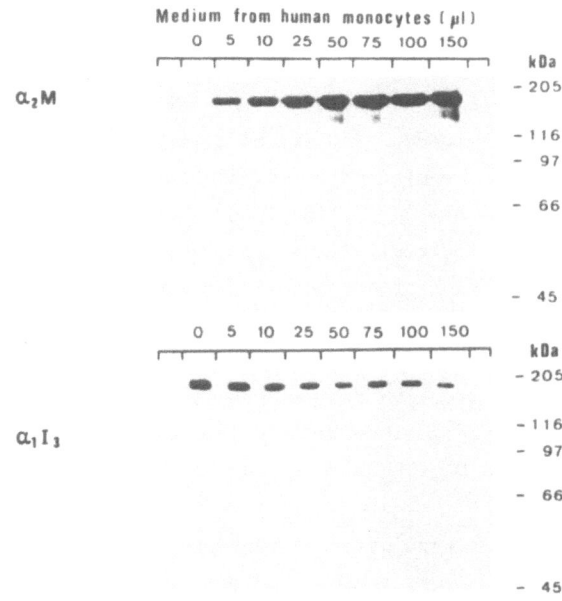

Fig. 1　Induction of α_2-macroglobulin and depression of α_1-inhibitor3 synthesis by supernatants from human monocytes

and the negative acute phase protein $\alpha_1 I_3$ is shown in Fig. 1. Similarly as **in vivo** during acute inflammation $\alpha_2 M$ synthesis is increased up to 100-fold and $\alpha_1 I_3$ as well as albumin (not shown) are decreased to about 30 - 40% of controls. Therefore, conditioned media from human monocytes and rat hepatocyte primary cultures represent a suitable system to investigate the regulation of acute-phase protein synthesis and of their mediators.

2. Induction of hepatocyte-stimulating factor in human monocytes and its discrimination from interleukin-1

Until now little is known about the regulation of HSF synthesis in human monocytes[6,7]. In our hands HSF was found to be synthesized spontaneously by human monocytes or rat Kupffer cells[8]. The use of selected fetal calf serum or human serum with low lipopolysaccharide (LPS) contamination led to the observation that monocytes cultured in these media synthesized no or only very small amounts of HSF. Using such functionally LPS-free sera we could now clearly demonstrate an LPS-dependent stimulation of HSF synthesis in human monocytes (Fig. 2A, filled circles).

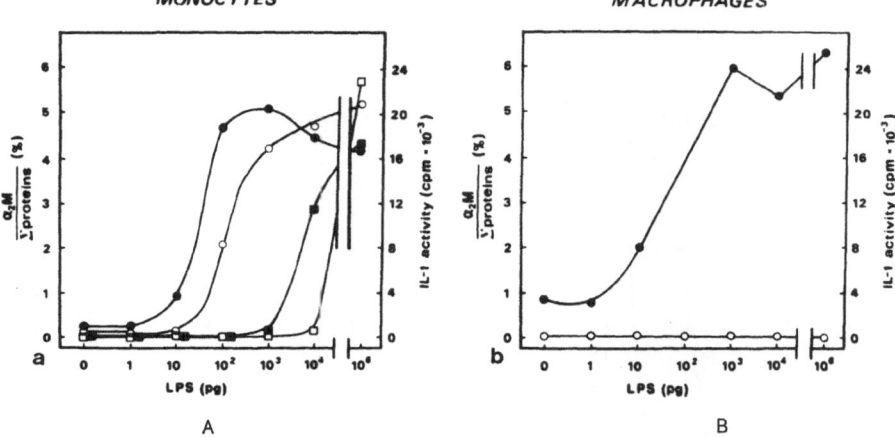

Fig. 2   Induction of hepatocyte-stimulating factor and interleukin-1 by lipopolysaccharide
HSF activity is expressed as the ratio of the radioactivity of immunoprecipitated $\alpha_2 M$ and the TCA-precipitable radioactivity of total proteins in the medium of hepatocytes.

Amounts of LPS as low as 100 pg/ml of medium were found to be suffi-cient to release a large amount of HSF activity. To further corroborate that it is LPS, which stimulates HSF production in monocytes, we pre-incubated the different amounts of LPS with 100 U/ml of the LPS inactivator polymyxin B prior to LPS addition to monocytes. It can be seen from Fig. 2A (filled squares) that polymyxin B totally inhibited the action of LPS up to a concentration of 1 ng/ml.

Since it is known that in monocytes interleukin-1 (IL-1) synthesis is stimulated by LPS, we determined IL-1 activity in the thymocyte stimulation assay as described by Northoff et al.[9]. Fig. 2A (open circles) shows that measurable Il-1 activity was induced by 10 pg of LPS per ml. When the same experiment was carried out with human monocytes after a preincubation period of 7 days in the presence of 5% human serum and in the absence of stimulant, the IL-1 production was completely turned off, whereas HSF synthesis was preserved (Fig. 2B). From this finding we conclude that HSF and IL-1 must be differently regulated biological entities and that HSF is not identical with IL-1.

This conclusion was further supported by results from gel permeation chromatography. LPS-stimulated supernatants from freshly prepared monocytes (Fig. 3,A) or from monocytes preincubated for 7 days in the absence of stimulant prior to the addition of LPS (Fig. 3B) were

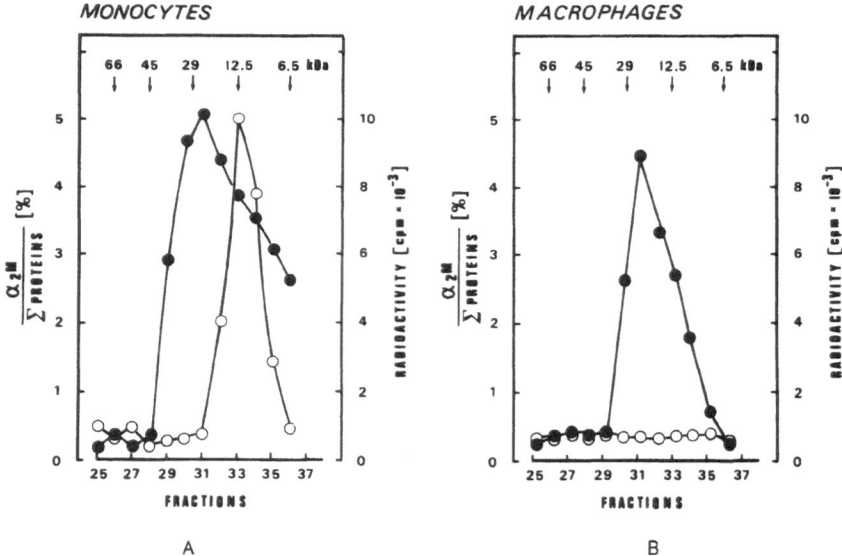

A   B

Fig. 3   Separation of hepatocyte-stimulating factor and interleukin-1 activities by gel permeation chromatography

subjected to a Superose 12HR10/30 column. HSF activity eluted with an apparent molecular weight of about 23 kDa in both cases (filled circles). Il-1 activity, which was only present in supernatants of fresh monocytes (Fig. 2A), eluted as a sharp peak with an apparent molecular weight of about 12.5 kDa (Fig. 3A, open circles). Based on our gel filtration experiments, we conclude that the high molecular weight inhibitors described in references[10-12] could not have caused the loss of Il-1 activity in the medium of preincubated monocytes. The same applies to low molecular weight inhibitors, such as thymidine[13], prostaglandin[14,15] and hydrogen peroxide[15]. From the gel permeation experiments we can, however, not exclude the possible existence of an Il-1 inhibitor with a molecular weight identical to that of Il-1. This latter possibility however, was also ruled out by the results of a mixing experiment, where different amounts of supernatants from precultured monocytes were added to different amounts of Il-1. Essentially no inhibition of Il-1 activity was observed (data not shown).

In conclusion, Il-1 and HSF are different biological entities, both of which can be elicited by stimulation of human monocytes with LPS. They are, however, differently regulated, since monocytes cultured for more than 24 h lose their ability to produce Il-1 in response to LPS, while HSF synthesis is preserved.

3. Discrimination of hepatocyte-stimulating factor from tumor necrosis factor

To study the effect of tumor necrosis factor on the synthesis of acute phase proteins, rat hepatocyte primary cultures were incubated with different amounts of recombinant human TNFα ranging from 1 pg to 10 µg per ml medium for 12 h at 37°C and proteins were labelled with $|^{35}S|$methionine. Three positive acute-phase proteins, α_2-macroglobulin (α_2M), α_1-acute-phase globulin (α_1APG) and α_1-proteinase inhibitor (α_1PI), known to show very strong, strong or weak increases during inflammation and the negative acute-phase protein albumin were immunoprecipitated from the media of hepatocytes incubated with TNFα. Fig. 4 A shows that recombinant TNFα did not affect the synthesis of any of the 4 selected rat acute-phase proteins.

Since recombinant TNFα may have properties different from the physiologically synthesized material, conditioned media from LPS-stimulated human monocytes containing naturally synthetesized TNFα were added to hepatocyte primary cultures. A concentration of 0.5 ng TNF /ml was

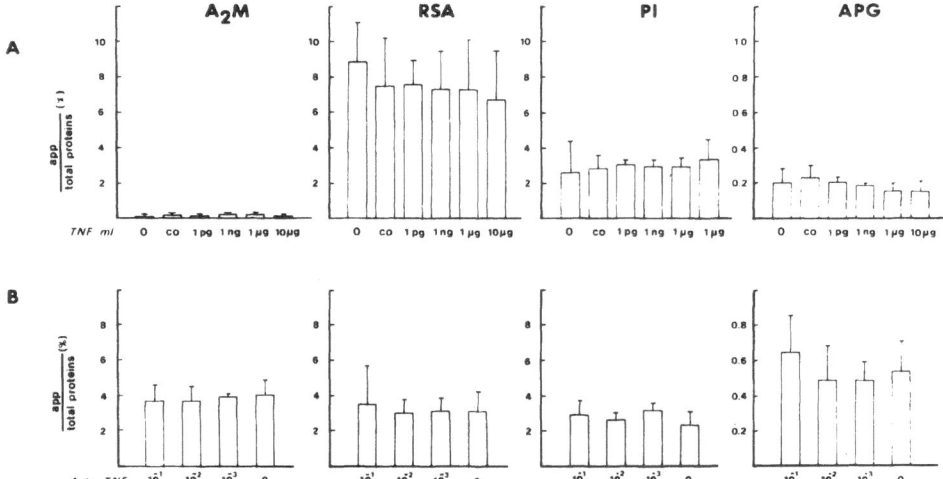

Fig. 4   Effect of human recombinant TNFα and conditioned medium from human monocytes on the synthesis of acute-phase proteins in rat hepatocyte primary cultures

determined in the monocyte media.

As shown in Fig. 4B, last columns, conditioned monocyte media stimulated α_2M synthesis 25-fold and α_1APG synthesis 2.5-fold, whereas α_1PI did not change. Albumin synthesis was decreased by 60%. These changes in the synthesis of the 4 acute-phase proteins are quite similar to those observed in vivo during inflammation. Thus, the conditioned medium of human monocytes contains inflammatory mediators leading to the induction of acute-phase protein synthesis.

Addition of a neutralizing antibody to TNFα should abolish the acute-phase response, if TNFα present in the monocyte medium would be the inducer of acute-phase protein synthesis in hepatocytes. We therefore preincubated the conditioned media with different amounts of anti-TNFα prior to the addition to hepatocytes. As shown in Fig. 4B, first 3 columns, no inhibition of the acute-phase induction was observed, although the antibody had neutralized TNFα activity completely (data not shown).

From the fact that human recombinant TNFα did not induce α_2M, α_1APG or α_1PI synthesis in rat hepatocytes, while conditioned medium from human monocytes was a potent stimulator, it must be concluded that TNFα is not a main mediator of the acute-phase response in rat hepatocyte primary cultures and is distinct from HSF.

4. Effects of different interferons on the induction of α_2-macroglobulin in rat hepatocyte primary cultures

Since monocytes synthesize interferon ß (IFN ß), we added different amounts of human purified IFNα and ß as well as recombinant human IFN ß and γ to our rat hepatocytes. After 12 h of incubation at 37°C, RNA was isolated from the hepatocytes, blotted on a gene screen membrane and hybridized with ^{32}P-labelled α_2M cDNA[16]. As shown in Fig. 5 high doses of purified IFN ß stimulated the synthesis of α_2M mRNA to a similar extend as conditioned medium from human monocytes and macrophages. Interestingly, neither human recombinant IFN ß or γ nor human natural IFNα were able to stimulate α_2M mRNA synthesis.

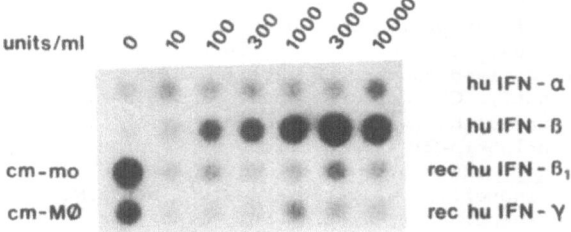

Fig. 5   Effect of different interferons on the synthesis of α_2-macroglobulin in rat hepatocytes

Hence we conclude that a protein, produced by human fibroblasts, but different from IFNα, ß$_1$ and γ can stimulate acute phase protein synthesis in rat hepatocyte primary cultures. This is in accordance with the recent finding of J. Gauldie. He found recombinant B cell stimulating factor 2 (BSF-2), which is identical to IFN ß$_2$ to act as an acute phase mediator (personal communication).

Acknowlegements

We thank H. Gottschalk for her help with the preparation of this manuscript. This work was supported by grants from the Deutsche Forschungsgemeinschaft, Bonn, and the Fonds der Chemischen Industrie, Frankfurt.

REFERENCES

1. C. A. Dinarello, 1984, Interleukin-1 and the pathogenesis of the acute-phase response, New Engl. J. Med. 1984. 311:1413.

2. J. D. Sipe, Cellular and humoral components of the early inflammatory reaction, in: The Acute Phase Response to Injury and Infection, A. Koj, and A.H. Gordon, eds., p. 23, Elsevier, Amsterdam (1985).

3. J. J. Oppenheim, E.J. Kovacs, K. Matsushima, and S.K. Durum, 1986, There is more than one interleukin 1, Immunol. Today 7:45.

4. I. Kushner, 1982, The phenomenon of the acute-phase response, Ann. N. Y. Acad. Sci. 389:39.

5. G. Schreiber, and G. Howlett, in: Plasma Protein Secretion by the Liver, H. Glaumann, T. Peters, and C. Redman, eds., p. 423, Academic Press, London (1983).

6. D. G. Ritchie, B.A. Levy, M.A. Adams, and G.M. Fuller, 1982, Regulation of fibrinogen synthesis by plasmin-derived fragments of fibrinogen and fibrin: An indirect feedback pathway, Proc. Natl. Acad. Sci. USA 79:1530.

7. D. G. Ritchie, and G.M. Fuller, 1983, Hepatocyte-stimulating factor: a monocyte-derived acute-phase regulatory protein, Ann. N.Y. Acad. Sci. 408:490.

8. J. Bauer, M. Birmelin, G.-H. Northoff, W. Northemann, T.-A. Tran-Thi, H. Ueberberg, K. Decker, and P.C. Heinrich, 1984, Induction of rat α2-macroglobulin **in vivo** and in hepatocyte primary cultures: synergistic action of glucocorticoids and a Kupffer cell-derived factor, FEBS Lett. 177:89.

9. H. Northoff, C. Carter, and J.J. Oppenheim, 1980, Inhibition of concanavalin A-induced human lymphocyte mitogenic factor (interleukin-2). Production by suppressor T-lymphocytes. J. Immunol. 125:125.

10. C. A. Dinarello, L.J. Rosenwasser, and S.M. Wolff, 1981, Demonstration of a circulating suppressor factor of thymocyte proliferation during endotoxin fever in humans, J. Immunol. 127:2517.

11. E. P. Amento, J.T. Kurnick, and S.M. Krane, 1982, Interleukin-1 production by a human monocyte cell line is induced by a T lymphocyte product, Immunobiology 163:276.

12. H. Fujiwara, and J.J. Ellner, 1986, Spontaneous production of a suppressor factor by the human macrophage-like cell line U937, J. Immunol. 136:181.

13. M. J. Stadecker, J., Calderon, M.L. Karnovsky, and E.R. Unanue, 1977, Synthesis and release of thymidine by macrophages, J. Immunol. 119:1738.

14. M. Gordon, M.A. Bray, and J. Morley, 1976, Control of lymphokine secretion by prostaglandins, Nature 262:401.

15. Z. Metzger, J.T. Hoffeld, and J.J. Oppenheim, 1980, Macrophage-mediated suppression. I. Evidence for participation of both hydrogen peroxide and prostaglandins in suppression of murine lymphocyte proliferation. J. Immunol. 124:983.

16. W. Northemann, M. Heisig, D. Kunz, and P.C. Heinrich, 1985, Molecular cloning of cDNA sequences for rat α2-macroglobulin and measurement of its transcription during experimental inflammation, J. Biol. Chem. 260:6200.

ASTROCYTES SYNTHESIZE AND SECRETE α2-MACROGLOBULIN: DIFFERENCES BETWEEN THE REGULATION OF α2-MACROGLOBULIN SYNTHESIS IN RAT LIVER AND BRAIN

Joachim Bauer, Peter-Joachim Gebicke-Haerter[+], Ursula Ganter, Isolde Richter and Wolfgang Gerok

Medizinische Klinik und [+]Pharmakologisches Institut der Universität, D-7800 Freiburg, West Germany

ABSTRACT

α2-Macroglobulin (α2M) is an important proteinase inhibitor both in the blood and in the interstitial space of many mammalian species. Recently, occurrence of α2M in human and fetal rat brain has been reported. However, its cellular origin remained obscure. Here it will be shown that astroglial cells cultured from newborn rats synthesize and secrete α2M. In addition to astrocyte primary cultures a rat astrocytoma cell line, C6-cells, also synthesize and secrete α2M.

In contrast to hepatocytes of the adult rat, where α2M is expressed as an acute-phase protein and where glucocorticoids and monocyte-derived factors are required for its synthesis, α2M synthesis in astrocytes of newborn rats is independent from both. It is therefore concluded that expression of α2M is differently regulated in the liver of the adult rat and fetal or neonatal brain.

INTRODUCTION

α2-Macroglobulin is a tetrameric glycoprotein consisting of four identical subunits with a molecular weight of 182 kDa each. Because of its broad specificity which is directed against proteinases of all four major classes, α2M is regarded as one of the most potent proteinase inhibitors (for review see [1]). In the rat, α2M synthesis was found in fetal liver, spinal cord, and brain[2-4]. Soon after birth α2M synthesis levels off and is induced in the adult animal only during gestation in the uterus and during the acute-phase response in the liver[3-9].

Until now, α2M synthesis in brain could not be attributed to a specific cell type. Here it will be shown for the first time that rat astrocytes, one of the major glial cell types of the central nervous system, produce α2M, thus establishing a new source of its synthesis.

MATERIALS AND METHODS

Chemicals

Monensin and tunicamycin were from Sigma Chemicals (St. Louis, MO). Dulbecco's modified Eagle's medium (DMEM), RPMI 1640, horse serum, fetal calf serum (FCS) was from Gibco (Karlsruhe, FRG). $|^{35}S|$methionine (600 Ci/mmol) was from the Radiochemical Center (Amersham, UK). Protosol was from New England Nuclear (Boston, MA), protein A-Sepharose CL-4B from Pharmacia (Freiburg, FRG). X-Ray films (XAR5) were from Kodak.

Astrocyte cultures

Astrocyte primary cultures from brain cortices of newborn Wistar rats were prepared according to Booher and Sensenbrenner[10]. Briefly, after removal of the meninges, forebrains were minced and gently disso-ciated by trituration in Hank's balanced salt solution (HBSS), then filtered through a 75 µm nylon sieve. Cells were collected by centrifu-gation at 200 x g for 10 min, resuspended in DMEM supplemented with 10% FCS, plated in 35 mm Falcon culture dishes (5 x 10^5 cells/dish) and incubated at 37°C in a humidified atmosphere of 95% air, 5% CO_2. Purity of the cells, as determined by glial fibrillary acidic protein (GFAP) immunohistochemical staining, was more than 85% as previously described by our laboratory[11].

Antibodies

Monospecific, polyclonal antibodies against rat α2M were raised in rabbits as previously described[12]. Anti-GFAP antibodies raised in pigs and rabbits were kindly supplied by L. F. Eng, Palo Alto, CA[13,14]. The mouse monoclonal antibody ED 1, directed against a macrophage-specific intracellular antigen, was a generous gift of C.D. Dijkstra, Amster-dam[15]. Fluorescin-isothiocyanate (FITC) and rhodamine (TRITC) conjugated second antibodies were from Cappel Laboratories (Cochranville, PA) and Dako Inc. (St. Barbara, CA), respectively.

Immunohistochemical staining

Indirect immunofluorescence histochemistry was done as described previously[16]. Briefly, cell cultures were incubated for 2 h in the presence of 10 µM monensin to inhibit protein secretion. After three washings in DMEM, 20% horse serum, 20 mM Hepes, pH 7.2, cells were fixed in ethanol/acetic acid (95:5, v/v) at -80°C for 12 min, washed again and incubated with the respective antibodies (α2M 1:200, GFAP 1:500, ED1 1:800, diluted in wash medium) for 1 h at room temperature. Fluorescent second antibodies were reacted at 1:50 dilutions for 30 min at room temperature. The respective first and second antibodies were added simultaneously. Using TRITC-conjugated goat anti-rabbit and FITC-conju-gated rabbit anti-pig second antibodies for double labeling, the rabbit anti-pig antibody was used last. Specimens were mounted in PBS/glyce-rol/n-propyl-gallate and viewed with a Zeiss fluorescence microscope under the appropriate wave lengths.

Radiolabeling and immunoprecipitation

Cell cultures were incubated in 1 ml methionine-free RPMI 1640 medium/dish at 37°C in the presence of $|^{35}S|$methionine (25 µCi/dish) for 2.5 h. After incubation, media were separated from cells. The cells from each dish were homogenized in 1 ml of 25 mM Tris/HCl, pH 7.5, 20 mM NaCl, 1% deoxycholate (Na^+) and 1% Triton X-100. Non-soluble material was removed by centrifugation at 12,000 x g for 10 min. The resulting supernatants and the culture media were used for immunoprecipitations.

For immunoprecipitations, cell homogenates and culture media were diluted in 2 volumes TNET-buffer (20 mM Tris/HCl, pH 7.6, 0.14 M NaCl, 5 mM EDTA, 1% Triton X-100). The respective soluble material was incubated with nonimmune rabbit control serum, non-specifically bound material was bound to protein A-Sepharose and processed as the specific immunoprecipitates. The specific immunoprecipitation was done in the presence of excess antibody. After incubation at 4°C overnight, immune complexes were bound to protein A-Sepharose, washed four times with TNET-buffer and twice with 50 mM sodium phosphate buffer, pH 7.5. Antigen and antibodies were resuspended in 0.1 M Tris/HCl, pH 6.8, 5% ß-mercaptoethanol, 5% SDS and 10% glycerol at 95°C for 4 min. Solubilized proteins were analyzed by SDS-PAGE[17] and subjected to fluorography according to Bonner and Laskey[18]. For quantification of the radioactivity incorporated into α2M, the corresponding bands were cut out from the gels, solubilized in Protosol/water (9:1, v/v) at 45°C overnight and counted in a liquid scintillation spectrometer.

RESULTS

Intracellular de novo synthesis and secretion of α2M by rat astrocyte primary cultures after 14 days in culture is shown in Fig. 1. Astrocyte cultures were metabolically radiolabeled for 2.5 h and α2M was immunoprecipitated from the cell homogenate (lane 2) and from the culture medium (lane 3). Material nonspecifically bound by preabsorption to a rabbit nonimmune serum is shown in lane 1 (cell homogenate) and lane 4 (medium).

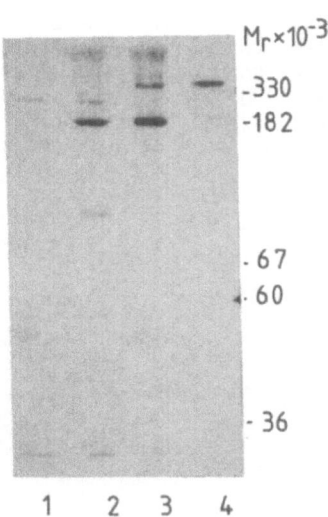

Fig. 1. Astrocyte primary cultures kept in culture for 14 days were metabolically labeled with |35S|methionine for 2.5 h. α2M (182 kDa) was immunoprecipitated from the cell homogenate (lane 2) and culture medium (lane 3) and subjected to SDS-PAGE and fluorography. The material preabsorbed by a rabbit nonimmune control serum from the cell homogenate (lane 1) and culture medium (lane 4) classifies the faint upper bands in lanes 2 and 3 as unspecifically bound material.

Dependence of α2M secretion from time in culture is plotted in Fig. 2. No astrocytic α2M secretion is detectable at day 1. Very low levels are found during day 3-6. Thereafter, secretion increases from 0.008% total secreted proteins at day 7 to 0.07% at day 12, finally to around 0.5% of total protein secretion at day 26.

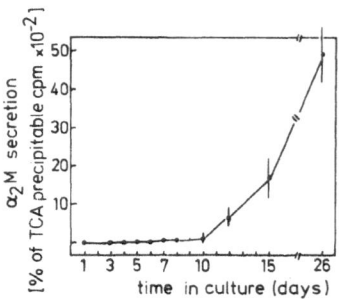

Fig. 2. Freshly prepared astrocytes after different times in culture as indicated in the figure were metabolically labeled with $|^{35}S|$methionine for 2.5 h. α2M was immunoprecipitated from equivalent TCA-precipitable total cpm (0.4×10^6) from the culture media, subjected to SDS-PAGE and fluorography. The radioactive 182 kDa α2M bands were excised from the gel, solubilized, counted and related to total TCA-precipitable protein.

In order to identify unequivocally the specific cell type synthesizing α2M, 14-day old astrocyte primary cultures were immunocytochemically double stained (phase bright optics Fig. 3,A) with antibodies directed against α2M (Fig. 3,B) and against the established astrocyte marker protein GFAP (glial fibrillary acidic protein) (Fig. 3,C). Fig. 3, B and C show co-staining for α2M and GFAP in the same cells.

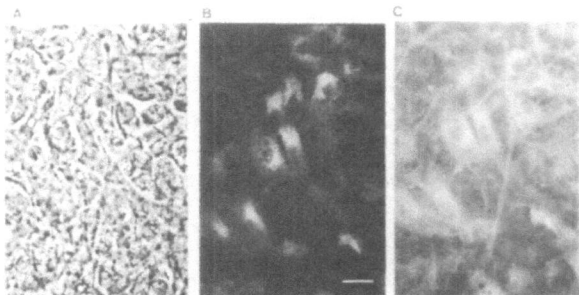

Fig. 3. A phase-contrast micrograph of a 2 week-old astrocyte culture is shown in A. An identical field is shown in B and C after double immunostaining for α2M (B) and GFAP (C). Bar, 20 μm.

α2M synthesis was also found in a rat astrocytoma cell line. After 2.5 h of |^{35}S|methionine incorporation into newly synthesized proteins, the 182 kDa α2M was immunoprecipitated from the cell homogenate (Fig. 4, lane 2) and culture medium (lane 3) of C6-cells. Fig. 4 (lane 1) shows material preabsorbed to a nonimmune control serum.

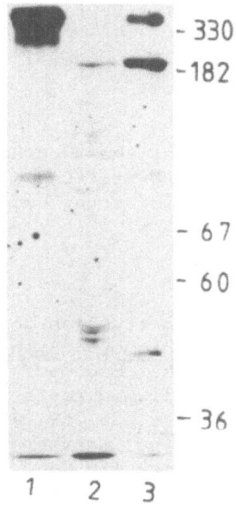

Fig. 4. 10^6 C6 astrocytoma cells were radiolabeled with |^{35}S|methionine for 2.5 h and α2M was immunoprecipitated from the cell homogenate (lane 2) and culture medium (lane 3), subjected to SDS-PAGE and autoradiographed. The additional band above the 182 kDa α2M band in lane 3 is classified as unspecific by demonstration of material preabsorbed to nonimmune rabbit control serum (lane 1).

The presence of glucocorticoids was not necessary for α2M synthesis in astrocyte primary cultures or in astrocytoma cells. When added in different concentrations, glucocorticoids did not affect α2M synthesis. α2M synthesis was slightly repressed in a dose-dependent manner when dialyzed supernatants of human monocytes were added to the astrocyte primary cultures (not shown).

DISCUSSION

The regulation of α2M expression differs in hepatocytes of the adult rat and in astrocytes of newborn rats. Liver-derived α2M in the adult rat is expressed as an acute-phase protein and requires the synergistic action of glucocorticoids and monocyte-derived factors[7-9]. In contrast, glucocorticoids are not required for α2M synthesis in rat astrocytes. Here, monocyte-derived factors act even slightly inhibitory. This could explain the lack of astrocytic α2M synthesis during the first days in culture, since large amounts of monocytic cells are present until their removal from the adherent astrocytes during the first medium changes, which therefore also serve as purifying steps. Detectable α2M synthesis only after several days of culture could therefore reflect a derepression.

What is the biological meaning of the ability of astrocytes to synthesize a strong proteinase inhibitor? α2M synthesis could be an important event during brain development where a balance between serine proteinases (around the neurite growth tips) and protease inhibitors was found to be essential for migration of neurons and for neurite out-growth[19-23]. In the adult brain, after injury astrocytic α2M synthesis may help to protect the adjacent tissue from proteolytic enzymes coming from the inflamed area.

ACKNOWLEDGEMENT

This work was supported by the Deutsche Forschungsgemeinschaft, Bonn, and the Müller-Fahnenberg-Stiftung, Freiburg. U. Ganter is a fellow of the Landesgraduiertenförderung Baden-Württemberg.

REFERENCES

1. O. Saksela. Plasminogen activation and regulation of pericellular proteolysis, Biochim. Biophys. Acta 823:35 (1985).
2. K. M. Dziegielewska, N.R. Saunders, and H. Soreq. mRNA from developing rat cerebellum directs in vitro synthesis of plasma proteins, Dev. Brain Res. 23:259 (1985).
3. V. Kodelja, M. Heisig, W. Northemann, P.C. Heinrich, and W. Zimmermann. α2-Macroglobulin gene expression during rat development studied by in situ hybridization, EMBO J. 5:3151 (1986).
4. D. E. Panrucker, P.C.W. Lai, and F.L. Lorscheider. Distribution of acute-phase α2-macroglobulin in rat fetomaternal compartments, Am. J. Physiol. 245:138 (1983).
5. H. E. Weimer, and D.C. Benjamin. Immunochemical detection of an acute phase protein in rat serum, Am. J. Physiol. 209:736 (1965).
6. A. H. Gordon. The α2-macroglobulins in rat serum, Biochem. J. 159:643 (1976).
7. J. Bauer, M. Birmelin, G.-H. Northoff, W. Northemann, T.-A. Tran-Thi, H. Ueberberg, K. Decker, and P.C. Heinrich. Induction of rat α2-macroglobulin in vivo and in hepatocyte primary cultures: synergistic action of glucocorticoids and a Kupffer-cell-derived factor, FEBS Lett. 177:89 (1984).
8. J. Bauer, W. Weber, T.-A. Tran-Thi, G.-H. Northoff, K. Decker, W. Gerok, and P.C. Heinrich. Murine interleukin 1 stimulates α2-macroglobulin synthesis in rat hepatocyte primary cultures, FEBS Lett. 190:271 (1985).
9. J. Bauer, T.-A. Tran-Thi, G.-H. Northoff, F. Hirsch, H.-J. Schlayer, W. Gerok, and P.C. Heinrich. The acute-phase induction of α2-macroglobulin in rat hepatocyte primary cultures, Eur. J. Cell Biol. 40:86 (1986).
10. J. Booher, and M. Sensenbrenner. Growth and cultivation of disso-ciated neurons and glial cells from embryonic chick, rat and human brain in flask cultures, Neurobiology 2:91 (1972).
11. M. Keller, R. Jakisch, A. Seregi, and G. Hertting. Comparison of prostanoid forming capacity of neuronal and astroglial cells in primary cultures, Neurochem. Int. 7:655 (1985).

12. W. Northemann, T. Andus, V. Gross, and P.C. Heinrich. Cell-free synthesis of rat α2-macroglobulin and induction of its mRNA during experimental inflammation, Eur. J. Biochem. 137:257 (1983).

13. L. F. Eng. The glial fibrillary acidic (GFS) protein, in: "Proteins of the Nervous System", R.A. Bradshaw, and D.M. Schneider, eds., Raven Press, New York (1980).

14. L. F. Eng, and S.J. De Armond. Glial fibrillary acidic (GFA) protein immunocytochemistry in development and neuropathology, in: "XIth International Congress of Anatomical-Glial and Neuronal Cell Biology", E.A. Vidrio, and S. Fedoroff, eds., Alan R. Liss, New York (1981).

15. C. D. Dijkstra, E.A. Doepp, P. Joling, and G. Kraal. The heterogeneity of mononuclear phagocytes in lymphoid organs: distinct macrophage subpopulations in the rat recognized by monoclonal antibodies ED 1, ED 2, and ED 3, Immunology 54:589 (1985).

16. P. J. Gebicke-Haerter, H.-H. Althaus, I. Rittner, and V. Neuhoff. Bulk separation and long-term culture of oligodendrocytes from adult pig brain. I. Morphological studies, J. Neurochem. 42:357 (1984).

17. J. King, and K. Laemmli. Polypeptides of the tailfibres of bacteriophage T4, J. Mol. Biol. 62:465 (1971).

18. W. M. Bonner, and R.A. Laskey. A film detection method for tritium-labelled proteins and nucleic acids in polyacrylamide gel, Eur. J. Biochem. 46:83 (1974).

19. A. Krystosek, and N.W. Seeds. Plasminogen activator secretion by granule neurons in cultures of developing cerebellum, Proc. Natl. Acad. Sci. USA 78:7810 (1981).

20. J. Guenther, H.P. Nick, and D. Monard. A glia-derived neurite-promoting factor with protease inhibitory activity, EMBO J. 4:1963 (1985).

21. P. H. Patterson. On the role of proteases, their inhibitors and the extracellular matrix in promoting neurite outgrowth, J. Physiol. (Paris) 80:207 (1985).

22. G. Moonen, M.P. Grau-Wagemans, and I. Selak. Plasminogen activator - plasmin system and neuronal migrations, Nature 298:752 (1982).

23. J. Lindner, J. Guenther, H.-P. Nick, G. Zinser, H. Antonicek, M. Schachner, and D. Monard. Modulation of granule cell migration by a glia-derived protein, Proc. Natl. Acad. Sci. USA 83:4568 (1986).

CHARACTERIZATION OF DIFFERENT FORMS OF DIPEPTIDYL PEPTIDASE IV FROM RAT LIVER AND HEPATOMA BY MONOCLONAL ANTIBODIES

Sabine Hartel, Christoph Hanski, Reinhard Neumeier
Reinhart Gossrau[+] and Werner Reutter

Institut für Molekularbiologie und Biochemie
[+]Institut für Anatomie der Freien Universität
Berlin
Arnimallee 22, [+]Königin-Luise-Str. 15
D-1000 Berlin 33 (Dahlem)

INTRODUCTION

Dipeptidyl peptidase IV (DPP IV, E.C.3.4.14.5) is a glyco-protein present in high activities in the plasma membranes of mammalian tissues. Although the enzyme has been investigated in detail for several years, its biological role remains still obscure. Because of the specificity for N-terminal Gly-Pro-dipep-tides, its possible role in collagen metabolism was suggested (1). It could be shown, that DPP IV is involved in the interaction of hepatocytes with the extracellular matrix (2). In Morris hepatoma 7777 the specific activity of plasma membrane DPP IV is reduced to about 3% (3), which is concomitant with a decreased cell-substratum adhesion of hepatoma cells (4).
In addition to the membrane-bound enzyme, about 15% of total DPP IV activity in liver and about 50% of the total DPP IV activity in hepatoma is found in the soluble fraction (5). This observation prompted the present study on the relationship and differences between the enzyme of liver and hepatoma. The comparison of DPP IV from various sources was carried out with monoclonal antibodies directed against different epitopes of the enzyme.

MATERIALS AND METHODS

Monoclonal antibodies were produced according to Köhler and Milstein (6). Balb/c mice were immunized with 0.7 U of partially purified DPP IV from liver plasma membranes (7) and the splenocytes were fused with the mouse myeloma cell line NS-1. Immunoglobulin G-producing hybridoma cells were selected in an enzyme-linked immunosorbent assay (ELISA). Protein A (1 ug/ml) (Sigma, Heidelberg, FRG) in carbonate buffer (0.05 M, pH 9.6) was adsorbed over night to microtiter wells (Nunc, Heidelberg, FRG) at room temperature. Culture supernatants were then incubated for 2 h at 37°C. Afterwards 30 mU/ml of purified DPP IV in PBS containing 0.1% sulphobetaine 14 were added. Each step

was followed by three washings with PBS, pH 7.2, containing 1%
bovine serum albumin, 0.1% Tween 20 and 0.2% sodium azide (washing buffer). Bound DPP IV was detected by determination of
the enzyme activity with Gly-Pro-p-nitroaniline-tosylate (Bachem, Bubendorf, Switzerland) as the substrate (8). Antibodies
were concentrated 5-fold from culture supernatants by $(NH_4)_2SO_4$
precipitation. The production of the rabbit antiserum against
DPP IV was described previously (9). Plasma membranes were isolated from liver and Morris hepatoma 7777 of male rats as described earlier (10). The soluble form of DPP IV was obtained
from liver homogenate as described (5). Immunoprecipitation was
carried out with protein A-Sepharose (Pharmacia, Freiburg,
FRG). The precipitated DPP IV was separated on SDS-polyacrylamide gels (11), blotted onto nitrocellulose and immunostained
(7). A modified dot blot assay of Freeman et al. (12) was used
to test whether the monoclonal antibodies recognize different
epitopes. Nitrocellulose (Schleicher and Schüll, Dassel, FRG)
was positioned into a dot blot chamber (Bio Rad, München, FRG)
and increasing amounts (20-50 mU/ml) of purified DPP IV were
pipetted into the wells. After the protein solution soaked into
the nitrocellulose, the paper was incubated with PBS containing
3% bovine serum albumine for 1 h at 37°C. The nitrocellulose
was then cut into discs and incubated over night at 4°C with
the concentrated hybridoma culture supernatants or with PBS.
After washing the discs were incubated over night at 4°C with
^{35}S-methionine-labelled hybridoma culture supernatants (13).
The nitrocellulose discs were washed until the radioactivity in
the washing buffer was reduced to background levels, dried and
counted. For histochemical studies 6 um thick cryostat sections
of liquid-nitrogen frozen male tissues were used. The acetone-chloroform pretreated sections were either incubated with
the substrate Gly-Pro-4-methoxy-2-naphthylamine and Fast Blue B
or Neufuchsine for simultaneous coupling (5) or with hybridoma
culture supernatants. The immunohistochemical staining was performed either with an anti-mouse-ß-galactosidase-linked antibody (Amersham, Braunschweig, FRG) or an alkaline phosphatase-
linked antibody (Dakopatts, Hamburg, FRG) as described (14).
Phospholipase C treatment of plasma membranes was carried out
according to Low et al. (15).

RESULTS AND DISCUSSION

1. Preparation and characterization of monoclonal antibodies

Nine hybridoma cell lines secreting DPP IV-specific monoclonal antibodies were obtained. All nine antibodies belong to
IgG class, isotype 1, 2a, 2b and 3. The use of a competitive
dot blot assay with purified DPP IV showed, that four different
epitopes, named A, B, C and D, of the DPP IV molecule were recognized by the antibodies (Table 1). The cross-reactivity with
hepatoma plasma membrane DPP IV as well as with the soluble
form of the liver was tested by immunoprecipitation. Eight monoclonal antibodies recognized all three forms in the native
state. The antibodies precipitated completely liver plasma membrane DPP IV but only 70% of DPP IV from hepatoma plasma membranes or the soluble form. One monoclonal antibody (25.8), recognizing epitope D, precipitated only DPP IV from liver plasma
membrane, indicating that the relevant epitope is neither accessible in hepatoma plasma membranes nor in the soluble form
(Table 1).

208

Table 1. Determination of common epitopes recognized by the monoclonal antibodies and their reactivity with the three DPP IV forms. The common epitopes were determined by competitive dot blot assay and the reactivity by immunoprecipitation.

Monoclonal antibody	Epitope	Reactivity with		
		Liver plasma membrane form	Liver soluble form	Tumor plasma membrane form
25.1, 25.2	A	+	+	+
25.3, 25.6, De 13.4	B	+	+	+
25.4, 25.5, 25.7	C	+	+	+
25.8	D	+	-	-

When DPP IV from rat liver plasma membranes was blotted onto nitrocellulose (Western blotting) only six of the antibodies recognized polypeptide bands. The remaining three antibodies are directed against an epitope (C) on the native glycoprotein which does not regain its original conformation after SDS-electrophoresis and transfer to nitrocellulose (Fig. 1). Incubation of DPP IV with the monoclonal antibodies and subsequent determination of the enzyme activity indicates, that none of the antibodies inhibits the enzyme activity (data not shown).

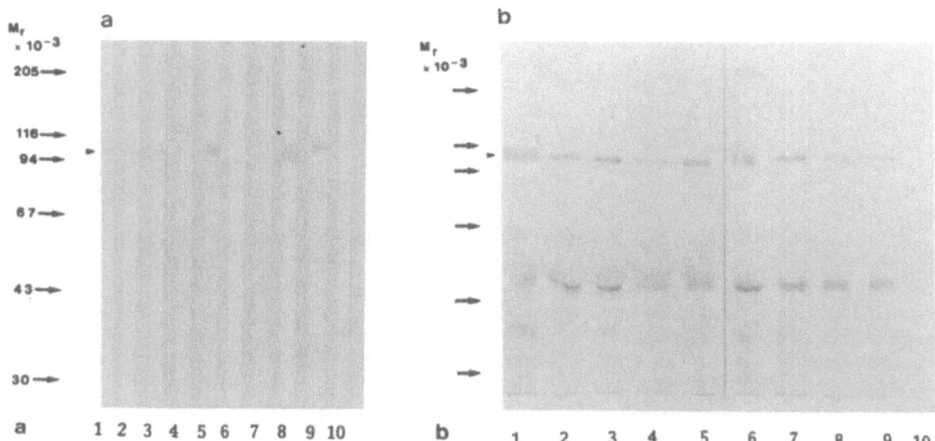

Fig. 1. Immunoblot of DPP IV from rat liver plasma membranes: a. After Western blotting and b. after precipitation of DPP IV with the monoclonal antibodies. Epitope A (1, 2), epitope B (3,5,9), epitope C (4,6,7), epitope D (8) and negative control without the first antibody (10), arrows: 105 kD

2. Immunolocalization of DPP IV in various rat tissues

The distribution of the enzyme was tested on acetone-chloroform pretreated cryostat sections by immunohistochemical staining and compared with DPP IV activity in these sections. In liver eight monoclonal antibodies recognize the bile canalicular region, where most of DPP IV activity is present. Some staining was also observed in the sinusoidal epithelium. Antibody 25.8 does not show any staining, indicating that the epitope D is not accessible on frozen sections of rat liver, but can be made accessible by solubilization (Fig. 2).

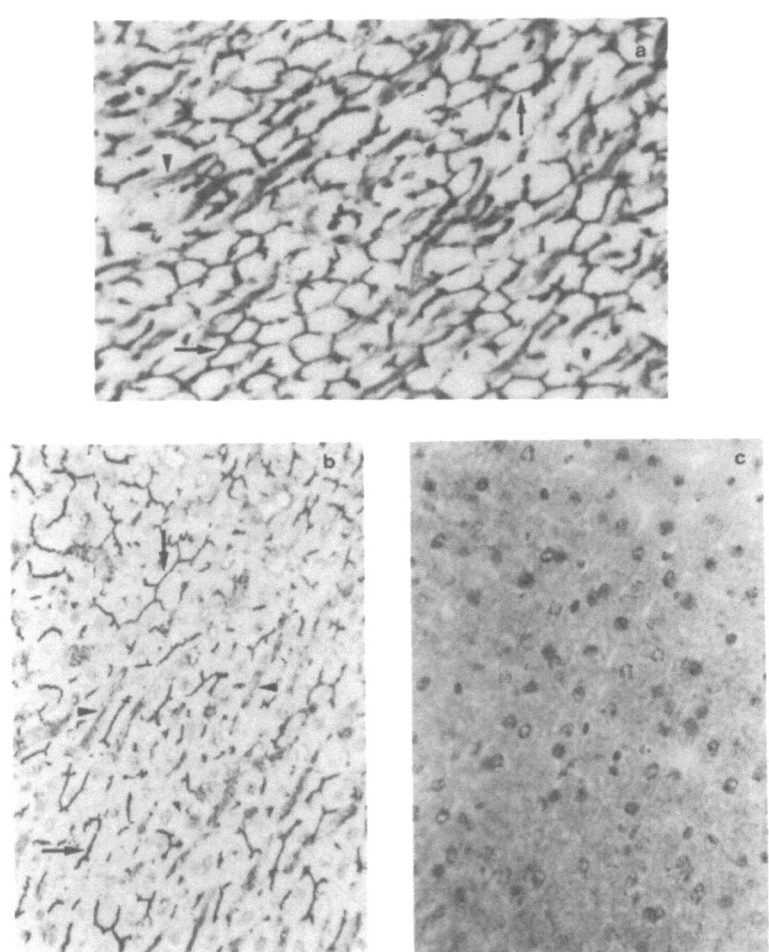

Fig. 2. Liver: a. Histochemical staining for DPP IV acitivity. b. Immunohistochemical staining for epitope A. Similar staining was obtained for epitopes B and C. c. Immunohistochemical staining for epitope D.
Arrows: bile capillaries, arrow heads: sinus endothelium; magnification x512

The comparison of the histochemical staining for DPP IV in most rat organs with the immunohistochemical staining in these tissues showed that the epitopes A, B and C were accessible to the antibodies in most investigated organs. In the glomeruli of the kidney immunostaining is weaker than in the proximal tubules, while the distribution of activity is similar in both structures (Fig. 3). Epitope D was not detected in most tissues, i.e. it was either absent or not accessible. However, it was found in acinar cells of the exocrine pancreas and special cells of the sublingual gland, in epithelial cells of seminal vesicle and epididymal duct and in certain cells of the ovary, testis, cerebrum, cerebellum, spleen and plexus myentericus (Fig. 4) (Hartel et al., manuscript in preparation).

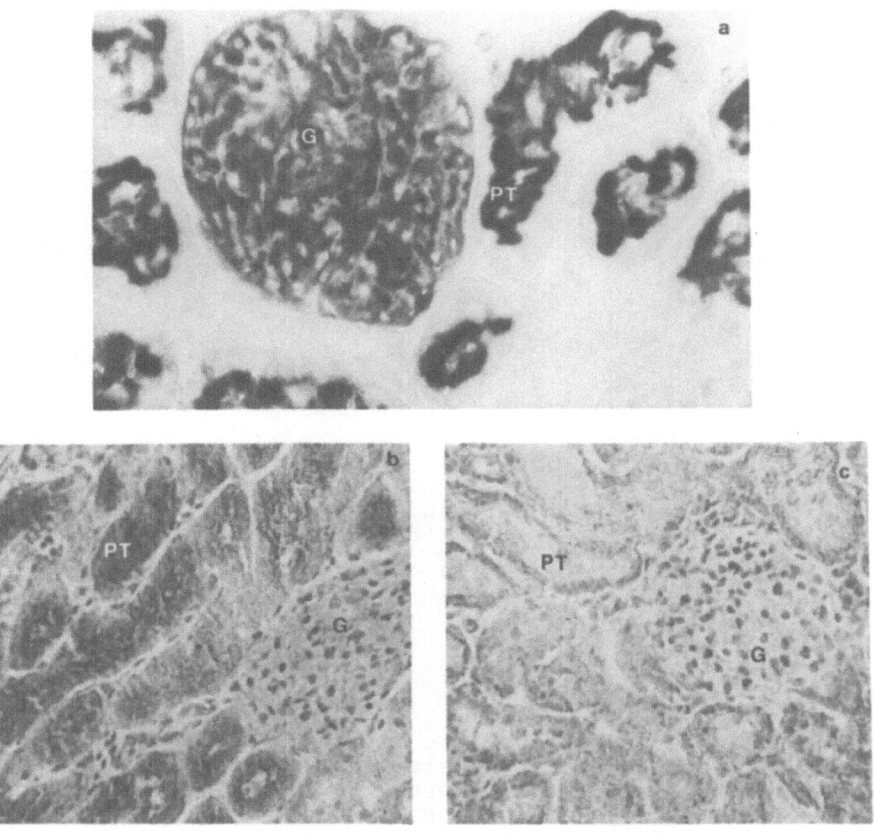

Fig. 3. Accessibility of DPP IV-epitopes in kidney of male rats and its correlation with the distribution of enzymatic activity: a. Histochemical staining of a glomerulus (G) and proximal tubules (PT). b. Immunohistochemical staining for epitope A in proximal tubules (PT), weak reactivity in glomerulus (G), similar staining was obtained for epitopes B and C. c. Absence of immunohistochemical staining for epitope D, G glomerulus, PT proximal tubulus; magnification x512

3. Release of DPP IV from rat liver plasma membrane by phospholipase C

The loss of DPP IV from hepatoma could be caused by defective or altered anchorage of the enzyme in tumor cell plasma membrane. To investigate this point the anchorage of DPP IV in liver and hepatoma was studied (manuscript in preparation). In brief, the methods used and results obtained are described. Plasma membranes from liver or hepatoma were incubated at 37°C with increasing amounts of phospholipase C. About 30% of the DPP IV activity from liver membrane were released within 2 h with 15 U/ml of the enzyme. The released activity did not increase after incubation with higher amounts of phospholipase C (Fig. 5). Macnair and Kenny reported, that in kidney plasma membrane DPP IV is anchored by a small hydrophobic domain (17). However, the present data suggest that in rat liver the DPP IV

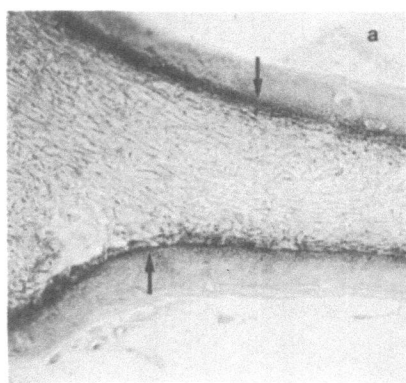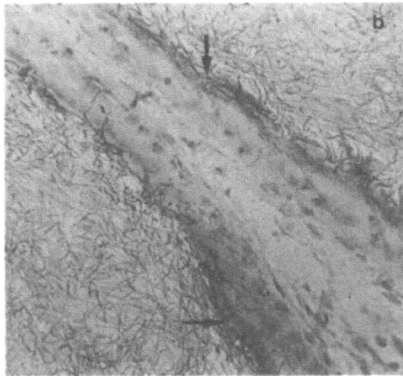

Fig. 4. Epididymal duct, arrows stereocilia: a. Histochemical staining. b. Immunohistochemical staining for epitope D; magnification x512.

molecule can be inserted by two different mechanisms, one of them being phosphlipase C-sensitive. Two different anchorage mechanisms have also been reported for the membrane glycoprotein N-CAM 120 (18, 19). No DPP IV-activity was released by phospholipase C from the hepatoma membrane, i.e. in this tumor the entire DPP IV population is phospholipase C-resistant, probably due to insertion into the membrane by a hydrophobic polypeptide. The monoclonal antibodies against the epitopes A, B and C precipitated equally well the intact and phospholipase C-treated DPP IV, indicating that these epitopes are located on the outside of the plasma membrane. However, the antibody against the epitope D did not precipitate DPP IV after treatment with phospholipase C, indicating that this epitope represents or is part of the anchoring domain of the enzyme.

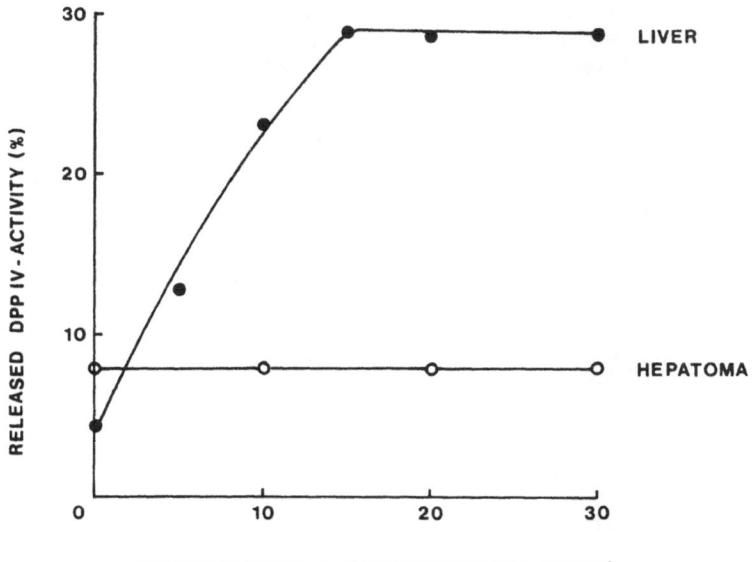

Fig.5. Release of DPP IV from plasma membranes by treatment
with phospholipase C.

SUMMARY

Nine monoclonal antibodies directed against DPP IV from
rat liver plasma membranes were obtained. They recognized four
different epitopes (A, B, C and D) of the enzyme. The epitopes
A, B and C were located on the outside of the hepatocyte plasma
membrane and were shared by DPP IV from hepatoma plasma membrane and the soluble form. Epitope D appeared to be partly inserted in the membrane and was found exclusively in the liver.
Epitopes A, B and C and DPP IV revealed by histochemical means
showed similar distribution patterns on frozen sections of various rat tissues, while epitope D did not show such a correlation. DPP IV is inserted in liver plasma membrane by two different mechanisms, one being phospholipase C-sensitive, while in
hepatoma the enzyme is anchored in this membrane by a phospholipase C-resistant mechanism only.

REFERENCES

1. V. K. Hopsu-Havu and T.O. Efkors, 1969, Distribution of a
   dipeptide naphthylamidase in rat tissues and its localization by using diazo coupling and labeled antibody
   techniques. Histochemie 17: 30
2. C. Hanski, T. Huhle and W. Reutter, 1985, Involvement of
   dipeptidylaminopeptidase IV in fibronectin-mediated adhesion on collagen. Biol. Chem. Hoppe Seyler 366: 1169

3. C. Hanski, T. Zimmer, R. Gossrau and W. Reutter, 1986, Increased serum levels of dipeptidylaminopeptidase IV in hepatoma-bearing rats coincide with the loss of the enzyme from the plasma membrane. Experientia 42: 826

4. R. Neumeier, J. Mauck, C. Hanski and W. Reutter, 1986, Comparison of adherent and non-adherent variants of hepatoma cells. Eur. J. Cell Biol. 42 (Suppl. 15): 46

5. C. Hanski, S. Hartel, C. Hoffmann, T. Zimmer, R. Neumeier, R. Gossrau and W. Reutter, 1987, Characterization of liver and hepatoma dipeptidyl peptidase IV. In: "Modulation of liver cell expression", W. Reutter, H. Popper, L. Junkee, P. C. Heinrich, D. Keppler and L. Landmann, eds., MTP Press Lancaster, in press.

6. G. Köhler and C. Milstein, 1975, Continuous cultures of fused cells secreting antibody of predefined specificity. Nature 256: 495

7. S. Hartel, C. Hanski, W. Kreisel, C. Hoffmann, J. Mauck, and W. Reutter, 1987, Rapid purification of dipeptidyl peptidase IV from rat liver plasma membrane. Biochim. Biophys. Acta, in press.

8. W. Kreisel, R. Heussner, B. Volk, R. Büchsel, W. Reutter and W. Gerok, 1982, Identification of the 110 000 M_r glycoprotein isolated from rat liver plasma membrane as dipeptidylaminopeptidase IV. FEBS Lett. 147: 85

9. C. Hanski, W. Reutter and W. H. Evans, 1984, Turnover of the protein and carbohydrate moieties of a 105kD glycoprotein of plasma membranes from mouse liver determined by immunoprecipitation. Eur. J. Cell Biol. 33: 123

10. R. Büchsel, D. Berger and W. Reutter, 1980, Routes of fucoproteins in plasma membrane domains. FEBS Lett. 113: 95

11. K. U. Laemmli, 1970, Cleavage of structural proteins during assembly of the head of bacteriophage T4. Nature 227: 680

12. J. W. Freemann, A. Chatterjee and H. Bush, 1985, Discrimination of epitopes identified by monoclonal antibodies binding to nitrocellulose bound antigens. J. Immunol. Methods 78: 259

13. A. Becker, R. Neumeier, C. Heidrich, N. Loch, S. Hartel and W. Reutter, 1986, Cell surface glycoproteins of hepatocytes and hepatoma cells identified by monoclonal antibodies. Biol. Chem. Hoppe-Seyler 367: 681

14. H. Stein, K. Gatter, H. Asbahr and D.Y. Mason, 1985, Use of freeze-dried paraffin-embedded sections of immunohistologic staining with monoclonal antibodies. Lab.Invest. 52: 676

15. M. G. Low and J. B. Fineau, 1978, Specific release of plasma membrane enzymes by a phosphatidylinositol specific phospholipase C. Biochim. Biophys. Acta 508: 565

16. R. D. C. Macnair and J. A. Kenny, 1979, Proteins of the kidney microvillar membrane. The amphipathic form of dipeptidyl peptidase IV. Biochem. J. 179: 379

17. H.-T. He, J. Barbet, J.-C. Chaix and C. Goridis, 1986, Phosphatidylinositol is involved in the membrane attachment of NCAM-120, the smallest component of the neural cell adhesion molecule. EMBO J. 5: 2489

18. J. J. Hemperly, G. M. Edelman and B. A. Cunningham, 1986, cDNA clones of the neural cell adhesion molecule (N-CAM) lacking a membrane-spanning region consistent with evidence for membrane attachment via a phosphatidylinositol intermediate. Proc.Natl.Acad.Sci. 83: 9822

NON − LYSOSOMAL, HIGH − MOLECULAR − MASS CYSTEINE PROTEINASES

FROM RAT SKELETAL MUSCLE

Burkhardt Dahlmann, Lothar Kuehn, Friedrich Kopp
Hans Reinauer and William T. Stauber*

Biochemische Abteilung, Diabetes Forschungsinstitut
Auf'm Hennekamp 65, D−4000 Düsseldorf F.R.G. and
* Department of Physiology, Medical Centre,
West Virginia University, Morgantown, WV 26506 U.S.A.

INTRODUCTION

It is well established that in protein metabolism protein synthesis as well as protein degradation are equally important for maintaining cellular viability. Thus, a basal level of intracellular proteolysis is measurable in living cell [1] , which comprises post-translational processing of newly synthesized proteins [2], degradation of missense proteins [3] as well as breakdown of inactivated proteins having lost their function [4,5]. In addition to the ´basal degradation` an ´accelerated proteolysis` takes place in the cells living under conditions of nutrition deprivation as well as hormone or amino acid deficiency [6,7]. For catalysis of these various intracellular proteolytic events multiple proteolytic pathways exist in a mammalian cell [8-11]. One of the major sites of intracellular proteolysis is the lysosomal compartment. However, since the lysosomal cathepsins have no free access to the protein substrates, initial rate−limiting steps of intracellular proteolysis probably proceed extra−lysosomally. The aim of our investigations is to identify, isolate and characterize the enzymes catalyzing steps of these extra−lysosomal pathways.

From the non−lysosomal compartment of skeletal muscle tissue we have isolated several neutral/alkaline proteinases. The present paper summarizes their purification and some of their properties.

ISOLATION PROCEDURE FOR THE PROTEINASES

Pooled skeletal muscles from rat hind legs were homogenized and extracted with 20 mM Tris/HCl/ 1 mM NaN3/ 0.1% 2−mercaptoethanol, pH 7.5, and were fractionated with $(NH_4)_2SO_4$ to obtain fractions with proteins precipitating between 45 − 65% saturation with $(NH_4)_2SO_4$ and one containing proteins precipitating at 65 − 90% saturation with respect to the salt. Further fractionation of these two groups of proteins is summarized in Fig.1.

Abbreviations used: Dip−F, di−isopropyl phosphorofluoridate; DTT, dithiothreitol; MCP, multicatalytic proteinase; NMec, 4−methyl−7−coumarylamide; 2NNap, 2−naphthylamide; n.d., not determined; Pms−F, phenylmethanesulphonyl fluoride; SDS, sodium dodecyl sulphate.

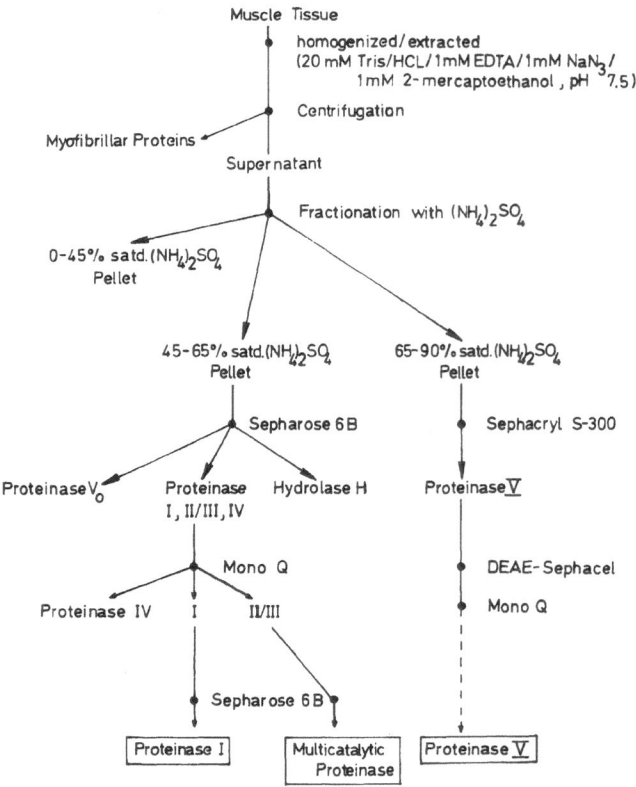

Fig. 1 Purification scheme for high molecular mass proteinases from rat skeletal muscle tissue.

Proteins precipitating between 45 – 65% saturation with the salt were subjected to gel filtration on Sepharose 6B. During this step the proteolytic
activity was resolved into a caseinolytic activity eluting in the void
volume of the column, indicating that this activity has a $M_r \geq 4000000$. The
enzyme, whose properties are presently under investigation, was designated Proteinase V_o. In the fractionation volume, a large peak of activity
was eluted from the column containing hydrolytic and proteolytic activities
against the substrates methylcasein, Z-Phe-Arg-NMec, Suc-Ala-Ala-Phe-NMec
and Bz-Val-Gly-Arg-NMec. By chromatography on a Mono Q column these activities were resolved further: at a concentration of 200 mM NaCl a proteinase was eluted from the column that was designated Proteinase IV, a proteinase eluting at 220 mM NaCl was designated Proteinase I, and a proteinase, designated Proteinase II/III, was eluted at about 310 mM NaCl [12].
Since the latter enzyme seems to contain more than one active site, it was
designated 'Multicatalytic Proteinase' [13].

A further enzyme, Proteinase $\underline{V}$, was identified in the fraction of proteins precipitating at 65 – 90% saturation with $(NH_4)_2SO_4$. It was purified
by gel filtration on Sephacryl S-300 and by chromatography on DEAE Sepha-
cel and Mono Q [14].

With the procedure described, Proteinase I and the Multicatalytic Proteinase were purified to apparent homogeneity [13,14], whereas the Proteinase $\underline{V}$ preparation still contained minor contaminants of proteolytically
inactive material.

Table 1.   Physicochemical Properties of the Proteinases

	M_r	pI	M_r (subunits)
Proteinase I	750 000	4.5	150 000
			90 000
			60 000
			40 000
MCP	650 000	5.1	32 000
			to
			25 000
			(eight peptides)
Proteinase $\underline{V}$	300 000		

PHYSICOCHEMICAL PROPERTIES OF THE PROTEINASES

According to gel filtration experiments Proteinase I has a M_r of 750 000, the Multicatalytic Proteinase has a slightly lower M_r whereas Proteinase $\underline{V}$ is a clearly smaller molecule with an M_r of 300 000. As shown in Table 1. we have found clear differences in pI values and subunit composition for the Multicatalytic Proteinase and Proteinase I [13,14].

In the electronmicroscope the Multicatalytic Proteinase is viewed as a cylinder-shaped molecule with a length of 15 nm and a diameter of 10 nm, and consisting of four rings [15].

SUBSTRATES OF THE PROTEINASES

Several synthetic peptide substrates and their rate of hydrolysis by the three proteinases are listed in Table 2. The Multicatalytic Proteinase hydrolyses nearly all of these substrates, those containing an arginine, a phenylalanine as well as that with a glutamic acid residue adjacent to the leaving group. However, the enzyme does not split Z-Phe-Arg-NMec. By contrast, this substrate is hydrolysed by Proteinase I and Proteinase $\underline{V}$. All three enzymes degrade methylinsulin; in addition, Proteinase I and the Multicatalytic Proteinase hydrolyse methylcasein [13,14].

Z-Phe-Arg-NMec is optimally hydrolysed at pH 8.5 and pH 9.5 - 10.5 by Proteinase I and Proteinase $\underline{V}$, respectively. Depending on the substrate tested, the Multicatalytic Proteinase showed different pH optima of activity: pH 7.2 with Suc-Ala-Ala-Phe-NMec, pH 10.5 for Bz-Val-Gly-Arg-NMec and pH 9.0 for methylcasein degradation [13].

INHIBITORS OF THE PROTEINASES

To clarify the three proteinases according to the type of the active site contained on the molecule, the effect of various compounds on enzyme activity was tested [16] (Table 3.). The activities of all three enzymes are strongly inhibited by p-hydroxymercuriphenylsulfonic acid and, with exception of Proteinase $\underline{V}$, by the mercury containing mersalyl acid. Since addition of cysteine (as well as of DTT, not shown) considerably enhances the activity of Proteinase I and Proteinase $\underline{V}$, we conclude that both enzymes belong to the family of cysteine proteinases. This probably holds

Table 2. Substrates of the Proteinases

Substrates	Enzyme Activity (pmol/ml x min)		
	Proteinase I	Proteinase $\underline{\text{V}}$	MCP
Bz-Arg-NMec	0	1	0
Z-Arg-Arg-NMec	0	13	37
Z-Ala-Arg-Arg-NMec	0	0	83
Z-Gly-Gly-Arg-NMec	0	1	156
Bz-Val-Gly-Arg-NMec	0	1	194
Tos-Pro-Arg-NMec	14	1	12
Z-Phe-Arg-NMec	19	20	0
Suc-Ala-Ala-Phe-NMec	0	0	34
Glt-Gly-Gly-Phe-NMec	0	0	33
Z-Leu-Leu-Glu-2NNap	0	n.d.	80

true for the Multicatalytic Proteinase, as well, although no significant activation of the enzyme was observed in the presence of these reducing compounds (Table 3.).

Interestingly, several inhibitors have varying or even antagonistic effects on the active site(s) of the Multicatalytic Proteinase (Table 4.). Since similar effects have been found with Ca^{++}, Na^+ and K^+ ions, it is assumed that the enzyme has more than one active site.

Table 3. Inhibitors of the Proteinases

Compound (final concn.)	Proteinase I	Proteinase $\underline{\text{V}}$	Multicatalytic Proteinase		
Substrate ...	Z-Phe-Arg-NMec	Z-Phe-Arg-NMec	Bz-Val-Gly-Arg-NMec	Suc-Ala-Ala-Phe-NMec	$[^{14}C]$-Methyl-casein
Dip-F (1 mM)	n.d.	n.d.	117	80	n.d.
Pms-F (1 mM)	89	100	100	95	n.d.
p-OH-mercuriphenyl-sulfonic acid (1 mM)	0	33	0	0	0
Mersalyl acid (1 mM)	40	95	0	0	0
Cysteine (15 mM)	1960	180	96	105	80
EDTA (5 mM)	136	n.d.	87	113	n.d.
Pepstatin (0.036 mM)	100	n.d.	135	100	n.d.

Table 4. The Effect of Various Compounds on the Activities of the Multicatalytic Proteinase

Compound	final concn. (mM)	Substrate...	Multicatalytic Proteinase Activity (% of control)			
			Bz-Val-Gly-Arg-NMec	Suc-Ala-Ala-Phe-NMec	Z-Leu-Leu-Glu-2NNap	[14C]-Methyl-casein
Leupeptin	0.125		11	67	96	102
Antipain	0.017		14	111	n.d.	n.d.
Chymostatin	0.125		19	49	128	322
ATP	5		30	100	n.d.	n.d.
CaCl$_2$	5		19	128	88	55

POSSIBLE REGULATORS OF PROTEINASE I AND THE MULTICATALYTIC PROTEINASE

Proteinase inhibitors are commonly found to regulate the function of proteolytic enzymes. However, the modulation of a proteinase like the Multicatalytic Proteinase, possibly containing more than one active site, may be difficult to achieve, especially, when inactivation of one catalytic site is paralleled by activation of another catalytic site of the enzyme (Table 4.). In the course of our studies we have found that the Multicatalytic Proteinase, when isolated from the tissue, is in a latent state of activity. Activation of the enzyme occurred when low concentrations of SDS were added to the enzyme solution [17]. In an attempt to find activators closer to the physiologic, we tested various free fatty acids, fatty acid esters and phospholipids. Long chain free fatty acids (C:18-C:22) were found to be the most effective activators [17] and oleic acid provoked a nearly 50-fold increase in caseinolytic activity (Table 5). No activation was found, when oleic acid was bound to albumin or when tested in the ester form as glycerol-, ethanol- or carnitine-ester (Table 5.). On the other hand, fatty acid thiol esters with Coenzyme A are still activators for the peptide hydrolysis by the proteinase, but do not stimulate the degradative activity against casein. Furthermore, different phospholipids, e.g. phosphatidylinositol and phosphatidylglycerol substantially increase the activities of the Multicatalytic Proteinase. The most abundant phospholipid in mammalian tissues, phosphatidylcholine, does not affect the activities of the Multicatalytic Proteinase (Table 5.).

Concerning Proteinase I, we know that the activity of this enzyme is very labile and we have found that ATP is able to stabilize the activity [18]. A stabilizing effect was also found with other nucleotides, e.g. CTP, GTP and UTP [14]. Since in the absence of ATP the Proteinase I activity decreases very rapidly, an alternative interpretation of these results would be that ATP has an activating effect on the activity of Proteinase I. As shown in Fig. 2 , after a preincubation of the enzyme for 60 min in the presence of ATP, the activity is about 4-fold as high as compared to the enzyme preinincubated without ATP.

Table 5.   Effect of Lipids on the Activities of the Multicatalytic Proteinase.

Substrate...	Multicatalytic Proteinase Activity (% of control)		
Compound [a]	Bz-Val-Gly-Arg-NMec	Suc-Ala-Ala-Phe-NMec	$[^{14}C]$ - Methyl-casein
Stearic acid	n.d.	n.d.	1384
Oleic acid	1578	450	4770
Oleic acid (bound to albumin)	100	100	n.d.
Triolein	95	105	26
Oleic acid ethyl-ester	95	95	0
Stearoyl-L-carnitine	100	125	39
Stearoyl-Coenzyme A	1047	215	0
Oleoyl-Coenzyme A	1673	560	14
Phosphatidyl-choline	93	100	161
Phosphatidyl-inositol	303	959	431
Phosphatidyl-D,L-glycerol (di-oleoyl)	1058	1063	293

[a] all compounds were tested at 100 μM final concentration

LOCALISATION OF THE PROTEINASES

To assess possible functions of the proteolytic enzymes, it is indispensible to know their location in the tissue. Immunohistochemical investigations with monospecific, polyclonal antibodies to Proteinase I have shown, that Proteinase I antibodies bind diffusly to cultured rat myoblasts [19], indicating that Proteinase I is myoblast associated. With the same approach but using Multicatalytic Proteinase antibodies a consistently positive reaction of the antibodies with cell nuclei was revealed (Stauber, Fritz, Maltin, Dahlmann, unpublished). Since a similar locale was demonstrated in muscle sections from adult rats, we conclude that the Multicatalytic Proteinase has a nuclear association.

CONCLUDING REMARKS

Skeletal muscle tissue contains a spectrum of high molecular mass, non-lysosomal cysteine proteinases, three components of which were isolated in our laboratory. One of these enzymes, designated Proteinase I, is activated and stabilized by ATP and thus may be involved in the energy-dependent pathways of intracellular protein breakdown [9]. Since its activity is not dependent on ubiquitin, the enzyme is not involved in the ATP/ubiquitin-dependent proteolytic system [10] that has recently been identified in skeletal muscle, as well [20].

Proteinase II/III, designated Multicatalytic Proteinase, is considerably activated by micromolar concentrations of lipids and phospholipids. This property as well as its location within the cell nucleus disclose interesting views on possible functions of the enzyme [21,22]. The Multi-

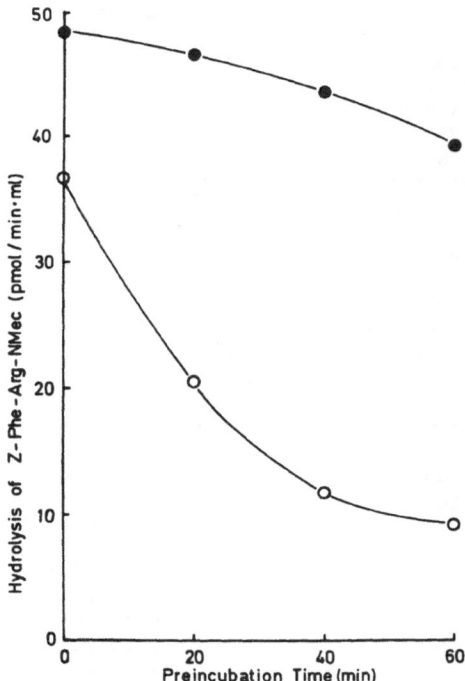

Fig. 2 The effect of ATP on the activity of Proteinase I in vitro.
Purified Proteinase I was incubated with (●) or without (O)
5 mM ATP at 37°C. At the times indicated aliquots were removed and the activity was tested with Z-Phe-Arg-NMec as
substrate.

catalytic Proteinase has been identified in [23] and isolated from various
mammalian tissues [24-35] and thus may play a basal role in intracellular
(or even intranuclear) proteolysis.

To date the amount of data on Proteinase $\overline{V}$ is too limited to judge
its identity and possible function in the muscle tissue.

Acknowledgements

This study was supported by the 'Deutsche Forschungsgemeinschaft',
Bonn, by the 'Ministerium für Wissenschaft und Forschung des Landes Nordrhein-Westfalen', Düsseldorf, and by the 'Bundesministerium für Jugend,
Familie, Frauen und Gesundheit', Bonn. Our international collaboration
was fostered by the award of a NATO Research Grant.

References

1. Amenta,J.S. & Brocher,S.C.: Mechanisms of protein turnover in cul-
     tured cells. Life Sciences 28: 1195-1208 (1981)
2. Dean,R.T. & Judah,J.D.: Post-translational proteolytic processing of
     polypeptides. Comp. Biochem. 19B: 233-298 (1980)
3. Klemes,Y., Etlinger,J.D. & Goldberg,A.L.: Properties of abnormal pro-
     teins degraded rapidly in reticulocytes. J.Biol.Chem. 256: 8436-
     8444 (1981)

4. Ballard,F.J.: Intracellular protein degradation. Essays Biochem. 13: 1-37 (1978)

5. Holzer,H. & Heinrich,P.C.: Control of proteolysis. Ann.Rev.Biochem. 49: 63-91 (1980)

6. Mortimore,G.E.: Mechanisms of cellular protein catabolism. Nutr. Rev. 40:1-12 (1982)

7. Ballard,F.J. & Gunn,J.M.: Nutritional and hormonal effects on intracellular protein catabolism. Nutr.Rev. 40: 33-42 (1982)

8. Glaumann,H., Ericsson,J.L.E. & Marzella,L.: Mechanisms of intralysosomal degradation with special reference to autophagocytosis and heterophagocytosis of cell organelles. Int.Rev.Cytol. 73: 149-182 (1981)

9. Hershko,A. & Ciehanover,A.: Mechanisms of intracellular protein breakdown. Ann.Rev.Biochem. 51: 335-364 (1982)

10. Ciehanover,A., Finley,D. & Varshavsky,A.: The ubiquitin-mediated proteolytic pathway and mechanisms of energy-dependent intracellular protein degradation. J.Cell.Biochem. 24: 27-53 (1984)

11. Pontremoli,S. & Melloni,E.: Extralysosomal protein degradation. Ann. Rev.Biochem. 55: 455-481 (1986)

12. Dahlmann,B.: High-M_r cysteine proteinases from rat skeletal muscle. Biochem.Soc.Trans. 13: 1021-1023 (1985)

13. Dahlmann,B., Kuehn,L., Rutschmann,M. & Reinauer,H.: Purification and characterization of a multicatalytic, high-molecular-mass proteinase from rat skeletal muscle. Biochem.J. 228: 161-170 (1985)

14. Dahlmann,B., Kuehn,L. & Reinauer,H.: Identification of two alkaline cysteine proteinases from rat skeletal muscle. in: Cysteine Proteinases and their Inhibitors. (Turk,V., ed.) pp. 133-146 Walter de Gruyter & Co. Berlin, New York (1986)

15. Kopp,F., Steiner,R., Dahlmann,B., Kuehn,L. & Reinauer,H.: Size and shape of the multicatalytic proteinase from rat skeletal muscle. Biochim.Biophys.Acta 872: 253-260 (1986)

16. Hartley,B.S.: Proteolytic enzymes. Ann. Rev.Biochem. 29: 45-72 (1960)

17. Dahlmann,B., Rutschmann,M., Kuehn,L. & Reinauer,H.: Activation of the multicatalytic proteinase from rat skeletal muscle by fatty acids or sodium dodecyl sulphate. Biochem.J. 228: 171-177 (1985)

18. Dahlmann,B., Kuehn,L. & Reinauer,H.: Identification of three high molecular mass cysteine proteinases from rat skeletal muscle. FEBS-Lett. 160: 243-247 (1983)

19. Stauber,W.T., Fritz,V.K., Dahlmann,B., Kay,J., Heath,R. & Mayer,M.: Alkaline proteinase localization in myoblasts. J.Histochem.Cytochem. 35: 83-86 (1987)

20. Fagan,J.M., Waxman,L. & Goldberg,A.L.: Skeletal muscle and liver contain a soluble ATP+ubiquitin-dependent proteolytic system. Biochem.J. 243: 335-343 (1987)

21. Dahlmann,B., Kopp,F., Kuehn,L., Reinauer,H. & Schwenen,M.: Studies on the multicatalytic proteinase from rat skeletal muscle. Biomed. Biochim.Acta 45: 1493-1501 (1986)

22. Dahlmann,B., Kuehn,L., Reinauer,H. & Kay.J.: Multicatalytic proteinase activity in skeletal muscle from starving rat. Biochem.Soc.Trans. (1987) in the press

23. Kuehn,L., Dahlmann,B. & Reinauer,H.: Tissue distribution of the multicatalytic proteinase in the rat: An immunological and enzymic study. Cienc.Biol. 11: 101-107 (1986)

24. Wilk,S. & Orlowski,M.: Cation sensitive neutral endopeptidase:Isolation and specificity of the bovine pituitary enzyme. J.Neurochem. 35: 1172-1182 (1980)

25. Edmunds,T. & Pennington,R.J.T.: A high molecular weight peptide hydrolase in erythrocytes. Int.J.Biochem. 14: 701-703 (1982)

26. Hardy,M.F., Mantle,D. & Pennington,R.J.T.: Characteristics of a new muscle protease. Biochem.Soc.Trans. 11: 348-349 (1983)

27. Ray,K. & Harris,H.: Purification of a neutral lens endopeptidase: close similarity to a neutral proteinase in pituitary. Proc.Natl.Acad.Sci. U.S.A. 82: 7545-7549 (1985)

28. Rivett,A.J.: Purification of a liver alkaline protease which degrades oxidatively modified glutamine synthetase. J.Biol.Chem. 260: 12600-12606 (1985)

29. Ishiura,S., Sano,M., Kamakura,K. & Sugita,H.: Isolation of two forms of the high molecular mass serine protease, ingensin, from porcine skeletal muscle. FEBS-Lett. 189: 119-123 (1985)

30. Ishiura,S., Yamamoto,T., Nojima,M. & Sugita,H.: Ingensin, a fatty acid activated serine proteinase from rat liver cytosol. Biochim.Biophys. Acta 882: 305-310 (1986)

31. Nojima,M., Ishiura,S., Yamamoto,T., Okuyama,T., Furuya,H. & Sugita,H.: Purification and characterization of a high molecular weight protease, ingensin, from human placenta. J.Biochem. 99: 1605-1611 (1986)

32. Ishiura,S. & Sugita,H.: Ingensin, a high molecular mass alkaline protease from rabbit reticulocyte. J.Biochem. 100: 753-763 (1986)

33. McGuire,M.J. & DeMartino,G.N.: Purification and characterization of a high molecular weight proteinase (macropain) from human erythrocytes. Biochim.Biophys.Acta 873: 279-289 (1986)

34. Tanaka,K., Ii,K., Ichihara,A., Waxman,L. & Goldberg,A.L.: A high molecular weight protease in the cytosol of rat liver. I. Purification, enzymological properties and tissue distribution. J.Biol.Chem. 261: 15197-15203 (1986)

35. Zolfaghari,R., Baker,C.R.F., Canizaro,P.C., Amirgholami,A. & Behal,F.J.: A high molecular mass neutral endopeptidase-24.5 from human lung. Biochem.J. 241: 129-135 (1987)

ROLE OF FACTORS DERIVED FROM ACTIVATED MACROPHAGES IN REGULATION OF

MUSCLE PROTEIN TURNOVER

V.E. Baracos

Department of Animal Science
University of Alberta
Edmonton, Alberta, T6G 2P5, Canada

INTRODUCTION

A marked loss of body weight and body protein, and a generalized wasting of skeletal muscle, frequently accompany infection, injury and certain neoplastic diseases (1,2). These conditions result in a negative nitrogen balance, which may exceed that seen in fasting (2-9). This net nitrogen loss may result from increased degradation of cell protein, reduced rates of protein synthesis, or both (2-9). Skeletal muscle, which comprises a major body protein reserve, would appear to be a primary site for this catabolic response (2-9). Experimental infections, injury or tumours result in a marked loss of muscle mass, and enhancement of protein breakdown in muscle (2,3,7,8). Similarly, in human patients, there is evidence of increased muscle proteolysis and net release of amino aids from this tissue (4,10). However, the factor(s) initiating muscle catabolism have still to be fully characterized.

Mononuclear phagocytes (Mϕ)(blood monocytes and fixed tissue macrophages) synthesize and release over 75 defined molecules, which act in a highly coordinated fashion to mediate the antimicrobial, antitumour and inflammatory activity of these cells (11). A summary of Mϕ secretory products is presented in Table 1. We propose that the activated Mϕ may produce factors which modulate the metabolism of protein, coordinately with primary host defense functions. It would seem advantageous that in addition to primary defense activities (such as activation of T and B cells), animals also be able to mobilize body protein reserves to provide amino acids for oxidation, gluconeogenesis and for biosynthetic processes associated with host defense. Thus our recent work has concentrated on the effects of Mϕ factors on protein metabolism in skeletal muscle. Prior studies showed that partially purified interleukin-1 (Il-1) from human monocytes activated protein degradation in isolated skeletal muscle, in a prostaglandin (PG)E$_2$ - dependant fashion (5,6). PGE$_2$ alone also stimulated muscle protein breakdown in vitro. Both interleukin-1 and PGE$_2$ are products of activated Mϕ. During sepsis or following administration of endotoxin to animals, there is an increase in muscle protein breakdown which is sensitive to inhibitors of PG production, suggesting the involvement of these factors in vivo (3,7).

The availability of supernatants from activated Mφ (12), and homogeneous natural and recombinant Mφ products (16-19,23) has allowed further investigation of the effects of these factors on muscle protein turnover. This report summarizes our studies which attempt to clarify the importance of Mφ factors in muscle catabolism.

Table 1. Secretory Products of Activated Macrophages

1. Lysosomal hydrolases

2. Lysozyme

3. Neutral Proteases

    -elastase

    -collagenase

    -plasminogen activator

    -cytolytic protease

4. Arachidonic acid metabolites

    -PGI_2, TXB_2, PGE_2, $PGF_{2\alpha}$,

    LTC_4, LTB_4

5. Reactive Oxygen

    -O_2^-, H_2O_2, OH·

6. Complement components

7. Monokines

    -Interleukin - 1

    -Interferon

    -Tumour necrosis factor

MATERIALS AND METHODS

Young male Sprage Dawley rats were maintained on Wayne lab chow and water ad libitum. Extensor digitorum longus (EDL) or soleus (SOL) muscles weighing 15 - 20 mg, or quarter diaphragm muscles weighing 20 - 25 mg were dissected as previously described (5,20,21). Tissues were incubated in 3.0 ml of Krebs Ringer Bicarbonate (KRB) medium, which was continuously bubbled with 5% CO_2/95% O_2. KRB was supplemented with 5mM glucose and 0.1 U/ml insulin. Muscles were routinely incubated at 36°C for 30 min, then transferred to fresh media and incubated for a further 3 h. When Mφ products were studied, these factors were added to the incubation medium only. Protein synthesis, protein degradation, net protein degradation and the production of PGE_2 were measured as previously described (5,20,21). Data are presented as means ± SEM. The results were compared by the Student's T test or analysis of variance.

Supernatants of activated porcine blood monocytes were generously prepared by Dr. J. Saklatvala (Strangeways Laboratories, Cambridge, England), as described previously (12). Briefly, porcine blood monocytes were collected and cultured in Dulbecco's Modified Eagle medium (DME). The release of monocyte secretory products was activated by the addition

TABLE 2. EFFECT OF MONOCYTE SUPERNATANT FR 38 ON PROTEIN TURNOVER

Addition	Net Protein Breakdown	Protein Synthesis
	(nmol phenylalanine/mg muscle/3h)	
None	0.259 ± 0.013	0.329 ± 0.023
Fr.38 1:7.5	0.402 ± 0.022*	0.139 ± 0.007*
Fr.38 1:15	0.367 ± 0.017*	0.219 ± 0.017*
Fr.38 1:30	0.319 ± 0.023	0.310 ± 0.029

Difference from no additions * $p<0.002$ (n=8)

Rat EDL muscles were preincubated in KRB medium for 30 min at 36°C. Tissues were then transfered to fresh medium containing the additions indicated, and incubated for a further 3 h.

TABLE 3. EFFECT OF MONOCYTE AND rINTERLEUKIN-1 AND TNF ON NET PROTEIN DEGRADATION IN RAT SOLEUS MUSCLE

Treatment	Net Protein Degradation (nmol tyrosine/mg muscle/3 h)	PGE$_2$ Production (pg/mg muscle/3 h)
No additions	0.199 ± 0.017	32 ± 3
Monocyte Il-1, human (1 μg/ml)	0.410 ± 0.008*	103 ± 9*
rIl-1β, human (1 μg/ml)	0.237 ± 0.030	29 ± 4
No additions	0.248 ± 0.008	
rIl-1β, human (10 μg/ml)	0.269 ± 0.015	
rTNFα, murine, human (10 μg/ml)	0.213 ± 0.020	

*$p<0.001$ (n=10)

Rat SOL muscles were preincubated in KRB medium for 30 min at 30°C. Muscles were then transferred to fresh medium containing the additions indicated, and were incubated for a further 3 h.

of Concanavalin A. After 40 h, culture supernatants were aspirated, and concentrated by ultrafiltration over a membrane of 10kd cutoff. The concentrated solution was clarified by centrifugation, and NaCl added to a final concentration of 0.5 M. It was then chromatographed at room temperature on a column of AcA54 (9 x 90 cm). Fractions (150 ml) were assayed for interleukin-1 activity (ability to stimulate release of proteoglycan from cartilge) (12).

Homogeneous porcine interleukin-1 α and β were also provided by Dr. Saklatvala. Recombinant human interleukin-1β, recombinant tumour necrosis factor, and partially purified human monocyte interleukin-1 were a gift of Dr. C.A. Dinarello (Tufts University, Boston).

RESULTS

EFFECTS OF MONOCYTE SUPERNATANTS ON PROTEIN TURNOVER IN RAT MUSCLE

Forty-two fractions of porcine monocyte supernatants were tested for their effects on net protein degradation in rat diaphragm muscle, as determined by the release of tyrosine into the incubation medium. Fractions were assayed at a dilution of 33 μl / ml of medium. Since monocytes secrete a number of neutral proteinases (11), we performed control studies to determine the spontaneous rate of tyrosine release in the KRB/monocyte fraction mixtures. Control incubations containing no muscles were incubated for 0 and 3 h at 36°C. Fractions eluting early from the column (Fr 1-8) showed significant proteolytic activity, while subsequent fractions (Fr 9-42) showed little or no spontaneous tyrosine release. When muscles were incubated in media containing the different fractions, the 3 h tyrosine level in the control incubation was used to correct the value obtained from the corresponding muscle incubations. As seen in Fig. 1, several fractions caused significant increases of up to 75% in net protein degradation in rat diaphragm muscle (p<0.01). These included Fr 12, corresponding to the peak of interleukin-1 activity, and two peaks eluting subsequently between Fr 22-28 and 36-42. Similar results were obtained with several independant preparations. Since previous work had suggested that the activity towards muscle might be prostaglandin - dependant (5-7), PGE_2 released into incubation medium was determined. As shown in Fig. 2, all fractions eluting early from the column (Fr 2-22) significantly increased muscle PGE_2 production (p<0.05).

Fractions were also tested for their effects on muscle protein synthesis. As seen in Fig. 1, most fractions had no significant effect on muscle protein synthesis. However, Fr 32-42 caused a significant inhibition of muscle protein synthesis of up to 50% (p<0.010). Thus the catabolic activity eluting last from the column, would appear to result primarily from a reduction in muscle protein synthesis. In contrast, the catabolic effect of the two earlier peaks would appear to result from an activation of protein degradation.

A preliminary characterization of the catabolic activity found in Fr 38 has been carried out. Addition of Fr 38 to rat EDL muscles caused a dose - dependant increase in net protein degradation, apparently by suppressing protein synthesis (Table 2). The activity was stable to heating (70°C for 20 min), however it was abolished by dialysis or by repeated freezing and thawing.

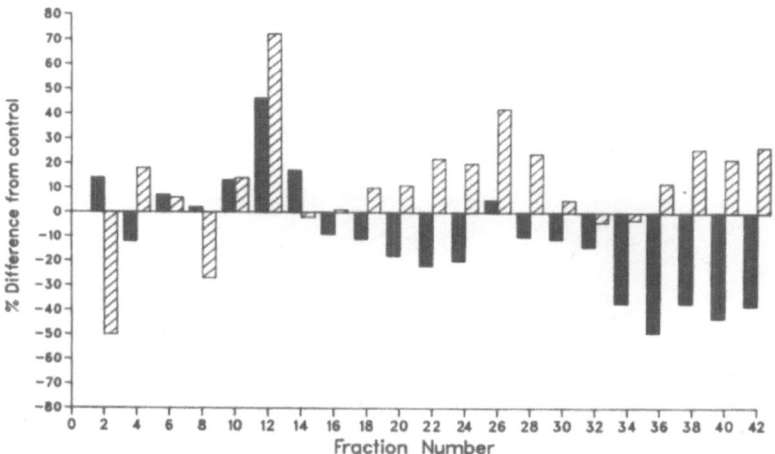

FIGURE 1. MUSCLE PROTEIN TURNOVER IN RAT MUSCLE TREATED WITH
FRACTIONS OF MONOCYTE SUPERNATANTS

Hatched bars, net protein degradation; Solid bars, protein synthesis
      Porcine blood monocytes were cultured and activated as described in
(12).   After chromatography on AcA54, fractions were diluted (33
$\mu l/ml$) in KRB medium.   Rat quarter diaphragm muscles were
preincubated for 30 min. in KRB, then transfered to fresh media
containing the fractions indicated, for 3 h. Control incubations
contained KRB alone. (n=8)

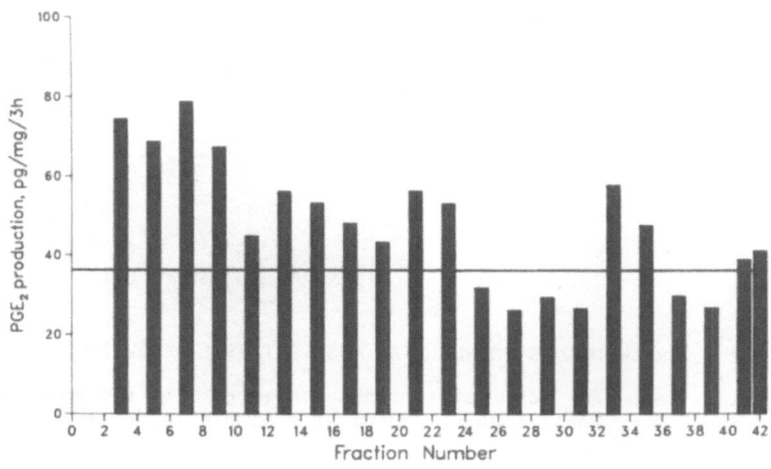

FIGURE 2. PGE$_2$ PRODUCTION OF RAT MUSCLES INCUBATED WITH
FRACTIONS OF MONOCYTE SUPERNATANTS

      Rat quarter diaphragms were incubated as described in the legend to
Fig. 1.   PGE$_2$ levels in the medium were determined after 3 h of
incubation by radioimmunoassay.   The horizontal bar represents the
mean value for muscles incubated in KRB alone (n=8).

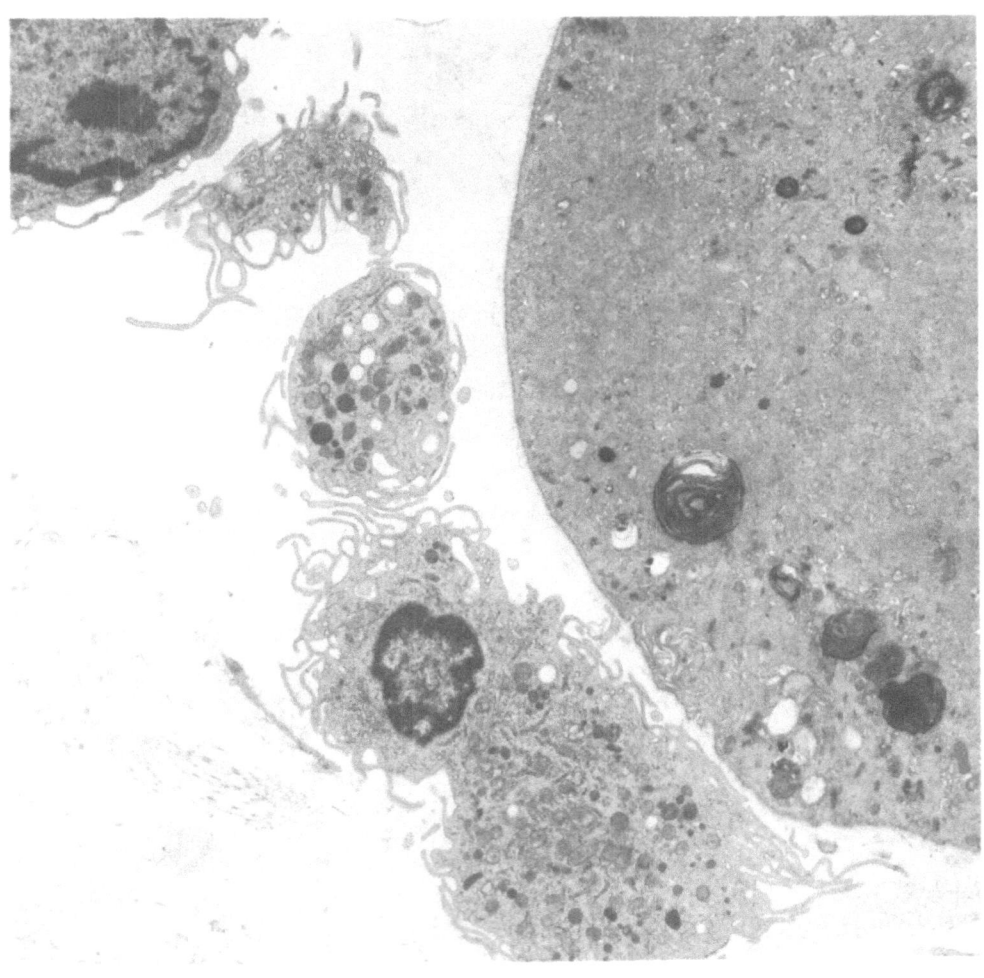

FIGURE 3. MACROPHAGES IN SKELETAL MUSCLE TISSUE AFTER LOCAL TRAUMA

Macrophage seen in connective tissue adjacent to a damaged muscle cell, 48 h after local trauma induced by a single impact to the medial gastrocnemius muscle. (X 50,000)(B.D. Fisher and VE Baracos, unpublished observations).

EFFECTS OF PURIFIED INTERLEUKIN-1 AND TUMOUR NECROSIS FACTOR

Activated monocytes produce a number of protein and nonprotein factors which could affect muscle protein synthesis and degradation. Our prior studies (5,6) had suggested that interleukin-1 may be involved. Also of interest, were the effects of tumour necrosis factor (TNF),a macrophage product that has been proposed to play a critical role in the pathogenesis of cachexia (15,28). In order to further clarify the involvement of interleukin-1 and tumour necrosis factor in muscle protein catabolism, we tested the effects of purified and recombinant (r)interleukin-1 and rTNF on muscle protein turnover in vitro. Porcine interleukin-1 α and β, purified to homogeneity from the same fractions used in these experiments (12), were tested on rat soleus muscles at 0.2, 20 and 200 ng / ml. To our surprise, neither form showed any effect on net protein degradation or PGE_2 production in vitro. As shown in Table 3, an interleukin-1 containing fraction obtained from gel filtration of human monocyte supernatants (14), caused stimulation of net degradation and PGE_2 production in rat soleus muscle (p<0.001), while rinterleukin-1 (18) had no effect on either process. Similarly, rTNFα (murine, 10 μg/ml) had no effect on net breakdown of muscle protein. These results suggest that these monokines have no action on skeletal muscle in their purified or recombinant forms.

DISCUSSION

When supernatants from activated Mφ were separated by gel filtration chromatography, several peaks of catabolic activity towards skeletal muscle were observed. When rat muscles were incubated with fractions showing interleukin-1 activity,net protein degradation increased by up to 75%, and the production of PGE_2 increased (Table 2, Fig. 1). A second peak eluting after the interleukin-1 - containing fractions also caused the net catabolism of muscle protein. A third and final peak eluting from the column strongly inhibited muscle protein synthesis. These results confirm our earlier observations that Mφ can release factors that are catabolic when added directly to skeletal muscles (5,6). It seems clear that partially purified fractions of Mφ supernatants can both stimulate net muscle catabolism in vitro (Table 2, Fig. 1)(4,5,8), and alter protein and amino acid metabolism when injected into rats (9,40,41). In related studies, we have also confirmed the observations of Clowes et al. (3), that circulating factors which can stimulate muscle proteolysis can be found in infected animals (Goldberg,AL et al., in preparation); these may originate from activated Mφ. However, the circulating product(s) that actually initiate protein breakdown and PGE_2 synthesis in muscle remain to be identified. Work to clarify the nature of these factors is in progress.

By contrast to the results noted above, highly purified porcine interleukin-1, or rinterleukin-1, failed to influence net protein degradation or PGE_2 production. In related studies (Goldberg AL, et al., in preparation), we tested purified and rinterleukin-1 from seven different laboratories; these failed to affect net protein degradation over a wide range of concentrations and experimental conditions. The failure of PGE_2 levels to rise drammatically when treated with rinterleukin-1 contrasts with our prior results with partially purified interleukin-1 - containing fractions (5,6), and with in vitro studies of muscles of rats injected with endotoxin or live bacteria (3,7). Thus the failure of rinterleukin-1 to promote muscle PGE_2 production may account for its inability to stimulate muscle protein degradation. Similarly, exhaustive testing of two other Mφ products, TNF and interferon (α, β, γ), showed no effects of these agents on net breakdown of muscle in vitro. Thus it remains to be clarified which Mφ factors affect muscle protein turnover in vitro or in vivo.

Mononuclear phagocytes, as a class, comprise a major first line of host defense, and are ubiquitous in their distribution in blood and tissues (11). They include peripheral blood monocytes, and fixed tissue macrophages found in all tissues, including skeletal muscle (Fig. 3). In addition, Mφ may be recruited to the site of a localized tissue injury, infection or inflammation. These cells may be activated in vivo or in vitro by a wide variety of stimuli including microorganisms, microbial products, tumour cells, antigens, inflammatory agents, plant lectins, lymphokines and other inducers (11). Net catabolism of skeletal muscle protein is a hallmark of diverse pathological states which are associated with Mφ activation. Although it is possible to speculate that activated phagocytes have a primary role in initiation of muscle catabolism of muscle protein, further work is required to identify the exact nature of the product(s) involved.

REFERENCES

1. Beisel, W.R. 1984. Metabolic effects of infection. Prog. Food Nutr. Sci. 8:43-75.

2. DeWys, W.D. 1982. Pathophysiology of cancer cachexia. Cancer Res. 42: 7215-7265

3. Fagan, J.M., and A.L. Goldberg. 1985. Muscle protein breakdown, prostaglandin E_2 production and fever following bacterial infection. In The Physiological, Metabolic and Immunologic Actions of Interleukin-1. (Kluger, M.J., J.J. Oppenheimer, and M.C. Powanda, editors). Alan R. Liss, New York, 201-210.

4. Clowes Jr., G.H.A., B.C. George, C.A. Villee Jr., and C. Saravis. 1983. Muscle proteolysis induced by a circulating peptide in patients with trauma or sepsis. New Engl. J. Med. 308:545-552.

5. Baracos, V.E., H.P. Rodemann, C.A. Dinarello, and A.L. Goldberg. 1983. Stimulation of muscle protein degradation and prostaglandin E_2 release by leukocytic pyrogen (Interleukin-1). New Engl. J. Med. 308:553-558.

6. Goldberg, A.L., Baracos, V., Rodemann, P., Waxman, L., and Dinarello, C. 1984. Control of protein degradation in muscle by prostaglandins, Ca^{2+}, and leukocytic pyrogen (Interleukin-1) Fed. Proc. 43:1301-1306.

7. Ruff, R., and Secrist, D. 1984. Inhibitors of prostaglandin synthesis or cathepsin D prevent muscle wasting due to sepsis in the rat. J. Clin. Invest. 73:1483-1486.

8. Jepson, M., M. Pell, P.C. Bates, and D.J. Millward. 1986. Effects of endotoxemia on protein metabolism in skeletal muscle and liver of fed and fasted rats. Biochem. J. 235:329-336.

9. Moldawer, L.L. 1986. Interleukin-1, tumor necrosis factor-α (cachectin) and the pathogenesis of cancer cachexia. Ph.D. thesis, Dept. of Surgery, Goteborgs Universitet, Sweden.

10. Benegard, K. 1982. Metabolic balance across the leg in weight - losing cancer patients, Cancer res. 42: 4293-4298

11. Adams, D.O., T.A. Hamilton. 1984. The cell biology of macrophage activation. Ann. Rev. Immunol. 2: 238-318.

12. Saklatvala, J., S.J. Sarsfeld, Y. Townsend. 1985. Pig interleukin-1. J. Exp. Med. 162: 1208-1222

13. Yang, R.D., L.L. Moldawer, A. Sakamoto, R.A. Keenan, D.E. Matthews, V.R. Young, R.W. Jr. Wannemacher, G.L. Blackburn, and B.R. Bistrian. 1983. Leukocyte endogenous mediator alters protein dynamics in the rat. Metabolism 32:654-660.

14. Dinarello, C. 1984. Interleukin-1. REv. Infect. Dis. 6:51-95.

15. Cerami, A., Y. Ykeda, N. LeTrang, P.J. Hotez, and B. Beutler. 1985. Weight loss associated with an endotoxin-induced mediator from peritoneal macrophages: the role of cachectin (tumor necrosis factor). Immun. Lett. 11:273-177.

16. Lomedico, P.T., V. Gubler, C.P. Hellman, M. Dukovitch, J.G. Giri, Y-CE Pan, and K. Collier. 1984. Cloning and expression of murine interleukin-1 cDNA in E.coli. Nature 312:456-462.

17. March, C.J., B. Mosley, A. Larsen, D.P. Cerretti, G. Braedt, V. Price, S. Gillis, C.S. Henney, S.R. Kronheim, K. Brabstein, P.J. Canlon, T.D. Hopp, and D. Cosenan. 1985. Cloning sequence, and expression of two distinct human interleukin-1 cDNAs, Nature 315:641-647.

18. Aueron, P.E., A.C. Webb, L.J. Rosenwasser, S.F. Mucci, A. Rich, S.M. Wolfe, and C.A. Dinarello. 1984. Nucleotide sequence of human monocyte interleukin-1 precursor mRNA Proc. Natl. Acad. Sci. USA 81:7907-7911.

19. Wingfield, P., M. Payton, J. Tavernier, M. Barnes, A. Shaw, K. Rose, M.G. Simona, S. Demczuk, K. Williamson, and J.M. Dayer. 1986. Purification and characterization of human interleukin-1β expressed in recombinant Escherichia coli. Eur. J. Biochem. 160:491-497.

20. Baracos, V.E., and A.L. Goldberg. 1986. Maintenance of normal length improves protein balance and energy status in isolated rate skeletal muscles. Am. J. Physiol. 251:C588-C596.

21. Tischler, M., M. Desautels, and A.L. Goldberg. 1982. Does leucine, leucycl tRNA or some metabolite of leucine regulate protein synthesis and degradation in skeletal and cardiac muscle. J. Biol. Chem. 257:1613-1621.

22. Dinarello, C.A., G.H.A. Clowes Jr., A.H. Gordon, C.A. Saravis, and S.M. Wolff. 1984. Cleavage of human Interleukin-1: Isolation

of a peptide from plasma of febrile humans and activated monocytes. J. Immunol. 133:1332-1338.

23. Beutler, B., and A. Cerami. 1986. Cachectin and tumor necrosis factor as two sites of the same biological coin. Nature 320:584-588.

24. Sobrado, J., L.L. Moldawer, C.A. Dinarello, G.L. Blackburn, and B.R. Bistrian. 1983. Effect of ibuprofen on fever and metabolic changes induced by leukocytic pyrogen (Interleukin-1) and endotoxin. Infect. Immun. 42:997-1005.

25. Tocco-Bradley, R., L.L. Moldawer, M. Georgieff, C.T. Jones, G.L. Blackburn, and B.R. Bistrian. 1986. Effects of recominant murine Interleukin-1 on protein dynamics in the rat. Proc. Soc. Exp. Biol. Med. 182: 263-271

RESPONSES OF LYSOSOMAL AND NON-LYSOSOMAL

PROTEASES TO UNLOADING OF THE SOLEUS

Erik J. Henriksen, Soisungwan Satarug
Marc E. Tischler, and Peter Fürst

Department of Biochemistry, University of Arizona
Tucson, AZ USA and Institute for Biological Chemistry
University of Hohenheim, FRG

INTRODUCTION

Atrophy of skeletal muscle can be induced by a number of interventions, including denervation of the hindlimb,[1] hindlimb immobilization,[2] and total mechanical unloading by either tail-cast suspension,[3] harness suspension,[4] or exposure to microgravity.[5] One probable causal factor for this loss of muscle mass is an increased rate of degradation of myofibrillar proteins. When coupled with an unchanged or decreased rate of protein synthesis, this leads to a net protein breakdown and thus muscle atrophy. While the exact mechanism for this intramuscular protein breakdown is still unclear,[6] it is most likely due to either an increase in the proteolytic capacity of the muscle cell, an increased susceptibility of substrate proteins to proteolytic enzymes, or a combination of the two.[6]

It is well established that denervation[7,8] or immobilization[9-11] of the hindlimbs leads to increases in the activity of several lysosomal enzymes, including proteases, in skeletal muscle. On the other hand, hindlimb muscle from rats unloaded by exposure to microgravity showed no changes in lysosomal acid protease activity, but did display increases in non-lysosomal Ca^{2+}-activated proteases and in possibly non-lysosomal tripeptidylaminopeptidases.[12] This suggested that the breakdown of contractile protein was non-lysosomal in origin.

The purpose of the investigations discussed herein was, therefore, to characterize the respective importance of lysosomal and non-lysosomal proteolysis in inducing the muscle atrophy and other attendant metabolic alterations in skeletal muscle unloaded by tail-cast suspension.

METHODS

Treatment of Animals

Female Sprague-Dawley rats (85-99g) maintained on food and water ad libitum were anesthetized with either Innovar-Vet (10 μl/100g body wt) or ether and subjected to tail-cast suspension[3] for periods of 1 to 6 days. Control animals were also tail-casted, but were allowed to remain weight-bearing during the entire treatment period. The tail casts were composed of Hexcelite orthopedic tape and Dow Corning Silastic medical-grade elastomer.

Measurement of Total Proteolysis

Muscle were placed in 3 ml of Krebs-Ringer bicarbonate (KRB) buffer (37°C, pH 7.4) equilibrated with 95% O_2:5% CO_2 and containing amino acids at normal rat plasma concentrations,[13] except for tyrosine, alanine, glutamate, and glutamine, which were omitted. Cyclohexamide (0.5 mM) was added to prevent protein synthesis. Following a 30 min preincubation, total proteolysis was measured as the release of tyrosine into fresh buffer during a 2 hr incubation period.[14] Tyrosine was assayed fluorometrically.[15]

Alkaline and Acid Proteases

Fresh muscle was homogenized in 50 mM phosphate buffer (pH 6.0 for acid protease or pH 8.0 for alkaline protease) containing 1 mM EDTA and 1 mM DTT, and centrifuged (5,000 X g for 10 min at 4°C). An aliquot of the supernatant (40 μl) was then added to 100 μl of incubation buffer (same as homogenization buffer, except pH 4.0 for acid protease assay) containing 150 nCi/ml [methyl-^{14}C] casein. The reaction was terminated after 1 hr by adding 700 μl of ice-cold 10% TCA. Following centrifugation (5,000 X g for 10 min at 4°C), 400 μl of the supernatant was counted for free ^{14}C-labeled peptides. One unit of protease activity was calculated as 1 dpm released per min at 37°C. The protease reactions were found to be linear over the 1 hr period at the protein concentration used.

Measurement of Glucose Metabolism

For measurement of glucose uptake, muscles were preincubated for 60 min in 1 mM glucose and 1% bovine serum albumin (BSA) and then incubated for 15 min with 1 mM 2-deoxy-D-[1,2-^{3}H] glucose (0.3 mCi/mmol) substituted for glucose in the absence and presence of insulin. The incubations were terminated by quickly blotting the muscles and solubilizing them in 0.5 N NaOH. The samples were then counted in 5 ml of aqueous scintillation cocktail. To permit for correction of the extracellular space, [carboxyl-^{14}C] inulin also was present during the incubation.

Oxidation of glucose and release of lactate and pyruvate were measured simultaneously in muscles incubated with 5 mM D-[U-^{14}C] glucose (10 μCi/mmol). Glucose oxidation was determined from the production of $^{14}CO_2$.[16] Lactate[17] and pyruvate[18] in the medium were assayed spectrophotometrically.

Insulin Binding Study

Insulin binding to the soleus was measured using a modification of the LeMarchand-Brustel et al.[19] Soleus muscles were preincubated at 22°C for 60 min in KRB buffer supplemented with 5 mM glucose, 0.2% BSA, and 2 mg/ml bacitracin to prevent insulin degradation. Muscles were then incubated for 4 hr at 22°C in fresh buffer containing 2 mM pyruvate, 0.2% BSA, 2 mg/ml bacitracin, 1 ng/ml ^{125}I-insulin, and varying concentrations of unlabeled insulin (0.17-33.5 nM). Non-specific binding was determined by incubation with unlabeled insulin (8.33 μM). After the incubations, muscles were washed 6 times (5 min/wash) with ice-cold isotonic saline (140 mM NaCl, 0.2% BSA, 10 mM Hepes, pH 7.6) to remove unbound insulin. Muscles were then dissolved in 2 ml of 1 N NaOH and counted in a gamma spectrophotometer.

Statistics

All values presented are means ± SEM for the number of animals indicated in each figure or table. Differences between unloaded and control muscles were evaluated using the Student's unpaired t-test. Significance of differ-

Table 1. Changes in Soleus Mass over a 6-day Unloading Period

Days of unloading	Soleus muscle wt (mg)/100 g body wt		
	1	3	6
Weight bearing	38.4 ± 1.3	40.4 ± 1.1	39.9 ± 1.5
Unloaded	37.2 ± 0.8	28.3 ± 1.1	23.2 ± 0.8
Difference	NS	−29%[a]	−42%[a]

Values are means ± SEM for 5 animals. [a] $p < 0.001$ vs. weight bearing group. NS, not statistically significant.

Table 2. Effects of Unloading and Insulin on Glucose Metabolism in Soleus Muscle

Muscle	Insulin	Glucose uptake	Glucose oxidation	Lactate and pyruvate release
		pmol/mg/15 min	nmol/mg/2 hr	
Weight bearing	−	246 ± 9	0.56 ± 0.02	11.9 ± 0.7
	+	430 ± 8	1.09 ± 0.05	22.2 ± 1.0
Unloaded	−	335 ± 16[a]	0.51 ± 0.04	11.3 ± 0.8
	+	715 ± 24[a]	1.74 ± 0.10[a]	33.5 ± 2.1[b]

Values are means ± SEM for 5 animals. Unloading period was 6 days. All insulin effects (0.1 mU/ml) are significant at $p < 0.001$. [a] $p < 0.001$, [b] $p < 0.01$ vs. weight bearing group. Data are from ref. 22.

ences between contralateral muscles for insulin effects was tested by using the Student's paired t-test.

RESULTS AND DISCUSSION

While there was no significant loss of mass by the soleus after 24 hr of unloading (Table 1), by day 3 this muscle was 29% smaller than controls. At day 6 of treatment, the soleus was 42% smaller than control muscle mass normalized to body weight. Coincident with the appearance of significant atrophy of the soleus was an increased sensitivity of this muscle to insulin for glucose metabolism. As shown in Table 2, on day 6 of unloading, glucose uptake by the soleus was 66% greater than the control muscle in the presence of physiological insulin (0.1 mU/ml). Three days of unloading have been shown

to be sufficient for this increased sensitivity to become significant.[20]
The increased insulin-stimulated uptake of glucose by the unloaded soleus
manifested itself in an increased metabolism of this sugar, as evidenced by
enhanced oxidation of glucose and release of lactate and pyruvate (Table 2).
A similar increase in sensitivity to insulin following 3 days or more of unloading has also been observed for protein metabolism, with a greater effect
of this hormone in inhibiting protein breakdown and stimulating protein synthesis.[21]

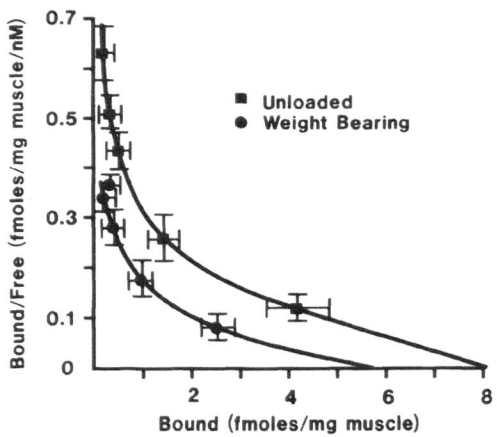

Fig. 1.   Scatchard plot of insulin binding to unloaded and weight
          bearing soleus muscles. The unloading period used was 6
          days. Each point represents the mean ± SEM for 4 muscles.
          Data are from ref. 22.

The greater increased insulin-stimulated uptake of glucose by the unloaded soleus can be overcome at maximal concentrations (> 10 mU/ml) of the
hormone.[22] This suggested that the underlying cause for this phenomenon was
receptor-mediated, rather than due to a post-binding alteration.[23] Therefore, an insulin-binding study was undertaken to investigate possible
changes in the binding capacity for insulin by the soleus as a result of unloading. Scatchard analysis revealed that the unloaded soleus did indeed
possess a greater capacity for insulin binding than the control muscle
(Fig. 1). However, the total amount of hormone bound per muscle did not
differ between the two groups (Table 3), despite the unloaded muscle being
33% smaller. This indicated that the increased binding capacity of the unloaded soleus could be completely accounted for by the loss of mass by this

Table 3. Comparison of Total Specifically Bound Insulin in
6-day Unloaded and Weight Bearing Soleus Muscles

Insulin	Total bound per muscle	
	Weight bearing	Unloaded
nM	fmol	
0.17	2.5 ± 0.2	2.5 ± 0.3
1.00	12.5 ± 0.8	11.3 ± 1.2
33.50	103.7 ± 16.7	99.0 ± 9.3

Values are means ± SEM for 4 muscles. Differences between
groups at a given insulin concentration are not statistically
significant. Data are taken from Fig. 1.

muscle. Unlike the myofibrillar proteins in the unloaded soleus,[24] the insulin receptor apparently did not undergo an increased net turnover during
the 6 days of treatment.

The insulin receptor, like other membrane receptor proteins, is recycled
between an intracellular storage site and the plasma membrane, and ultimately
is degraded in the lysosome.[25] Because of the above observation that there
was no enhanced degradation of the insulin receptor during unloading, it was
hypothesized that lysosomal proteolysis in general may not increase with unloading atrophy. To test this contention, total proteolysis was measured in
unloaded and control soleus muscles in the absence and presence of either
lysosomotropic agents (methylamine or leucine methyl ester) or mersalyl, an
inhibitor of Ca^{2+}-activated protease (Table 4). While the lysosomotropic
agents had no effect on the increased rate of proteolysis due to unloading,
mersalyl completely abolished the difference between the unloaded and control
muscles. In contrast, the enhanced proteolysis of the denervated soleus was
significantly reduced by both methylamine or leucine methyl ester, and to a
much lesser degree by mersalyl.[26] These results strongly suggest that with
total mechanical unloading, non-lysosomal proteolysis, perhaps Ca^{2+}-activated
protease activity, plays a major role in inducing the observed atrophy,
whereas atrophy due to denervation is primarily lysosomal in origin, with a
small contribution by non-lysosomal proteolytic mechanisms.

In support of the primary role of non-lysosomally mediated proteolysis
during unloading, the activity of alkaline protease was found to increase in
the atrophying soleus (Table 5). A significantly greater activity of this
non-lysosomal protease was seen after just 24 hr of unloading. This suggested that this enzyme could perhaps play a causal role in the instigation of
the muscle atrophy that becomes significant after more extended periods of
unloading. By day 6, the activity of alkaline protease was 73% greater than
controls. At the same time, acid protease, a lysosomal enzyme, displayed
only a 31% greater activity than controls (11.4 ± 0.6 vs. 8.7 ± 0.3 U/mg
muscle, p < 0.005), which could be accounted for simply by the loss of muscle
mass. Additionally, this activity of acid protease was 36% less than that of
alkaline protease in the same 6-day unloaded muscle. Taken together, these
data indicate that while the contribution of lysosomal acid protease cannot

Table 4. Effect of Lysosomotropic Agents and Mersalyl on the
Increased Proteolysis due to Unloading of the Soleus

	None	Methylamine	LME	Mersalyl
	(nmol tyrosine/mg muscle/2 hr)			
Control	0.334 ± 0.009	0.299 ± 0.029	0.294 ± 0.009	0.348 ± 0.020
	(percent difference from control)			
Unloaded	+ 18 ± 6	+ 20 ± 10	+ 20 ± 6	- 2 ± 9

Values are means ± SEM for 5 animals. Concentrations of inhibitors
used were: methylamine (10 mM), leucine methyl ester (LME, 10 mM),
and mersalyl (0.2 mM).

Table 5. Effect of Unloading on Activity of Alkaline Protease in
Whole Muscle Homogenates of Soleus

	Alkaline protease activity (U/mg muscle)		
Days of unloading	1	3	6
Weight bearing	10.0 ± 0.4	9.9 ± 0.3	10.2 ± 0.4
Unloaded	13.3 ± 0.4	13.9 ± 0.5	17.7 ± 0.8
Difference	+ 32%[a]	+ 41%[a]	+ 73%[a]

Values are means ± SEM for 5 animals. [a] $p < 0.001$ vs. weight
bearing group.

be ignored, the non-lysosomal alkaline protease in all likelihood is of importance in contributing to muscle atrophy due to unloading.

CONCLUSION

The alterations in protein turnover and the resulting changes in other
metabolic processes due to unloading of the soleus are summarized in Table 6.
In addition, the effects of denervation atrophy in this muscle are presented.
Unloading atrophy, believed to result from increased cytosolic proteolysis,
does not alter the turnover of the insulin receptor. The resulting increase
in binding capacity for insulin (expressed relative to muscle mass or protein) leads to an increased sensitivity to this hormone for both protein and
glucose metabolism. In contrast, denervation atrophy is believed to be
largely lysosomal in origin, and leads to a proportionate loss of myofibril-

Table 6. Comparison of Alterations in Metabolic Processes due to Unloading or Denervation Atrophy

Metabolic Alteration	Unloading	Denervation
Loss of myofibrillar proteins	Increased	Increased
Loss of the insulin receptor	No change	Increased
Sensitivity to insulin	Increased	Decreased
	(receptor-mediated)	(post-binding defect)
Primary locus of increased proteolysis	Non-lysosomal	Lysosomal

lar and insulin receptor proteins. The decreased sensitivity to insulin in denervated muscle is thought to arise from alterations distal to the insulin receptor.

The results presented highlight the fact that while unloading and denervation both induce atrophy of the soleus, the underlying mechanisms for this negative protein balance and the attendant metabolic changes may be quite different between the two models.

AKNOWLEDGEMENTS

This work was supported in part by a grant from the German Academic Exchange Service to E. J. Henriksen and Grant NAG2-384 from the National Aeronautics and Space Administration to M. E. Tischler.

REFERENCES

1. D. F. Goldspink, The effects of denervation on protein turnover of rat skeletal muscle, Biochem. J. 156:71 (1976).
2. F. W. Booth and J. R. Kelso, Production of rat muscle atrophy by cast fixation, J. Appl. Physiol. 34:404 (1973).
3. S. R. Jaspers and M. E. Tischler, Atrophy and growth failure of rat hindlimb muscles in tail-cast suspension, J. Appl. Physiol. 57:1472 (1984).
4. X. J. Musacchia, D. R. Deavers, G. A. Meininger, and T. P. Davis, A model for hypokinesia: effects on muscle atrophy in the rat, J. Appl. Physiol. 48:479 (1980).
5. E. J. Henriksen, M. E. Tischler, S. Jacob, and P. H. Cook, Muscle protein and glycogen responses to recovery from microgravity and unloading by tail-cast suspension, Physiologist 28:S193 (1985).
6. B. Dahlman, L. Kuehn, and H. Reinauer, Proteolytic enzymes and enhanced muscle protein breakdown, Adv. Exptl. Med. Biol. 167:505 (1982).
7. M. S. Pollack and J. C. W. Bird, Distribution and particle properties of acid hydrolase in denervated muscle, Am. J. Physiol. 215:716 (1968).
8. R. Libelius, I. Lundquist, S. Tagerud, and S. Thesleff, Endocytosis and lysosomal enzyme activities in dystrophic muscle: The effect of denervation, Acta Physiol. Scand. 113:259 (1981).

9. S. R. Max, R. F. Mayer, and L. Vogelsang, Lysosomes and disuse atrophy of skeletal muscle, Arch. Biochem. Biophys. 146:227 (1971).

10. F. A. Witzmann, J. P. Troup, and R. H. Fitts, Acid phosphatase and protease activities in immobilized rat skeletal muscles, Can. J. Physiol. Pharmacol. 60:1732 (1982).

11. H. Miyazawa, S. Ishiura, K. Yonemoto, A. Takagi, and H. Sugita, Effect of hindlimb immobilization on the lysosomal enzyme activity in the rat skeletal muscle, Biomed. Res. 4:597 (1983).

12. D. A. Riley, S. Ellis, G. R. Slocum, T. Satyanarayana, J. C. W. Bain, and F. R. Sedlak, Morphological and biochemical changes in soleus and extensor digitorum longus muscles of rats orbited in Spacelab 3, Physiologist 28:S207 (1985).

13. L. E. Mallette, J. H. Exton, and C. R. Park, Control of glucose from amino acids in the perfused rat liver, J. Biol. Chem. 244:5713 (1969)

14. R. M. Fulks, J. B. Li, and A. L. Goldberg, Effects of insulin, glucose, and amino acids on protein turnover in rat diaphragm, J. Biol. Chem. 250:290 (1975).

15. T. P. Waalkes and S. Udenfriend, A fluorometric method for the estimation of tyrosine in plasma and tissues, J. Lab. Clin. Med. 50:733 (1967).

16. T. W. Chang and A. L. Goldberg, The origin of alanine produced in skeletal muscle, J. Biol. Chem. 253:3677 (1978).

17. I. Gutmann and A. W. Wahlefeld, L-(+)-Lactate determination with lactate dehydrogenase and NAD, in: "Methods of Enzymatic Analysis," H. U. Bergmeyer, ed., Verlag Chemie International, Deerfield Beach (1981).

18. R. Czok and W. Lamprecht, Pyruvate phosphoenolpyruvate, and D-glycerate-2-phosphate, in: "Methods of Enzymatic Analysis," H. U. Bergmeyer, ed., Verlag Chemie International, Deerfield Beach (1981).

19. Y. LeMarchand-Brustel, B. Jeanrenaud, and P. Freychet, Insulin binding and effects in isolated soleus muscle of lean and obese mice, Am. J. Physiol. 234:E348 (1978).

20. E. J. Henriksen and M. E. Tischler, Acute unloading of the soleus results in a diminished capacity for glucose uptake, Fed. Proc. 46:326 (1987).

21. M. E. Tischler and S. Satarug, unpublished observations.

22. E. J. Henriksen, M. E. Tischler, and D. G. Johnson, Increased response to insulin for glucose metabolism in the 6-day unloaded rat soleus muscle, J. Biol. Chem. 261:10707 (1986).

23. C. R. Kahn, Insulin resistance, insulin insensitivity, and insulin unresponsiveness: a necessary distinction, Metabolism 27:1893 (1978).

24. S. R. Jaspers and M. E. Tischler, The role of glucocorticoids in the response of rat leg muscles to reduced activity, Muscle Nerve 9: 554 (1986).

25. K. A. Heidenreich and J. M. Olefsky, The metabolism of insulin receptors: internalization, degradation, and recycling, in: "Molecular Basis of Insulin Action," M. P. Czech, ed., Plenum Press, New York (1985).

26. M. E. Tischler, S. Satarug, and E. J. Henriksen, unpublished data.

CATHEPSIN B AND D ACTIVITY IN HUMAN SKELETAL MUSCLE IN DISEASE STATES

Gianfranco   Guarnieri,   Gabriele   Toigo,   Roberta   Situlin
Maria Alessandra Del Bianco, and Lucia Crapesi

Istituto di Patologia Medica
University of Trieste
Trieste, Italy

The effects of nutrition and disease state on whole body protein metabolism have been widely studied in recent years, but the distribution of changes among the various body tissues is less known. It is, therefore, important to establish the contribution made by skeletal muscle to whole body protein metabolism in disease state, because skeletal muscle is quantitatively and metabolically the most important body protein "store"[1] and   remains a good marker of protein-energy malnutrition (PEM) when the protein pools of other tissues and organs become metabolically un-available[1-3].

The results of in vitro studies, in which tissue specimens are studied independently of circulation and the nervous and hormonal influences, should be evaluated in the intact animal [4, 5]. Moreover, model studies in animals should be evaluated in humans, because animals may differ from man in body composition and in the contribution of each tissue to whole body changes.

The aim of this study is to evaluate in human skeletal muscle the usefulness of proteinase activity determination for the study of muscle protein degradation in disease states.

PROTEINASES AND MUSCLE PROTEIN DEGRADATION

The mechanisms of protein degradation in skeletal muscle are still not well understood[6-10]. Many proteolytic systems are active in skeletal muscle in different conditions. Basal degradative processes may differ from increased proteolysis during fasting or in disease state, such as in trauma/sepsis or in diabetes[11,12]. The mechanisms responsible for the selective degradation of protein are also poorly understood[9,12].

It is not clear if lysosomes are important for basal proteolysis[9]. After a cytoplasmic initiation of proteolysis, by limited and specific protein degradation, a "cascade effect" might follow in the lysosomes [13]. Indeed, many studies suggest that the lysosomal cathepsins B and D are involved in skeletal muscle proteolysis (Table I and II) and are capable of degrading myofibrillar proteins [14]. According to Furuno and Goldberg,[11] the basal degradative process in muscle does not involve lysosomes or thiol proteinases, and muscle can enhance protein breakdown by two mechanisms: lack of insulin and nutrients enhances a lysosomal process in muscle, as in other cells, whereas Ca^{2+} and muscle injury activate a distinct pathway involving thiol proteinases.

The non lysosomal ATP-dependent pathway probably hydrolyzes the "abnormal" proteins [37]. Non-lysosomal calcium-dependent proteinases might play a specific role in the turnover of structural proteins, such as cytoskeletal, myofibrillar or neurofilament proteins and in the proteolytic modification of specific receptors or enzyme proteins [38]. They could initiate the turnover of myofibrillar protein by disassembling these proteins from their highly organized structure in the myofibril, but almost certainly they are not involved in general metabolic turnover of the sarcoplasmic or stroma muscle proteins [39]. The chymotripsin-like serine-proteinases are active at alkaline pH ("alkaline proteinases") and probably degrade myofibrillar and sarcoplasmic proteins [40]. It is not clear if the alkaline proteinases are produced only in the mast-cells [40-42]. Their physiological importance is probably limited, but in disease state, with increased mast-cell infiltration, their contribution to muscle protein degradation might increase [40]. Muscle alkaline proteinase activity is increased in hypercatabolic patients and in benign prolonged starvation [43]. Total parenteral nutrition completely reversed the effect of benign prolonged starvation [43]. The tripsin-like serine proteinases ("human neutral proteinases") appear to have all the capacity necessary for initiating the

Table I. Muscle total or free cathepsin D activity in different conditions

---

INCREASED ACTIVITY

- ageing[15,16]
- prolonged running[15]
- denervation[17]
- muscle dystrophy[18]
- starvation[19]
- dexamethasone (free)[20,21]
- thyroid hormones[22,23]
- cancer[24,25]
- experimental burn injury, cold stress[23,26]

DECREASED ACTIVITY

- reinnervation[17]
- insulin (free)[27,28]
- anabolic agents (free)[21,29]
- dietary protein deficiency[30]

244

Table II. Some conditions increasing muscle cathepsin B activity

---

- thyroid hormones[22]
- muscle dystrophy[18]
- interleukin 1, fever, sepsis, burn injury, and a peptide present in trauma and sepsis[26,31-35]
- myopathy from experimental vitamin E deficiency[36]

---

catabolism of muscle proteins at neutral pH[40]. Alterations in neutral proteinase activity have been observed in muscular distrophy[44], immobilization[45], denervation[46] and during the first few days after experimental thermal injury[47]. In two recent studies, instead, the accelerated muscle protein degradation after experimental thermal injury was mediated by lysosomal acid proteinases (cathepsins B and perhaps H and L)[26,35]. By contrast, the calcium stimulated neutral protease activity was unaffected by burn injury[26]. A time course shows parallel increases in whole muscle proteolysis and acid protease activity after experimental burn injury[26]. Apparently the increase in lysosomal cathepsins is sufficient to account for the burn-induced increase in muscle protein breakdown[26]. Prostaglandins are probably not involved in the accelerated muscle proteolysis induced by burn, trauma or sepsis[35,48-51], as suggested by other studies[31-34].

Studies in cell cultures suggest that the rate of cellular growth is regulated by the rate of lysosomal protein degradation[52,53], but increased proteolytic degradation may occur both in anabolic conditions of muscle growth and in catabolic states[30].

Endogenous proteinase inhibitors might protect the cells from inappropriate endogenous or external proteolysis and/or might be involved in the control mechanisms responsible for intracellular or extracellular protein breakdown[54].

In vitro measurements of total enzyme activities in muscle homogenates, in optimal assay conditions, with changed concentrations of possible activators and/or inhibitors acting in vivo, may not reflect the actual enzyme activities in the intact animal. However, it is important to point out that proteinase activities measured in vitro in tissue extracts have been shown to follow rates of degradation in vivo and in incubated tissue fragments[25,55].

From these studies we may conclude that lysosomal processes are importantly involved in muscle proteolysis, but firm evidence that they play a rate-limiting or regulatory role in determining the rate of degradation is currently lacking.

EXPERIMENTAL RESULTS OF MUSCLE CATHEPSIN B AND D ACTIVITIES IN DISEASE
STATE

To better define the changes of muscle synthesis and degradation in
disease state, we directly measured the activities of cathepsins B and/or
D in specimens of human skeletal muscle obtained by needle-biopsy. In the
tissue fragments we also determined the concentration of DNA, RNA and ASP
(alkali-soluble, non collagen, proteins). Muscle DNA content in adults
is considered a satisfactory reference standard and in protein-energy
malnutrition the total number of nuclei in muscle is slightly affected [1].
The ASP:DNA ratio is an index of a hypothetical muscle cell size, i.e.
the imaginary volume of cytoplasm managed by a single nucleus [1]. Total
muscle RNA content, which is about 80% ribosomal RNA, and the RNA:DNA and
RNA:ASP ratios are a measure of the capacity of the cell to synthesize
proteins [1]. Starvation and refeeding are accompanied by progressive changes in muscle RNA content and in muscle protein synthesis and content [1,56].

The methods for the determination of cathepsin D activity and metabolite content were previously reported [56-58] and are outlined in Figure
1. Cathepsin B activity was measured by a modification of the method of
Baici et al.[59], using a highly specific fluorimetric substrate (N- -Cbz-L-
-Arg-7-amido-4-methylcoumarin from Bachem, Switzerland).

Chronically uremic patients

At the First International Congress on Proteases we reported significantly lower cathepsin D activities in patients on chronic hemodialysis[58] (Table III). Also the RNA to DNA and ASP to DNA ratios were significantly lower in patients than in controls, and were highly correlated
(r = 0.79 and r = 0.83) with total or free cathepsin D activity. Therefore in this chronic situation the indices of muscle protein synthesis
and degradation showed the coordinate-parallel changes described in whole

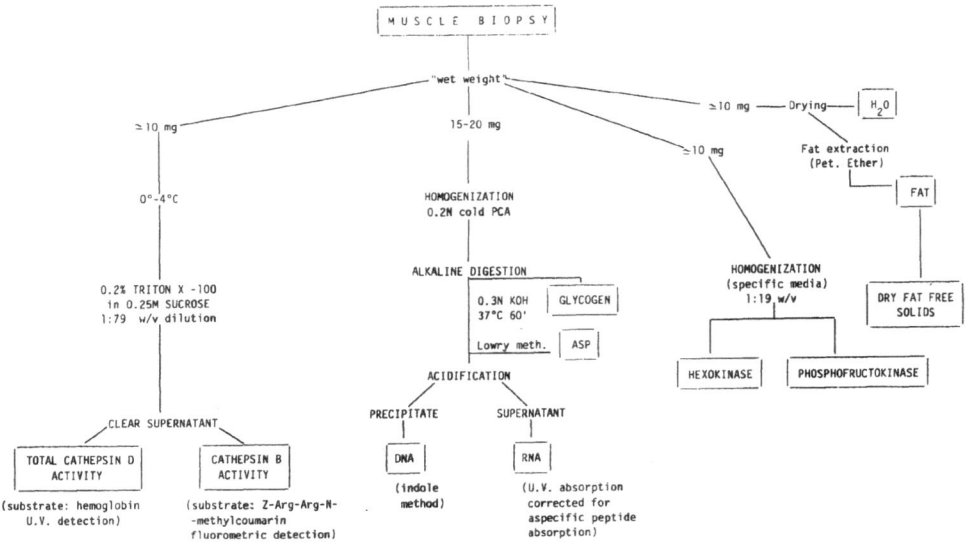

Figure 1. A schematic illustration of methods used.

Table III. Muscle cathepsin D activity in controls and in patients

	Tot. activ. (mU/100 mg wet wt)	Free activ. (mU/100 mg wet wt)	Free:total (%)	Total activity:DNA	Free activity/DNA
Controls (n = 8)	98 + 15	43 + 8	44 + 9	2.88 + 0.68	1.26 + 0.36
Hemodialysis patients (n = 10)	63 + 13[a]	29 + 7[b]	47 + 8	1.33 + 0.46[a]	0.62 + 0.26[b]
Patients with early renal failure (n = 11)	85 + 16			2.44 + 0.50	
Patients with advanced renal failure (n = 8)	83 + 25			2.39 + 0.99	
Patients with chronic pancreatitis (n = 11)	78 + 20[c]	41 + 13	53 + 10	1.50 + 0.40[a]	0.80 + 0.20[b]

Student's t-test (patients vs controls): [a]$p < 0.001$; [b]$p < 0.005$; [c]$p < 0.05$.

animals by Millward et al.[23,30] by means of different techniques. Cathepsin D activity was also inversely correlated with dietary protein and energy intake and with serum urea nitrogen[58]. Also granulocyte proteinase activity is reduced in chronically uremic patients treated with diet or regular dialysis[60].

Subsequent studies[1,61] could not demonstrate a significant decrease of cathepsin D acitivity in patients with early or advanced renal failure on protein-restricted diets (Table III).

Patients with chronic relapsing pancreatitis

In 14 patients with chronic relapsing pancreatitis muscle cathepsin D activity (Table III), and the RNA to DNA and ASP to DNA ratios were significantly lower than in controls[62]. These changes were probably related to malnutrition, which was present in our patients[62], although increased levels of muscle protease inhibitors might not be excluded. Our patients ingested only 26 kcal/kg ideal body wt/d and 0.9 g of proteins/ /kg ideal body wt/d, which is a borderline level of intake in disease. According to Millward et al.[23,30,55] a reduced nutrient supply decreases muscle protein synthesis, and the initial decrease in breakdown is consi-

Table IV. Some highly significant correlations present in patients with chronic relapsing pancreatitis

		r	$p <$
- basal insulin	free cathepsin D	- 0.99	0.01
- peak insulin	free/total cathepsin D	- 0.90	0.01
- basal C-peptide	free/total cathepsin D	- 0.91	0.01
- peak C-peptide	free/total cathepsin D	- 0.91	0.01
- insulin area	free cathepsin D	- 0.95	0.01
- C-peptide area	free/total cathepsin D	- 0.92	0.01
- peak insulin	3 methyl HIS	- 0.88	0.01

Table V. Effects of multiple trauma and total parenteral nutrition (TPN) on muscle cathepsin activities and metabolite content in humans

| | CONTROLS | TRAUMA PATIENTS | | | |
		p<	before TPN	p<	after TPN
Tot. cathepsin D, mU/100 mg w.wt	98 + 15	0.005	65 + 19	0.005	150 + 71
Free cathepsin D, mU/100 mg w.wt	43 + 8	0.01	29 + 9	0.005	70 + 14
Tot. cathepsin D/DNA, mU/µg	2.88 + 0.68	0.005	1.60 + 0.64		1.90 + 0.20
Free cathepsin D/DNA, mU/µg	1.26 + 0.36	0.01	0.73 + 0.28		0.90 + 0.17
Tot. cathepsin D/ASP, mU/mg	5.60 + 1.37	,0.05	3.60 + 2.07	0.02	8.30 + 5.02
Free cathepsin D/ASP, mU/mg	2.43 + 0.60	0.05	1.60 + 0.88		3.90 + 2.16
Cathepsin B, U/mg w.wt			24 + 9	0.01	77 + 44
Cathepsin B/DNA, U/µg			0.58 + 0.22		0.92 + 0.43
Cathepsin B/ASP, U/mg			1.25 + 0.61	0.05	3.93 + 1.72
DNA, mg/100 g w.wt	34 + 8		39 + 10	0.001	80 + 10
RNA, mg/100 g w.wt	77 + 23		89 + 15		98 + 20
ASP, g/100 g w.wt	18 + 2.1		19.7 + 6.0		19.8 + 5.7
RNA/DNA, mg/mg	2.32 + 0.59		2.38 + 0.66	0.005	1.23 + 0.21
ASP/DNA, mg/mg	554 + 109		518 + 168	0.005	247 + 67

p: level of significance of Student's t test for paired or unpaired data

dered adaptive to changes in synthesis. The mean muscle cathepsin D activity was lower than in controls, but in the range of measured activities we observed a highly significant inverse correlation with serum insulin and C-peptide levels both in basal conditions and during a glucose load (Table IV). Serum insulin levels were also inversely correlated with urinary 3 methyl-histidine excretion. Insulin plays an important role in stimulating muscle protein synthesis, but the effects of insulin and diabetes on muscle protein degradation in vivo are not clearly defined and, of course, these correlations are not an index of a cause-effect relationship between insulin secretion and muscle protein degradation. Anyway, insulin has been reported to acutely reduce muscle free cathepsin D activity and to inhibit lysosomal protein degradation[6,11,27,63-70].

Patients with severe accidental trauma

We examined 10 patients admitted to the Intensive Care Unit of our Hospital for severe accidental injuries (category 1, according to the classification of the American College of Surgeons[71]). A muscle needle biopsy was taken from quadriceps femoris by Bergstrom's needle on the 3rd-4th day after trauma and then the patients were treated with TPN[14]. Energy intake was about 25% above measured energy expenditure, 50% as lipids (Intralipid[R]), and 50% as glucose[14]. Nitrogen intake was 250-300 mg/kg ideal body wt/d as cristallin amino acids. Patients often received corticosteroids for head injury. In our patients a second muscle biopsy was taken after 2-3 weeks of parenteral nutrition.

Muscle DNA, RNA and ASP contents were normal on the first assessment (Table V). Muscle DNA concentrations were significantly higher, and the

RNA to DNA and ASP to DNA ratios significantly lower than in controls on the second assessment, in spite of 2-3 weeks of adequate nutritional support. Cathepsin D activities (Table V) were decreased, in comparison with controls, on the 3rd-4th day after trauma and increased at the end of parenteral nutrition, 2-3 weeks after trauma. Similar changes were observed for cathepsin B activity (Table V). Cathepsin B and D activities were significantly correlated (Table VI) and also in these patients, as in those with pancreatitis, cathepsin D activity was inversely correlated with insulin secretion evaluated by serum C-peptide levels.

Our patients were previously healthy and were admitted for severe accidental injuries. Shock, endogenous and exogenous hormone levels, and nutrient intake could all affect muscle metabolism in our patients. During the first days after injury they were resuscitated, treated for shock, fractures etc., and artificially ventilated. Their energy intake was actually very low and no nitrogen was supplied. According to Millward et al.[55], muscle protein degradation falls in protein deficiency and in fasting adults and, interestingly, the changes in degradation occur in concert with changes in lysosomal proteinases. Also ischemia, muscle paralysis (as in our artificially ventilated patients), anoxia, and acidosis are known to inhibit proteolysis[72,73]. Moreover, it is interesting to recall that glutamine, which is definitely decreased in muscle hypercatabolic states, in physiologic concentrations has an inhibitory effect on protein degradation rate in cultured skeletal muscle cells[74]. A two--stage response is apparent from our results: in muscle, early changes after injury were often opposite to late changes, after TPN and corticosteroid administration, when muscle RNA:DNA and ASP:DNA ratios (namely muscle protein synthesis and content) were low and muscle cathepsin B and D activities (possibly an index of muscle protein degradation) were increased. Although being temptative to speculate that these changes of proteinase activity are an effort to spare some muscle protein in early stages and to provide substrates later on, other possible explanations exist. For the early changes, we should mention loss of enzyme from the sarcoplasm, enzyme inhibition, decreased enzyme synthesis or increased enzyme degradation. Besides, it is not clear from these preliminary results whether changes present at the end of treatment were related to an inadequate nutrient intake, to the metabolic abnormalities of trauma, not reversed by the medical and nutritional treatment, or to corticosteroid administration. Opposite changes of muscle breakdown have been reported also by Waterlow et al.[75] during experimental starvation: an adaptive de-

Table VI. Some highly significant correlations present in patients after severe accidental trauma

	BEFORE TPN	r	p<
total cathepsin D	free cathepsin D	0.80	0.01
free cathepsin D	cathepsin B	0.80	0.01
	AFTER TPN		
total cathepsin D/ASP	serum C-peptide	-0.81	0.01

crease in breakdown was followed by an increase after many days of insufficient nutrient intake, when muscle proteins were probably drawn upon to provide substrates for energy generation. Also according to Hasselgren et al.[76], in experimental trauma complicated by sepsis, muscle protein synthesis was reduced, and degradation was increased. Other studies, measuring A-V amino acid difference, reported lower muscle protein synthesis after trauma[4,8] and are in agreement with our direct RNA measurements. Two-phase changes of muscle protein degradation were reported also by Rennie et al.[4] after trauma and surgery, by measuring 3-methyl--histidine efflux from the leg. The results of these changes of muscle protein synthesis and degradation were the extremely low muscle protein content (evaluated by the ASP to DNA ratio) and muscle wasting, evaluated by anthropometry[14]. In fact, the rates of protein synthesis and catabolism are much higher than dietary protein intake, and therefore limited changes of their rates induce large variations of whole body balance[77].

CONCLUSIONS

Skeletal muscle is the largest body protein store and it supplies AA for the synthetic and energy-producing processes. Therefore, muscle protein degradation is strictly regulated and several hormones and disease-related factors enhance AA release by cleavage of muscle proteins. However, the role of specific proteinases in muscle catabolism and the factors that regulate their activities are as yet largely unknown[42].

We do not know if the increase in lysosomal cathepsins is sufficient to account for muscle wasting of our patients, as suggested by the experimental studies of Odessey et al.[26] , and we ignore whether lysosomal proteinases do have in vivo a regulatory role on the increase of muscle protein degradation in disease state, in response to nutritional and hormonal signals, or whether the changes of their activity with changing protein catabolism are merely an epiphenomenon, namely a pathological response in deteriorating tissues. As reported above, muscle proteinase activities have been shown to follow rates of muscle protein degradation both in vivo and in vitro[22,55] . Probably muscle cathepsin activity is, at least, an index of muscle protein degradation rate.

According to Millward et al.[23] the effects of dietary protein deficiency and stress on muscle lysosomal proteinases could be mediated by thyroid hormones, which should be more frequently measured in future studies. We also believe that in future investigations changes of plasma and tissue activators and inhibitors of muscle proteinases should be evaluated and the enzyme protein should be also measured immunologically.

REFERENCES

1. G. F. Guarnieri, G. Toigo, R. Situlin, M. A. Del Bianco, L. Crapesi, and A. Zanettovich, Direct biochemical analysis of human muscle tissue in hospital malnutrition, Proc. Int. Workshop on Nutritional Assessment in Hospital Malnutrition, Venice (Italy) April 26-27, 1985, JPEN 11 (1987) (in press)

2. S. B. Heymsfield, C. McManus, V. Stevens, and J. Smith, Muscle mass: reliable indicator of protein-energy malnutrition severity and outcome, Am. J. Clin. Nutr. 35:1192-1199 (1982)
3. S. B. Heymsfield, V. Stevens, R. Noel, C. McManus, J. Smith, and D. Nixon, Biochemical composition of muscle in normal and semistarved human subjects: relevance to anthropometric measurement, Am. J. Clin. Nutr. 36:131-142 (1982)
4. M. J. Rennie, Muscle protein turnover and the wasting due to injury and disease, Br. Med. Bull. 41:257-264 (1985)
5. C. I. Harris, C. A. Maltin, R. M. Palmer, P. J. Reeds, and A. B. Wilson, Biochemical and morphological observations of skeletal muscles incubated in vitro, in: "Progress in Clinical and Biological Research, vol. 180: Intracellular Protein Catabolism," E. A. Khairallah, J. S. Bond, J. W. C. Bird, eds., Alan R. Liss, Inc., New York (1985)
6. M. E. Tischler, Hormonal regulation of protein degradation in skeletal and cardiac muscle, Life Sci. 28:2569-2576 (1981)
7. P. J. Reeds, and W. P. T. James, Protein turnover, Lancet March 12: 571-574 (1983)
8. M. J. Rennie, and R. Harrison, Effects of injury, disease, and malnutrition on protein metabolism in man, Lancet February 11:323-325 (1984)
9. E. A. Khairallah, J. S. Bond, and J. W. C. Bird, eds., "Progress in Clinical and Biological Research, vol. 180: Intracellular Protein Catabolism," Alan R. Liss, Inc., New York (1985)
10. W. H. Hörl, and A. Heidland, eds., "Proteases - Potential Role in Health and Disease," Plenum Press, New York and London (1984)
11. K. Furuno, and A. L. Goldberg, The activation of protein degradation in muscle by calcium or muscle injury does not involve a lysosomal mechanism, Biochem. J. 237:859-864 (1986)
12. J. F. Dice, J. M. Backer, P. Miao, L. Bourret, and M. A. McElligott, Regulation of catabolism of ribonuclease a microinjected into human fibroblasts, in: "Progress in Clinical and Biological Research, vol. 180: Intracellular Protein Catabolism," E. A. Khairallah, J. S. Bond, and J. W. C. Bird, eds., Alan R. Liss, Inc., New York (1985)
13. E. A. Khairallah, J. S. Bond, and J. W. C. Bird, "Regulation of protein turnover," in: "Progress in Clinical and Biological Research, vol. 180: Intracellular Protein Catabolism," E. A. Khairallah, J. S. Bond, and J. W. C. Bird, eds., Alan R. Liss, Inc., New York (1985)
14. G. Guarnieri, G. Toigo, R. Situlin, A. Del Bianco, L. Crapesi, A. Zanettovich, E. Romano, F. Iscra, and G. Mocavero, Muscle-biopsy studies on protein metabolism in traumatized patients, in: "Clinical Nutrition and Metabolic Research. Proc. 7th Congr. ESPEN, Munich 1985," G. Dietze, T. Grünert, A. Kleinberger, and S. Wolfram, eds., Karger, Basel (1986)
15. A. Salminen, and V. Vinko, Effects of age and prolonged running on proteolytic capacity in mouse cardiac and skeletal muscle, Acta Physiol. Scand. 112:89-95 (1981)
16. K. Lundholm, and T. Schersten, Leucine incorporation into proteins and cathepsin D activity in human skeletal muscles. The influence of the age of the subject, Exp. Geront. 10:155-159 (1975)
17. L. D. Sellin, R. Libelius, I. Lundquist, S. Tagerud, and S. Thesleff,

Membrane and biochemical alterations after denervation and during reinnervation of mouse skeletal muscle, Acta Physiol. Scand. 110: 181-186 (1980)

18. J. B. Li, Protein synthesis and degradation in skeletal muscle of normal and dystrophic hamsters, Am. J. Physiol. 239:E401-E406 (1980)

19. A. M. Samarel, E. A. Ogunro, A. G. Fercuson, P. Allenby, and M. Lesch, Regulation of cathepsin D metabolism in rabbit heart. Evidence for a role for precursor processing in the control of enzyme activity, J. Clin. Invest. 69:999-1007 (1982)

20. A. F. Clark, and P. J. Vignos Jr., The role of proteases in experimental glucocorticoid myopathy, Muscle Nerve 4:219-222 (1981)

21. B. G. Vernon, and P.J. Buttery, The effect of the growth promoter trenbolone acetate, dexamethasone and thyroxine on skeletal muscle cathepsin D (EC 3.4.4.23) activity, Proc. Nutr. Soc. 40:13A (1980)

22. R. S. Decker, and K. Wildenthal, Lysosomal alterations in heart, skeletal muscle, and liver of hyperthyroid rabbits. J. Lab. Invest. 44:455-465 (1981)

23. D. J. Millward, P. C. Bates, J. G. Brown, M. Cox, R. Giugliano, M. Jepson, and J. Pell, Role of thyroid, insulin and corticosteroid hormones in the physiological regulation of proteolysis in muscle, in: "Progress in Clinical and Biological Research, vol. 180: Intracellular Protein Catabolism," E. A. Khairallah, J. S. Bond, and J. W. C. Bird, eds., Alan R. Liss, Inc., New York (1985)

24. T. Schersten, and K. Lundholm, Lysosomal enzyme activity in muscle tissue from patients with malignant tumor, Cancer 30:1246-1251 (1972)

25. K. Lundholm, A.-C. Bylund, J. Holm, and T. Schersten, Skeletal muscle metabolism in patients with malignant tumor, Europ. J. Cancer 12: 465-473 (1976)

26. R. Odessey, Burn-induced stimulation of lysosomal enzyme activity in skeletal muscle, Metabolism 35:750-757 (1986)

27. D. E. Rannels, R. Kao, and H. E. Morgan, Effect of insulin on protein turnover in heart muscle, J. Biol. Chem. 250:1694-1701 (1975)

28. J. B. Li, S. R. Rannels, M. E. Burkart, and L. S. Jefferson, Effects of insulin on protein degradation and lysosomal cathepsin D in perfused skeletal muscle, Fed. Proc. 34:535 (1975)

29. P. A. Sinnett-Smith, N. W. Dumelow, and P. J. Buttery, Protein turnover in sheep treated with trenbolone acetate and zeranol, Proc. Nutr. Soc. 42:58A (1983)

30. D. J. Millward, P. C. Bates, J. G. Brown, S. R. Rosochacki, and M. J. Rennie, Protein degradation and the regulation of protein balance in muscle, in: "Protein Degradation in Health and Disease," Ciba Symposium No. 75:307-329 (1980)

31. G. H. Clowes Jr., B. C. George, C. A. Villee Jr., and C. A. Saravis, Muscle proteolysis induced by a circulating peptide in patients with sepsis or trauma, N. Engl. J. Med. 308:545-552 (1983)

32. V. Baracos, H. P. Rodemann, C. A. Dinarello, and A. L. Goldberg, Stimulation of muscle protein degradation and prostaglandin E_2 release by leukocytic pyrogen (interleukin-1). A mechanism for the increased degradation of muscle proteins during fever. N. Engl. J. Med. 308:553-558 (1983)

33. A. L. Goldberg, V. Baracos, P. Rodemann, L. Waxmann, and C. Dinarello, Control of protein degradation in muscle by prostaglandins, Ca^{2+},

and leukocytic pyrogen (interleukin-1). Fed. Proc. 43:1301-1306 (1984)

34. R. Ruff, and D. Secrist, Inhibitors of prostaglandin synthesis or cathepsin B prevent muscle wasting due to sepsis in the rat. J. Clin. Invest. 73:1483-1486 (1984)

35. A. S. Clark, R. A. Kelly, and W. E. Mitch, Systemic response to thermal injury in rats. Accelerated protein degradation and altered glucose utilization in muscle, J. Clin. Invest. 74:888-897 (1984)

36. A. M. Spanier, and J. W. C. Bird, Endogenous cathepsin B inhibitor activity in normal and myopathic red and white skeletal muscle, Muscle Nerve 5:313-320 (1982)

37. J. D. Etlinger, H. McMullen, R. F. Rieder, A. Ibrahim, R. A. Janeczko, and S. Marmorstein, Mechanisms and control of ATP-dependent proteolysis, in: "Progress in Clinical and Biological Research, vol. 180: Intracellular Protein Catabolism," E. A. Khairallah, J. S. Bond, and J. W. C. Bird, eds., Alan R. Liss, Inc., New York (1985)

38. G. N. DeMartino, and D. E. Croall, Calcium-dependent proteases from liver and heart, in: "Progress in Clinical and Biological Research, vol. 180: Intracellular Protein Catabolism," E. A. Khairallah, J. S. Bond, and J. W. C. Bird, eds., Alan R. Liss, Inc., New York (1985)

39. D. E. Goll, T. Edmunds, W. C. Kleese, S. K. Sathe, and J. D. Shannon, Some properties of the Ca^{2+}-dependent proteinase," in: "Progress in Clinical and Biological Research, vol. 180: Intracellular Protein Catabolism," E. A. Khairallah, J. S. Bond, and J. W. C. Bird, eds., Alan R. Liss, Inc., New York (1985)

40. J. Kay, R. Heath, B. Dahlmann, L. Kuehn, and W. T. Stauber, Serine proteinases and protein breakdown in muscle, in: "Progress in Clinical and Biological Research, vol. 180: Intracellular Protein Catabolism," E. A. Khairallah, J. S. Bond, and J. W. C. Bird, eds., Alan R. Liss, Inc., New York (1985)

41. B. Dahlmann, L. Kuehn, M. Rutschmann, and H. Reinauer, High molecular mass cysteine proteinases from rat skeletal muscle tissue, in: "Progress in Clinical and Biological Research, vol. 180: Intracellular Protein Catabolism," E. A. Khairallah, J. S. Bond, and J. W. C. Bird, eds., Alan R. Liss, Inc., New York (1985)

42. M. Mayer, Regulation of myofibrillar protease, plasminogen activator and protein degradation in cultured myoblasts, in: "Progress in Clinical and Biological Research, vol. 180: Intracellular Protein Catabolism," E. A. Khairallah, J. S. Bond, and J. W. C. Bird, eds., Alan R. Liss, Inc., New York (1985)

43. O. Z. Lernau, S. Nissan, B. Neufeld, and M. Mayer, Myofibrillar protease activity in muscle tissue from patient in catabolic conditions, Eur. J. Clin. Invest. 10:357-361 (1980)

44. N. C. Kar, and C. M. Pearson, A calcium-activated neutral protease in normal dystrophic human muscle. Clin. Chim. Acta 73:293 (1976)

45. A. Jakubiec-Puka, and W. Drabikowski, Changes in proteolytic activity in muscle of rat after immobilization, in: "Structure and Function of Normal and Diseased Muscle and Peripheral Nerve," I. Hausmanowa--Petrusewicz, and H. Jedrzejowska, eds., Polish Medical Publ., Warsaw (1974)

45. A. Jakubiec-Puka, and W. Drabikowski, Influence of denervation and reinnervation on autolytic activity and on protein composition of

skeletal muscle in rat, Enzyme 23:10 (1978)

47. J. J. Newman, D. R. Strome, C. W. Goodwin, A. D. Mason Jr., and B. A. Pruitt, Neutral proteinase activity in skeletal muscle from thermally injured rats, J. Surg. Res. 35:515-519 (1983)

48. R. Odessey, Effect of inhibitors of proteolysis and arachidonic acid metabolism on burn-induced protein breakdown, Metabolism 34:616-620 (1985)

49. D. J. Loegering, and J. Turinsky, Prostaglandin E_2 production and protein metabolism in septic rats, Fed. Proc. 44:1377 (1985)

50. P.-O. Hasselgren, M. Talamini, R. LaFrance, J. H. James, H. C. Peters, and J. E. Fischer, Effect of indomethacin on proteolysis in septic muscle, Ann. Surg. 202:557-562 (1985)

51. H. R. Freund, J. H. James, R. LaFrance, L. S. Gallon, U. O. Barcelli, L. L. Edwards, S. N. Joffe, S. Bjornson, and J. E. Fischer, The effect of indomethacin on muscle and liver protein synthesis and on whole-body protein degradation during abdominal sepsis in the rat, Arch. Surg. 121:1154-1158 (1986)

52. A. M. Spanier, W. A. Clark Jr., and R. Zak, Replacement perfusion of cultured eucaryotic cells: a method for the accurate measurement of the rates of growth, protein synthesis and protein turnover, in: "Progress in Clinical and Biological Research, vol. 180: Intracellular Protein Catabolism," E. A. Khairallah, J. S. Bond, and J. W. C. Bird, eds., Alan R. Liss, Inc., New York (1985)

53. U. Pfeifer, and R. Dunker, Inhibition of protein degradation and of cellular autophagy in growing as compared to density inhibited fibroblasts, in: "Progress in Clinical and Biological Research, vol. 180: Intracellular Protein Catabolism," E. A. Khairallah, J. S. Bond, and J. W. C. Bird, eds., Alan R. Liss, Inc., New York (1985)

54. V. Turk, J. Brzin, B. Lenarčič, P. Ločnikar, T. Popović, A. Ritonja, J. Babnik, W. Bode, and W. Machleidt, Structure and function of lysosomal cysteine proteinases and their protein inhibitors, in: "Progress in Clinical and Biological Research, vol. 180: Intracellular Protein Catabolism," E. A. Khairallah, J. S. Bond, and J. W. C. Bird, eds., Alan R. Liss, Inc., New York (1985)

55. D. J. Millward, J. G. Brown, and B. Odedra, Protein turnover in individual tissues with special emphasis on muscle, in: "Nitrogen metabolism in man," P. Waterlow, and J. Stephen, eds., Applied Science Publ., London (1981)

56. G. F. Guarnieri, G. Toigo, R. Situlin, L. Crapesi, and M. A. Del Bianco, Proteinase activity, as well as DNA, RNA and protein content in human skeletal muscle in malnutrition and disease states, in: "Proc. of the Workshop: Wertigkeit und Aussagefähigkeit metabolischer Parameter in der parenteralen Ernährung, Fuschl, March 6-7, 1987," W. Zuckschwerdt, ed., Verlag, München (1987) (in press)

57. G. Guarnieri, G. Toigo, R. Situlin, L. Faccini, U. Coli, S. Landini, G. Bazzato, F. Dardi, and L. Campanacci, Muscle biopsy studies in chronically uremic patients: evidence for malnutrition, Kidney Int. 24:S-187-S-193 (1983)

58. G. F. Guarnieri, G. Toigo, R. Situlin, L. Faccini, R. Rustia, and F. Dardi, Muscle cathepsin D activity, and RNA, DNA and protein content in maintenance hemodialysis patients, in: "Proteases: Potential Role in health and Disease," W. H. Hörl, and A. Heidland, eds., Plenum Publ. Corp. (1984)

59. A. Baici, M. Gyger-Marazzi, and P. Sträuli, Extracellular cysteine proteinase and collagenase activities as a consequence of tumor-host interaction in the rabbit V2 carcinoma, Invasion Metastasis 4:13-27 (1987)

60. A. Heidland, W. H. Hörl, N. Heller, H. Heine, S. Neumann, and E. Heidbreder, Proteolytic enzymes and catabolism: enhanced release of granulocyte proteinases in uremic intoxication and during hemodialysis, Kidney Int. 24:S-27-S-36 (1983)

61. G. Guarnieri, G. Toigo, R. Situlin, L. Crapesi, M. A. Del Bianco, A. Zanettovich, L. Faccini, A. Lucchesi, L. Oldrizzi, C. Rugiu, and G. Maschio, Nutritional assessment in patients with early renal insufficiency on long-term low protein diet, in: "Contr. to Nephrol., vol. 53", G. M. Berlyne, and S. Giovannetti, eds., Karger, Basel (1986)

62. G. Guarnieri, G. Toigo, R. Situlin, L. Crapesi, M. A. Del Bianco, A. Zanettovich, E. Mandero, and G. Resetta, Muscle biopsy studies on protein-energy malnutrition in patietns with chronic relapsing pancreatitis, Infusionsther. 13:166-171 (1986)

63. J. F. Dice, and C. D. Walker, The general characteristics of intracellular protein degradation in diabetes and starvation, in: "Protein Turnover and Lysosome Function," H. L. Segal, and D. J. Doyle, eds., Academic Press, New York San Francisco London (1978)

64. B. Draznin, and M. Trowbridge, Inhibition of intracellular proteolysis by insulin in isolated rat hepatocytes, J. Biol. Chem. 257:11988--11993 (1982)

65. V. M. Pain, E. C. Albertse, and P. J. Garlick, Protein metabolism in skeletal muscle, diaphragm, and heart of diabetic rats, Am. J. Physiol. 245:604-610 (1983)

66. D. J. Millward, P. C. Bates, J. G. Brown, M. Cox, R. Giugliano, M. Jepson, and J. Pell, Role of thyroid, insulin and corticosteroid hormones in the physiological regulation of proteolysis in muscle, in: "Progress in Clinical and Biological Research, vol. 180: Intracellular Protein Catabolism," E. A. Khairallah, J. S. Bond, and J. W. C. Bird, eds., Alan R. Liss, Inc., New York (1985)

67. P. J. Garlick, V. R. Preedy, and P.J. Reeds, Regulation of protein turnover in vivo by insulin and amino acids, in: "Progress in Clinical and Biological Research, vol. 180: Intracellular Protein Catabolism," E. A. Khairallah, J. S. Bond, and J. W. C. Bird, eds., Alan R. Liss, Inc., New York (1985)

68. R. I. Inculet, R. J. Finley, J. H. Duff, R. Pace, C. Rose, A. C. Groves, and L. I. Woolf, Insulin decreaes muscle protein loss after operative trauma in man, Surgery 99:752-758 (1986)

69. J. M. Fagan, S. Satarug, P. Cook, and M. E. Tischler, Rat muscle protein turnover and redox state in progressive diabetes, Life Sci. 23:783-790 (1987)

70. M. Goodman, Myofibrillar protein breakdown in skeletal muscle is diminished in rats with chronic streptozocin-induced diabetes, Diabetes, 36:100-105 (1987)

71. D. D. Trunkey, Overview of trauma, Surg. Clins. N. 62:3-7 (1982)

72. B. Chua, R. L. Kao, D. E. Rannels, and H.E. Morgan, Inhibition of protein degradation by anoxia and ischemia in perfused rat hearts. J. Biol. Chem. 254:6617-6623 (1979)

73. V. E. Baracos, and A. L. Goldberg, Ca^{2+}, interleukin-1 and failure to maintain normal length stimulate protein degradation in isolated skeletal muscle, in: "Progress in Clinical and Biological Research, vol. 180: Intracellular Protein Catabolism," E. A. Khairallah, J. S. Bond, and J. W. C. Bird, eds., Alan R. Liss, Inc., New York (1985)

74. R. J. Smith, Regulation of protein degradation in differentiated skeletal muscle cells in monolayer culture, in: "Progress in Clinical and Biological Research, vol. 180: Intracellular Protein Catabolism," E. A. Khairallah, J. S. Bond, and J. W. C. Bird, eds., Alan R. Liss, Inc., New York (1985)

75. J. C. Waterlow, P. J. Garlick, and D. J. Millward, Protein turnover in mammalian tissues and in the whole body, Elsevier/North-Holland Biomedical Press, Amsterdam (1978)

76. P.-O. Hasselgren, R. Jagenburg, L. Karlström, P. Pedersen, and T. Seeman, Changes of protein metabolism in liver and skeletal muscle following trauma complicated by sepsis, J. Trauma 24:224-228 (1984)

77. F. J. Ballard, and L. C. Read, Coordinate regulation of protein synthesis and breakdown in cultured cells, in: "Progress in Clinical and Biological Research, vol. 180: Intracellular Protein Catabolism," E. A. Khairallah, J. S. Bond, and J. W. C. Bird, eds., Alan R. Liss, Inc., New York (1985)

HORMONAL REGULATION OF MUSCLE PROTEIN CATABOLISM IN ACUTELY UREMIC RATS: EFFECT OF ADRENALECTOMY AND PARATHYROIDECTOMY

R.M. Schaefer[1], M. Moser[1], P. Kulzer[1], G. Peter[2]
A. Heidland[1], W.H. Hörl[3], and S.G. Massry[4]

Department of [1]Internal Medicine and [2]Dermatology
University of Wuerzburg, FRG
[3]Department of Internal Medicine, University of
Freiburg, FRG
[4]Department of Internal Medicine, University of
Southern California, School of Medicine, Los
Angeles, California

INTRODUCTION

Clinically prominent skeletal muscle dysfunction and wasting are found not only in uremia (1) but also in patients suffering from both primary hyperparathyroidism (2, 3) and glucocorticoid excess (4, 5). Markedly elevated serum levels of parathyroid hormone are frequently observed in uremic patients (6). In addition, Wernze reported a twofold increase of plasma corticosterone 6 and 24 hours after bilateral nephrectomy in the rat (7). This finding corresponded well with former observations by Bondy et al., who were able to demonstrate a decrease of blood urea nitrogen after adrenalectomy in acutely uremic rats (8). From a metabolic point of view, net degradation of skeletal muscle proteins with a concomitant excessive release of amino acids has been reported in uremia (9). Moreover, enhanced muscle protein breakdown was also found to be caused by high levels of parathyroid hormone (10) as well as glucocorticoids (11 - 13).

Since uremia, hyperparathyroidism, and glucocorticoid excess display some common clinical as well as metabolic alterations, and since uremia might be associated with enhanced circulating levels of both parathyroid hormone and glucocorticoids, a relationship between muscle protein breakdown and those hormones seems possible. For these reasons, we have investigated the potential contributions of both parathyroid hormone and glucocorticoids to the enhanced protein catabolism of acutely uremic rats.

METHODS

Male Wistar rats (Ivanovas, Kisslegg/Allgäu, FRG) weighing 200 - 250 g were used for the experiments. The animals were kept at constant humidity and temperature with a controlled 12 hour light-on and light-off cycle. The rats had free access to tap water and standard rat chow (Altromin, Lippe/ Westfalen, FRG) ad libitum. After an adaption period of 5 days, the animals underwent sham operation (SHAM), bilateral nephrectomy (BNX), or bilateral nephrectomy and adrenalectomy (BNX/ADX). A further group of rats underwent parathyroidectomy (PTX). Parathyroidectomy was performed by electrocautery after exposure of the parathyroid glands through the strap muscles. Adequacy of parathyroidectomy was confirmed by demonstrating a serum calcium level of less than 2.0 mmol/l 72 hours following surgery. Then, these animals received water, containing 8 g of calcium-gluconate per liter, to maintain their serum calcium levels in the normal range. 10 days after PTX, these animals underwent bilateral nephrectomy. For all surgical procedures, rats were anaesthetized with hexobarbital (50 mg/kg b. w.) intraperitoneally. SHAM rats were deprived of food, but were allowed to drink ad libitum. Uremic animals had neither access to food nor water. Forty-eight hours after sham operation or induction of uremia, blood was obtained from the abdominal aorta under hexobarbital anaesthesia and the gastrocnemius muscle was excised.

Glucose, electrolytes, urea nitrogen and creatinine were assayed using a Technikon Autoanalyzer. Blood pH and bicarbonate were determined with an AVL-Gas check analyzer (AVL-GmbH, Bad Homburg, FRG). Urea-nitrogen appearance, an indicator of net protein breakdown, was determined during the 48 hour period following surgery according to Flügel-Link et al. (14). For determination of N^t-methylhistidine, plasma samples were deproteinized with 10 % TCA, centrifuged and washed 4 times with diethylether. Amino acid assays were performed on an amino acid analyzer LC 4010 (Biotronik, Maintal, FRG). The column (DC 6A, Biotronik) had a length of 200 mm and a lithium buffer system was used. The column temperature was maintained at 60°C for separation of the basic amino acids.

The determination of the alkaline myofibrillar proteinase in homogenates from skeletal muscle was performed according to Mayer et al. (15). In brief, gastrocnemius muscle of one leg was homogenized at 4°C in 5 v/w of 0.01 M potassium phosphate (pH 7.7) containing 0.05 M KCl, with an Ultra-Turrax^R blender (Janke & Kunkel, Staufen, FRG). The homogenate was passed through two layers of cheese cloth and centrifuged at 5,000 g for 10 min at 4°C. The sediment was washed twice and finally resuspended in the homogenization buffer. This suspension was used as the alkaline proteinase preparation. To measure its proteolytic activity, this fraction was incubated in 0.1 M glycine-NaOH buffer (pH 9.1) containing 1.2 mM KCl for 60 min at 37°C. The amount of tyrosine equivalents, released into the supernatant, was taken as a measure of the proteolytic activity. Proteolysis was stopped by addition of 2.5 ml of 5 % (w/v) TCA. After centrifugation at 1,000 g for 10 min, tyrosine equivalents were determined in the supernatant by the method of Lowry et al. (16).

All data are given as mean + SEM. Statistical assessments were made using Student's t-test or the analysis of variance (ANOVA) as appropriate.

RESULTS

Forthy-eight hours after bilateral nephrectomy the animals displayed a considerable state of azotemia with high serum levels of urea-N, creatinine, potassium and phosphate. Glucose was unchanged and serum calcium was decreased. There was frank metabolic acidosis with low serum bicarbonate levels (Table 1). In animals, which underwent both nephrectomy and removal of their adrenal glands, an identical degree of renal insufficiency was observed as compared with those who were only nephrectomized. Thus, serum creatinine levels were comparable in both groups of animals. As to urea-N, BNX/ADX-animals had significantly lower serum levels in comparison to BNX-rats. Moreover, BNX/ADX rats developed marked hypoglycemia (33 + 4.0 mg/dl). The metabolic acidosis was more pronounced as can be seen by the lower serum bicarbonate levels (BNX/ADX: 10.1 + 0.6 vs. BNX: 14.0 + 0.5 mmol/l) and marked hyperkalemia was found in these animals (BNX/ADX: 8.2 + 0.2 vs. BNX: 7.3 + 0.2 mmol/l). Parathyroidectomy in combination with calcium supplementation prior to the induction of uremia resulted in an identical pattern of metabolic alterations, as compared with animals that underwent only bilateral nephrectomy. Only serum calcium levels tended to be lower in PTX/BNX animals (PTX/BNX: 1.7 + 0.05 vs. BNX: 1.8 + 0.05 mmol/l).

When urea-N appearance, as a measure for ureogenesis, was calculated following picture emerged. The urea-N appearance was markedly increased in rats 48 h after bilateral nephrectomy (370 + 26 mg/48 h) as compared to SHAM rats (220 + 15 mg/48 h). With the removal of the adrenal glands this increment of urea-N appearance in the uremic state was almost completely reversed (238 + 20 mg/48 h). Parathyroidectomy in combination with calcium supplementation prior to removal of the kidneys caused no reduction in the rate of urea-N production (Fig. 1).

In order to investigate as to wether these changes in urea generation were reflected in similar alterations of skeletal muscle protein degradation, we followed plasma N^t-methylhistidine as an parameter of myofibrillar protein breakdown (Table 2). Compared to animals which underwent only bilateral nephrectomy, BNX/ADX rats displayed significantly lower plasma values of N^t-methylhistidine. On the other hand parathyroidectomy prior to induction of uremia caused no reduction in the elevated plasma levels of this amino acid (Table 2).

Table 1. Blood chemistries of rats 48 h following sham operation (SHAM), bilateral nephrectomy (BNX) alone, or in combination with adrenalectomy (BNX/ADX) or parathyroidectomy (PTX/BNX)

	SHAM	BNX	BNX/ADX	PTX/BNX
Urea-N mg/dl	29.0 $\pm$ 1.0	253 $\pm$ 6.5[a]	176 $\pm$ 6.0[b]	255 $\pm$ 6.4
Creatinine mg/dl	0.5 $\pm$ 0.01	6.5 $\pm$ 0.3[a]	6.1 $\pm$ 0.3	6.7 $\pm$ 0.4
Glucose mg/dl	122 $\pm$ 10	127 $\pm$ 5	33 $\pm$ 4.0[b]	129 $\pm$ 7
Potassium mmol/l	4.0 $\pm$ 0.1	7.3 $\pm$ 0.2[a]	8.2 $\pm$ 0.2[b]	7.5 $\pm$ 0.5
Calcium mmol/l	2.4 $\pm$ 0.03	1.8 $\pm$ 0.05[a]	2.0 $\pm$ 0.05	1.7 $\pm$ 0.05
Phosphate mmol/l	2.4 $\pm$ 0.1	8.4 $\pm$ 0.3[a]	6.5 $\pm$ 0.3[b]	8.5 $\pm$ 0.6
Bicarbonate mmol/1	24.8 $\pm$ 0.5	14.0 $\pm$ 0.5[a]	10.1 $\pm$ 0.6[b]	14.7 $\pm$ 0.8

Data are presented as the mean $\pm$ SE, obtained from 8 animals in each group.

[a] $p < 0.01$ for SHAM vs. BNX;

[b] $p < 0.01$ for BNX vs. BNX/ADX;

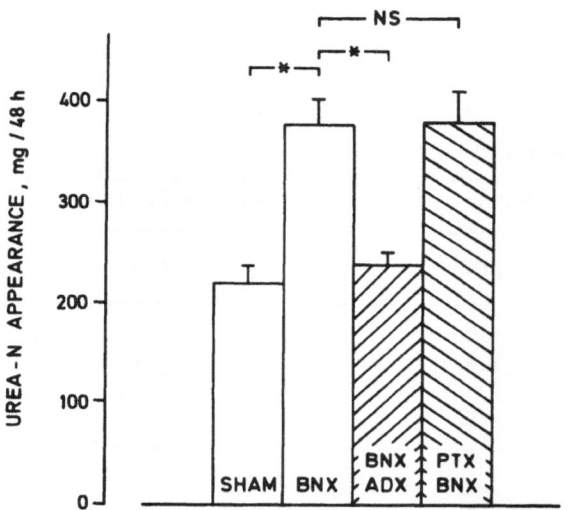

Fig. 1. Effect of sham operation (SHAM), bilateral nephrectomy (BNX) alone or in combination with adrenalectomy (BNX/ADX) or parathyroidectomy (PTX/BNX) on urea-N appearance. Data are given as mean ± SE obtained from 8 animals in each group.

*p < 0.01 and NS denotes not significant.

Finally, the activity of the myofibrillar proteinase (pH 9.1) in skeletal muscle homogenates was determined. As can be seen in Fig. 2, uremia caused a considerable increase in the activity of this proteinase (125 + 9 μg tyr. eq./mg prot./60 min) as compared to SHAM rats (96 + 4 μg tyr. eq./mg prot./60 min). This increment, which appears to be due to uremia, was totally reversed by the removal of the adrenals at the time of nephrectomy, resulting in a normalization of this proteolytic activity (BNX/ADX: 94 + 6 μg tyr. eq./mg prot./60 min). Parathyroidectomy prior to the induction of uremia had no demonstrable effect on the enhanced activity of the myofibrillar proteinase in the uremic rat (127 + 6 μg tyr. eq./mg prot./60 min).

Table 2. Plasma N^t-methylhistidine levels in bilaterally nephrectomized (BNX) rats alone or in combination with adrenalectomy (BNX/ADX) or parathyroidectomy (PTX/BNX)

	BNX	BNX/ADX	PTX/BNX
N^t-methyl-histidine μg/ml	9.6 + 0.4	6.0 + 0.3*	9.9 + 0.1

Data are presented as the mean + SE, obtained from 8 animals in each group.
*p <0.01 for BNX vs. BNX/ADX.

DISCUSSION

Acute uremia in rats, achieved by bilateral nephrectomy, displayed metabolic disturbances which were comparable to those observed in patients suffering from acute renal failure. Acute uremia resulted in a clearcut increase of net protein catabolism as indicated by an enhanced urea generation (Fig. 1). That this increased rate of ureogenesis was, at least partly, due to an increment in myofibrillar protein breakdown is suggested by the enhanced activity of the myofibrillar proteinase (pH 9.1) of skeletal muscle homogenates (Fig. 2). In order to investigate the role of glucocorticoids in acute uremia one group of rats underwent bilateral nephrectomy and adrenalectomy simultaneously. In those animals

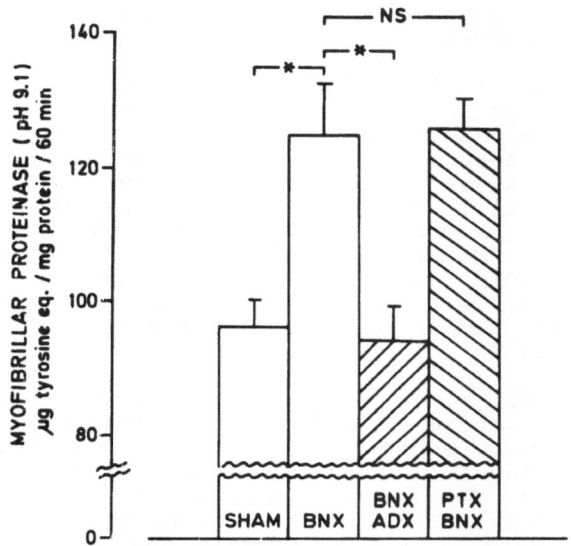

Fig. 2. Effect of sham operation (SHAM), bilateral
nephrectomy (BNX) alone or in combination with adrenalectomy
(BNX/ADX) or parathyroidectomy (PTX/BNX) on the activity
of the myofibrillar proteinase of skeletal muscle.
Data are given as mean ± SE obtained from 8 animals in each
group.
*p <0.01 and NS denotes not significant.

net urea-N appearance was considerably reduced, compared to uremic rats with intact adrenals, and almost reached the level of SHAM rats (Fig. 1). That this was due to a reduced rate of muscle protein degradation, was indicated by a decrease in plasma N^t-methylhistidine levels (Table 2) and a reduction of the activity of the myofibrillar proteinase (Fig. 2). These findings correspond well with those obtained by Bondy and coworkers (8). These authors also described reduced serum urea nitrogen levels in acutely uremic rats following adrenalectomy. Moreover, recent data on the effect of the antiglucocorticoid RU 38486, also indicated that glucocorticoids contribute considerably to the enhanced catabolism in acutely uremic rats (17). It is of note that the removal of the adrenal glands caused marked hypoglycemia in the uremic animals (Table 1), suggesting that glucocorticoids play a key role in maintaining blood glucose levels by an enhanced release of amino acids from skeletal muscle in acute uremia. These observations are comparable to those made by Fröhlich and coworkers (18) who demonstrated that acute uremia exerted a stimulating effect on gluconeogenesis of the isolated perfused rat liver. Employing the same model, Lacy (19) was able to show that amino acid uptake and urea formation by the liver was markedly enhanced in acutely uremic rats. The hypoglycemia was more pronounced in uremic rats following adrenalectomy (Table 1) as compared to those uremic rats being treated with the antiglucocorticoid (17), suggesting that by ablation of the adrenal glands a considerable source of catecholamines is removed and that a lack of these hormones also contributes to hypoglycemia in acutely uremic animals.

In a further set of experiments the effect of parathyroidectomy on the enhanced protein catabolism was investigated. As can be seen in Table 1 and 2 blood chemistries, including serum urea-N levels and plasma values of N^t-methylhistidine were identical in acutely uremic rats with or without parathyroidectomy. Thus, urea-N appearance (Fig. 1) was the same in both groups of animals and the activity of the myofibrillar proteinase was also unchanged in parathyroidectomized uremic rats (Fig. 2). These findings are in contrast to those reported by Garber (10), who found that high doses of PTH in vitro induced an excessive release of amino acids from normal muscle, whereas muscle from chronically uremic rats was unresponsive to this hormone. For these reasons, it was suggested that PTH may contribute to the enhanced muscle protein breakdown in uremia. However, both studies are not readily comparable, owing to different types of uremia and different study designs (one being an in vitro experiment, whereas our study was performed in intact animals).

Taken as a whole, our findings evidence a clearcut role for glucocorticoids in the enhanced protein catabolism of acutely uremic rats, whereas PTH seems not to contribute to this metabolic alteration.

SUMMARY

To examine the role of glucocorticoids and PTH on the enhanced protein catabolism of acute uremia, rats were rendered uremic and had their adrenals or their parathyroid glands concomitantly removed. Adrenalectomy resulted in a marked reduction of urea generation in uremic animals due to decrease of myofibrillar protein breakdown as indicated by lower serum levels of N^t-methylhistidine and a reduction in the activity of the myofibrillar proteinase from skeletal muscle. This reduced urea formation was accompanied by marked hypoglycemia. Parathyroidectomy, on the other hand, caused no change of those paramters of protein catabolism, suggesting that PTH does not account for the protein degradation observed in acutely uremic rats.

ACKNOWLEDGEMENTS

We are indebted to A. Goj and M. Roeder for their excellent technical asistance. The secretarial help of E. Hammer and Ch. Winter is greatly appreciated.

REFERENCES

1. M. Floyd, D. Ayyar, D. Barwick, P. Hudgson and D. Weightman, Myopathy in chronic renal failure, Q. J. Med. 53: 509 (1974)
2. R. Smith and G. Stern, Myopathy, osteomalacia and hyperparathyroidsm, Brain 90: 593 (1967)
3. B. Frame, E. Heinze, M. Block and G. Manson, Myopathy in primary hyperparathyroidsm. Observations in three patients, Ann. Intern. Med. 68: 1022 (1968)
4. R. Müller and E. Kugelberg, Myopathy in Cushing's syndrome, J. Neurol. Neurosurg. Psychiat. 22: 314 (1959)
5. G.T. Perkoff, R. Silber, F. Tyler, G. Cartwright and M. Wintobe, Studies in disorders of muscle: XII. Myopathy due to the administration of therapeutic amounts of 17-hydroxycorticosteroids, Am. J. Med. 26: 891 (1959)
6. E. Slatopolsky, K. Martin and K. Huska, Parathyroid hormone metabolism and its potential as a uremic toxin, Am. J. Physiol. 239: Fl (1980)
7. H. Wernze, Changes in plasma renin substrate in physiological and pathophysiological states, In: Krause, Hummerich, Poulsen, Radioimmunoassay renin-angiotensin, pp 91 (Thieme Stuttgart, 1978)
8. P. Bondy, F. Engel, B. Farrar, The metabolism of amino acids and protein in the adrenalectomized nephrectomized rat, Endocrinology 44: 476 (1949)
9. A. Garber, Skeletal muscle protein and amino acid metabolism in experimental chronic uremia in the rat, J. Clin. Invest. 62: 623 (1978)
10. A. Garber, Effects of parathyroid hormone on skeletal muscle protein and amino acid metabolism in the rat. J. Clin. Invest. 71: 1806 (1983)
11. C. Long, B. Katzin and E. Fry, The adrenal cortex and carbohydrate metabolism, Endocrinology 26: 309 (1940)
12. F. Tomas, H. Munro and V. Young, Effect of glucocorticoid administration on the rate of muscle protein breakdown in vivo in rats, as measured by urinary excretion of N^t-methylhistidine, Biochem. J. 178 (1979)
13. S. Rannels and L. Jefferson, Effects of glucocorticoids on muscle protein turnover in perfused rat hemicorpus. Am. J. Physiol. 238: E 564 (1980)

14. R. Flügel-Link, J. Salusky, M. Jones and J. Kopple, Protein and amino acid metabolism in posterior hemicorpus of acutely uremic rats, <u>Am. J. Physiol.</u> 244: E 615 (1983)

15. M. Mayer, R. Amin and E. Shafrir, Rat myofibrillar protease: enzyme properties and adaptive changes in conditions of muscle protein degradation, <u>Archs. Biochem. Biophys.</u> 161: 20 (1974)

16. O. Lowry, N. Rosebrough, A. Farr and R. Randall, Protein measurement with the Folin phenol reagent, <u>J. Biol. Chem.</u> 193: 265 (1951)

17. R. Schaefer, M. Teschner, P. Kulzer, J. Leipold, G. Peter and A. Heidland, Evidence for reduced catabolism by the antiglucocorticoid RU 38486 in acutely uremic rats. <u>Am. J. Nephrol.</u> 7: 127 (1987)

18. J. Fröhlich, J. Schölmerich, G. Hoppe-Seyler, K. Maier, H. Talke, O. Schollmeyer and W. Gerok, The effect of acute uremia on gluconeogenesis in isolated perfused rat livers, <u>Eur. J. Clin. Invest.</u> 4: 453 (1974)

19. W. Lacy, Effect of acute uremia on amino acid uptake and urea production by perfused rat liver, <u>Am. J. Physiol.</u> 216: 1300 (1969)

RELATION BETWEEN URINARY PROTEINASES AND PROTEINURIA IN RATS WITH A

GLOMERULAR DISEASE

Jean-Claude Davin, Malcolm Davies, Jean-Michel Foidart,
Jacqueline B. Foidart, Charles A. Dechenne, and Philippe
R. Mahieu

Department of Medicine, Pediatrics and Gynecology,
State University of Liege, Liege, Belgium, and
Kidney Research Unit Foundation for Wales, Welsh
National Scholl of Medicine, Royal Infirmary,
Cardiff, Wales, United Kingdom

INTRODUCTION

Electronegative charges of the glomerular filtration barrier play
probably an important role in its impermeability to anionic proteins [1-9].
Sialic acid [10] (present in laminin) and glycosaminoglycans [11] seem to be the
major molecules responsible for this propriety. Neutral proteinases synthe-
tized either by mesangial cells or by monocytes and neutrophils, infiltrating glomeruli in some pathologic states, are able to degrade GBM glycoproteins in vitro [12-14].

In the present study, we have tried to provide a support to the
concept that neutral proteinases may also play a role in the damage to
the GBM in vivo. In this purpose, we have measured the urinary neutral
proteinase activity and the urinary excretion of GBM components, that is,
laminin and type IV collagen, in normal rats and in proteinuric rats with
either a non-proliferative glomerular disease or a GN associated with an
infiltration of glomeruli by mononuclear phagocytes.

METHODS

The experimental models of glomerular diseases

The accelerated model of nephrotoxic nephritis was induced in Sprague
Dawley rats (group A) as previously described [15]. Briefly, female rats
weighing 100–120 g were pre-immunized with 1 mg of rabbit IgG in 0,5 ml
of complete Freund's adjuvant injected intraperitoneally. One week later,
the rats received 1 mg of rabbit anti-GBM IgG intravenously. A nephrotic
syndrome was induced [16] in another group of Sprague Dawley rats (group B)
weighing 100–120 g by the administration of one single intraperitoneal

injection of 150 mg per kg of aminonucleoside of puromycin (Sigma Chemical Co, St Louis, Mo, USA). Animals were sacrificed three to 24 days after administration of aminonucleoside or of anti-GBM immunoglobulin.

Assessment of urinary neutral proteinase activity

The methods decribed here were detailed extensively elsewhere [17]. Briefly, 24 hour urine collections were dialyzed at 4°C against distilled water, were sterilized by filtration through Millipore filters (Millipore Benelux, Brussels) and were then stored at -20°C until use. Neutral proteolytic activity against azocasein (Azocoll, Calbiochem-Behring Corp., La Jolla, California, USA)or $|^3H|$-labelled acetyl-casein was determined according to the method described by Starkey[18].
In order to characterize the enzymes responsible for the degradation of substrates, the following inhibitors of proteinases were tested : EDTA 2 mM or 15 mM, Cystein 2 mM, Soybean trypsin inhibitor (SBTI), 100 and 500 μg/ml, aprotinin, 500 μg and 1 mg/ml, and phenylmethylsulfonylfluoride (PMS-F) 1 mM.
The effect of neutral proteinases on BM components was established by incubation of dialyzed urine samples (0.2 ml) with $|^{14}C|$ - labelled type IV collagen or $|^3H|$ - laminin[17]. In the latter experiments, enzyme inhibition profiles were also obtained[17].
The apparent molecular weight of neutral proteinases was determined by chromatography on a Sephacryl S-200 SF column[17] (Pharmacia, Uppsala, Sweden) whereas isoelectric points were assessed by chromatofocusing[17].

Assessment of proteinuria and of urinary laminin or type IV collagen excretion

The protein concentration of 24 hour urines was measured by the method of Kingsbury and Clarck[19]. The urinary excretion of laminin and of type IV collagen was determined by a solid phase radioimmunoassay[17].

Morphological studies of the kidney sections

The kidneys were examined by light and immunofluorescence microscopy at various periods of time according to classical technics[17]. The kidney sections were also stained with colloïdal iron in order to study the distribution of glomerular anionic charges.

RESULTS

Morphological aspects

In A rats, microscopic examination of the kidneys revealed a glomerular hypercellularity resulting from an infiltration by mononuclear cells and a proliferation of intrinsic glomerular cells[17]. In B rats, no glomerular hypercellularity was observed ; the only abnormalities noticed in that model were multiple small eosinophilic droplets in glomerular epithelial cells and eosinophilic hyaline casts in some Bowman's spaces and tubular lumens[17]. In each rat with a proteinuria, a decrease and alteration in the staining intensity for colloïdal iron was apparent when compared with kidney sections from normal rats.

Proteinuria

In A rats, a proteinuria started on day two after the injection of anti-GBM IgG and reached mean maximum values on day four (240 mg/24h). The proteinuria decreased thereafter to reach about 50 mg/24h three weeks later. The proteinuria of B rats appeared six days after the administration of PAN,

Table 1 : Urinary excretion of neutral proteinase activity, laminin and type IV collagen
in NTN and PAN nephrosis rats.

Rats	Urinary neutral proteinase activity (U/24h ; mean +/- SEM)		Urinary laminin (nannograms/24h ; mean +/- SEM)		Urinary type IV collagen (nannograms/24h ; mean +/- SEM)	
Normal controls	4.69 +/- 0.60	(N = 9)	1,154 +/- 325	(N = 10)	306 +/- 36.5	(N = 8)
NTN *	38.55 +/- 8.66	(N = 11)	12,371 +/- 2,205	(N = 11)	1,495 +/- 370	(N = 11)
PAN Nephrosis **	42.17 +/- 7.92	(N = 9)	15,878 +/- 3,038	(N = 9)	2,011 +/- 288	(N = 9)

* NTN rats were tested four days after the injection of anti-GBM IgG. The results obtained were significantly
higher than in normal rats for each parameter considered (P < 0.001).

** PAN nephrosis rats were tested twelve days after administration of PAN. The results obtained were significantly
higher than in normal rats for each parameter considered (P < 0.001).

increased to a mean maximum value of 180 mg per 24 h on day 12 and declined
to reach near normal values by 22 to 24 days.

Urinary excretion of neutral proteinase activity, laminin and type IV collagen

The proteinase activity against azocasein and $|^3H|$- labelled acetyl-
casein was present in urines of normal rats as well as of A and B rats.

This activity was destroyed by heating at 60°C for 30 minutes and was
maximal in the neutral range of Ph. The values obtained in A and B rats were
however significantly higher than those of normal rats (Table 1). Moreover,
those values were parallel to the proteinuria and were maximal on day four
for A rats and on day 12 for B rats (data not shown).
For the urinary excretion of laminin or of type IV collagen, comparable data
were obtained (Table 1).
In both groups of nephrotic rats, significant direct linear correlations
were demonstrated between the following parameters : 1) the proteinuria and
the urinary proteinase activity (A rats : r = 0.83, P< 0.001 ; B rats :
r = 0.89, P<0.001) ; 2) the proteinuria and the urinary excretion of
laminin (A rats : r = 0.509, P < 0.001 ; B rats : r = 0.876, P< 0.001) ;
3) the urinary neutral proteinase activity and the urinary excretion of
either laminin (A rats : r = 0.86, P < 0.001 ; B rats : r = 0.84, P< 0.01)
or type IV collagen (A rats : r = 0.83, P < 0.001 ; B rats : r = 0.76,
P < 0.01).

Characterization of urinary neutral proteinases (data not shown)

In normal, A and B rats, an inhibition of 50% of the enzymatic activity
was determined by the serine-proteinase inhibitors (PMS-F, SBTI and aprotinin). On the contrary, the metallo-proteinase inhibitors (EDTA and Cysteine)
inhibited only the urinary proteinase activity of A and B rats. Gel chromatography of normal or nephrotic urine showed an active neutral proteinase
activity peak with an apparent molecular weight of 30,000 daltons. The isoelectric point of active enzymes determined by chromatofocusing was about
9 in normal as well in nephrotic rats. $|^{14}C|$-labelled type IV collagen and
$|^3H|$laminin were degraded by incubation in vitro with urines of normal, A
and B rats. However, the percentage of degradation was three times higher
with the urines of nephrotic rats than with those of normal rats. The inhibitory experiments showed comparable results than above.

DISCUSSION

The functionnal role of the anionic sites in the impermeability of the
glomerular filtration barrier has been demonstrated several years ago [1,8].
The loss of these sites has been demonstrated in experimental models of
proteinuria [5,9] as well as in human nephrotic syndromes [20,21]. Sialic acid
present at the surface of endothelial and epithelial cells and in GBM glycoproteins (laminin, for example) is one of the components of the glomerular
filtration barrier, responsible for its electronegativity.

In the present study, we have shown in nephrotic rats that : 1) the
urinary laminin is increased ; 2) this increase is temporally related with
the loss of the glomerular colloïdal iron staining ; and 3) the urinary
excretion of laminin runs parallel with that of type IV collagen. Those
results may therefore suggest that laminin molecules are removed from the
glomerular filtration barrier. However, one cannot exclude that some laminin
and type IV collagen antigens found in urine might come from other basement
membranes than the GBM since the plasma levels of laminin and type IV
collagen were not measured in the present work.

Neutral proteinases generated by mesangial cells, monocytes and neutro-phils may be implicated in the degradation of GBM glycoproteins in vitro 12, 13, 22. We have shown in both experimental models of proteinuric rats studied that : 1) the urinary excretion of proteinases is increased ; 2) those proteinases are able of degrading laminin and type IV collagen in vitro ; 3) the urinary excretion of those enzymes is significantly correlated with the proteinuria and with the urinary excretion of laminin and type IV collagen ; 4) urine of proteinuric rats is able to degrade to a greater extent than urine of normal rats, labelled laminin and type IV collagen in vitro ; 5) proteinases found in urine from normal or nephrotic rats are neutral proteinases since their activity is maximum at pH 7.4 ; 6) in normal rats, those enzymatic activities mainly result from serine-type proteinases (inhibition by PMS-F, SBTI and aprotinin) whereas in proteinuric rats, they also result from metallo-proteinases (inhibition by EDTA and Cysteine).

Although the association between the proteinuria and the increased urinary proteinases excretion may simply be a coincidence, our date support the concept that neutral proteinases may play a role in the damage of the GBM also in vivo.

The origin of those urinary neutral proteinases is unknown. However, the following possible origins of urinary proteinases have to be considered: the plasma, the phagocytic cells infiltrating the glomeruli and the intrinsic kidney cells.

An eventual role of plasma proteinases cannot be evaluated from this study since the latters were not measured. However, is must be recalled that it is well-known that the plasma contains serine-type proteinases and that some authors have recently demonstrated the presence of a plasma metallo-proteinase in patients with acute and chronic renal failure[23]. Those enzymes might theoretically degrade the GBM in vivo despite the fact that their action could be counteracted by circulating plasma inhibitors [14].

Phagocytic cells may be implicated in A rats but not in B rats. Indeed, in the latter rats, no phagocytic cells are demonstrated in glomeruli [16]. On the contrary, in A rats, mononuclear phagocytes infiltrate the glomeruli two to five days after the injection of anti-GBM IgG and are in close contact with the GBM [24]. However, the latter cells cannot be responsible for the presence of urinary metallo-proteinases since they only secrete serine-type proteinases [14] in vitro.

The last possible origin of those proteinases is intrinsic kidney cells. Indeed, lysosomes are increased in glomerular and proximal tubular cells during the aminonucleoside nephrosis [16] and the mesangial cells can produce metalloproteinases capable of degrading purified rat GBM, in vitro, at physiological pH [12].

Our findings have therefore shown that the proteinuria, the urinary excretion of basement membrane components and the urinary excretion of neutral proteinases are significantly correlated in both experimental models of proteinuria studied. It is thus suggested that metalloproteinases and/or serine-type proteinases from different sources may play a role in the development of proteinuria by their action on the glomerular filtration barrier.

Acknowledgemts

We are indebted to Ms M-A Delavignette for her secretarial assistance and to Mrs Y. Pirard and A. Desoroux for their technical assistance.

REFERENCES

1. RENNKE HG, COTRAN RS, VENKATACHALAM MA, Role of molecular charge in glomerular permselectivity : Tracer studies with cationized ferritins, J Cell Biol 67 : 638 (1975).
2. RENNKE HG, VENKATACHALAM MA, Glomerular permeability : In vivo tracer studies with polyanionic and polycationic ferritins, Kidney Int 11 : 44 (1977).
3. RENNKE HG, VENKATACHALAM MA, Structural determinants of glomerular permselectivity, Fed Proc 36 :2619 (1977).
4. KANWAR YS, FARQUHAR MG, Anionic sites in the glomerular basement membrane : In vivo and in vitro localization in the laminae rarae by cationic probes, J Cell Biol 81 : 137 (1979).
5. MICHAEL AF, BLAU E, VERNIER RL, Glomerular polyanion alteration in amionucleoside nephrosis, Lab Invest 23 : 649 (1970).
6. BOHRER MP, BAYLIS C, ROBERTSON CR, BRENNER BM, Mechanisms of puromycin-induced defects in the transglomerular passage of water and macromolecules, J Clin Invest 60 : 152 (1977).
7. CHIN J, SMITH M, DRUMMOND KN, Proteinuria and glomerular sialprotein (GSP) in experimental renal disease, (abstract), Clin Res 21 : 681, (1973).
8. CHANG RLS, DEEN WM, ROBERTSON CR, BRENNER BM, Permselectivity of the glomerular capillary wall : III. Restricted transport of polyanions, Kidney Int 8 : 212 (1975).
9. BENNETT CM, GLASSOCK RJ, CHANG RLS, DEAN WM, ROBERTSON CR, BRENNER BM, Permselectivity of the glomerular capillary wall : Studies of experimental glomerulonephritis in the rat using dextran sulfate, J Clin Invest 57 : 1287 (1976).
10. KANWAR YS, FARQUHAR MG, Detachment of endothelium and epithelium from the glomerular basement membrane produced by perfusion with neuraminidase, Lab Invest 42 : 375 (1980).
11. KANWAR YS, LINKER A, FARQUHAR MG, Increased permeability of the glomerular basement membrane to ferritin after removal of glycosaminoglycans (heparan sulfate) by enzyme digestion, J Cell Biol 86 : 688 (1980).
12. LOVETT DH, STERZEL B, KASHGARIAN M, RYAN JL, Neutral proteinase activity produced in vitro by cells of the glomerular mesangium, Kidney Int 23 : 342 (1983).
13. DAVIES M, BARRET AJ, TRAVIS J, SANDERS E, COLES GA, The degradation of human glomerular basement membrane with purified lysosomal proteinases : Evidence for the pathogenic role of the polymorphonuclear leucocytes in glomerulonephritis, Clin Sci Mol Med 54 : 233 (1978).
14. DAVIES M, COLES GA, HUGHES KT, Glomerular basement membrane injury by neutrophil and monocyte neutral proteinases, Renal Pysiol 3 : 106 (1980).
15. DUBOIS CH, FOIDART JB, HAUTIER MB, DECHENNE CA, LEMAIRE MJ, MAHIEU PR, Proliferative glomerulonephritis in rats : Evidence that mononuclear phagocytes infiltrating the glomeruli stimulate the proliferation of endothelial and mesangial cells, Eur J Clin Invest 11 : 91 (1981).
16. RYAN BG, KARNOVSKY MJ, An ultrastructural study of the mechanism of proteinuria in aminonucleoside nephrosis, Kidney Int 8:219 (1975)
17. DAVIN JC, DAVIES M, FOIDART JM, FOIDART JB, DECHENNE CA, MAHIEU PR, Urinary excretion of neutral proteinases in nephrotic rats with a glomerular disease, Kidney Int 31 : 32 (1987).
18. STARKEY PM, Proteinases in mammalian cells and tissues, edited by BARRET AJ North Holland, Amsterdam, p.57 (1977).
19. KINGSBURY FB, CLARCK CP, The rapid determination of albumin in urine, J Lab Clin Med 11 : 981 (1926).

20. BRIDGES CR, MYERS BD, BRENTER BM, DEEN WM, Glomerular charge alterations in human minimal change nephropathy, Kidney Int 22 : 677 (1982).
21. VERNIER RL, KLEIN DJ, SISSON PS, MAHAN JD, OEGEMA TR, BROWN DM, Heparan sulfate-rich anionic sites in the human glomerular basement membrane : decreased concentration in congenital nephrotic syndrome, N Engl J Med 309 : 1001 (1983).
22. SANDERS E, COLES GA, DAVIES M, Lysosomal enzymes in human urine : Evidence for polymorphonuclear leucocyte proteinase involvement in the pathogenesis of human glomerulonephritis, Clin Sc Mol Med 54 : 667 (1978).
23. HÖRL WH, WANNER C, THAISS F, SCHOLLMEYER P, Detection of a metallo-proteinase in patients with acute and chronic renal failure, Am J Nephrol 6 : 6 (1986).
24. SCHREINER GF, COTRAN RS, PARDO V, UNANUE ER, A mononuclear cell component in experimental immunological glomerulonephritis, J Exp Med 147 : 369 (1978).

CHARACTERIZATION AND CLINICAL ROLE OF GLOMERULAR AND TUBULAR PROTEASES FROM HUMAN KIDNEY

Jürgen E. Scherberich, Gunter Wolf, Claudia Stuckhardt
Peter Kugler * and Wilhelm Schoeppe

Dept.Nephrology, University Hospital, Frankfurt am Main
Dept. Anatomy, University Würzburg*

INTRODUCTION

As shown by histochemical studies glomeruli as well as the tubule apparatus exhibit characteristic distribution patterns of proteases (Kugler 1982, 1985, Lojda et al. 1979). However, these data were predominantly obtained through investigating tissue sections from mice, rats or rabbits. Up to the present only few studies were performed on material from normal human kidney (Kugler et al.1985, Scherberich et al.1984). In addition, little is known about the physiological role of renal proteases, although some of them might be involved in cleavage of peptides and proteins, and subsequent tubular reabsorption of break down products (Baumann 1981, Kinne 1985). Recent studies presented evidence that a glomerular aminopeptidase (aminopeptidase A) of human kidney acts as an angiotensinase (angiotensinase A) possibly involved in maintaining the intrarenal balance of renin-angiotensin system (Kugler et al. 1985, Scherberich et al. 1986, Wolf et al. 1986, 1987). In the present paper we report on further histo/biochemical properties of proteases from human kidney and emphasize will be placed on their possible role under normal and pathologcal conditions (e.g. Wanner et al. 1986).

MATERIAL AND METHODS

Human kidneys were obtained within two hours after accidental causes, perfused with Ringer solution, and further used for histological and biochemical investigations: Plasma-membranes were prepared by differential and sucrose density gradient centrifugation as reported earlier (Scherberich et al. 1984). For histochemical studies small tissue blocks were snap frozen in isopenthane cooled by liquid nitrogen and 7 u cryosections were prepared (Dittes-Duspiva system). Staining of tissue sections for glomerular and tubular peptidases was performed using Fast Blue B as a coupling salt and a panel of the following peptide-4-methoxy-2-naphthylamide-(MNA)-substrates (Gossrau 1981, Lojda et al. 1979, Kugler et al. 1985)): Aminopeptidase M (APM): Ala-MNA, aminopeptidase A (APA): α-Glu-MNA, gamma-glutamyltranspeptidase (GGT): γ-Glu-MNA, Dipeptidylaminopeptidase I (DAP I): Gly-Arg-

275

MNA, Dipeptidylaminopeptidase IV (DAP IV): Gly-Pro-MNA.
APA inhibition studies were performed applying Angiotensin I, II, and III in different concentrations (tissue sections. determination of Km in total homogenates, Wolf et al. 1987). APA actitivity in urine specimens of healthy persons and patients with kidney diseases was estimated as reported previously (Scherberich et al. 1986).

RESULTS AND DISCUSSION

Studies on the distribution patterns of renal proteases disclosed the following results:

Aminopeptidase M : In the fetal kidney APM activity was restricted to few developing tubular segments (fig. 1A). The distribution of APM- positive tubules was nearly identical with that of epithelia reacting with an antibody raised against brush border membranes of the normal adult human kidney (Scherberich 1984). Cryosections of normal adult human kidney revealed strong APM activity of the brush border region of the proximal tubule (fig.1 B), whilst distal tubules as well as glomeruli were completely negative. In kidney biopsies (cases with chronic GN and interstitial involvement; n =12) irregular distribution of APM-immunoreactive tubules indicated a different capability of epithelia to synthesize the enzyme(-protein) under pathological conditions (fig.1 C). In addition shedding of brush border membranes into the tubular lumen might result in APM- positive obstructing blebs and cast formation (Scherberich 1984).
APM was identified to be partially integrated within a multienzyme complex of the brush border membrane surface. After limiting proteolytic digestion of membranes with papain increase of APM activity into the supernatant paralleled cleavage of 5 nm globular particles, which were characterized by high affinity towards immobilized Con A and WGA respectively (Scherberich et al. 1984).

Gammaglutamyl transpeptidase activity in the human kidney was similarly distributed as described for APM. Analyzed in detail previously, also contraluminal plasma-membranes of the proximal tubule (basal labyrinth) disclosed a weak but significant staining reaction for GGT (Kugler et al. 1985).

Dipeptidylaminopeptidase IV (DAP IV) was found to constitute a marker enzyme of the brush border, similar to GGT and APM (fig. 2 A). However, DAP IV appeared also as an integral component of apical vesicles of tubule epithelia. In contrast to DAP IV, DAP I was detected in various segments of the tubular system, with a rather strong activity in the proximal convolut; in addition, the epithelia of Bowman`s capsule gave a positive reaction for DAP I (fig. 2 B). Apparently DAP I is localized within lysosomes (Kugler et al. 1985).

Aminopeptidase A (APA) was a predominant marker of glomeruli, as shown in fig. 3 A; to lesser extent also the proximal tubule was positive for APA. Further studies indicated that APA is localized in endothelia and podocytes (Kugler et al. 1985). APA activity was also observed in developing glomeruli of the fetal kidney (fig. 4). In contrast to data obtained in mice, in the human kidney APA was not found in cells of Goormaghthig`s plasmodium (Kugler 1982).

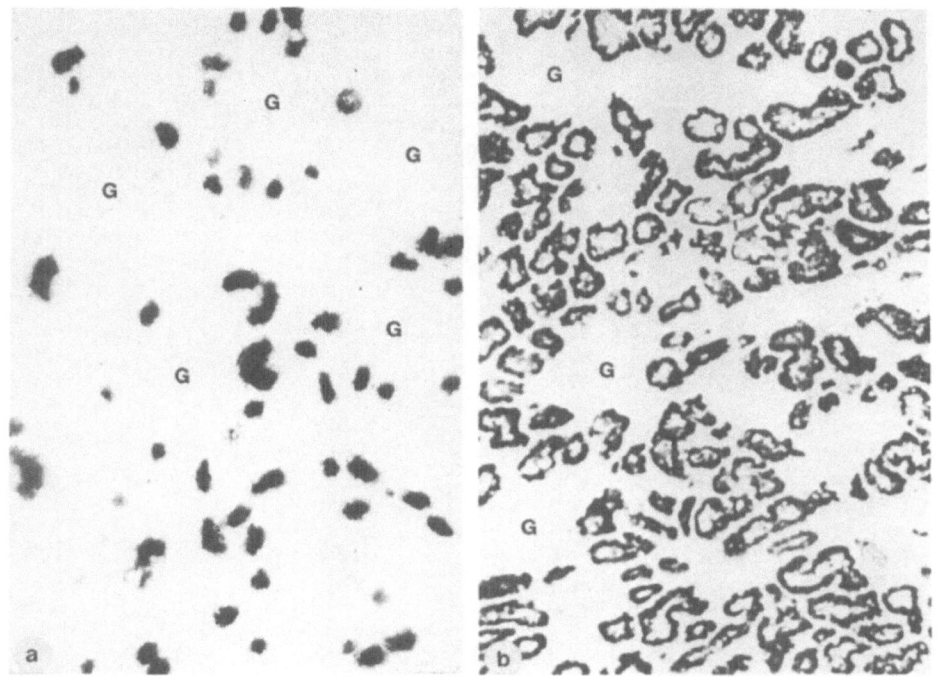

FIG.1. Histochemical demonstration of Aminopeptidase-M (APM) on tissue
sections from normal human kidney. 1 A: Distribution of APM activity on a
5 μ cryosection from human fetal kidney (24th week of gestation); high
APM activity is restricted to few developing tubules; enlarged distance
between APM-positive segments of the nephron as compared to adult normal
kidney, where a more homogenous distribution pattern was found(1 B). Both
sections were made from material of the cortical region. G = glomerulus,
magnification x 40 .

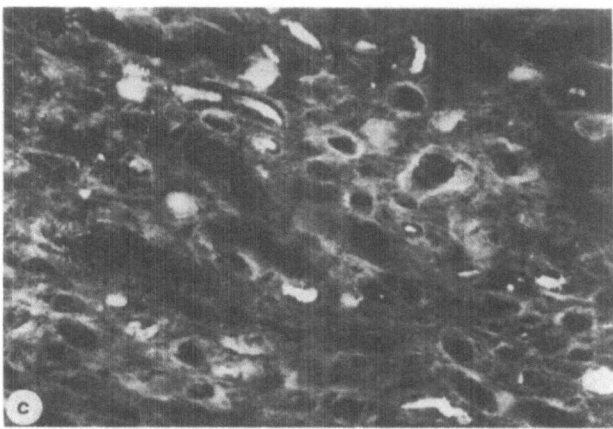

FIG.1 C: Cryosection of a kidney from a patient with interstitial nephro-
pathia disclosing irregular distribution of immunoreactive aminopeptidase M.
1st antibody directed against APM purified from brush border membranes of
human kidney; 2nd antibody: FITC-labelled-anti-rabbit-gammaglobulin.
magnification x 30; for details see text.

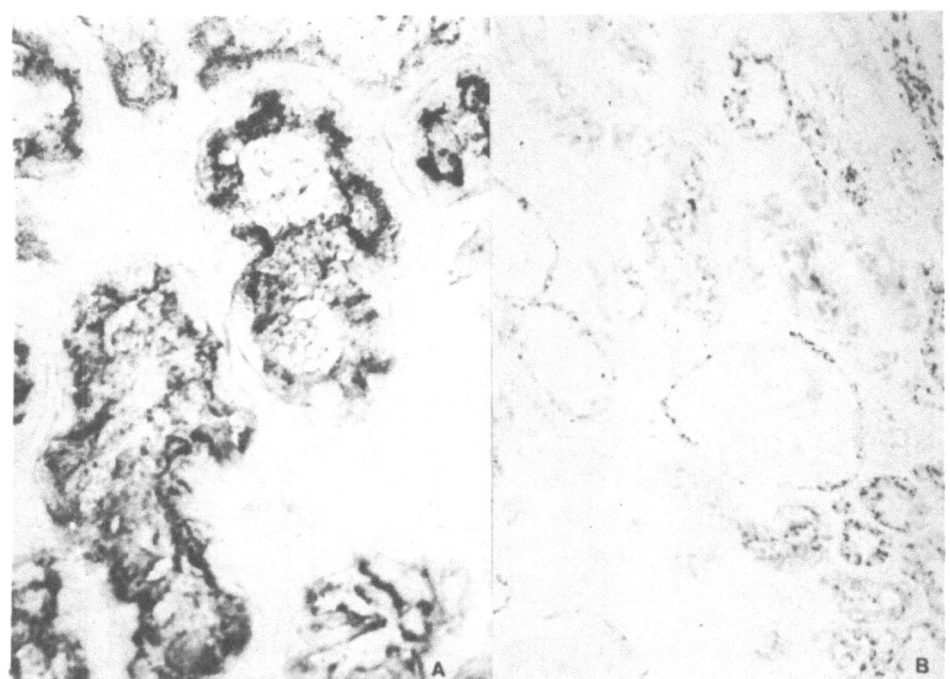

FIG 2: Histochemical demonstration of Dipeptidylaminopeptidase IV and I
(DAP IV, DAP I) in normal adult human kidney. Fig.2 A : DAP IV is a marker enzyme of the brush border membrane from the proximal tubule.
Fig 2 B: DAP I is predominantly localized in tubule lysosomes, in addition,
the epithelia of Bowman`s capsule exhibit a positive staining reaction.
magnification fig.2A: x 120; fig. 2B: x 80 .

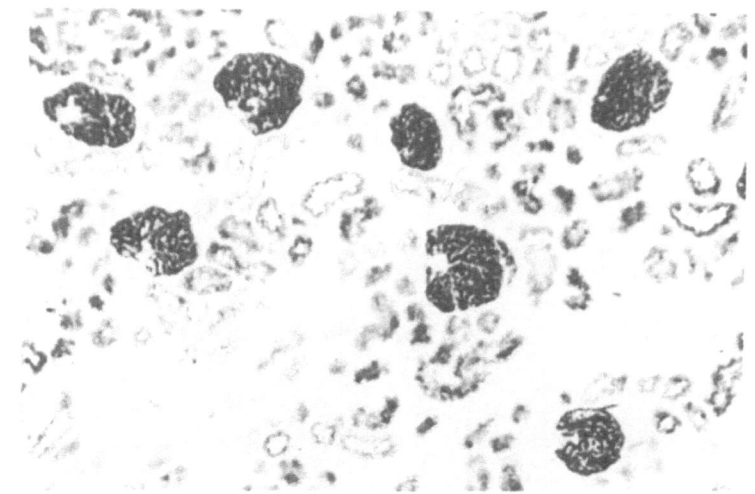

FIG.3. Histochemical demonstration of Angiotensinase A (Aminopeptidase A)
on a cryosection from normal adult human kidney. APA activity is found on
glomeruli (endothelia, podocytes) and, to lesser extend, on proximal tubular segments; magnification x 60 .

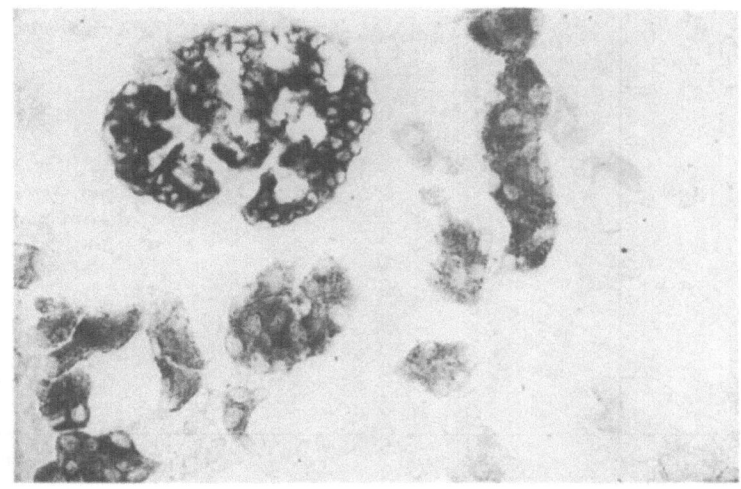

FIG. 4 . Histochemical distribution of Angiotensinase A (Aminopeptidase A)
on a cryosection from normal fetal kidney (20th week of gestation; corti-
comedullar region). Strong APA activity of developing glomerulus and tubular epithelia; magnification x 140 .

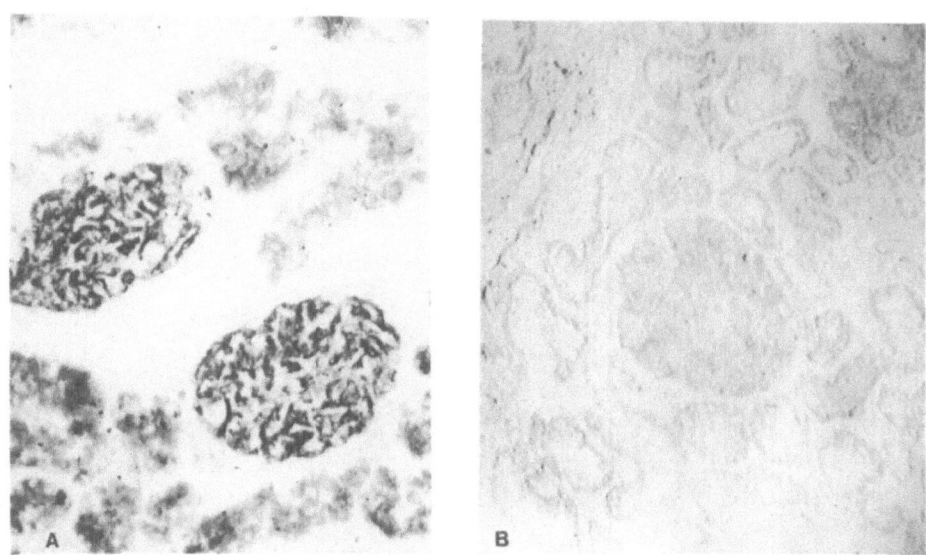

FIG.5 A . Strong histochemical Angiotensinase A (Aminopeptidase A) activity of glomeruli and, to lesser extend, of proximal tubules. 5 μ cryosection
of normal adult kidney (control).
FIG 5 B : Kidney cryosection pretreated with 0.05 mMol Angiotensin II for
30 min and subsequently stained for Angiotensinase A activity, revealing a
total inhibition of tissue enzyme activity. (magnification x 100)

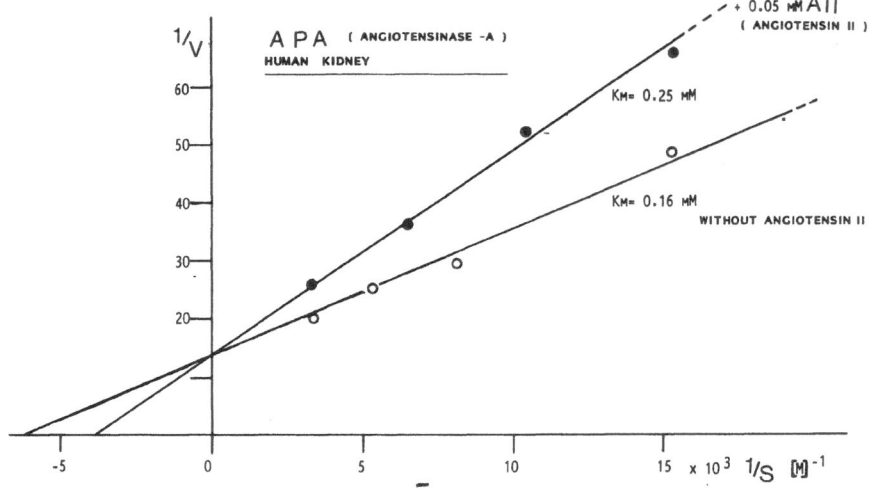

FIG. 6 . Determination of Michaelis –Menten constant as a function of substrate concentration for Angiotensinase A activity according to the Line-
weaver –Burk plot. o = substrate dependent APA activities (Km = 0.16).
● = APA activities after adding 0.05 mMol angiotensin II (A II); Km =
0.25). APA of human kidney is competitively inhibited by Angiotensin II,
(total homogenate of human kidney cortex).

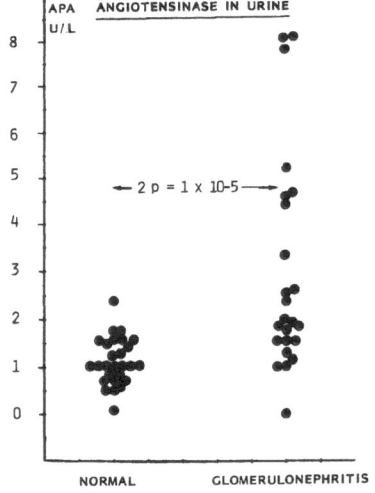

FIG.7. Excretion of Angiotensinase A in urine of healthy persons (n=27)
and patients suffering from glomerulonephritis of various courses (n=24).
Urine specimens of " 2nd morning urine " were assayed for APA activity
(U/L) . As compared to the controls excretion rate of APA is significantly higher in patients with glomerulonephritis (Mann –Whitney test).

Preincubation of tissue sections with angiotensin II (0.05 mM/L) revealed nearly a total inhibition of APA activity ascompared to the controls (fig. 5). Studies performed in kidney tissue fractions showed that APA was inhibited by angiotensin II in a competitive manner, whilst APM was not influenced (Wolf et al. 1987). Further details are shown in fig.6. APA was activated by Ca^{++} but not by Mn^+, Mg^{++} or Zn^{++}. Sucrose density gradient centrifugation (20-50 % w/w) of crude membrane fractions from human kidney cortex disclosed maximum of APA activity in the density range of d = 1.2 which was also identical with that of angiotensin-converting enzyme activity (Scherberich et al.1986). These data indicate that APA is an angiotensin II splitting re- ceptor of the glomerular tuft, and may act as a Ca^{++} dependent modulator of the intrarenal renin-A-II system. A hypothetical model demonstrating the possible integral function of APA in the control of local glomerular hemodynamic has been presented else- where (Scherberich et al. 1986, Wolf et al.1986, 1987).

Kidney- related APA as assessed in sponateously voided urine spe- cimens of healthy persons revealed the following levels: APA 1.13 ± 0.43 U/L, SEM: 0.09; median 1.04 U/L; n =27). In patients with glomerulonephritis the urinary excretion of APA was significantly higher compared to the controls, indicating possible damage to glomerular cells under these conditions (fig.7).
     If whether or not enhanced excretion of Angiotensin II splitting APA is related to arterial hypertension is presently under work.

References

Baumann,K., 1981, Renal transport of proteins, Eds.R.Greger, F. Lang, S. Silbernagl, p 118, Springer, New York

Gossrau, R., 1981, Investigations of proteinases in the digestive tract using 4-methoxy-2-naphthylamine (MNA) substrates, J.Histochem.Cytochem.29, 464

Kinne,R., 1985 , Biochemical aspects of tubular transport, in: Renal tubular disorders, pp 1 , Eds. H.Gonick, V.Buckalew, Marcel Dekker, New York

Kugler,P.,1982 , Histochemistry of angiotensinase A in the glomerulus and the juxtaglomerular apparatus, Kidney Int. 22,44

Kugler,P.,Wolf,G.,Scherberich,J.E., 1985 , Histochemical demon- stration of peptidases in the human kidney, Histochemi- stry,83,337

Lojda,Z., Gossrau,R. Schiebler,T.H., 1979 , Enzyme histochemi- stry, Springer, Berlin, Heidelberg, N.York

Scherberich,J.,E., Gauhl,C., Heinert,G.,Mondorf,W.,Schoeppe,W., 1984 , Characterization and clinical significance of membrane bound proteases from human kidney cortex, Adv.Exp.Med. 167,179

Scherberich,J.E.,Kinne,R.,Gauhl,C.,Mondorf,W.,Schoeppe,W, 1982, Isolation and characterization of basal-lateral and lumi- nal plasma-membranes from the proximal tubule of human kidney, Prot.Biol.Fluids, 29, 139

Scherberich,J.E.,Stuckhardt,C.,Wolf,G.,Kugler,P., 1986, Angio-
    tensinase A of human kidney: histo-/biochemical studies of
    an enzyme possibly involved in the regulation of intrare-
    nal renin-angiotensin system, Nieren-u.Hochdruckkrankh.,
    15,386

Wanner,Ch., Schollmeyer,P.,Hörl,W.H., 1986, Urinary proteinase
    activity in patients with multiple traumatic injuries,
    sepsis or acute renal failure, J.Lab.Clin.Med.

Wolf,G., Stuckhardt,C, Scherberich,J.E., 1986, Biochemical and
    immunohistochemical studies on angiotensinase A of human
    kidney and renal adenocarcinoma, Immunobiology, 173, 445

Wolf,G.,Scherberich,J.E.,Stuckhardt,C,Schoeppe,W., 1987, Angio-
    tensin-degrading aminopeptidase A (APA) of human kidney,
    Acta Endocrinol. 114, 159

Wolf,G., Scherberich,J.E., Stuckhardt,C., Schoeppe,W, 1987 ,
    Characterization of angiotensin receptors from human kid-
    ney, Prot.Biol.Fluids (Pergamon), 35, 289

EFFECT OF GLOMERULAR PROTEINURIA ON THE ACTIVITIES OF LYSOSOMAL

PROTEASES IN ISOLATED SEGMENTS OF RAT PROXIMAL TUBULE

C.J.Olbricht

Abteilung Nephrologie, Medizinische Hochschule
D3000 Hannover
Federal Republic of Germany

INTRODUCTION

Numerous renal diseases are accompanied by glomerular proteinuria. This syndrom is characterized by increased glomerular protein filtration and increased urinary protein excretion. Under normal conditions the proximal tubule absorbs most of the filtered proteins. In glomerular proteinuria the proximal reabsorption of proteins is increased (1). The uptake into the proximal tubule cells occurs by endocytosis. The uptake process is characterized by a high capacity (as compared with normal filtered loads), and a low affinity. Absorbed proteins are transferred into the lysosomal system. Within the heterolysosomes the proteins are catabolized to amino acids as shown in Figure 1 .

Only limited information is currently available concerning the proteolytic lysosomal enzymes involved in normal protein catabolism and in the increased protein digestion that occurs in glomerular proteinuria. The lysosomal proteinases Cathepsin B and L are abundant in the kidney (2,3). The highest activity was found in segments S1 and S2 of the proximal tubule (Fig.2) where under normal conditions about 70% of the filtered protein load is absorbed. In more distal nephron segments where protein uptake is minimal under normal conditions, a comparatively low activity of those proteolytic enzymes is found. This raises the possibility that cathepsins B and L are involved in intralysosomal degradation of absorbed proteins (3). Hence, it is our hypothesis that activation of cathepsins by increased protein uptake may account for the increased protein catabolism observed in glomerular proteinuria. To explore this hypothesis four different experimental models of glomerular proteinuria were applied in rats. Following induction of proteinuria we measured cathepsin B and L activities in the microdissected segments S1, S2, and S3 of the proximal tubule (Fig.2).

METHODS

Experimental Protocol

Seven experimental groups of Sprague-Dawley rats were used and fed standard rat chow (Purina). Three of the groups that were studied included young female rats weighing 95-163 g, old female rats of 9-11 mo of age weighing 254-335 g, and old male rats of 7-9 mo of age weighing 364-439 g.

Both groups of older rats had glomerular proteinuria due to glomerulo-
sclerosis (4). The fourth group of animals was composed of female rats
weighing 180-237 g. Each animal was made proteinuric by eight 1-g
intraperitoneal injections of bovine serum albumin (Sigma, fraction V)
dissolved in 4 ml of 0.9% saline. The interval between injections was 12 h.
A fifth group of animals was composed of female rats weighing 175-252 g and
served as the control group for the albumin-loaded rats. These animals
received the same number of intraperitoneal injections of the vehicle only.
Both groups were studied 24 h following the last injection. A sixth group of

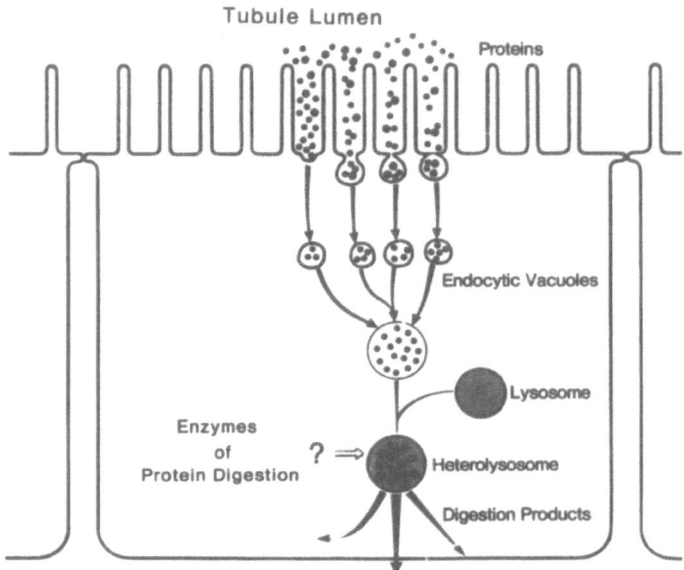

Fig. 1   Schematic representation of endocytic uptake of filtered
          proteins in the proximal tubule cell. Proteins are
          segregated in endocytic vacuoles. These vacuoles then
          migrate to the cell interior, where they fuse with
          lysosomes. Digestion of proteins occurs within the
          heterolysosomes.

7 female rats weighing 127-185 g received i.v. injections in the tail vein
with a single dose of puromycin aminonucleoside (PAN, Sigma) amounting to 15
mg/100g body weight as a 2% solution in saline. Another 6 control rats (body
weight 143-171 g) received an i.v. injection of the same volume of 0.9%
saline only. Experiments were performed 4 to 8 days following injection. All
animals in these seven groups were maintained in a metabolic cage for a 24 h
urine collection prior to the experiments. The animals were sacrificed by
stunning or by cervical dislocation, and blood was withdrawn by cardiac
puncture. Endogenous creatinine in urine and plasma was determined using a
modified Jaffes picric acid method (Beckman Creatinine Analyzer 2). Protein
in urine was determined by the Lowry method (5).

Determination of Cathepsin B and L Activities

The activities of cathepsin B and L in microdissected nephron segments were measured by ultramicroassays developed recently (3). Immediately after sacrifice the left kidney was removed and perfused with ice-cold collagenase solution. Pieces of tissue were sliced along the cortico-medullary axis and incubated for 45 min in the same collagenase solution gassed with 100% oxygen at 37°C. Collagenase was used to facilitate dissection of the proximal tubule segments, and did not alter cathepsin B and L activities (3). Individual segments of proximal tubules were dissected as described previously (3). Based on morphology and function three different segments of the proximal tubule , S1, S2, and S3 can be identified and dissected separately (Fig. 2). The length of the dissected segments was measured using an eyepiece micrometer and did not exceed 1.5 mm. The segments were transferred into reaction vials containing hypotonic buffer solution and

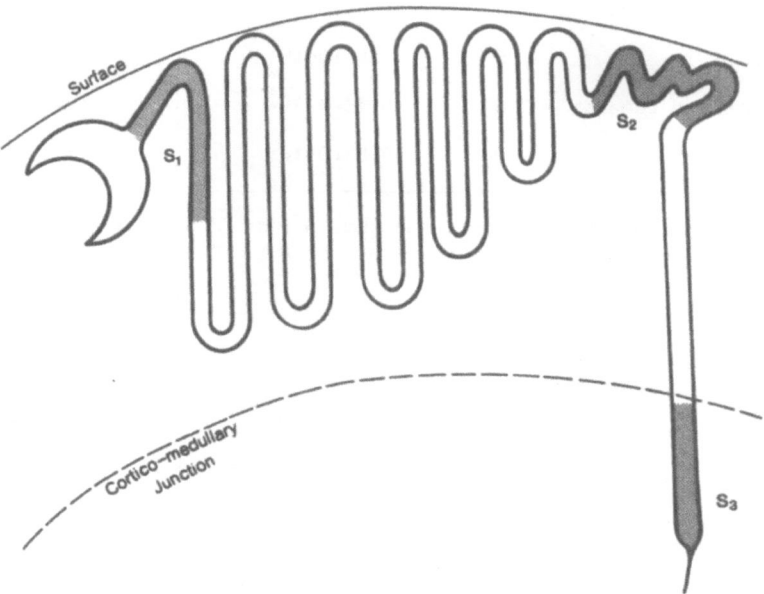

Figure 2   The segments S1, S2, and S3 of the proximal tubule as identified and dissected in the present study.

were subjected to freezing and thawing. Next, the substrate Z-phenylalanyl-arginine-7-amido-4-methylcoumarin for cathepsin B and L together, and the substrate Z-arginyl-arginine-7-amido-4-methylcoumarin for cathepsin B were added to the vials. Following 45 min incubation at 37°C, the reaction was stopped by sodium monochloroacetate. Cathepsin B and cathepsin B and L together cleave the respective substrates and release 7-amino-4-methylcoumarin (NMec). The concentration of NMec was determined by fluorometry. The enzyme activity was calculated as pmol of NMec generated per mm tubule length per min incubation time from a standard curve using NMec. The assay was linear with respect to incubation time and tubule length. In each rat between three and ten samples of each segment were analyzed. The significance of differences was tested by a two-sample t test for unpaired data. P values ≤ 0.05 were considered significant.

Table 1   Urine protein excretion and creatinine clearance

Experimental Group	n	Creatinine Clearance ml/min/100g BW	Protein Excretion mg/24 h
Young females	7	0.98+0.09	0.68+0.1
Older females	9	0.60+0.03*	10+6.3§
Older males	6	NM	22+6§
Albumin-loaded females	5	1.04+0.04	411+134*
Controls	6	0.93+0.15	2.7+0.4
PAN-injected females	6	0.45+0.31*	192+141*
Controls	7	0.85+0.14	1.5+0.8

Values are means+SEM; NM, not measured. § P < 0.05 vs young females. BW, body weight. * P < 0.05 vs control.

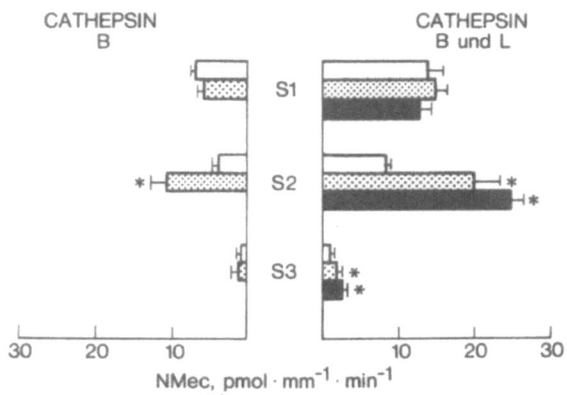

Fig. 3 Cathepsin B and combined cathepsin B and L activity in proximal tubule segments of young female rats (open bars), of older female rats (stippled bars), and of older male rats (solid bars). Values are mean+SD. Asterisks indicate p < 0.05 vs young female rats.

RESULTS

Urine Protein Excretion and Creatinine Clearance

The urine protein excretion and the creatinine clearances are depicted
in table 1. Total urine protein excretion per 24 h was significantly greater
in older animals of both sexes compared with the young female rats. In rats
injected with albumin and in rats following injection of PAN, the urine
protein excretion was dramaticly increased compared with the respective
controls. Older female rats and rats injected with PAN had slightly reduced
values of the creatinine clearance.

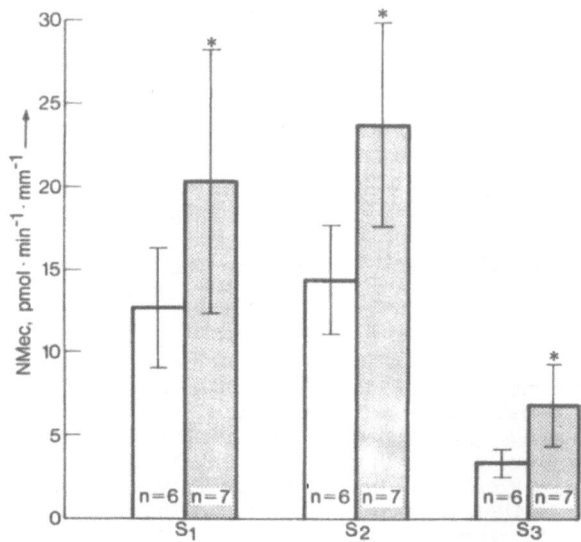

Fig. 4   Cathepsin B and L activity in proximal tubule
segments from albumin loaded rats, represented
by shaded bars, and from controls (open bars).
Values are Mean+SD. Asterisks indicate $p < 0.05$
vs control values. n represents number of rats.

Cathepsin B and L Activities

The activities of cathepsin B and of cathepsin B and L in individual
nephron segments dissected from young females and from older females and
males are depicted in Fig. 3. In S1 segments no differences were found for
cathepsin B and for cathepsin B and L. The combined cathepsin B and L
activitiy in S2 segments increased significantly in the older proteinuric
female and male rats in comparison with the young female rats ($p < 0.001$).
The activity of cathepsin B in S2 segments was nearly three times higher in
older females compared with younger female animals ($p < 0.005$). The
activities of cathepsin B and L in S3 segments were also significantly
higher in the two proteinuric groups ($p < 0.05$). No differences in the S3
activities were found for cathepsin B.

In albumin-loaded rats the cathepsin B and L activities were increased in all three segments of the proximal tubule (Fig. 4). The results were characterized by a large scatter of the data. The level of significance was $p < 0.05$ for S1 and S2, and $p < 0.01$ for S3, respectively. In PAN injected rats the combined cathepsin B and L activities were increased in S2 and in S3 segments (Fig. 5) in comparison with sham injected controls.

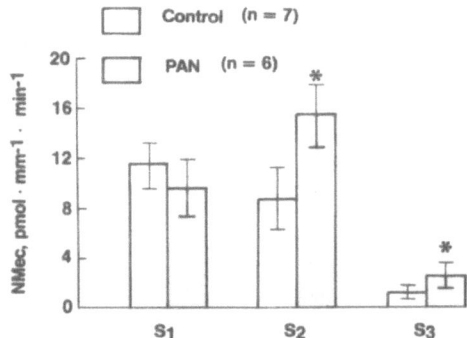

Fig. 5   Cathepsin B and L activities in proximal tubule segments of control rats and of rats 4 to 8 days following PAN injection. Values are mean+SD. Asterisks denote $p < 0.05$ vs control values. n indicates the number of animals.

Discussion

In the present study we observed high levels of cathepsin activity in the S1 and S2 segments of the proximal tubule in all animals examined. In addition, in the four groups of rats with proteinuria there was a significant increase in the levels of cathepsin activity in the S2 and S3 segments. In albumin-loaded rats enhanced cathepsin activity was also present in S1 segments. Factors that must be considered in the interpretation of these results include the presence of proteinuria, the age and sex of the animals, the effect of PAN, and the possible role of enhanced endogenous protein turnover. Certainly in other organs it is well documented that cathepsins B and L are responsible for a portion of the turnover of cell proteins (6).

The demonstration of increased cathepsin activity in the proximal tubule in animals made proteinuric with exogenous albumin injections and in rats with proteinuria following PAN injection in a pattern similiar to both groups of older rats with age-related proteinuria makes it unlikely that the increased cathepsin activity is an age-related phenomenon.

Likewise, the presence of elevated cathepsin activity in both male and female animals would argue strongly against the sex of the animal playing a major role in our findings. Furthermore, PAN does not stimulate cathepsin B and L activities in proximal tubule segments as recently demonstrated (7). Finally, although endogenous cell protein turnover may well contribute to basal levels of cathepsin activity in both proteinuric and nonproteinuric animals, it does not appear likely that the enhanced activity can be explained by this mechanism. The single uniform feature that was present in the four groups of animals that demonstrated increased cathepsin activities in segments of proximal tubule was the presence of proteinuria.

In our study both groups of older rats, the younger female rats injected with albumin, and the rats injected with PAN had significantly

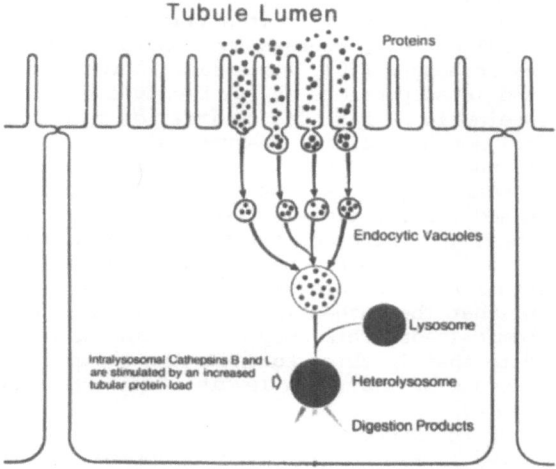

Fig. 6   Schematic representation of the hypothesis. Enhanced
concentration of proteins in the tubule lumen is
accompanied by increased protein uptake. Increased
amounts of protein within the lysosomal system
stimulate cathepsin B and L activities.

higher levels of urine protein excretion compared with young nonproteinuric females and with the two nonproteinuric control groups. Their is considerable evidence that the proteinuria in older rats is due to increased protein filtration (4). Likewise, it is well established that the proteinuria in albumin-loaded animals and in rats following PAN injection is of glomerular origin (8). Hence, it seems safe to conclude that in all four proteinuric groups the protein load per nephron was increased. We believe the elevation in cathepsin B and L activities in proximal tubule segments is secondary to an increase in endocytosis of proteins that occurs as a consequence of the increase in the load of protein delivered to the proximal tubule.

Although direct support for this hypothesis is not provided by the results of the present study, Park et al (9) have demonstrated an increase in protein uptake by the proximal tubule as the protein load was increased. The potential of acute increases in the load of protein delivered to the nephron to stimulate the activity of intralysosomal proteinases has been demonstrated by previous studies (10,11). The injection of lysozyme in mice induced a significant increase in proteolytic cathepsin activity in homogenates of renal cortex , whereas the nonproteolytic acid phosphatase activity remained unchanged (10). A similiar increase of cathepsin D-like activity was found in suspensions of rat kidney cortex tubules after induction of proteinuria by PAN. The activity of fucosidase, a non-proteolytic intralysosomal enzyme measured simultaneously , was not affected by proteinuria (11). These results support the present data inasmuch as it appears quite likely that the stimulation of lysosomal enzymes by increased tubular protein levels may selectively affect the proteolytic enzymes. The increased cathepsin B and L activities in S2 and S3 segments of rats made proteinuric by PAN injection may account for the increased intrarenal albumin catabolism observed in rats with PAN nephrosis (12). Our finding that the cathepsin activity in the S1 segment was increased in albumin-loaded rats only but not in the other proteinuric groups may indicate that in this region of the nephron enhanced endocytosis occurs but in presence of very high levels of protein concentration.

The results of the present study are consistent with the hypothesis schematicly depicted in Fig. 6. The proximal protein uptake by endocytosis is enhanced in the presence of glomerular proteinuria. The increased amount of proteins that is sequestered in the lysosomal system stimulates the activity of the intralysosomal proteolytic enzymes cathepsin B and L. The intralysosomal protein catabolism is enhanced and increased amounts of amino acids are released from the lysosomes into the cell and in the circulation. Due to this adaptation of cathepsin B and L the steady state between protein uptake and protein catabolism is well maintained in spite of increased endocytosis of proteins in the proximal tubule. This mechanism would provide a certain compensation for the increased glomerular loss of plasma proteins. However, it should be emphasized that not proteins are conserved but only amino acids.

REFERENCES

1. T. Maack, C.H.Park, and M.J.F. Camargo, Renal filtration, transport, and metabolism of proteins, in : "The Kidney: Physiology and Pathophysiology", D.W. Seldin and G. Giebisch eds, Raven Press, New York (1985).
2. E. Kominami, T.Tsukuhara, Y.Bando, and N.Katunuma, Distribution of cathepsin B and H in rat tissue and peripheral blood cells, J Biochem 98:97 (1985).

3. C.J. Olbricht, J.K.Cannon, L.C.Garg, and C.C.Tisher, Activities of cathepsin B and L in isolated nephron segments from proteinuric and nonproteinuric rats, Am J Physiol , 250:F1055 (1986).
4. J.M. Alt, H.Hackbarth, F.Deerberg, and H.Stolte, Proteinuria in rats in relation to age-dependant renal changes, Lab Animals ,14:95 (1980).
5. O.H. Lowry, N.J. Rosebrough, A.L. Farr, and R.J. Randall, Protein measurement with the Folin phenol reagent, J Biol Chem , 193:265 (1951)
6. J.S. Amenta and S.C.Brocher, Mechanisms of protein turnover in cultured cells, Life Science , 28:1195, (1981).
7. C.J. Olbricht, J.C.Cannon, and C.C.Tisher, Cathepsin B and L in nephron segments of rats with puromycin aminonucleoside nephrosis, Kidney Inter , 32:354 (1987).
8. D.E. Oken, S.C.Cotes, and C.W.Mende, Micropuncture study of tubular transport of albumin in rats with aminonucleoside nephrosis, Kidney Inter , 7:87 (1977).
9. C.H. Park and T. Maack, Albumin absorption and catabolism by isolated perfused proximal convoluted tubules of the rabbit, J Clin Invest, 73:767 (1984).
10. T. Maack, Changes in the activity of acid hydrolases during renal reabsorption of lysozyme, J Cell Biol , 35:268 (1967).
11. W.H. Baricos and S.V. Shah, Increased cathepsin D-like activity in cortex, tubules, and glomeruli isolated from rats with experimental nephrotic syndrome, Biochem J , 223:393 (1984).
12. J. Katz, G. Bonorris, and A.L.Sellers, Albumin metabolism in amino-nucleoside nephrotic rats. J Lab Clin Med , 62:910 (1963).

MEPRIN PHENOTYPE AND CYCLOSPORIN A TOXICITY IN MICE

Jane F. Reckelhoff[1], Shirley S. Craig[2], Robert J. Beynon[3] and Judith S. Bond[1]

Departments of [1]Biochemistry & Molecular Biophysics and of [2]Anatomy, Virginia Commonwealth University, Richmond, Virginia 23298-0001, USA; [3]Department of Biochemistry, University of Liverpool, Liverpool L69 3BX, UK

INTRODUCTION

Meprin is a membrane-bound metallo-proteinase that is present in high concentrations in the kidney brush border of mice and rats (Bond and Beynon, 1986). The enzyme has been purified and found to exist as a tetrameric glycoprotein that contains one mol zinc and three mol calcium per 85,000 molecular weight subunit (Beynon et al., 1981; Butler et al., 1987). Meprin hydrolyzes a variety of peptide and protein substrates (e.g., insulin B chain, glucagon, angiotensins I and II, bradykinin, hemoglobin, casein, aldolase, and phosphorylase kinase); the substrate used most often to assay activity is azocasein, a good general substrate for neutral and alkaline proteinases.

A heritable deficiency of meprin activity occurs in many inbred mouse strains (Bond et al., 1984). The low-activity strains (phenotype **b**; specific activities less than 10% of high activity, phenotype **a** strains) are derived from several different genealogical lines and include C3H, CBA, AKR, MaMy and RF strains. Genetic studies have led to the conclusions that the low meprin activity trait is inherited as an autosomal recessive allele and to the localization of the gene that controls expression of meprin activity (the *Mep-1* gene). The *Mep-1* gene is linked to the major histocompatability complex (MHC) on chromosome 17 in the mouse (Reckelhoff et al., 1985).

Toxicological studies of cyclosporin A (CyA), a cyclic undecapeptide of fungal origin, have shown that certain inbred strains of mice were more susceptible to toxic effects of this peptide (weight loss, calcium deposits in heart tissue, increased serum urea and creatinine levels) compared to other strains (Borland et al., 1985). The susceptible strains, CBA and C3H mice, were low-meprin activity strains and those resistant to side effects of the drug were high-meprin activity strains,

BALB/c, C57BL/10, A mice. Furthermore, it was found that
susceptibility to cyclosporin A toxicity is inherited as an
autosomal recessive trait as is low-meprin activity. It seemed
possible that mice containing high activities of this potent
membrane-bound proteinase may have a biological advantage in the
presence of administration of large amounts of a peptide drug
such as cyclosporin A. To determine definitively the
relationship between meprin phenotype and cyclosporin A
toxicities, we conducted a series of experiments, presented
herein. The effects of administration of cyclosporin A on
mortality, calcium accumulation in heart and other tissues, and
biochemical and morphological parameters of kidney were examined
in high- and low-meprin activity strains. Because cyclosporin A,
a potent immunosuppressive agent, is being used after organ
transplantation and also has serious toxic side effects in man,
these studies are relevant to clinical problems.

The results of this study indicate that there are genetic
factors that predispose some strains of mice to the toxic effects
of cyclosporin A and that the expression of toxicity may be
different in individual strains. Individual strains differ in
their response to the drug as determined by mortality, cardiac
calcification or renal damage. There was no simple correlation
between meprin phenotype and toxic responses. There are probably
multiple factors that determine the response of individual
strains to the drug.

MATERIALS AND METHODS

Inbred strains of mice were obtained from Jackson
Laboratory, Bar Harbor, Maine (BALB/c, CBA, CBA/Ca, C3H/HeJ, C58,
C57 BL/6, C57BR/cd, DBA/2); Charles River Laboratories, Dublin,
Virginia (C3H/HeN); Dr. Frank Lilly, Albert Einstein School of
Medicine, Bronx, New York (BALB.K); or from Dr. Chella David,
Immunogenetic Mouse Colony of the Mayo Medical School, Rochester,
Minnesota (B6/Kh, B6.K1/FlDv, B6.K2/FlDv, B6.H-2^k/BeDv). The mice
(male, 20 - 25 g) were maintained at 23°C on a 12 h dark/ 12 h
light cycle with free access to Purina Lab Chow (Ralston Purina,
St. Louis, Missouri) and water.

Cyclosporin A (CyA) was purchased as Sandimmune IV (the
solubilised commercial product provided for patient use).
Sandimmune IV contains 50 mg CyA, 650 mg Cremophore-EL, and 32.9%
ethanol per ml of solution. The Sandimmune was further diluted to
10 or 20 mg CyA/ml with sterile 0.9% NaCl. The control vehicle
(ethanol vehicle) contained the same concentrations of Cremophore-
EL, ethanol and NaCl as the diluted Sandimmune. Mice were
injected intraperitoneally with 50 mg CyA per kg body weight
except where noted. They were weighed every 2 to 3 days and doses
were adjusted accordingly. Animals were killed by cervical
dislocation. Blood samples were subjected to centrifugation at
10,000 x g, for 2 min, to obtain the serum.

Serum creatinine concentrations were determined by a
colorimetric assay kit (Sigma Chem. Comp., No. 555) based on the
method of Heinegard and Tiderstrom (1973). Serum urea nitrogen
was determined by the method of Crocker (1967) using a
colorimetric assay kit (Sigma, No 535). Creatine phosphokinase
(CPK) was determined by the method of Hughes (1962) using a
colorimetric assay kit (Sigma, No. 520). Glutamic-oxaloacetic

transaminase (SGOT) was assayed using Sigma kit No. 505 according to the method of Reitman and Frankel (1957).

For metal analyses kidneys and hearts were homogenized in 3% trichloroacetic acid in a glass tissue grinder with a Teflon pestle. The 10% (wt/vol) homogenates were subjected to centrifugation at 10,000 x g for 5 min and supernatant fractions were used for flame photometry (Video 22 AA/AE spectrophotometer, Instrumentation Laboratories, Boston, Massachusetts). Samples were diluted 10-, 20-, or 100-fold depending on tissue and metal being tested. Calcium concentrations in serum were determined using a Calcette (Precision Instruments).

To obtain tissue for electron microscopy, mice were anesthetized with sodium pentabarbital (50 mg/kg body wt). Blood was removed from tissues by perfusion through the left ventricle with heparinized Ringer's solution and then fixative (2% paraformaldehyde and 2% glutaraldehyde in 0.1 M phosphate buffer, pH 7.2). Hearts and kidneys were removed and immersed in the fixative for 4 h at 25°C. Tissues were washed with phosphate buffer and postfixed in 2% osmium tetroxide in 0.1 M phosphate buffer at 4°C for 2 h. Samples were dehydrated in ethanol and embedded in Fluka Durcupan ACM. Thin sections were cut on a Sorvall MT-2 ultramicrotome. The sections were collected on copper grids, stained with uranyl acetate (Watson, 1958) and lead citrate (Reynolds, 1963) and were examined with an Hitachi HU-12 electron microscope.

RESULTS

The first parameter that was studied was mortality. Five strains of mice were used for these studies including two high-meprin strains (C57BL/6 and BALB/c) and three low-meprin strains (C3H/HeN, CBA and C3H/HeJ). The mice were injected with 50 to 250 mg cyclosporin A per kg body weight per day for 9 days (Fig. 1). At the 50 mg/kg dose, C3H and C57BL/6 mice survived until day 10; one of the ten BALB/c mice died and 2 of the 10 CBA mice died. The 100 mg/kg dose proved to be the best for discriminating between strains. At the latter dose, C57BL/6 (a high-meprin strain) and C3H/HeJ (a low-meprin strain) had the highest survival rates while C3H/HeN, CBA (both low-meprin) and BALB/c (high-meprin) had less resistance to the drug. At cyclosporin A doses of 150 mg/kg or higher, all mice died by 6 days and there was very little difference between strains. These data indicated that there was no correlation between meprin phenotype and cyclosporin A-induced mortality.

To determine the effects of the drug on kidney function, the sera of mice were tested for creatinine and urea nitrogen (Table 1). The serum creatinine levels were elevated in the three high-meprin strains tested but not in the three low-meprin strains. However, the elevated creatinine levels were all within normal values of 1.11 mg/dl serum (Crispens, 1975). Urea nitrogen levels were slightly elevated in 5 of the 6 strains; only the CBA/J mice showed no significant increase in response to the drug. These results indicate that the drug causes some impairment of renal function in mice but gave no indication of greater renal damage in low-meprin mice.

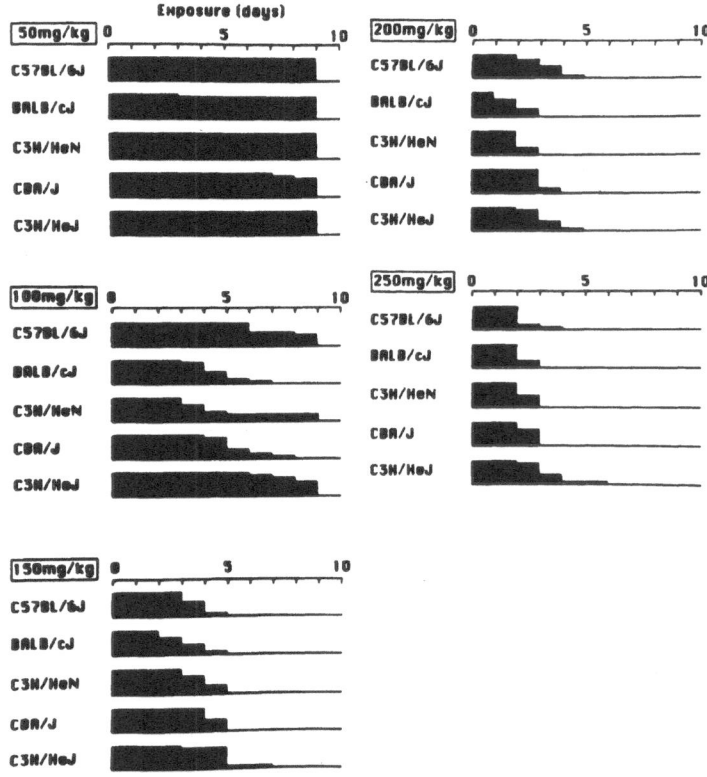

Fig.1. Mortality profiles of mouse strains administered various doses of cyclosporin A. Mice
were injected intraperitoneally, beginning with
groups of 10 per group at each dose level for 9
days. CyA dosages are noted in the boxes.
Thickness of the vertical bars indicates the
portion of animals surviving at each day during
the 9 day course of CyA administration.

Serum glutamic-oxalacetic transaminase (SGOT), which
reflects liver damage and serum creatine phosphokinase (CPK)
which reflects myocardial damage, were measured in control and
cyclosporin A-treated mice. CPK was not elevated significantly
over controls in 6 strains. Typical values for CPK were 30.0 ±
4.7 U/ml for controls, compared with 25.5 ± 1.9 U/ml for
cyclosporin A-treated mice. Typical values for SGOT were 72.3 ±
4.3 U/ml for controls, compared with 91.1 ± 7.2 U/ml for drug-
treated animals. The increase in SGOT in drug-treated mice was
significant for all strains of mice tested but there was no
difference in liver damage in high- and low-meprin mice.

Calcium concentrations were determined in serum, kidney and
heart tissue of cyclosporin A-treated and control mice (Tables 2
and 3). There were no differences in serum or kidney calcium
concentrations between control and drug-treated mice in any of
the strains tested. Heart calcium concentrations were elevated,
quite markedly in some instances, in several strains of mice
including low-meprin strains (C3H/HeJ, C3H/HeN, B6.H-2^k/BeDv,
CBA/J) and high-meprin strains (C57BL/6J, DBA/2J and B6/Kh).

Table 1. Serum creatinine and urea nitrogen in mice injected with cyclosporin A.

Strain	meprin phenotype	serum creatinine		serum urea nitrogen	
		control	CyA-treated	control	CyA-treated
C57BR/cdJ	a	0.39 ± 0.07 (4)	1.05 ± 0.22 (7)*	16.5 ± 2.71 (4)	20.9 ± 1.46 (7)*
C58/J	a	0.88 ± 0.12 (2)	1.06 ± 0.13 (8)*	26.0 ± 1.75 (2)	30.4 ± 1.50 (8)*
DBA/2J	a	0.68 ± 0.12 (4)	0.99 ± 0.24 (6)*	8.10± 0.61 (3)	15.3 ± 2.40 (6)*
C3H/HeJ	b	0.48 ± 0.10 (2)	0.43 ± 0.05 (2)	14.7 ± 1.46 (5)	27.4 ± 6.07 (6)*
C3H/HeN	b	0.92 ± 0.36 (2)	1.11 ± 0.11 (6)	20.0 ± 0.10 (2)	25.3 ± 2.25 (6)*
CBA/J	b	0.33 ± 0.09 (5)	0.34 ± 0.09 (5)	19.5 ± 2.07 (5)	24.4 ± 3.60 (6)

Mice were injected with 50 mg CyA/kg body wt/day for 14 days. Controls received the ethanol vehicle. Values represent mg/dl serum (mean ± S.E.). The number of animals in each group (n) is listed parenthetically. Meprin 'a' phenotype has high meprin activity; meprin 'b' phenotype low meprin activity. *$p < 0.05$.

Thus these data indicate that there are genetic factors that determine calcium accumulation in the heart in mice but there is no indication of a correlation with meprin phenotype.

Zinc, lead and magnesium concentrations were also determined in hearts and kidneys of cyclosporin A-treated and control mice (Table 4 and 5). No lead was detected in either tissue from any strain (the lower limit of detection was 0.01 μmol/g). There were no changes in zinc concentrations in heart and kidney of drug-treated mice compared to controls. Magnesium concentrations were unaffected in kidney but showed significant decreases in hearts of 5 out of seven strains tested. There was no correlation between meprin phenotype and decreases in heart magnesium, nor was there a correlation between calcium increases and magnesium decreases in response to cyclosporin A treatments. In addition, manganese concentrations in kidneys of the strains listed in Table 5 were not altered by CyA-treatment (data not shown).

Electron microscopic examination of kidney sections from mice treated with cyclosporin A revealed that all strains studied had increased vacuolization and disruption of the brush border membrane (Fig. 2). Electron micrographs demonstrated varying

Table 2. Calcium concentrations in serum and kidney of mice injected with cyclosporin A

Strain	meprin phenotype	serum		kidney	
		control	CyA-treated	control	CyA-treated
C58/J	a	8.61 ± 0.44 (2)	7.87 ± 0.21 (3)	5.80 ± 0.30 (2)	6.10 ± 0.01 (4)
C57BR/cdJ	a	8.50 ± 0.26 (4)	7.98 ± 0.21 (7)	4.60 ± 0.20 (4)	5.00 ± 0.30 (3)
DBA/2J	a	13.4 ± 0.27 (5)	12.5 ± 0.41 (6)	6.60 ± 0.55 (5)	6.67 ± 0.49 (6)
C3H/HeN	b	7.70 ± 1.52 (2)	7.70 ± 0.28 (6)	4.90 ± 1.20 (2)	6.30 ± 0.10 (6)
CBA/J	b	8.08 ± 0.42 (5)	7.95 ± 0.30 (6)	4.60 ± 0.20 (5)	4.90 ± 0.08 (6)
C3H/HeJ	b	13.8 ± 0.21 (5)	13.7 ± 0.28 (6)	5.00 ± 0.55 (5)	5.42 ± 0.99 (6)

Mice recived 50 mg CyA/kg body/wt for 14 days. Serum values are mg calcium per dl (mean ± S.E.; n for each group is given parenthetically). Serum values were determined by Calcette. Kidneys were homogenized in 3% trichloracetic acid (10% wt/vol), and calcium was determined in acid extracts by flame photometry. Kidney values are given as μmol calcium per g wet wt. No values in either tissue from treated animals were statistically different from those of control animals.

Table 3. Concentrations of calcium in hearts of mice injected with cyclosporin A

Strain	meprin phenotype	control	CyA-treated
C58/J	a	0.28 ± 0.14 (2)	0.26 ± 0.05 (6)
C57BL/6J	a	0.26 ± 0.02 (5)	0.40 ± 0.05 (5)*
C57BR/cdJ	a	0.07 ± 0.01 (4)	0.08 ± 0.01 (6)
DBA/2J	a	4.00 ± 0.95 (5)	14.0 ± 3.49 (6)*
B6/Kh	a	0.57 ± 0.03 (3)	0.70 ± 0.08 (4)*
BALB/cJ	a	0.37 ± 0.11 (5)	0.49 ± 0.04 (5)
C3H/HeJ	b	0.43 ± 0.08 (4)	5.72 ± 2.90 (5)*
C3H/HeN	b	0.30 ± 0.02 (8)	29.6 ± 7.60 (5)*
B6.K1/F1Dv	b	0.67 ± 0.03 (3)	0.66 ± 0.04 (5)
B6.K2/F1Dv	b	0.77 ± 0.17 (3)	0.73 ± 0.05 (4)
B6.H-2^k/BeDv	b	0.55 ± 0.05 (2)	2.08 ± 1.38 (4)*
CBA/J	b	0.27 ± 0.03 (5)	0.52 ± 0.05 (5)*
CBA/CaJ	b	1.02 ± 0.15 (5)	1.24 ± 0.25 (5)
BALB.K	b	0.78 ± 0.06 (5)	0.84 ± 0.84 (5)

Mice received 50 mg CyA/kg body wt/day for 14 days. Controls received the ethanol vehicle. Hearts were homogenized in 3% trichloroacetic acid (10% wt/vol), and calcium was determined in the acid extracts using flame photometry. Values represent μmol calcium/g wet wt heart (mean ± S.E.; n for each group is given parenthetically). *$p < 0.5$.

degrees of nephrotoxicity in all strains. There was some indication that damage by the drug to the membrane and underlying cell structures was more pronounced in low-meprin strains (C3H/HeN and CBA/J) compared to high-meprin mice (C57Bl/6J and C58/J).

Heart sections from cyclosporin A-treated mice were also examined by electron microscopy (Fig. 3). There was some disruption of cellular structure and loss of density in C57BL/6J, C58/J (high-meprin mice) and CBA/J (low-meprin mice) treated with cyclosporin A compared with controls. However, comparisons between heart sections from control and drug-treated C3H/HeN (low-meprin) mice showed, in addition to muscle cell disruption, large dense bodies (Fig. 4). These deposits stained positively for calcium, using the Von Kossa calcium staining method (Pearse, 1985). In addition, x-ray spectra indicated that the deposits were composed of calcium (A. LeFurgey, personal communication). The micrographs corroborate the large increases in heart calcium found by atomic absorption.

DISCUSSION

This work supports the hypothesis that there are genetic factors that determine susceptibility to CyA toxicity in mice. The reason for the strain-specific mortality is unknown. Serious toxic side effects of the drug in humans and animal models include nephrotoxicity, hypertension and neurotoxicity. One or more of these processes may lead to death. Administration of CyA to mice leads to accumulation of CyA in virtually all tissues. In

Table 4. Concentrations of magnesium and zinc in hearts of mice injected with cyclosporin A.

	Strain	meprin phenotype	control	CyA-treated
magnesium	C57BR/cdJ	a	7.02 ± 0.42 (3)	7.07 ± 0.58 (3)
	DBA/2J	a	7.90 ± 0.19 (3)	6.37 ± 0.41 (3)*
	BALB/cJ	a	3.24 ± 0.42 (4)	2.72 ± 0.58 (4)*
	C3H/HeJ	b	7.24 ± 0.03 (3)	6.66 ± 0.01 (3)*
	CBA/J	b	3.00 ± 0.13 (5)	2.42 ± 0.09 (5)*
	CBA/CaJ	b	2.92 ± 0.09 (5)	1.84 ± 0.23 (5)*
	BALB.K	b	2.42 ± 0.20 (5)	2.54 ± 0.08 (5)
zinc	C57BR/cdJ	a	0.21 ± 0.01 (4)	0.20 ± 0.05 (6)
	C58/J	a	0.11 ± 0.02 (2)	0.17 ± 0.02 (6)
	DBA/2J	a	0.10 ± 0.02 (5)	0.11 ± 0.03 (6)
	CBA/J	b	0.19 ± 0.03 (5)	0.21 ± 0.03 (6)
	C3H/HeJ	b	0.09 ± 0.02 (5)	0.15 ± 0.01 (6)
	CeH/HeN	b	0.24 ± 0.12 (5)	0.15 ± 0.02 (6)

Mice received 50 mg Cya/kg body wt/day for 14 days. Controls received the ethanol vehicle. Hearts were homogenized in 3% trichloracetic acid (10% wt/vol), and metal concentrations were determined in acid extracts by flame photometry. Values are μmol/g wet wt heart (mean ± S.E., when n for each group is listed parenthetically). *$p<0.05$.

Table 5. Concentrations of magnesium and zinc in kidneys of mice injected with cyclosporin A.

	Strain	meprin phenotype	control	CyA-treated
magnesium	C58/J	a	17.40 ± 1.40 (2)	18.70 ± 2.10 (3)
	C57BR/cdJ	a	16.30 ± 0.79 (3)	15.60 ± 0.65 (3)
	DBA/2J	a	9.64 ± 0.28 (5)	9.77 ± 0.63 (6)
	C3H/HeN	b	18.00 ± 2.20 (2)	16.20 ± 0.99 (3)
	C3H/HeJ	b	9.82 ± 0.60 (5)	9.62 ± 0.30 (6)
	CBA/J	b	15.40 ± 0.87 (3)	16.30 ± 0.76 (3)
zinc	C58/J	a	0.23 ± 0.01 (2)	0.24 ± 0.01 (3)
	C57BR/cdJ	a	0.26 ± 0.01 (3)	0.26 ± 0.01 (3)
	DBA/2J	a	0.26 ± 0.01 (5)	0.27 ± 0.01 (6)
	C3H/HeN	b	0.19 ± 0.03 (2)	0.23 ± 0.01 (4)
	C3H/HeJ	b	0.27 ± 0.03 (2)	0.29 ± 0.01 (6)
	CBA/J	b	0.20 ± 0.02 (3)	0.23 ± 0.01 (3)

Treated mice were injected intraperitoneally with 50 mg CyA per kg body wt per day for 14 days. Controls received the ethanol vehicle. Kidney tissue was homogenized in 3% trichloroacetic acid (10% wt/vol), and metal concentrations were determined in supernatants of acid extracts. VAlues are μmol/g wet wt kidney (mean ± S.E.). The number of tissues assayed is shown parenthetically.

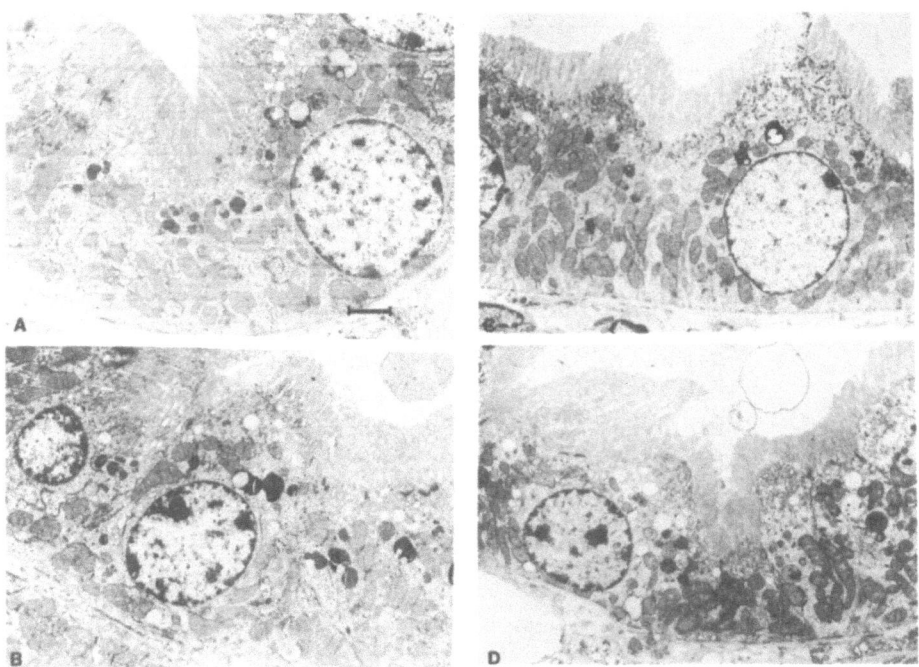

Fig.2.Electron micrographs of proximal tubule cells in kidneys
of mice treated with CyA (B and D) and control mice (A and
C). Mice were injected with 50 mg CyA/kg/day for 14 days;
controls were injected with the ethanol vehicle. (A)
C57BL/6J control, (B) C57BL/6J CyA-treated, (C) C3H/HeJ
control, (D) C3H/HeJ CyA-treated. C57BL/6J mice have high
meprin activity, C3H/HeJ have low meprin activity. (bar,
1μm).

one mouse study high concentrations of the drug were found, in
order of decreasing amounts, in thymus, spleen, kidney, lung,
liver, intestine and brain; the cause of death was attributed to
neurotoxicity (Boland et al., 1984). It is possible that some
strains are able to metabolize or excrete the drug more readily
than other strains, thus preventing toxic effects. However, very
little is known about the metabolism of CyA in animals; only
hydroxylation of some of the methyl groups has been observed
(Maurer et al., 1984). In addition, it is not known whether the
parent compound is more or less toxic than the metabolites or
what the mechanisms are for the beneficial or detrimental effects
of the drug.

This work supports the hypothesis that there are genetic
factors that determine susceptibility to CyA toxicity in mice.
The reason for the strain-specific mortality is unknown. Serious
toxic side effects of the drug in humans and animal models
include nephrotoxicity, hypertension and neurotoxicity. One or
more of these processes may lead to death. Administration of CyA
to mice leads to accumulation of CyA in virtually all tissues. In
one mouse study high concentrations of the drug were found, in

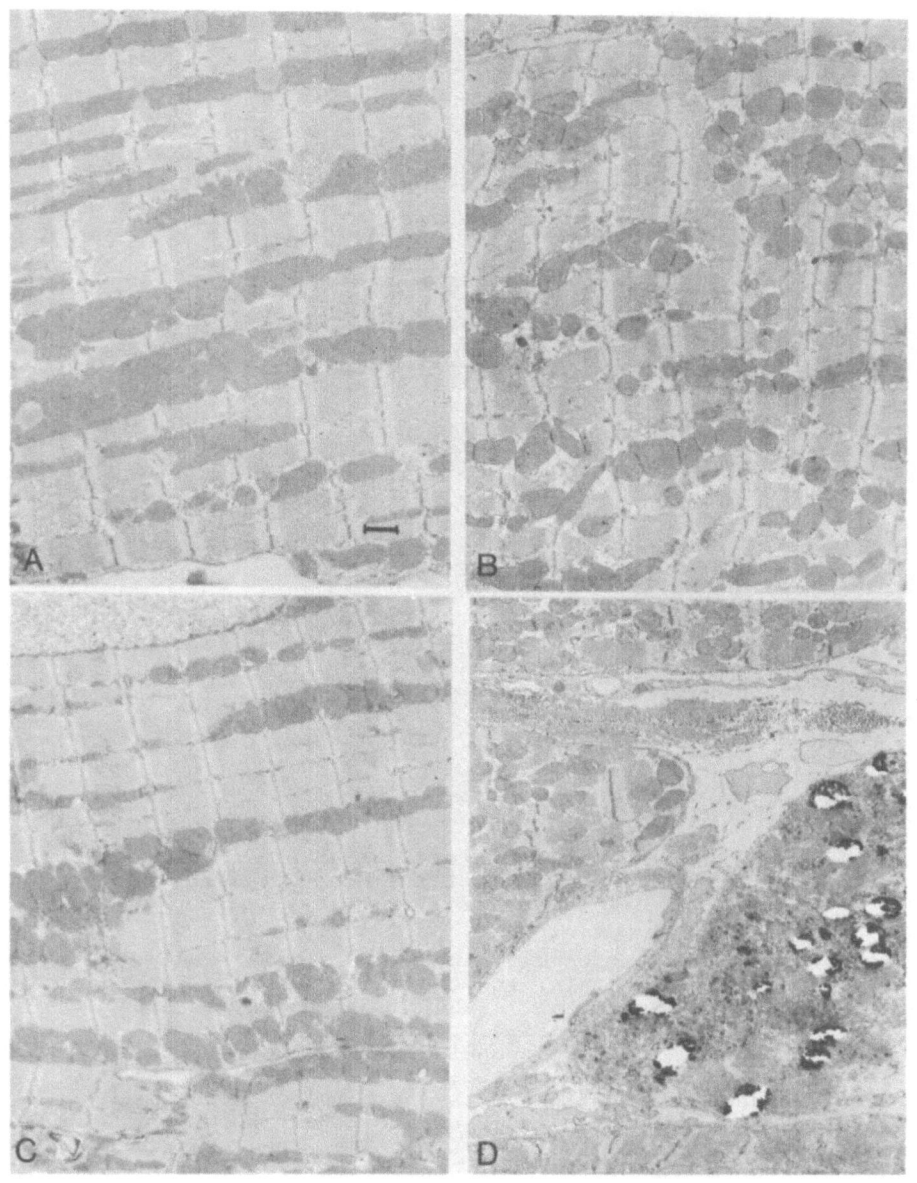

Fig.3. Electron micrographs of heart tissue of mice treated with
CyA (B, D and E) and control mice (A and C). Mice were
injected with 50 mg CyA/kg/day for 14 days; controls were
injected with the ethanol vehicle. (A) C57BL/6J control,
(B) C57BL/6J CyA-treated, (C) C3H/HeN control, (D)
C3H/HeN CyA-treated. C57BL/6J mice have high meprin
activity, C3H/HeN mice have low meprin activity.(bar,
1μm).

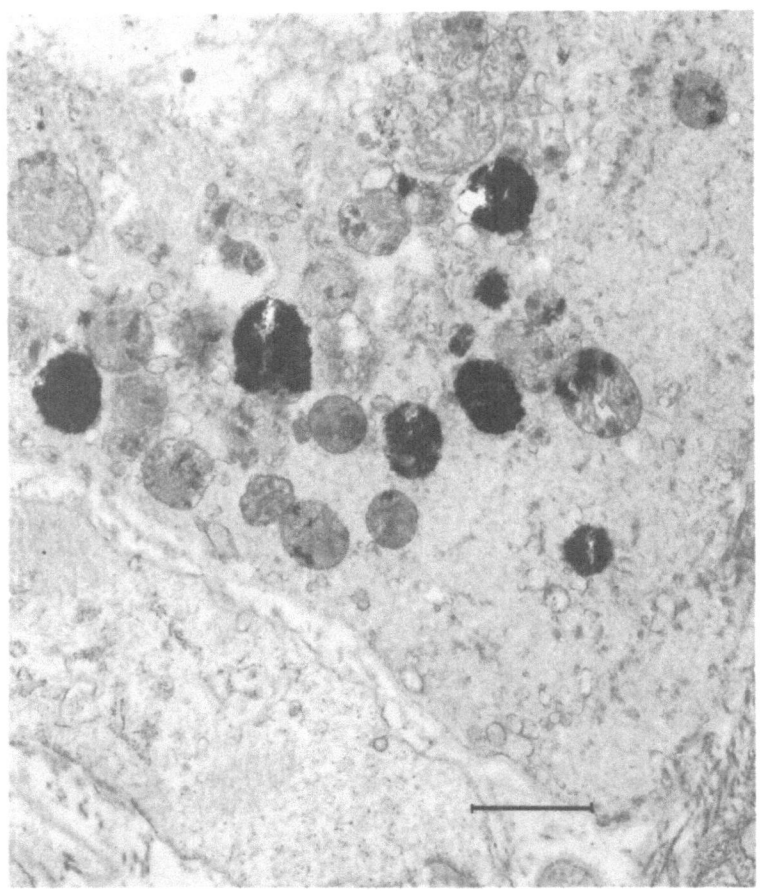

Fig.4. Electron micrograph of C3H/HeN heart tissue
after CyA administration. Mice were injected
with 50 mg CyA/kg/day for 14 days. (bar, 1 μm).

order of decreasing amounts, in thymus, spleen, kidney, lung,
liver, intestine and brain; the cause of death was attributed to
neurotoxicity (Boland et al., 1984). It is possible that some
strains are able to metabolize or excrete the drug more readily
than other strains, thus preventing toxic effects. However, very
little is known about the metabolism of CyA in animals; only
hydroxylation of some of the methyl groups has been observed
(Maurer et al., 1984). In addition, it is not known whether the
parent compound is more or less toxic than the metabolites or
what the mechanisms are for the beneficial or detrimental effects
of the drug.

The DBA/2J strain is abnormal in this respect; myocardial
calcification occurs in 48% of these mice (male and female) at 12
to 16 months in the absence of drug adminnistration (Crispens,
1975). The C3H mice were found to be particularly susceptible to
heart calcification in response to CyA admininstration in the
present studies (10 to 100-fold increases in heart calcium were

observed). These calcium effects were tissue specific (e.g., serum and kidney calcium concentrations were unaffected) and ion specific (zinc and manganese concentrations in heart were unaffected while magnesium concentrations decreased if altered). The heart cells that did accumulate large amounts of calcium appeared to be severely disrupted and it is possible that the increased calcium concentrations were a secondary result of dying cells. CyA has been found to bind to calmodulin and cyclophilin (a cytosolic binding protein) and this interaction has been suggested to be important for the immunosuppresive effects of the drug (Colombani et. al., 1985). It is possible that interactions of CyA with calcium binding proteins leads to accumulation of the ion in tissues as well.

These studies confirm that CyA is nephrotoxic and causes vacuolization and disruption of the brush border membrane. Serum urea nitrogen and creatinine concentrations were only mildly elevated in these studies but electron microscopic studies revealed more severe damage occured in several strains. Mice with the low-meprin phenotype were somewhat more suceptible to renal disruption than those with the high-meprin phenotype. It is possible that the rate of metabolism of the drug differs in this tissue and that meprin is involved in binding/sequestering or degradation of the drug in kidney.

ACKNOWLEDGMENTS

We are grateful to Dr. Ann LeFurgey, Duke University, for the x-ray analyses of heart tissue. This work was supported by NIH Grant DK 19691, Biomedical Research Support Grant #5S07RR05430-24, and NATO Research Grant 0574(82)264/TT.

REFERENCES

Beynon, R.J., Shannon, J.D., and Bond, J.S., 1981, Purification and characterization of a metalloendoproteinase from mouse kidney, Biochem. J., 199: 591-598.

Boland, J. Atkinson, K., Britton, K., Darveniza, P., Johnson, S., and Biggs, J., 1984, Tissue distribution and toxicity of cyclosporine A in the mouse, Pathol., 16: 117-123.

Bond, J.S., and Beynon, R.J., 1986, Meprin: a membrane-bound metallo-endopeptidase, Curr. Top. Cell. Reg., 28: 263-290.

Bond, J.S., Beynon, R.J., Reckelhoff, J.F. and David, C.S., 1984, Mep-1 gene controlling a kidney metalloendopeptidase is linked to the major histocompatibility complex in mice, Proc. Natl. Acad. Sci. USA, 81: 5542-5545.

Borland, I.A,, Sells, R.A., Gosney, J.R., and Beynon, R.J., 1985, Strain-specific calcification of the heart in cyclosporine-treated mice, Transplant. Proc., 17: 1345-1348.

Butler, P.E., McKay, M.J., and Bond, J.S., 1987, Characteristics of meprin, a mouse kidney metalloendopeptidase, Biochem. J., 241: 229-235.

Colombani, P.M., Robb, A., Hess, A.D., 1985, Cyclosporin A binding to calmodulin: a possible site of action on T lymphocytes, Science, 228: 337-339.

Crispens, C.G., Jr., 1975, "Hanbook on the Laboratory Mouse", C.C. Thomas, Springfield, Illinois.

Crocker, C.L., 1967, Rapid determination of urea nitrogen in serum or plasma with deproteinization, Am. J. Med. Technol., 33: 361-365.

Heinegard,D., and Tiderstrom, G., 1973, Determination of serum
    creatinine by direct colorimetric method, <u>Clin. Chim. Acta</u>,
    43: 305-310.
Hughes, B.P., 1962, A method for the estimation of serum creatine
    kinase and aldolase activity in normal and pathological
    sera, <u>Clin. Chim. Acta</u>, 7: 597-603.
Maurer, G., Loosli, H.R., Schreier, E., and Keller, B., 1984,
    Disposition of cyclosporine A in several animal species and
    man: 1. Structural elucidation of its metabolites, <u>Drug
    Metab. Disp.</u>, 12: 120-126.
Pearse, A.G.E., 1985, "Histochemistry Theoretical and Applied",
    Vol. 2, 4th edition, Churchill Livingstone, New York.
Reitman, S., and Frankel, S., 1957, A colorimetric method for the
    determination of serum glutamic oxalacetic and glutamic
    pyruvic transaminases, <u>Am. J. Clin. Pathol.</u>, 28: 56-63.
Reynolds, E.S., 1963, Use of lead citrate at high pH as an
    electron-opaque stain in electron microscopy. <u>J. Cell Biol.</u>,
    17: 208-212.
Watson, M.L., 1958, Staining of tissue sections for electron
    microscopy with heavy metals, <u>J. Biophys. Biochem. Cytol.</u>,
    4: 475-497.

POTENTIAL ROLE OF LYSOSOMAL PROTEASES IN GENTAMICIN NEPHROTOXICITY

C.J. Olbricht, E. Gutjahr, M. Fink, and K.M. Koch

Abteilung Nephrologie, Medizinische Hochschule
D 3000 Hannover 61, Konstanty-Gutschow-Str. 8
Federal Republic of Germany

INTRODUCTION

Gentamicin is an antibiotic with a broad spectrum of antibacterial activity. It is particularly valuable in severe infections. Nephrotoxicity, however, limits its use in just those critically ill patients in whom it is most needed. Nephrotoxicity has been related to a selective accumulation of gentamicin in proximal tubules. It is generally believed that gentamicin enters proximal tubule cells by endocytosis at the luminal cell membrane. Subsequently, the endocytic vacuoles fuse with primary lysosomes to form heterolysosomes (1) (Fig. 1).

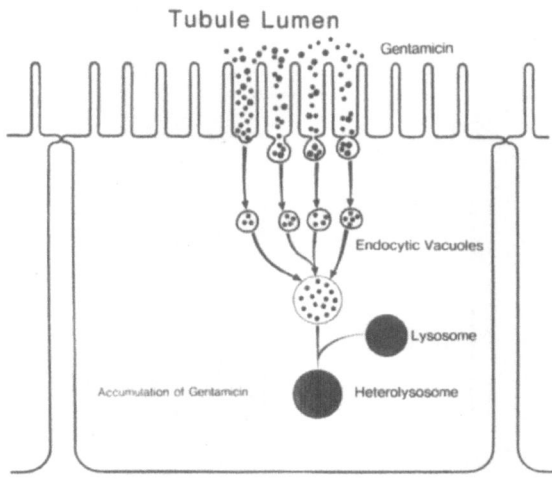

Fig. 1 Schematic representation of gentamicin (GM)
        uptake by proximal tubule cells. GM enters
        the cell at the luminal aspect by endo-
        cytosis and accumulates in heterolysosomes.

However, the mechanism of toxicity is unknown. Since gentamicin accumulates within lysosomes, and since the earliest identifiable lesions seen in gentamicin nephrotoxicity are myeloid bodies within the lysosomes (2), it has been suggested that lysosomal dysfunction may be the prime mechanism of nephrotoxicity (1). An important function of lysosomes is the catabolism of proteins as a part of physiologic intracellular protein turnover (3). Key enzymes involved in lysosomal protein degradation are the cathepsins B and L (4). It is conceivable that gentamicin affects these proteases and thereby interferes with physiologic cellular protein turnover. To evaluate this hypothesis activities of cathepsin B and L were measured in isolated segments of rat proximal tubule following injection of gentamicin.

METHODS

Nine female Sprague-Dawley rats weighing 160-200 g received four subcutaneous injections of gentamicin in a dose of 100 mg/kg body weight, respectively. 11 control rats weighing 165-203 g received injections of the solvent only. The intervals between injections were 24 h. The rats were kept in metabolic cages for 24 h urine collections. The animals were sacrificed 24 h following the last injection. Blood was sampled by cardiac puncture. The left kidney was removed and sliced along the cortico-medullary axis. The segments S1, S2, and S3 of the proximal tubule were isolated from the slices by microdissection as previously described (5). In the present study, S1 was identified as the first 3 mm of the proximal tubule attached to the glomerulus, while S2 included the last 1 mm of the proximal convoluted tubule, and S3 was identified as the last 1 mm of the pars recta of the proximal tubule.

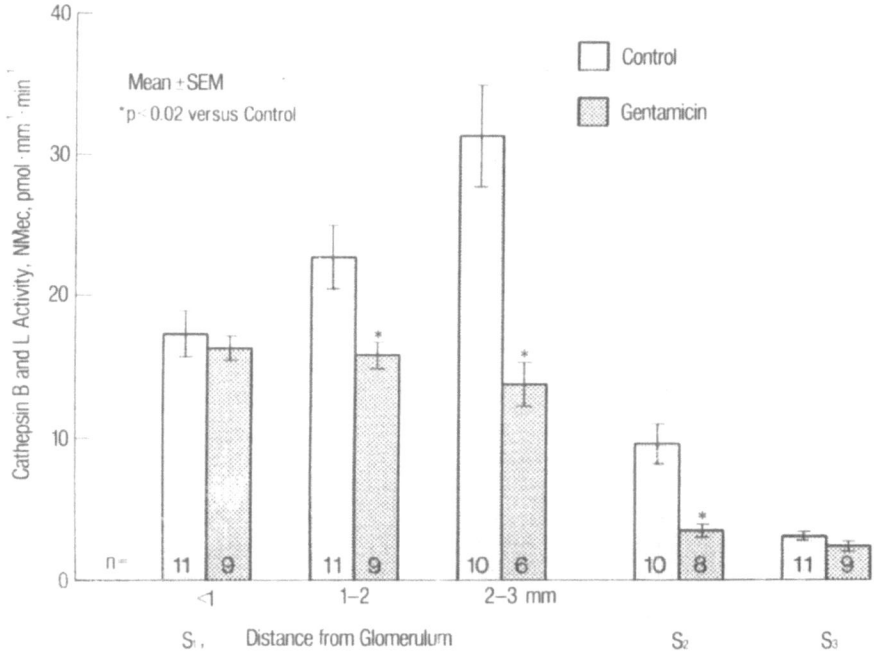

Fig. 2 Cathepsin B and L activity in individual segments of proximal tubule. Comparison between control rats and gentamicin injected rats. n indicates number of animals.

306

The combined activities of cathepsin B and L were measured in individual segments by means of a fluorometric ultramicro-assay as previously described (5). Z-phenylalanyl-arginine-7-amido-4-methylcoumarin (Z-Phen-Arg-NMec) was used as substrate for cathepsin B and L together. The enzymes cleave the substrate and release 7-amino-4-methylcoumarin (NMec) which is highly fluorescent. The enzyme activity was expressed in pmol of released reaction product NMec per mm tubule length per min incubation time. In each animal between 3 and 7 samples of the individual segments were analyzed. Creatinine was determined in serum and urine, and the creatinine clearance was calculated. The in vitro effect of gentamicin on cathepsin B and L was evaluated by incubating individual S1 segments of four rats with the following concentrations of gentamicin (mg/ml): 0.3; 0.6; 1.0; 5.0 .

RESULTS

Control rats had an urine volume of 11 ± 2 ml/24h and a creatinine clearance of 0.71 ± 0.1 ml/min 100g body weight. In gentamicin treated rats urine volume was increased to 24 ± 11 ml/24h and the creatinine clearance was decreased to 0.56 ± 0.1 ml/min/100 g body weight ($p < 0.05$, respectively). In gentamicin injected animals the cathepsin B and L activity was significantly reduced in the second and third mm of the proximal tubule (1-2 mm, $p < 0.02$; 2-3 mm, $p < 0.005$) and in the S2 segment ($p < 0.005$) (Fig. 2). In the first mm of the proximal tubuel and in the S3 segment, the cathepsin activity remained unchanged. The in vitro incubation of S1 segments with gentamicin revealed a concentration dependant inhibition of cathepsin B and L by gentamicin (Fig.3).

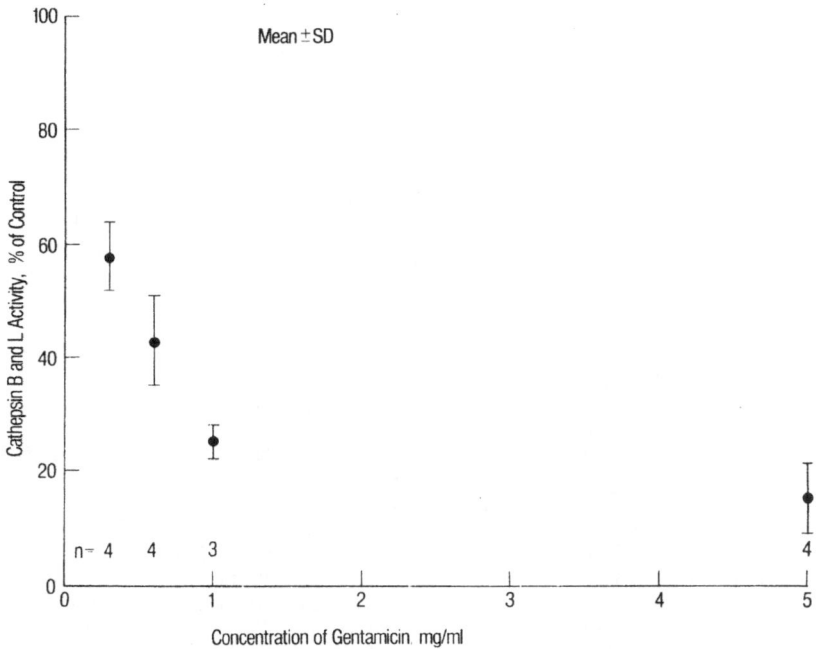

Fig. 3 Concentration dependant inhibition of cathepsin B and L by gentamicin, measured in individual S1 segments of rat proximal tubule in vitro. n indicates the number of rats. In each rat between 5 and 7 S1 segments were analyzed.

DISCUSSION

Gentamicin decreased the activity of the lysosomal proteolytic enzymes cathepsin B and L in the proximal tubule of rat kidney. Factors that must be considered in the interpretation of these data include enhanced exocytosis of phagolysosomes with release of lysosomal enzymes into the tubular lumen, frank proximal tubular cell necrosis, and inhibition of cathepsins by gentamicin. Increased urinary excretion of lysosomal enzymes following gentamicin injection has been described (1) and may partly contribute to the reduced cathepsin activities in the present study. Substantial tubular necrosis as an important factor involved in cathepsin depletion is unlikely since morphologic changes during the first 4 days following gentamicin injection are rather minimal (1). It is our interpretation of the data that inhibition of cathepsin B and L is a major factor contributing to the reduced level of enzyme activity in S1 and S2 segments of proximal tubule. This interpretation is supported by the in vitro data demonstrating a concentration dependant inhibition of cathepsin B and L by gentamicin. The failure of gentamicin to inhibit the cathepsin B and L activity in the first mm of the S1 segment and in the S3 segment of the proximal tubule in vivo may reflect differences in gentamicin uptake along the proximal tubule (6,7).

The results suggest that in proximal tubules gentamicin may interfere with intracellular protein catabolism by inhibition of lysosomal proteases, since cathepsin B and L are key proteases in cellular protein turnover. This may be an important mechanism of gentamicin nephrotoxicity. However, the connection between enzyme inhibition and cell damage remains unknown . It may be similiar to lysosomal storage disease where the reduced activity or the deficiency of one lysosomal enzyme causes serious cell damage.

ACKNOWLEDGEMENT

This work was supported by DFG (German Research Fund), grant C.J. Olbricht 1789

REFERENCES

1. G.J.Kaloynides and E.Pastoriza-Munoz, Aminoglycoside nephrotoxicity, Kidney Inter , 18:571 (1980).
2. J.C.Kosek, R.I.Mazze, and M.J.Cousins, Nephrotoxicity of gentamicin, Lab Invest , 30:48 (1974).
3. J.S.Amenta and S.C.Brocher, Mechanism of protein turnover in cultured cells, Life Science , 28:1195 (1981).
4. A.J.Barrett, Cathepsins and other thiol proteinases, in, "Proteinases in mammalian cells and tissues", A.J.Barrett ed, 181-208, Elsevier/North-Holland Biomedical Press, Amsterdam (1977).
5. C.J.Olbricht, J.K.Cannon, L.C.Garg, and C.C.Tisher, Activities of cathepsins B and L in isolated nephron segments from proteinuric and nonproteinuric rats, Am J Physiol , 250:F1062 (1986).
6. J.P.Morin, G.Viotte, A.Vanderwalle, F.VanHoof, P.Tulkens, and J.P.Fillastre, Gentamicin-induced nephrotoxicity: A cell biology approach, Kidney Inter , 18:583 (1980).
7. R.P.Wedeen, V.Batuman, C.Cheeks, E.Marquet, and H.Sobel, Transport of gentamicin in rat proximal tubule, Lab Invest , 48:212 (1983).

URINARY PROTEINASE ACTIVITY IN PATIENTS WITH ACUTE RENAL FAILURE

AFTER TRAUMA AND KIDNEY TRANSPLANTATION

Christoph Wanner[1], Stephan Greiber[1], Günther Kirste[2], Peter Schollmeyer[1], and Walter H. Hörl[1]

Department of Medicine[1], Division of Nephrology and Department of Surgery[2], University of Freiburg, Hugstetterstr. 55, 7800 Freiburg/FRG

INTRODUCTION

A pathophysiological stress that produces a "hypercatabolic" metabolic state often will cause an increased basal energy expenditure in addition to eccessive loss of body protein. This condition is further characterized by severe muscle wasting and increased urinary nitrogen excretion. Both protein synthesis and breakdown are stimulated in severe skeletal trauma (Birkhahn et al., 1981), sepsis (Long et al., 1977) and after burn (Kien et al. 1978). Under these conditions enhanced proteolytic activity in plasma ultrafiltrates and urine fractions of patients with severe trauma and acute renal failure has been reported (Hörl and Heidland, 1980; Hörl et al., 1981; 1982a; 1983; 1986). Until now, the effect of postischemic acute renal failure on urinary protease excretion has not been investigated. Therefore, the present study was undertaken to compare urinary protease activity in patients with acute renal failure after cadaveric kidney transplantation and following multiple traumatic injuries. Furthermore, the proteolytic enzymes involved were partly characterized.

PATIENTS AND METHODS

Seven severely traumatized patients (2 female, 5 male) aged 52.0 ± 9.0 years were investigated. The results were compared to those of 6 patients (3 female, 3 male; age 34.5 ± 1.8 years) after cadaveric kidney transplantation. All patients were suffering from posttraumatic or postischemic acute ·renal failure. Patients with oligo-anuric acute renal failure were characterized by a urine production of less than 500 ml/day. Kidney failure persisted for a mean of 12.0 ± 1.3 days after kidney transplantation. For controls the following groups of patients were studied: Ten healthy subjects aged 36.9 ± 4.1 years, 10 trauma patients with normal kidney function and 10 trauma patients with normal kidney function and sepsis. Traumatized patients were studied for nine days after the admission to the hospital, whereas patients after kidney transplantation were followed up to 35 days postoperatively. Urine samples were collected on ice (4°C) daily between 8 and 9 AM. Urine cultures were done in all patients and a bacterial contamination of $> 10^5$ microorganisms/ml were considered significant. In addition microscopical examination of the urine specimen was performed in most of the patients. The day of admission to the hospital or the day of transplantation was defined as day 0. Urine fraction obtained from the patients and healthy controls were incubated with azocasein (Serva, Heidelberg, FRG), phosphorylase kinase, or both by the methods previously described (Hörl et al., 1981; 1982 a and b; 1983; 1986). Phosphorylase kinase from rabbit skeletal muscle was isolated as described by the method of Cohen (1973) modified according to Jennissen and Heilmeyer (1975). Polyacrylamide gel electrophoresis in the presence of sodium dodecyl sulfate was carried out according to Porzio and Pearson (1977). Cytochrome C (Sigma, Chemical Co. Munich) was used as a standard protein. Protein concentration of the urine and urine supernatant after TCA precipitation were determined by the method of Lowry et al. (1951).

Statistics

All values are given as mean ± SEM. Wilcoxon's test for paired samples or the non-parametric Mann-Whitney U-test were used for analysis of significance.

RESULTS

In the present study 7 patients with multiple traumatic injuries and 6 patients after cadaveric kidney transplantation were investigated. Urinary proteinase activity was measured using phosphorylase kinase as substrate. Compared with healthy controls (CO), significantly higher proteolytic activity, measured with azocasein as a substrate, was found in trauma patients, patients with trauma and sepsis, patients with trauma, sepsis and acute renal failure as well as patients with acute renal failure after cadaveric renal transplantation (*p<0.01; Fig.1).

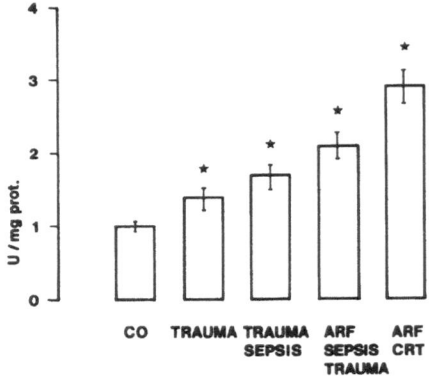

Fig. 1. Azocasein hydrolyzing activity in urine fractions of 10 healthy controls (CO), 15 trauma patients with normal kidney function, 7 patients with trauma complicated by sepsis, 7 patients with trauma, sepsis and acute renal failure (ARF) as well as 6 patients with oligo-anuric acute renal failure after cadaveric renal transplantation (CRT). Data expressed as Units/mg protein are given as mean ± SEM. * p<0.01.

Figure 2a shows the effect of urine fraction on the subunit structure of phosphorylase kinase 2, 5, 9 and 16 days following successful cadaveric renal transplantation. Two days postoperatively the alpha polypeptide chain of the enzyme was partly degraded. On the other hand, the enzyme was stable for 24 hours in the presence of urine fractions at day 9 and day 16. Figure 2b demonstrates the effect of an acute rejection episode on phosphorylase kinase degradation in vitro. The proteolytic digestion of the alpha and gamma subunits of the enzyme was associated with an increase of plasma creatinine from 4.2 mg/dl to 4.6 mg/dl. Administration of methylprednisolone caused an inhibition of urinary proteinase activity and an improvement of kidney function (plasma creatinine 0.9 mg/dl at day 12). The subunit structure of phosphorylase kinase was completely degraded after 24 hours of incubation of a urine fraction of a patient with trauma, sepsis and acute renal failure (Fig. 3).

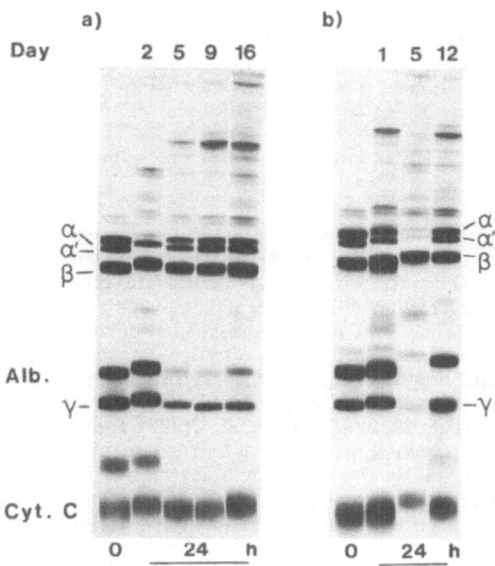

Fig. 2a. Effect of urine fractions obtained from a patient after successful kidney transplantation during uncomplicated postoperative follow-up on the subunit structure of phosphorylase kinase.

Fig. 2b. Effect of urine fractions obtained from a patient after successful kidney transplantation complicated by a severe rejection episode on the subunit structure of phosphorylase kinase.

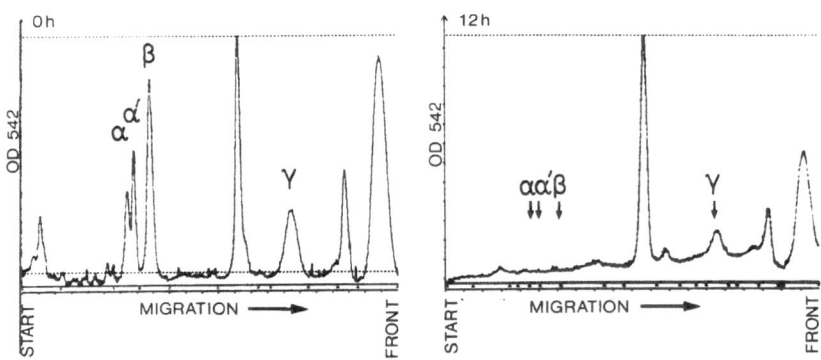

Fig. 3. Effect of urine fraction obtained from a patient with multiple traumatic injuries, sepsis, and acute renal failure on the subunit structure (α, α', β and γ) of phosphorylase kinase. Twenty-eight micrograms of enzyme were incubated for 12 hours at 37° C.

DISCUSSION

In order to investigate urinary proteinase activity in patients with ARF two groups have been selected. Patients after posttraumatic ARF complicated by sepsis were compared to patients with postischemic ARF following cadaveric renal transplantation. When urinary proteinase activity was calculated on protein basis significantly elevated proteinase activity following trauma and sepsis and markedly higher following trauma, sepsis and acute renal failure was found. The highest proteinase activity was observed after cadaveric renal transplantation (fig 1). Until now the origin of urinary proteinases is unclear. Of major interest are the kidney and the circulation. The kidney, particulary the cortex was found to be one source of proteolysis (Hörl et al. 1982). Davin et al. (1987) found activity of neutral metalloproteinases in urine of rats with a proliferative and a non proliferative glomerulonephritis. The proteolytic substances found in the urine could also derive from injured tissue and accumulate in the plasma before they are excreted into the urine. Potential sources are the striated muscle in rhabdomyolysis and after multiple traumatic injuries (Hörl and Heidland 1980, Hörl et al. 1981 and 1982), the lung (Heidland and Hörl 1984), the pancreas (Borgström and Ohlsson 1978), white blood cells (Hörl et al. 1983b; Heidland et al. 1984; Hörl and Heidland 1984) and a variety of bacterial strains in patients with septicemia (Tai et al. 1976; Goldberg and Dice 1974, Goldberg and St.John 1976; Morihara and Tzuzuki 1977)). Different strains of Pseudomonas aeruginosa were identified in blood cultures obtained from patients with septicemia, showing proteolytic digestion of elastin by pseudomonas elastase (Heidland et al. 1987). Elevated proteolytic activity in plasma of patients with trauma or sepsis and in patients who had undergone elective surgery were described by Clowes et al. (1983) and Beisel and Wannemacher (1980) found a putative humoral substance in the plasma of those patients. Since now no data are available about proteolysis in an organ crippled by cold storage time, immunological determinants and potential ciclosporin nephrotoxicity. Patients after kidney transplantation can also be considered catabolic since high doses of steroids are administered postoperatively. In summary, elevated proteinase activity was found early in the clinical coures in patients admitted with multiple traumatic injuries. Sepsis or acute renal failure as well as kidney transplantation induced a further increase of urinary proteolytic activity. These data indicate that different proteinases are present in urine fractions of traumatized patients and those with sepsis, acute renal failure or kidney transplantation.

ACKNOWLEDGEMENTS

This work was supported by the Deutsche Forschungsgemeinschaft, Ho 781/3-4. The excellent technical assistance of Miss U. Schaeffer is greatly appreciated.

REFERENCES

Beisel, W.R., Wannemacher, R.W. Jr., 1980, Gluconeogeneses, ureagenesis, and ketogenesis during sepsis. J. Parenter. Enteral. Nutr., 4:277

Birkhahn, R.H., Long, C.L., Fitkin. D., Jeevanandam, M., and Blakemore, W.S., 1981, Whole body protein metabolism due to trauma in man as estimated by L-(15N)ala-nine, Am J Physiol., 241:E64.

Borgström, A., and Ohlsson, K., 1978, Studies on the turnover of endogenous cathodal trypsinogen in man, Eur J Clin Invest., 8:379.

Clowes, G.H.A. Jr., George, B.C., Villee, C.A., and Saravis, C.A., 1983, Muscle proteolysis induced by a circulating peptide in patients with sepsis or trauma, N Engl J Med., 308:545.

Cohen, P., 1973, The subunit structure of rabbit-skeletal-muscle phosphorylase kinase and the molecular basis of its activation reaction, Eur J Biochem., 34:1.

Davin, J-C., Davis, M., Foidart, J-M, Foidart, J.B., Dechenne, C.A., and Mahieu,P.R., 1987, Urinary excretion of neutral proteinases in nephrotic rats with a glomerular disease, Kidney int., 31:32.

Goldberg, A.L., and Dice, J.F., 1974, Intracellular protein degradation in mammalian and bacterial cells: part I, Annu Rev Biochem., 43:835.

Goldberg, A.L., and St. John, A.C., 1976, Intracellular protein degradation in mammalian and bacterial cells: part II, Annu Rev Biochem., 45:747.

Heidland, A., and Hörl. W.H., 1984, Contribution of proteases to hypercatabolism in acute renal failure, in: Nephrology, Robinson, R.R., ed., Springer Verlag, NewYork, 763.

Heidland, A., Hörl, W.H., and Heller, N., 1984, Granulocyte lysosomal factors and plasma elastase in uremia: a potential factor of catabolism, Klin Wochenschr., 62,218.

Heidland, A., Schaefer, R.M., Weipert, J., Heidbreder, E., Teschner, M., Peter, G., and Hörl, W.H., 1987, Catabolism in acute renal failure: importance of glucocorticoids and lysosomal enzymes, in: Acute renal failure, Amerio, A., Coratelli, P., Campese, V.M., and Massry, S.G., eds., Plenum Press New York, London, 41.

Hörl, W. H., and Heidland, A., 1980, Enhanced proteolytic activity - cause of protein catabolism in acute renal failure, Am J Clin Nutr., 33:1423.

Hörl, W. H., Stepinski, J., Gantert, C., Hörl, M,. and Heidland, A., 1981, Evidence for the participation of proteases on protein catabolism during hypercatabolic renal failure, Klin Wochenschr., 59:751.

Hörl, W. H., Gantert, C., Auer, I. O., and Heidland, A., 1982, In vitro inhibition of protein catabolism by α_2-macroglobulin in plasma from a patient with posttraumatic acute renal failure, Am J Nephrol., 2:33.

Hörl, W.H., Schäfer, R.M., and Heidland, A., 1982, Role of urinary α_1-antitrypsin in Padutin (kallikrein) inactivation, Eur J Clin Pharmacol., 22:541.

Hörl, W. H., Stepinski, J., Schaefer, R., Wanner, C., and Heidland, A., 1983a Role of pro-teases in hypercatabolic patients with renal failure, Kidney Int., 24 (suppl.16):S-37.

Hörl, W.H., Jochum, M., Heidland, A., and Fritz, H., 1983b, Release of granulocyte proteinase during hemodialysis, Am J Nephrol., 3:213.

Hörl, W.H., and Heidland, A., 1984, Evidence of the participation of granulocyte proteinases on intradialytic catabolism, Clin Nephrol., 21:314.

Hörl, W. H., Wanner, C., Thaiss, F., and Schollmeyer, P., 1986, Detection of a metallo-proteinase during hemodialysis in patients with acute and chronic renal failure, Am J Nephrol., 6:6.

Jennissen, H.P., and Heilmeyer, L.M.G.jr., 1975, General aspects of hydrophobic chromatography. Absorption and elution characteristics of some skeletal muscle enzymes, Biochemisty, 14:754.

Kien, C.L., Young, V.R., Rohrbough, D.K., and Burke, J.F., 1978, Increased rates of whole body protein synthesis and breakdown in children recovery from burns, Ann Surg., 187:383.

Long, C.L., Jeedvanandam, M., Kim, B.M., and Kinney, J.M., 1977, Whole body protein synthesis and catabolism in septic man, Am J Clin Nutr., 30:1340.

Lowry, O.H., Rosebrough, N.J., Farr, A.L., and Randall R.J., 1951, Protein measurements with the Folin phenol reagent, J Biol Chem., 193:265.

Morihara, K., and Tzuzuki, H., 1977, Production of protease and elastase by Pseudomonas aeruginosa strains isolated from patients, Infect Immun., 15:679.

Porzio, M.A., and Pearson, A.M., 1977, Improved resolution of myofibrillar proteins with sodium dodecyl sulfate-polyacrylamide gel electrophoresis, Biochem Biophys Acta, 490:27.

Tai, J.Y., Kortt, A.A., Lu, T.Y., and Elliot, S.D., 1976, Primary structure of streptococcal proteinases. III. Isolation of cyanogen bromide peptides: complete covalent structure of the polypetide chain, J Biol Chem., 251:1955.

MECHANISMS FOR ACTIVATION OF PROTEOLYSIS IN UREMIA

William E. Mitch

Renal Division
Emory University School of Medicine
Atlanta, GA

INTRODUCTION

Both acute (ARF) and chronic (CRF) renal failure are commonly accompanied by evidence of abnormal protein metabolism, including negative nitrogen balance and anthropometric changes, consistent with loss of lean body mass. It would be advantageous to identify the causes and mechanisms underlying defects in protein metabolism which are associated with renal failure because defective protein conservation is one of the major factors causing morbidity in CRF and the high mortality rate in patients with ARF due to sepsis and/or trauma.[1] Evidence that we and others have gathered indicates that stimulation of catabolic pathways by the metabolic abnormalities resulting from loss of renal function is likely to be the most important cause of abnormal protein turnover in uremic patients.

The amount of protein being synthesized and degraded each day is large compared to dietary nitrogen. Consequently, factors affecting the turnover of endogenous proteins become critical in determining lean body mass if they persist for extended periods. For example, it can be estimated that a normal 70 kg adult is synthesizing and degrading protein at a rate equivalent to 320 g/day. Consequently, even a small percentage change in protein synthesis and/or degradation lasting for several months would lead to a severe loss of lean body mass in patients with CRF.[2]

CATABOLIC FACTORS IN UREMIA

Certain hormonal abnormalities associated with uremia, such as insulin resistance,[3,4] could change the control of protein synthesis and/or degradation and cause loss of lean body mass. In muscle, for example, the ability of insulin to stimulate amino acid uptake could be impaired, reducing intracellular pools of amino acids or causing an abnormal distribution of amino acids, leading to a secondary impairment of protein synthesis. Maroni et al. examined the impact of uremia on amino acid uptake by measuring the function of the three major transporters of neutral amino acids in muscle, Systems A, ASC and L.[5,6] By using methylaminoisobutyric acid, a probe that is transported specifically by System A, we found that transport by System A in muscle is impaired by ARF if the transporter is not stimulated by insulin.[5] However, the major property of System A is its sensitivity to hormones and this property was preserved; the percent increase in System A transport above the basal level with physiologic concentrations of insulin was the same in muscles of control and ARF. It also was shown that uptake through the other neutral amino acid transporters, Systems ASC and L, was unimpaired by ARF.[6] Thus, in spite of the presence of severe uremia with clearly defined defects in glucose and protein metabolism,[4,7] it appears that impaired amino acid transport, per se, does not play a major role in abnormal muscle protein turnover in acute uremia.

These findings suggest that the abnormal muscle protein synthesis associated with acute and chronic uremia is caused by a defect in the response to insulin.[7,8] Unfortunately, the mechanism by which insulin binding initiates a stimulation of protein synthesis in muscle is not understood. There seems to be no abnormality of insulin binding. When the dose-response relationships between insulin concentration and different parameters of muscle glucose metabolism including glucose uptake, glycogen synthesis and glucose oxidation were analyzed, May et al. found that the insulin concentration causing half-maximal stimulation of these metabolic functions was not different in muscles from control and ARF rats.[4] On the other hand, two findings did suggest the presence of post-receptor defects in insulin action. First, the dose-response curves indicated that the maximal rate of each metabolic function was lower in muscles of ARF rats, consistent with a post-receptor defect.[9] Second, insulin activation of glycogen synthase in muscle of ARF rats was clearly subnormal. An impaired ability of insulin to activate this cytosolic enzyme when insulin binding is intact also is consistent with the presence of a post-receptor defect. Consequently, it seems likely that defective insulin-stimulated protein synthesis, which is present at all levels of insulin,[7] is likely to be post-receptor in nature. Unfortunately, the subnormal rate of protein synthesis measured in the absence of insulin affects the interpretation of changes in the dose-response curve,[9] so the presence of a post-receptor defect in protein synthesis cannot be proven. It should be emphasized, however, that in both ARF and CRF, the degree of the abnormality in muscle protein synthesis was small compared to the increase in protein degradation.[7,8]

Acute and chronic uremia stimulate protein breakdown in muscle. In ARF, muscle protein degradation is associated with abnormalities in insulin-stimulated glucose metabolism, suggesting that defective glucose metabolism stimulates protein degradation. We found that in ARF, an increase in the proportion of muscle glucose uptake that is shunted into lactate production is closely associated with an increase in protein degradation.[7,8] A similar relationship was found when muscles of fed and starved normal rats, and rats responding to the catabolic stress of a thermal injury were studied.[10,11] To uncover a mechanism for this relationship, we used two methods of changing glucose metabolism to determine if there would be a concomitant change in protein degradation. First, we incubated muscles from normal rats with vanadate because we found that this compound exerted insulin-like effects, including a stimulation of glucose uptake and glycolysis.[12] However, vanadate did not change the proportion of glucose uptake that was shunted into glycolysis and therefore, according to our previous results, proteolysis should not be stimulated. In fact, neither protein synthesis nor protein degradation were changed. In a second study, Clark et al. used dichloroacetate to take advantage of its ability stimulate pyruvate dehydrogenase and alter lactate production. Although dichloroacetate did change the proportion of muscle glucose uptake shunted into glycolysis (or lactate production), protein degradation was unaffected.[13] These results indicate that a change in glycolysis may not directly stimulate proteolysis, although it remains possible that the association between glycolysis and protein metabolism seen in muscles of normal and catabolic rats is caused by activation of parallel pathways linked to a defect in insulin-mediated metabolism.

Other indirect evidence for a link between abnormal energy metabolism in uremia and defects in the control of protein turnover is found in the report of Li and Wassner.[14] They studied pair-fed CRF rats and noted that fat stores were reduced compared to control rats. In addition, they found a close correlation between fat stores and the weight of the epididymal fat pad, permitting them to study whether protein metabolism and fat stores are related. Using the isolated perfused hinquarter technique, they found that diminished fat stores in uremia were closely related to an increase in net protein degradation in muscle. Possibly, the relationship was caused by utilization of lipids to provide the energy necessary for protein breakdown in muscle.

An alternative explanation would be that a common abnormality in many tissues, including muscle and adipocytes, occurs in uremia leading to defective metabolism. McCaleb et al. showed that insulin resistance could be induced in normal adipocytes by incubating them with serum obtained from dialysis patients.[15] Moreover, Druml et al.[16] recently found that amino acid transport in adipocytes of CRF rats is subnormal and closely correlated with defective cation transport. Uremia-induced defective ion transport is reminiscent of the findings of Welt and associates who reported in 1964 that cation transport in erythrocytes of certain dialysis patients was abnormal[17]. The demonstration that the abnormality is also present in adipocytes and skeletal muscle and that cation transport abnormalities are closely linked to depressed amino acid transport may mean that defects in membrane potential, ion transport and/or the regulation of intracellular ion concentrations play an important role in the control of the metabolism of protein, lipids and other cellular constituents in uremia.

Recently, the cause of the defect in cellular ion transport in uremia has become more clear. Kelly and associates found that plasma of normal humans contains factors that exhibit digitalis glycoside-like activity; these factors inhibit NaK-ATPase activity in vitro, bind to the enzyme, displace ouabain from NaK-ATPase, and they impair normal cation flux in in erythrocytes.[18] In a subsequent report, Kelly et al. showed that unsaturated fatty acids and lysophospholipids present in plasma could account for all of these changes in NaK-ATPase.[19] The same factors are present at somewhat higher concentrations in the plasma of uremic patients.[20] Indirect data suggest that the inhibitory effect of plasma lipids involves a change in membrane composition and function rather than a direct effect of lipid at the cardiac binding site of NaK-ATPase. If this postulate is confirmed, then changes in membrane function might provide the apparent link between abnormal cation transport and the metabolic function of adipocytes.[16] Moreover, a membrane defect could involve transporters other than NaK-ATPase, leading to intracellular acidosis which would explain the increase in muscle protein breakdown that occurs in uremia.

A major factor causing excessive muscle proteolysis in experimental uremia is metabolic acidosis. In a study of rats with normal renal function, May et al. found that feeding ammonium chloride or hydrochloric acid stimulated both glucocorticoid production and muscle protein degradation.[21] Stimulation of muscle protein breakdown required both glucocorticoid production and metabolic acidosis because proteolysis was not increased if rats were first adrenalectomized, and then rendered acidotic, nor was it increased if normal muscles were incubated in acidified media. The importance of this experiment to uremia is that rats with CRF and mild metabolic acidosis (plasma bicarbonate <21 mM) exhibited an increase in muscle protein breakdown that was dependent on metabolic acidosis.[8] This defect in the control of protein turnover could not be attributed to the higher rate of glucocorticoid production because CRF rats that were pair-fed the diet supplemented with bicarbonate did not develop metabolic acidosis, and there was no increase in muscle protein breakdown even though corticosterone excretion rate was supranormal.

A potential mechanism causing abnormal protein turnover in uremia could involve excessive catabolism of leucine or its ketoacid, α-ketoisocaproate (KIC). In incubated muscles, it can be shown that leucine stimulates protein synthesis while KIC inhibits protein breakdown.[22] The anabolic effect of KIC may have clinical relevance since infusion of KIC into starving obese subjects was found to improve their nitrogen balance.[23] The improvement was attributable almost entirely to decrease in the rate of urea production, suggesting that KIC inhibited net protein degradation. A possible link between the anabolic effect of KIC and abnormal protein turnover in uremia is that metabolic acidosis by itself stimulates the activity of branched-chain ketoacid dehydrogenase in muscle and therefore, increases the oxidative degradation of KIC and branched-chain amino acids.[24] Metabolic acidosis occurring as a complication of CRF also increases KIC catabolism. Compared to pair-fed, sham-operated rats, CRF rats with mild metabolic acidosis (plasma bicarbonate <21 mM) exhibited accelerated decarboxylation of both valine and leucine in muscle.[25] When the metabolic acidosis was corrected by feeding bicarbonate, the accelerated decarboxylation was eliminated, and plasma and muscle levels of leucine actually rose. The link to the control of protein turnover is that correction of metabolic acidosis also blocked the increase in muscle protein degradation associated with CRF.

In summary, there is evidence in experimental uremia that the control of nitrogen metabolism is impaired and that muscle protein catabolism is accelerated. Two causes for the catabolic effect of uremia have been identified: an abnormality in insulin-stimulated protein synthesis, and an increase in muscle protein degradation. The latter can be eliminated by correcting the metabolic acidosis of uremia, but the mechanism for the defect in insulin-stimulated protein synthesis is unknown. Further work will be necessary to determine if the abnormalities in amino acid metabolism associated with uremia or other mechanisms account for these defects.

REFERENCES

1. W.E. Mitch, and M. Walser. Nutritional therapy of the uremic patient, in: The Kidney, Edition III, B.M. Brenner, and F.C. Rector, eds., W.B. Saunders, New York (1986).
2. M.A. Holliday. Protein intake, renal function and growth in chronic renal failure, in: The Progressive Nature of Renal Disease, W.E. Mitch, ed., Churchill-Livingstone, New York (1986).
3. R.A. Defronzo, A. Alvestrand, D. Smith, R. Hendler, E. Hendler, and J. Wahren. Insulin resistance in uremia. J. Clin. Invest. 67:563 (1981).
4. R.C. May, A.S. Clark, A. Goheer, and W.E. Mitch Identification of specific defects in insulin-mediated muscle metabolism in acute uremia. Kidney Int. 28:490 (1985).

5. B.J. Maroni, G. Karapanos, and W.E. Mitch. System A amino
    acid transport in incubated muscle: Effects of insulin and
    acute uremia. Am. J. Physiol. 251:F74 (1986).
6. B.J. Maroni, G. Karapanos, and W.E. Mitch. System ASC and
    Na-independent neutral amino acid transport in muscle of
    uremic rats. Am. J. Physiol. 251:F81 (1986).
7. A.S. Clark, and W.E. Mitch. Muscle protein turnover and
    glucose uptake in acutely uremic rats: Effects of insulin
    and the duration of renal insufficiency. J. Clin. Invest.
    72:836 (1983).
8. R.C. May, R.A. Kelly, and W.E. Mitch. Mechanisms for
    defects in muscle protein metabolism in rats with chronic
    uremia: The influence of metabolic acidosis. J. Clin.
    Invest 79:1099 (1987).
9. C.R. Kahn. Insulin resistance, insulin sensitivity and
    insulin unresponsiveness: a necessary distinction.
    Metabolism 27(Suppl 2):1893 (1978).
10. W.E. Mitch, A.S. Clark, and R.C. May. Relationships
    between protein degradation and glucose metabolism in
    skeletal muscle, in: Proc Vth International Congress on
    Proteolysis, E. Khairallah, J. Bond, and W.S. Bond, eds.,
    Alan R. Liss, New York (1985).
11. A.S. Clark, R.A. Kelly, and W.E. Mitch. Systemic
    response to thermal injury in rats: increased protein
    degradation and altered glucose utilization in muscle.
    J. Clin. Invest. 74:888 (1984).
12. A.S. Clark, J.M. Fagan, and W.E. Mitch. Selectivity of
    the insulin-like actions of vanadate on glucose and
    protein metabolism in skeletal muscle. Biochem. J.
    232:273 (1985).
13. A.S. Clark, W.E. Mitch, M.N. Goodman, J.M. Fagan, A.
    Goheer, and R.T. Curnow. Dichloroacetate augments
    insulin-stimulated glycogen synthesis and inhibits
    glycolysis in rat skeletal muscle. J. Clin. Invest.
    79:588 (1987).
14. J.B. Li, and S.J. Wassner. Protein synthesis and
    degradation in skeletal muscle of chronically uremic
    rats. Kidney Int. 29:1136 (1986).
15. M.L. McCaleb, R. Mevorach, R.B. Freeman, M.S. Izzo, and
    D.H. Lockwood. Induction of insulin resistance in normal
    adipose tissue by uremic human serum. Kidney Int.
    25: 416 (1984).
16. W. Druml, R.A. Kelly, R.C. May, and W.E. Mitch.
    Mechanisms for defective cation flux in chronic renal
    failure. Proc. Xth Int. Congress. Nephrology (In Press).
17. L.G. Welt, J.R. Sacks, and T.J. McManus. An ion
    transport defect in erythrocytes from uremic patients.
    Trans. Assoc. Am. Phys. 77:169 (1964).
18. R.A. Kelly, D.S. O'Hara, W.E. Mitch, and T.W. Smith.
    Characterization of digitalis-like factors in human
    plasma: interactions with NaK-ATPase and crossreactivity
    with cardiac glycoside-specific antibodies. J. Biol.
    Chem. 260:11396 (1985).
19. R.A. Kelly, D.S. O'Hara, W.E. Mitch, and T.W. Smith.
    Identification of digitalis-like factors in human plasma.
    J. Biol. Chem. 261:11704 (1986).
20. R.A. Kelly, D.S. O'Hara, W.E. Mitch, T.I. Steinman, R.C.
    Goldszer, H.S. Solomon, and T.W. Smith. Endogenous
    digitalis-like factors in hypertension and chronic renal
    insufficiency. Kidney Int. 30:723 (1986).

21. R.C. May, R.A. Kelly, and W.E. Mitch. Metabolic acidosis stimulates protein degradation in rat muscle by a glucocorticoid-dependent mechanism. J. Clin. Invest. 77:614 (1986).
22. W.E. Mitch, and A.S. Clark. Specificity of the effect of leucine and its metabolites on protein degradation in skeletal muscle. Biochem. J. 222:579 (1984).
23. W.E. Mitch, M. Walser, and D.G. Sapir. Nitrogen sparing induced by leucine compared with that induced by its keto analogue, α-ketoisocaproate, in fasting obese man. J. Clin. Invest. 67:553 (1981).
24. R.C. May, Y. Hara, R.A. Kelly, K.P. Block, M.G. Buse, and W.E. Mitch. Branched-chain amino acid metabolism in rat muscle: Abnormal regulation in acidosis. Am. J. Physiol. (In Press).
25. Y. Hara, R.C. May, R.A. Kelly, and W.E. Mitch. The influence of acidosis on branched-chain amino acid metabolism in chronic uremia. Kidney Int. (In Press).

EVIDENCE FOR THE ROLE OF PROTEINASES IN UREMIC CATABOLISM

Roland M. Schaefer[1], Markus Teschner[1], Gernot
Peter[2], Jochen Leibold[1], Peter Kulzer[1], and August
Heidland[1]

Departments of [1]Medicine and [2]Dermatology
University of Wuerzburg, FRG

INTRODUCTION

Characteristical feature of patients with acute renal
failure is a negative nitrogen balance with often dramatic
wasting of skeletal muscle (1). In any major catabolic
event, it is likely, that skeletal muscle provides an increased supply of amino acids for the enhanced metabolic activity
of the liver. This loss of tissue may represent an adaptive
response helping to meet the metabolic needs of the stressed
organism. On the other hand, this loss of protein stores
is probably a major contributing factor of the persistantly
high mortality in patients with hypercatabolic acute renal
failure.

The mechanism, underlying the accelerated protein
degradation, observed in uremia is not well understood.
Underlying illnesses commonly associated with acute renal
failure, such as septicemia (2), shock, trauma or uremia
per se might promote net protein breakdown (3). In order
to further elucidate the role of proteinases in the catabolic
state of acute uremia, we investigated the effect of proteinase inhibition on muscle protein breakdown. Acutely
uremic rats were treated with leupeptin, a low-molecular
weight proteinase inhibitor produced by streptomyces species
(4) a compound, which has been successfully employed in
animal experiments to slow accelerated muscle proteolysis
in several catabolic conditions (5, 6). This has been interpreted as a consequence of inhibition of muscular proteinases, both of lysosomal and cytosolic localisation (7).
Thus, leupeptin should prove to be a helpful means to clarify, whether proteinases are involved in the catabolic state
of acute uremia.

METHODS

Male Wistar rats (Ivanovas, Kisslegg/Allgäu, FRG) weighing 200 - 250 g, were used for the experiments. After an adaptive period of 5 days, the animals underwent sham operation and bilateral nephrectomy (BN). The surgical procedure was performed under anesthesia using hexobarbital (50 mg/kg b.w.) i.p. The acutely uremic animals were randomly assigned to two different subgroups (8 animals each). The experimental group received leupeptin (Peptide Institute, Osaka, Japan) intraperitoneally 6 times per day in a cummulative dose of 180 mg/kg b.w./24 hours. Sham operated animals as well as the binephrectomized controls received only the vehicle. 24 hours after surgery, the animals were anesthetized with hexobarbital, the abdominal aorta was exposed and blood samples were obtained. In addition, tissue from gastrocnemius muscle was removed.

The serum concentration of glucose, electrolytes, ureanitrogen and creatinine were assayed by means of a Technicon Autoanalyzer. Blood-pH and bicarbonate were determined by Astrup methods with an AVL-Gas Check analyzer (AVL-GmbH, Bad Homburg, FRG). N^t-methylhistidine was measured as previously described (8), the alkaline myofibrillar proteinase according to the method of Mayer et al. (9). The enzyme activity is given as µg of tyrosine equivalents, released per mg of myofibrillar protein.

Statistical analysis was performed by the Mann-Whithney rank sum test. Significance was defined as a P value of less than 0,05 taking into account the number of analyses performed. All results are given as mean + SEM.

RESULTS

Table 1 depicts the blood chemistries of the sham-operated and binephrectomized rats, either treated with leupeptin or not. According to the levels of potassium, phosphate, creatinine, SUN and acid base status, BN-animals became uremic 24 hours after operation. When acutely uremic rats were treated with leupeptin, serum urea levels were significantly reduced (-32%) as compared to untreated BN rats. This decrease was not due to a less severe form of azotemia, since creatinine, potassium and the degree of metabolic acidosis were comparable in BN rats with or without leupeptin treatment. It is of note that serum glucose levels were lower (-35%) in leupeptin-treated BN animals than in untreated BN rats, thereby suggesting a reduction in hepatic gluconeogenesis.

N^t-methylhistidine was measured in the plasma of these animals, to investigate, whether the leupeptin-induced changes in net protein breakdown could be correlated to alterations in the rate of myofibrillar protein degradation. In contrast to untreated BN rats, administration of leupeptin caused a significant decrease (-35%) in plasma levels of N^t-methylhistidine in acutely uremic animals (Tab. 2).

Table 1. Effect of leupeptin on blood chemistries 24 hours after sham operation (SHAM) or bilateral nephrectomy (BN).

Parameters	SHAM 24 hr	BN 24 hr	BN 24 hr + Leupeptin
SUN (mg/dl)	26 $\pm$ 1,0[a]	139 $\pm$ 3	95 $\pm$ 4[b]
Creatinine (mg/dl)	0,7$\pm$ 0,02[a]	4,1 $\pm$ 0,1	3,9$\pm$ 0,1
Glucose (mg/dl)	113$\pm$ 6	123 $\pm$ 12	80 $\pm$ 5[b]
Sodium (mmol/l)	137$\pm$ 0,5	140 $\pm$ 0,5	141$\pm$ 1,5
Potassium (mmol/l)	4,0$\pm$ 0,2[a]	5,7 $\pm$ 0,3	6,5$\pm$ 0,4
ion.Calcium (mmol/l)	1,01$\pm$ 0,02[a]	0,93$\pm$ 0,01	0,77$\pm$ 0,03[b]
Phosphate (mmol/l)	2,4$\pm$ 0,1[a]	5,3 $\pm$ 0,3	6,7$\pm$ 0,2[b]
Bicarbonate (mmol/l)	24,7$\pm$ 0,6[a]	14,8$\pm$ 1,0	14,0$\pm$ 2,1
pH	7,40$\pm$ 0,01[a]	7,23$\pm$ 0,02	7,21$\pm$ 0,03

Mean values $\pm$ SEM, obtained from 10 animals.
[a]$p < 0,01$ for SHAM vs. BN and [b]$p < 0,01$ for BN + Leupeptin vs. BN.

Table 2. Effect of leupeptin on plasma levels of N^t-methyl-histidine in acutely uremic rats.

	BN 24 hr	BN 24 hr + Leupeptin	p
N^t-Methylhistidine µg/dl	426 $\pm$ 30	276 $\pm$ 15	0,01

Mean values $\pm$ SEM obtained from 10 animals.

Table 3 demonstrates the results of the alkaline myofibrillar proteinase in skeletal muscle. The activity of this proteinase was significantly increased in acutely uremic rats in comparison to sham-operated controls. This increment was significantly reduced by administration of leupeptin.

Table 3: Effect of leupeptin on muscle alkaline protease (pH 9,1) activity in acutely uremic rats.

Assay	SHAM 24 hr	BN 24 hr	BN 24 hr + leupeptin
autolytic activity µg tyrosine eq./ mg protein/60 min	76 $\pm$ 3	123 $\pm$ 9[*]	93 $\pm$ 9

Mean values $\pm$ SEM, obtained from 8 animals. [*]$p < 0,01$, for $\overline{BN}$ vs. SHAM or $\overline{BN}$ vs. BN + Leupeptin.

DISCUSSION

Leupeptin displays a broad inhibitory spectrum against a variety of proteinases. Thus, lysosomal proteinases, namely cathepsin B, H and L (10), and non-lysosomal proteinases, such as the calcium activated proteinase (5) and the alkaline proteianse (11), both found in the cytosol of skeletal muscle, have been reported to be inhibited by this compound.

As suggested by lower serum levels of SUN and glucose, leupeptin administration resulted in a reduced urea- and gluconeogenesis in acutely uremic rats (Table 1). This finding already indicates, that leupeptin was able to curtail skeletal muscle breakdown of BN animals, since the extent of urea- and gluconeogenesis in acute uremia depends at least partly, on the availability of glugoplastic precursor amino acids from skeletal muscle. This is further underlined by the fact that leupeptin treatment caused a significant decrease in the plasma levels of N^t-methylhistidine, a sensitive indicator of actomyosin breakdown, 24 hours after bilateral nephrectomy, as compared to untreated uremic rats. These findings demonstrates reduced ureagenesis in leupeptin treated BN animals to be due to attenuated degradation of myofibrillar proteins.

Measurements of plasmatic parameters can, however, only be indirect indicators of reduced proteolysis in skeletal muscle. Thus, we determined, whether leupeptin diminishes the alkaline proteinase activity of skeletal muscle, which has been found to be closely related to actomyosin breakdown (13). An increased activity of this proteolytic enzyme in muscle tissue has been found in various catabolic conditions, such as inflammatory diseases, malignant tumors, prolonged starvation, glucocorticoid excess, diabetes and nephrotic syndrome (14).

It has been suggested that the alkaline protease activity of skeletal muscle might be of mast cell origin (15, 16). However, more recently, Mayorek et al. (17) provided evidence that the alkaline protease of skeletal muscle is different from mast cell proteinases and that it represents an endogenous activity of myocytes. Also, even if this proteinase was derived from mast cells, once released it could still degrade muscle protein. Hence, we believe that it is justified to interpret this enzyme as an important representative of proteolytic activity of muscle tissue.

In our study, the activity of this proteinase in acuteley uremic rats was significantly enhanced as compared to sham operated controls (table 3). With the administration of leupeptin this increased proteolytic activity of the uremic muscle could be significantly decreased. Hence, the capability of this agent, to reduce net protein breakdown in acutely uremic rats, provides good evidence that proteolytic enzymes are activated in uremia and do contribute to the catabolism of this state.

SUMMARY

Enhanced muscle protein breakdown has been demonstrated in acutely uremic rats by numerous authors. In order to investigate the pathogenetic role of skeletal muscle proteinases leupeptin, a low-molecular weight proteinase inhibitor, was administered intraperitoneally to acutely uremic rats. Twenty-four hours after bilateral nephrectomy, leupeptin-treated animals displayed significantly lowered serum urea levels (-32%), as compared to untreated uremic rats. As a sign of muscle protein breakdown, plasma levels of N^t-methylhistidine, an indicator of myofibrillar protein degradation, were also decreased (-35%) in the uremic animals treated with leupeptin as compared to untreated uremic rats. Finally, leupeptin treatment resulted in a significant inhibition of the myofibrillar alkaline proteinase activity, a porteinase which has been related to various catabolic conditions. These findings suggest that the increased muscle protein breakdown in uremia is caused by enhanced activity of muscular proteinases and that anti-proteolytic agents display favourable effects on the enhanced protein degradation observed in acute uremia.

REFERENCES

1. E.I. Feinstein, M.J. Blumenkrantz, M. Healy, A. Koffler, H. Silberman, S.G. Massry and J.D. Kopple, Clinical and metabolic responses to parenteral nutrition in acute renal failure, Medicine 60: 124 (1981)

2. G.H.A. Clowes, B.C. George, C.A. Villee and C.A. Saravis, Muscle proteolysis induced by a circulating peptide in patients with sepsis or trauma, New Engl. J. Med. 308: 545 (1983)

3. E.I. Feinstein, J.D. Kopple, H. Silberman and S.G. Massry, Total parenteral nutrition with high or low nitrogen intakes in patients with acute renal failure, Kidney Int. 24: 319 (1983)

4. H. Umezawa and T. Aoyagi, Activities of proteinase inhibitors of microbial origin, in Proteinases in Mammalian Cells and Tissues, edited by Barrett, A.J., Amsterdam, North-Holland, pp 637 (1977)

5. A. Stracher, E.B. McGowan and S.A. Shafiq, Muscular dystrophy: Inhibition of degeneration in vivo with protease inhibitors, Science 200: 50 (1978)

6. B. McCallister, W.W. Lacy and P.E. Williams, Leupeptin inhibits adrenocorticotropic hormone-induced protein breakdown in the conscious dog, J. Clin. Invest. 83: 390 (1983)

7. P. Libby and A.L. Goldberg, Leupeptin, a protease inhibitor decreases protein degradation in normal and diseased muscles, Science 199: 534 (1978)

8. R.M. Schaefer, M. Teschner, P. Kulzer, J. Leibold, G. Peter and A. Heidland, Evidence for reduced catabolism by the antiglucocorticoid RU 38486 in acutely uremic rats, Am. J. Nephrol. 7: 127 (1987)

9. M. Mayer, R. Amin and E. Schafrir, Rat myofibrillar protease: Enzyme properties and adaption changes in conditions of muscle protein degradation, Arch. Biochem. Biophys. 161; (1974)

10. P. Bohley, H. Kirschke, J. Langner, M. Miche, S. Riemann, Z. Salama, E. Schon, B. Wiederanders and S. Ansorge, Intracellular protein turnover, in Biochemical Functions of Proteinases, edited by Holzer H., Tschesche, H. Berlin, Heidelberg, New York, Springer, pp 17 (1979)

11. H. Reinauer and B. Dahlmann, Alkaline proteinases in sekeltal muscle, in Biochemical Functions of Proteinases, edited by Holzer H., Tschesche, H., Berlin, Heidelberg, New York, Springer, pp 94 (1979)

12. V.R. Young and H.N. Munro, N-Methylhistidine (3-methyl-histidine) and muscle protein turnover: an overview, Fed. Proc. 37: 2291 (1978)

13. B. Dahlmann, C. Schroeter, L. Herbertz and H. Reinauer, Myofibrillar protein degradation and muscle proteinases in nromal and diabetic rats, Biochem. Med. 21: 33 (1979)

14. O.Z. Lernau, S. Nissan, B. Neufeld and M. Mayer, Myo-fibrillar protease activity in muscle tissue from patients in catabolic conditions, Eur. J. Clin. Invest. 10: 357 (1980)

15. D.C. Park, M.E. Parson and R.J. Pennington, Evidence for mast cell origin of proteinase in skeletal muscle homogenates, Biochem. Soc. Trans. 1: 730 (1973)

16. T. Noguchi and M. Kandatsu, Some properties of alkaline protease in rat muscle compared with that in peri-toneal cavity cells, Agric. Biol. Chem. 40: 927 (1976)

17. N. Mayorek, A. Pinson and M. Mayer, Intracellular pro-teolysis in rat cardiac and skeletal muscle cells in culture, J. Cell. Physiol. 98: 587 (1979)

EGLIN C FAILS TO REDUCE CATABOLISM IN ACUTELY UREMIC RATS

M. Teschner[1], R.M. Schaefer[1], Ch. Rudolf[1], P. Kulzer[1], G. Peter[2], and A. Heidland[1]

Departments of[1]Medicine and [2]Dermatology, University of Wuerzburg, FRG
D-8700 Wuerzburg

INTRODUCTION

Eglin C, a proteinase inhibitor originally being isolated from the leech Hirudo medicinalis, is a potent inhibitor of the granulocyte proteinases elastase and cathepsin G (1). Thus, the agent enabled us to investigate the role of granulocytic proteinases in the pathogenesis of catabolism in acute renal failure.

In catabolic situations like acute uremia, skeletal muscle provides an increased supply of gluconeogenic amino acids for the enhanced metabolic activity of the liver (2). The resulting loss of skeletal muscle may partly represent an adaptive response helping to meet the metabolic needs of the stressed organism. This loss of protein stores, however, is probably a major contributing factor of the persistantly high mortality in patients with hypercatabolic acute renal failure (3).

The factors which promote skeletal muscle wasting in acute renal failure have not been fully defined yet. Increased release of proteinases, namely elastase from polymorphonuclear (PMN) leukocytes has been incriminated to be involved in the pathogenesis of enhanced protein breakdown (4). Interestingly, also human cathepsin G from PMN leukocytes has been reported to cleave myosin (5) indicating a possible role for this enzyme in uremic catabolism.

Aim of the present study was to investigate, whether the inhibition of eglin C-sensitive proteinases, like elastase and cathepsin G might result in a reduction of protein breakdown in acutely uremic rats.

METHODS

Male Wistar rats (Ivanovas, Kisslegg/Allgäu, FRG)
weighing 200 - 250 g, were used for the experiments. After an
adaptive period of 5 days, the animals underwent sham operation (sham) and bilateral nephrectomy (BN). The surgical procedure was performed under anesthesia using hexobarbital
(50 mg/kg b.w.) i. p.. The acutely uremic animals were randomly assigned to two different subgroups (8 animals each).
The experimental group received eglin C as a bolus injection
(5 mg/kg b.w.) and continously at a rate of 0,1 mg/kg b.w./h
by way of a subcutanously implanted Alzet model 2 mll osmotic
pump (ALZA Corp.) which was implanted according to the instructions of the manufacturer. Sham operated animals as well
as the binephrectomized controls received only the vehicle.
48 hours after surgery, the animals were anesthetized with
hexobarbital, the abdominal aorta was exposed and blood samples were obtained. In addition, tissue from gastrocnemius
muscle was removed.

The serum concentration of glucose, electrolytes, urea
nitrogen and creatinine were assayed by means of a Technicon
Autoanalyzer. Blood-pH and bicarbonate were determined by
Astrup methods with an AVL-Gas Check analyzer (AVL-GmbH, Bad
Homburg, FRG). N^t-methylhistidine was measured as previously
described (6), the alkaline myofibrillar proteinase was
determined according to the method of Mayer et al. (7).

Statistical analysis was performed by the Mann-Whithney
rank sum test. Significance was defined as a P value of less
than 0.05 taking into account the number of analyses performed. All results are given as mean $\pm$ SEM.

RESULTS

Table 1 depicts the blood chemistries of the sham-operated and binephrectomized rats, either treated with eglin C
or not. According to the levels of potassium, phosphate,
creatinine, SUN, N^t-methylhistidine and acid base status, BN-
animals became uremic 48 hours after operation. As far as
serum levels of SUN and glucose are concerned, there was
no difference among the eglin C-treated animals and the
binephrectomized controls. Similarily, plasma values of
N^t-methylhistidine did not differ.

Table 2 demonstrates the results of the alkaline myofibrillar proteinase of skeletal muscle. The activity of this
proteinase was significantly increased in acutely uremic rats
in comparison to sham-operated controls. This increment was
not influenced by administration of eglin C.

Table 1. Effect of eglin C on blood chemistries and N^t-methylhistidine 48 hours after bilateral nephrectomy (BN)

Parameter	SHAM 48 h	BN 48 h	BN 48 h + egline C
SUN (mg/dl)	22 ± 1*	272 ± 8,2	267 ± 15
Creatinine (mg/dl)	0,6 ± 0,01*	7,6 ± 0,3	7,0 ± 0,6
Glucose (mg/dl)	120 ± 6	112 ± 4	116 ± 3
Sodium (mmol/l)	144 ± 0,3	139 ± 1	142 ± 3
Potassium (mmol/l)	3,7 ± 0,3*	7,3 ± 0,2	7,5 ± 0,4
ion. Calcium (mmol/l)	1,0 ± 0,03	0,8 ± 0,05	0,67 ± 0,04
Phosphate (mmol/l)	1,9 ± 0,2*	8,8 ± 0,2	8,2 ± 0,5
Bicarbonate (mmol/l)	24,6 ± 0,4*	13,8 ± 0,9	13,1 ± 0,65
pH	7,41 ± 0,01*	7,2 ± 0,1	7,3 ± 0,2
N^t-methylhistidine (mg/dl)	0,12 ± 0,01*	0,99 ± 0,08	0,99 ± 0,05

Values are given as means + SEM of 8 animals
*p < 0,01 for SHAM versus BN

Table 2. Effect of eglin C on muscle alkaline protease (pH 9.1)
         activity in acutely uremic rats 48 hours after BN

Assay	SHAM 48 h	BN 48 h	BN 48 h + eglin C
autolytic activity μg tyrosine eq./mg protein/60 min	73 + 3	132 + 10,5*	112 + 9

Mean values + SEM, obtained from 8 animals.   *p < 0,01, for BN vs. SHAM

DISCUSSION

Eglin C has been successfully applied in animal experiments in order to attenuate the consumption of plasma proteins like antithrombin III and factor XIII in septicemia,
probably due to the inhibition of unspecific proteolysis, mediated by leukocytic lysosomal proteinases such as
elastase (8).
Hence, this agent is a valuable tool to investigate the possible role of granulocyte elastase in the pathogenesis of catabolism in acutely uremic rats. As it is demonstrated by serum levels of SUN and glucose, eglin C was not able to reduce
urea- and gluconeogenesis in acutely uremic rats. The extent
of urea- and gluconeogenesis in acute uremia depends,at least
partly,on the availability of glucoplastic precursor amino
acids from skeletal muscle(9).This finding might serve
as a hint, that eglin C was not able to reduce skeletal muscle breakdown of BN-animals. This is in accordance with the
results, that eglin reduced neither plasma levels of the amino acid N^t-methylhistidine (table 1), a sensistive indicator
of actomyosin breakdown (10), nor the stimulated activity of
the alkaline myofibrillar proteinase in comparison to binephrectomized controls. This finding is in contrast to results
obtained with the low molecular weight proteinase inhibitor
leupeptin (MG 476), which is able to enter muscle cells (11),
thereby exerting its antiproteolytic effect in acute uremia
as indicated by an attenuated activity of the alkaline myofibrillar proteinase resulting in decreased catabolism in
acutely uremic rats (6).

Probably, the proteinase-inhibitor eglin C with a molecular weight of 8 100 daltons cannot act intracellularily and
hence failes to reduce stimulated proteolytic activity in
skeletal muscle. Besides increased activity of intracellular
proteinases, extracellular ones namely elevated plasma-levels
of elastase, have also been incriminated to take part in the
pathogenesis of uremic catabolism (4). The missing effect,
however, of the elastase-inhibitor eglin C on uremic catabolism, in contrast to leupeptin, indicates that intracellular acting proteinases are responsible for the accel-
lerated protein degradation in acute uremia rather than
an imbalance of the circulating proteinase-antiproteinase system due to an enhanced release of granulocyte proteinases.

SUMMARY

Increased release of proteinases from polymorphonucelar
leukocytes, namely elastase, has been incriminated to take
part in the pathogenesis of enhanced muscle protein breakdown
in acute renal failure. In order to investigate, whether inhibition of the granulocyte proteinase elastase and cathepsin
G would have a beneficial effect on the extent of muscle protein degradation, eglin C, a potent inhibitor of the granulocyte proteinase elastase and cathepsin G, was administered
intraperitoneally to acutely uremic rats. 48 hours after bilateral nephrectomy, eglin C-treated animals displayed no
significant difference, as far as serum levels of SUN, glucose and N^t-methylhistidine are concerned. Similarily, eglin C
treatment failed to reduce the stimulated activity of the al-

kaline myofibrillar proteinase in comparison to binephrectomized controls. Hence, according to these results, granulocyte proteinases do not seem to be an important mediator of
uremic catabolism, since their inhibition by eglin C does not
reduce enhanced protein breakdown in acutely uremic rats.

ACKNOWLEDGEMENTS

We wish to express our thanks to Prof. H.P. Schnebli, Ciba
Geigy LTD., CH-4002 Basel, Switzerland, for providing us with
Eglin C.

REFERENCES

1. U. Seemüller, M. Meier, K. Ohlsson, H.P. Müller and H.
   Fritz, Isolation and characterization of a low molecular weight inhibitor (of chymotrypsin and human granulocytic elastase and cathepsin G) from leeches, Hoppe-
   Seyler's Z. Physiol. Chem. 358: 1105 (1977)

2. R.M. Flügel-Link, I.B. Salusky, M.R. Jones and J.D.
   Kopple, Protein and amino acid metabolism in posterior
   hemicorpus of acutely uremic rats, Am. J. Physiol. 244:
   E 615 (1983)

3. J.D. Kopple and B. Cianciaruso, The role for nutrition in
   acute renal failure; in Andreucci, Acute Renal Failure
   (ARF), Pathophysiology, prevention and treatment, chap.
   22 (Nijhoff, S'-Gravenhage, 1984)

4. A. Heidland, W.H. Hörl, N. Heller, H. Heine, S. Neumann,
   R.M. Schaefer and E. Heidbreder, Granulocyte lysosomal
   factors and plasma elastase in uremia: a potential factor of catabolism, Klin. Wschr. 62: 218 (1984)

5. J. Travis, P.J. Giles, L. Procelli, C.F. Reilly, R. Baugh
   and J. Powers, Protein degradation in health and disease, Excerpta medica, Ciba Found. Symp. 75: 51 (1980)

6. R.M. Schaefer, M. Teschner, G. Peter, J. Leipold, P.
   Kulzer and A. Heidland, Reduction of muscle protein
   degradation by leupeptin in acutely uremic rats, Min.
   Electr. Metab. (submitted for publication)

7. M. Mayer, R. Amin and E. Shafrir, Rat myofibrillar protease: Enzyme properties and adaptive changes in conditions of muscle protein degradation, Arch. Biochem.
   Biophys. 161: 20 (1974)

8. M. Jochum, H.F. Welter, M. Siebeck and H. Fritz, Proteinase inhibitor therapy of severe inflammation in pigs:
   first results with eglin, a potent inhibitor of granulocyte elastase in cathepsin G. In Proteinases in Inflammation and Tumor Invasion, pp 53 - 59, Walter de
   Gruyter & Co., Berlin - New York (1986)

9. J. H. Exton, Gluconeogenesis, _Metabolism_ 21: 945 (1972)

10. V.R. Young and H.N. Munro, N^t-Methylhistidine(3-methylhistidine) and muscle protein turnover: an overview. _Fed. Proc._ 37: 2291 (1978)

11. P. Libby and A.L. Goldberg, Leupeptin, a protease inhibitor decreases protein degradation in normal and diseased muscles, _Science_ 199: 534 (1978)

EVIDENCE FOR PROTEIN SPLIT PRODUCTS IN PLASMA OF PATIENTS WITH ACUTE
RENAL FAILURE

Marianne Haag, Helmut E. Meyer, Peter Schollmeyer
and Walter H. Hörl

Department of Medicine, Division of Nephrology
University of Freiburg

INTRODUCTION

Major trauma is followed by the loss of body nitrogen. It was suggested
that skeletal muscle is the major site of nitrogen loss. Leg balance studies
in patients with burns, trauma, and sepsis all demonstrate a marked augmen-
tation in the net release of amino acids from leg muscle. Several authors
have demonstrated a marked acceleration of protein turnover, with the increase in protein breakdown exceeding the increase in protein synthesis of
these patients (for review see (1)).

Trichloroacetic acid (TCA) is widely used as a protein coagulant in the
preparation of protein-free sera (2). The mean values of the serum supernatant following TCA precipitation determined by the Lowry method (3) are
246 ± 19 µg/ml in healthy controls and increase to $1,650 \pm 110$ µg/ml in
patients with posttraumatic acute renal failure (ARF) (4). Patients on
regular hemodialysis displayed a value of $1,036 \pm 29$ µg/ml in the supernatants (5). Similar observations were made in urinary samples obtained
from patients following trauma, sepsis or acute renal failure (6). It was
suggested that these supernatants consist of protein split products resulting
of protease-induced protein breakdown and is therefore a good indicator of
catabolism (7).

The aim of this study was to find out whether there is any evidence
for protein split products.

PATIENTS AND METHODS

Blood samples were obtained in EDTA from five healthy persons aged
32 ± 14 years and from five patients with acute renal failure (ARF) aged
58 ± 5 years. In controls and ARF patients we found following mean values
of blood biochemistry: creatinine: 0.82 ± 0.06 mg/dl vs 6.6 ± 1.05 mg/dl;
urea: 22 ± 1.58 mg/dl vs 218 ± 10.2 mg/dl; uric acid: 3.46 ± 0.28 mg/dl vs
8.5 ± 1.59 mg/dl; hemoglobin: 13.5 ± 0.49 g/dl vs 9.6 ± 0.49 g/dl. ARF was
complicated by sepsis in four cases. Three patients suffered from postoperative ARF, one patient was admitted with multiple traumatic injuries and one
patient with postrenal ARF due to stricture of the urethra. All patients were
treated with furosemid and dopamin according to Graziani et al. (8). Informed
consent was obtained from all patients. Parenteral nutrition consisted of

40% glucose and insulin combined with 1g/kg/day of essential and non-essential amino acids.

Plasma was precipitated with 50% TCA at a final concentration of5%. Protein was determined in plasma and in the supernatants after precipitation with TCA according to the method of Lowry (3). Free amino acid analysis was performed in plasma and supernatants after TCA precipitation with a amino acid analyzer LC 4010 (Biotronic, D-8000 München) using an ion-exchange column DC 6A (Durrim, Palo Alto, California/USA). We used Li-Citrate-buffer for elution according to the method of Kedenburg (9).

Fast protein liquid chromatography was undertaken with these supernatants. These supernatants were purified from TCA and equilibrated with the starting buffer using Sephadex G 10 column. Ion exchange chromatography was performed with these purified materials using the Mono Q HR 5/5 (Pharmacia) prepacked strong anion-exchange column. The conditions for this ion exchange chromatography were as follow: Mono Q HR 5/5; buffer A: 20 mM ammonium carbonate, pH 8,5, buffer B: 980 mM NaCl in buffer A, gradient: 0-30% in 18,7 ml, flow rate: 1 ml/min, detection: 280 nm at 0.05 or 0.1 AUFS.

Phenylthiocarbamyl-amino acid (PTC-amino acid) analysis was performed in the supernatants after precipitation with 50% TCA and in the peaks after performing FPLC. Hydrolysis was carried out with 6 N HCl at 150 C for one hour. Amino acid analysis was performed by precolumn derivatization with PITC (10). The PTC-amino acids were separated by reverse phase HPLC using the buffer system C of Heinrikson and Meredith (11) on a 3 μm Spherisorb ODS-2 column (125 x 4,6 mm). PTC-amino acids were detected at 260 nm using a Kratos photometer. Standard deviation does not exceed 10% for each amino acid.

We added to a supernatant after TCA precipitation of a healthy control urea, uric acid, creatinine, guanidine and dopamine HCl in different concentrations and performed protein determination according to the method of Lowry.

Guanidine was quantified in controls and ARF-patients by reversed-phase high-performance liquid chromatography (HPLC) according to the method of Gehrke (12).

RESULTS

Fig.1 shows the plasma concentration of free amino acids and the protein concentration in the supernatants after precipitation with 50% TCA at a final concentration of 5% in controls and patients with acute renal failure. The plasma levels of free amino acids were not significantly different in both groups. It was even lower in patients with acute renal failure. We found a markedly increase of "Lowry protein" in the supernatant of ARF-patients compared to healthy controls. Mean values for "Lowry protein" concentration were 982 $\pm$ 73 μg/ml in ARF-patients and 270 $\pm$ 43 μg/ml in healthy controls.

Using PTC-amino acid analysis for protein determination in the supernatant we found also a significant, but only slight increase in ARF-patients. The protein concentration was there 398 $\pm$ 35 μg/ml in ARF-patients and 207 $\pm$ 30 μg/ml in healthy controls.

If we take the percentage of PTC-amino acids of "Lowry protein", we found that in healthy controls PTC-amino acids represent 74.5 $\pm$ 10.9%, in ARF-patients, however, only 42.2 $\pm$ 5.2%.

In order to investigate several substances which may interfere with "Lowry protein" determination TCA supernatants were incubated with different

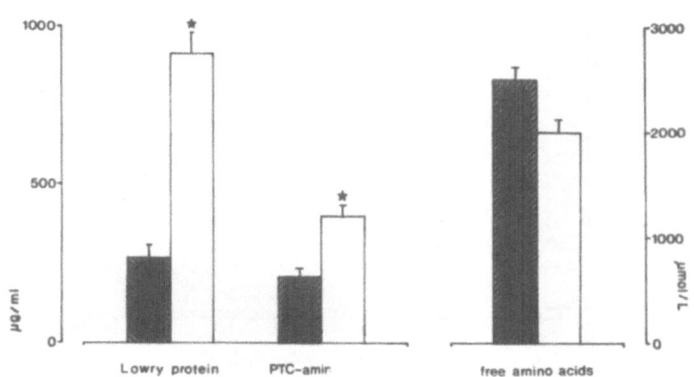

Fig. 1. "Lowry protein" determination and PTC-amino acid
analysis in supernatants after precipitation
with 50% TCA at a final concentration of 5% in
healthy controls and patients with acute renal
failure

concentrations of urea (10-500 mg/dl), uric acid, creatinine (1-20 mg/dl)
and dopamine. Urea and creatinine showed no effect on "Lowry protein" determination and also uric acid in concentrations measured in our patients with
acute renal failure. Dopamine and furosemide in concentrations according to
the plasma levels of the ARF patients did not influence "Lowry protein"
measurement. High doses of uric acid (50 mg/dl) and dopamine (80 µg/dl) induced an increase of "Lowry protein" levels in vitro experiments (uric acid:
526.2 vs. 971.1 µg/ml; dopamine: 502.2 vs. 988.2 µg/ml).

In separate experiments plasma guanidine levels were determined in controls and ARF patients in order to find out whether these substances are responsible for the elevated "Lowry protein" in TCA supernatants of ARF patients.
We could show that guanidine in ARF patients was lower than in controls
(data not shown).

Figures 2 and 3 demonstrate the results after fast protein liquid chromatography of a TCA supernatant of a healthy control and a patient with acute
renal failure. In both chromatographs we found the same peaks. The only
difference between the chromatograph of controls and ARF patients was, that
peak 5 was higher in ARF patients than in controls. Performing PTC-amino
acid analysis of all peaks we could confirm the results before. Nearly all
protein is found in peak 5 and protein concentration in peak 5 is significantly higher in ARF-patients.

DISCUSSION

During the last decades numerous investigations gave evidence for an
accelerated protein breakdown in acute renal failure and sepsis. The first
data concerning a stimulation of proteolysis were reported by Richet and
coworkers, who observed an increased tripeptidase activity in plasma of
hypercatabolic patients with renal failure (13,14). Hörl et al. (4, 15-18)
and Wanner et al. (6) have shown that plasma ultrafiltrates and urine frac-

341

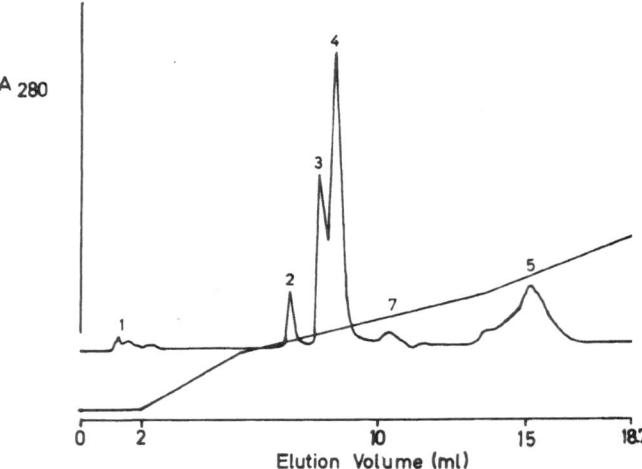

Fig. 2. Performing of FPLC of a supernatant of a
control after precipitation with 50% TCA
Conditions: sample 200 µl, column Mono Q HR
5/5; buffer A: 20 mM ammonium carbonate,
pH 8.5; buffer B: 980 mM NaCl in buffer A;
gradient: 0-30% in 18.7 ml; flow rate:
1 ml/min; detection: 280 nm, a.u.f.s.: 0.1

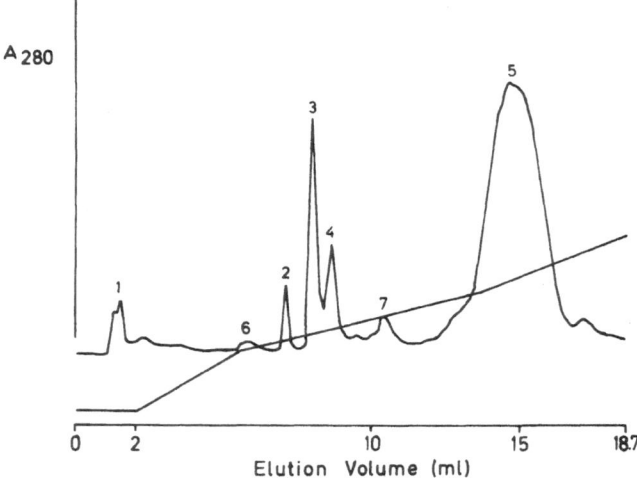

Fig. 3. Performing of FPLC of a supernatant of an
ARF-patient after precipitation with 50% TCA
Conditions: sample 200 µl diluted 1:2 in
buffer A, column Mono Q HR 5/5; buffer A:
20 mM ammonium carbonate, pH 8.5; buffer B:
980 mM NaCl in buffer A; gradient: 0-30%
in 18.7 ml; flow rate: 1 ml/min; detection:
280 nm, a.u.f.s.: 0.1

tions of patients admitted with multiple traumatic injuries, sepsis and acute renal failure are proteolytic. Baracos and coworkers (19) have demonstrated that a polypeptide synthesized by human monocytes (leukocytic pyrogen or interleukin-1) acts on skeletal muscle to promote protein breakdown by increasing the production of prostaglandin E. Clowes et al. (20) found that a factor in the plasma of patients with sepsis could stimulate protein degradation in muscle. As a result of these observations one would expect an increase of protein degradation products, especially in patients with impaired renal function.

The present study confirms earlier investigations that non-TCA precipitable "Lowry protein" markedly increases in patients with posttraumatic ARF (4). A further elevation of these supernatants was reported in patients with acute renal failure complicated by sepsis (21). This increase of non-TCA precipitable "Lowry protein" results partly from non protein related factors. Indeed, PTC-amino acid analysis of the supernatant peaks following FPLC confirmed significantly elevated values in patients with ARF compared to healthy controls. However, these peptide and amino acid analyses document that increased "Lowry protein" in TCA-supernatants is caused only partly by protein degradation products.

It may be argued that increased PTC-amino acid concentrations are induced by parenteral nutrition. However, we want to emphasize that plasma free amino acid levels were not significantly different in controls and in ARF patients, but even lower in ARF-patients.

In summary, "Lowry protein" in the supernatants after TCA precipitation is significantly higher in ARF patients compared to healthy controls. PTC-amino acid analyses of the peaks after FPLC revealed also significantly higher levels in ARF-patients than in controls. Nevertheless, a part of the TCA supernatants obtained from the patients with posttraumatic acute renal failure with and without sepsis are not related to protein degradation.

REFERENCES

1. R.A. Gelfand, R.A. DeFronzo and R. Gusberg, Metabolic alterations associated with major injury or infections, in: "New Aspects of Clinical Nutrition", G. Kleinberger and E. Deutsch, eds., Karger, Basel (1983)
2. F.H. Grimbleby and H. Ntailianes, Binding of trichloroacetic acid by protein. Nature: 189:835 (1961)
3. O.H. Lowry, N.J. Rosebrough, A.L. Farr and R.J. Randall, Protein measurements with the folin phenol reagent, J Biol Chem 193:265 (1951)
4. W.H. Hörl, J. Stepinski, C. Gantert, M. Hörl and A. Heidland, Evidence for the participation of proteases on protein catabolism during hypercatabolic renal failure, Klin Wochenschr 59:751 (1981)
5. W.H. Hörl, H.B. Steinhauer and P. Schollmeyer, Plasma levels of granulocyte elastase during hemodialysis: effects of different dialyzer membranes, Kidney Int 28:791 (1985)
6. C. Wanner, P. Schollmeyer and W.H. Hörl, Urinary proteinase activity in patients with multiple traumatic injuries, sepsis or acute renal failure, J Lab Clin Med 108:224 (1986)
7. A. Heidland and W.H. Hörl, Contribution of proteases to hypercatabolism in acute renal failure, in: "Nephrology", R.R. Robinson, ed., Springer Verlag, New York (1984)
8. G. Graziani, A. Cantaluppi, S. Casati, A. Citterio, A. Scalamagna, R. Silenzio, D. Brancaccio and C. Ponticelli, Dopamine and furosemid in oliguric acute renal failure, Nephron 37:39 (1984)

9. C.P. Kedenburg, A Lithium buffer system for accelerated single-column amino acid analysis in physiological fluids. Anal Biochem 40:35 (1971)

10. G.E. Tarr, Manual Edman sequencing system, in: "Methods of protein microcharacterization: A practical handbook", J.E. Shiveley, ed., Human Press, New Jersey (1986)

11. R.L. Heinrikson and S.C. Meredith, Amino acid analysis by reverse-phase high performance liquid chromatography: precolumn derivatization with phenylisothiocyanate. Anal Biochem 136:65 (1984)

12. C.W. Gehrke, K.C. Kuo and R.W. Zumwalt, Chromatography of nucleoside, J Chromatogr 188:129 (1980)

13. G. Richet, H. Villiers and R. Ardaillou, L'activité tripeptidasique du plasma au cours de l'insuffisance rénale. Contribution à l'étude du catabolisme protidique, Rev Franc Etudes Clin Biol 2:808 (1957)

14. G. Richet and R. Ardaillou, L'activité tripeptidasique du plasma au cours des affections sévères. Contribution à l'étude de l'hyper-catabolisme protidique, Presse Medicale 30:1229 (1959)

15. W.H. Hörl and A. Heidland, Enhanced proteolytic activity - cause of protein catabolism in acute renal failure, Am J Clin Nutr 33:1423 (1980)

16. W.H. Hörl, C. Gantert, I.O. Auer and A. Heidland, In vitro inhibition of protein catabolism by alpha-macroglobulin in plasma from a patient with posttraumatic acute renal failure, Am J Nephrol 2:33 (1982)

17. W.H. Hörl, J. Stepinski, R.M. Schaefer, C. Wanner and A. Heidland, Role of proteases in hypercatabolic patients with renal failure, Kidney Int 24 (suppl 16):S-37 (1983)

18. W.H. Hörl, C. Wanner, F. Thaiss and P. Schollmeyer, Detection of a metalloproteinase during hemodialysis in patients with acute and chronic renal failure, Am J Nephrol 6:6 (1986)

19. V. Baracos, H.P. Rodemann, C.A. Dinarello and A.L. Goldberg, Stimulation of muscle protein degradation and prostaglandin E release by leukocytic pyrogen (interleukin-1), N Engl Med 308:553 (1983)

20. G.H.A. Clowes, Jr., B.C. George, C.A. Villee, Jr. and C.A. Saravis, Muscle proteolysis induced by a circulating peptide in patients with sepsis or trauma, N Engl Med 308:545 (1983)

21. R.M. Schaefer, A. Heidland and W.H. Hörl, Role of leukocyte proteinases and proteinase inhibitors in the catabolism of acute renal failure, Kidney Int (in press)

PROTEASES AND ANTIPROTEASES AT DIFFERENT VASCULAR SITES IN RENAL FAILURE

Kurt Bausewein, Klaus Schafferhans, Rüdiger Götz,
Ulrich Gilge,Ekkehart Heidbreder,and August Heidland

Department of Medicine, Division of Nephrology
University of Wuerzburg, FRG

INTRODUCTION

A potential role of proteases in the pathogenesis of the catabolic state in patients with post-traumatic acute renal failure (ARF) has been discussed[1-3]. It was suggested that active proteases might play a role in protein wasting observed in uremic conditions[4]. Since trauma itself might lead to increased plasma proteolysis, the present study was performed both in patients with non-traumatic ARF and chronic renal failure (CRF) to examine potential imbalances between proteases and antiproteases in plasma. Some components of plasma proteolytic enzyme systems were also studied. All parameters were measured in plasma taken at different vascular sites in order to establish as to whether there were any arteriovenous differences across various organs indicating local extraction or release.

PATIENTS AND METHODS

34 patients were examined: 12 control persons with coronary heart disease who underwent a diagnostic heart catheter examination, 12 patients with non-traumatic ARF and 10 patients with CRF just before starting hemodialysis treatment.

Most of the patients were catheterized via the femoral vein before hemodialysis with blood being taken from the hepatic vein, the renal and iliac veins and from the femoral artery.

Table I demonstrates baseline clinical data. It is shown that there were no relevant differences between patients with ARF and CRF.

Plasma levels of elastase in complex with α_1-protease inhibitor (E-α_1PI) were determined by a highly sensitive enzyme-linked immunoassay[5]. Plasma concentrations of α_1-protease inhibitor (α_1PI), α_2-macroglobulin (α_2M), complement component C3c and fibronectin were evaluated by laser-nephelometric analysis using standardized antisera (Behringwerke AG, Marburg, FRG). The inhibitory activities of α_1PI and α_2M were measured by addition of trypsin using CHROMOCYM TRY^R (Boehringer, Mannheim, FRG) and BAPNA as substrates. Plasma prekallikrein and plasminogen levels were assayed by using chromogenic peptide substrates (S-2302, S-2251, Kabi, München, FRG).

TABLE 1. Baseline clinical data of controls, ARF and CRF patients

	control persons (n = 12)	ARF (n = 12)	CRF (n = 10)
BUN (mg/100 ml)	29 + 6	95 + 40	114 + 41
Creatinine (mg/100 ml)	1,2 + 0,5	9,3 + 5	13,7 + 5
Sodium (mmol/1)	140 + 3	138 + 6	139 + 5
pH	7,39+ 0,04	7,34+ 0,06	7,27+ 0,1
Systolic blood pressure (mmHg)	138 + 8	154 + 24	156 + 33
Diastolic blood pressure (mmHg)	85 + 9	88 + 10	88,9 + 9,2
Age (years)	55 + 8	49 + 13	51 + 12

Data are given as mean values + SEM

Results are expressed as mean values $\pm$ SEM. The significance of differences between paired data was assessed by the Wilcoxon paired signed rank-sum test. Unpaired group data were evaluated with the Mann-Whitney U-test. Significance was defined as a value of less than 0.05.

RESULTS

Determination of arterial plasma E-α_1PI complex in non-traumatic ARF patients showed a significant increase to 318,9 $\pm$ 42,0 ng/ml as compared to control values (100,1 $\pm$ 13,3 ng/ml) and values of CRF patients (151,9 $\pm$ 15,5 ng/ml, p$<$0,05) (Table 2).

α_1PI concentration and activity were markedly increased in ARF and CRF as compared with the controls. Moreover, the values of patients with ARF were significantly higher than those measured in CRF. However, the α_1PI activity/concentration ratio was reduced in ARF. There was no significant difference in this ratio between controls and CRF patients.

TABLE 2. Levels of E-α_1PI, α_1PI, α_2M, prekallikrein, plasminogen, C$_3$c, and fibronectin in arterial and renal venous plasma

		control (n = 12)	ARF (n = 12)	CRF (n = 10)
E-α_1PI (ng/ml)	femoral artery	100,1 + 13,3	318,9 + 42,0[a]	151,9 + 15,5[b]
	renal vein	102,1 ∓ 12,7	329,3 ∓ 49,8[a]	150,5 ∓ 24,3[b]
α_1PI (mg/dl)	femoral artery	199,4 + 16,0	413,5 + 23,7[a]	288,4 + 26,7[a,b]
	renal vein	199,6 + 13,7	414,8 + 25,2[a]	296,7 + 26,3[a,b]
α_1PI activity (% of standard plasma pool	femoral artery	89,9 + 5,5	162,4 + 10,8[a]	113,6 + 10,3[a,b]
	renal vein	100,5 + 4,3	148,2 + 9,6[a]	127,4 + 10,0[a,b]
α_1PI activity/ concentration ratio	femoral artery	1,02+ 0,05	0,86+ 0,03[a]	0,89+ 0,07[b]
	renal vein	1,15+ 0,06	0,79+ 0,04[a]	0,97+ 0,06[b]
α_2M (mg/dl)	femoral artery	104,0 + 7,7[c]	124,3 + 10,5	156,7 + 13,4[a,b]
	renal vein	108,0 + 8,1	126,3 + 10,3	164,3 + 13,7[a,b]
α_2M activity (% of standard plasma pool)	femoral artery	68,4 + 5,2	76,7 + 7,3	95,8 + 9,9[a]
	renal vein	66,4 + 4,9	77,3 + 7,7	91,7 + 6,3[a]
α_2M activity/ concentration ratio	femoral artery	1,02+ 0,04	0,95+ 0,04	0,93+ 0,04[a]
	renal vein	0,96+ 0,05	0,94+ 0,05	0,88+ 0,03[a]
Prekallikrein (% of standard plasma pool)	femoral artery	84,0 + 4,8	38,5 + 3,8[a]	60,2 + 5,0[a,b]
	renal vein	86,6 ∓ 4,5	37,3 ∓ 3,9[a]	59,5 ∓ 4,6[a,b]
Plasminogen (% of standard plasma pool)	femoral artery	96,0 + 10,2	83,7 + 4,0	82,2 + 3,8
	renal vein	95,6 + 7,0	81,5 + 4,6	86,7 + 5,4
C$_3$c (% of standard plasma pool)	femoral artery	68,9 + 5,1	79,4 + 5,2	59,4 + 3,4[b]
	renal vein	72,9 + 5,2	78,8 + 4,2	60,9 + 3,2[b]
Fibronectin (mg/dl)	femoral artery	31,1 + 1,9	41,0 + 5,5	38,2 + 2,2[a]
	renal vein	32,3 + 2,4	42,5 + 5,7	39,6 + 2,7

Mean values ± SEM in arterial and renal venous plasma.
[a] = $p < 0.05$ ARF, CRF versus controls; [b] = $p < 0.05$ CRF versus ARF
[c] α_2M in healthy controls (n=11) 165,0 ± 7,1 mg/dl
α_2M in standard plasma pool 154,0 mg/dl

α_2M concentration and activity were reduced in ARF patients. The activity/concentration ratio was not altered. In CRF, however, there was a significant decrease in this ratio due to a decreased α_2M activity while the concentration of this inhibitor was within the normal range. Surprisingly, there was a simultaneous decrease of both concentration and activity of α_2M in control persons undergoing coronary angiography. It is of note that in the case of α_2M there was a marked difference between the values of the control group and those obtained from a standard plasma pool of healthy subjects both with regard to concentration and activity. This could be due to the coronary heart disease or could be caused by the administration of the contrast agent.

Plasma prekallikrein was reduced to 38,5 ± 3,8 % (p ⊂ 0,05) in ARF and to 60,2 ± 5,0 (p⊂0,05) in CRF.

Plasminogen values were also lower in ARF and CRF, while C$_3$c was only diminshed in CRF.

Plasma fibronectin was moderately increased in ARF and CRF.

Furthermore, arteriovenous differences of the above mentioned parameters were determined across the iliac and hepatic vascular beds. No significant differences of iliac and hepatic vein concentrations were observed (data not shown), indicating that neither extraction nor release occurred in these organs.

DISCUSSION

The present study demonstrates that imbalances of proteases-antiproteases in non-traumatic ARF are similar to those reported in post-traumatic ARF.

The plasma concentration of E-α_1PI complex was found to be increased in non-traumatic ARF as compared to CRF. Elevated plasma levels of E-α_1PI complex have been reported in patients with post-traumatic ARF and septicemia[6,7]. Although there was an increase in α_1PI concentration, α_1PI activity was decreased. Therefore, in post-traumatic ARF a protease-antiprotease imbalance was suggested[4].

Establishing an activity/concentration ratio of the two most important plasmatic proteinase inhibitors there were differences in the inhibitory capacity in ARF and CRF. Even though there was an increase of both concentration and activity of α_1PI in ARF, the activity/concentration ratio was reduced. In CRF this ratio was not altered.

As to α_2M there was a decrease in concentration and activity in non-traumatic ARF. The activity/concentration ratio was not altered, whereas in CRF there was a significant decrease in this ratio resulting from reduced activity.

The constellation of increased granulocyte elastase and decreased activity/concentration ratios of the major plasmatic proteinase inhibitors in renal failure together with marked alterations in some key components of plasma proteolytic enzyme systems suggests that imbalances of proteases and their inhibitors might interfere with the main proteolytic cascades of plasma such as clotting, fibrinolysis, kinin generation and complement system.

ACKNOWLEDGEMENT

The excellent technical assistance of Mrs. M. Röder as well as the secretarial help of Miss Ch. Winter is greatly appreciated.

REFERENCES

1. W.H. Hörl and A. Heidland, Enhanced proteolytic activity - cause of protein catabolism in acute renal failure, Am. J. Clin. Nutr. 33: 1423 (1980)

2. W.H. Hörl, J. Stepinski and A. Heidland, Further evidence for the participation of proteases in protein catabolism during hypercatabolic renal failure, in: "Acute renal failure", H.E. Eliahou, ed., John Libbey, London (1982)

3. W.H. Hörl, J. Stepinski, R.M. Schaefer, C. Wanner and A. Heidland, Role of proteases in hypercatabolic patients with renal failure. Kidney Int. 24: S-37 (1983)

4. W.H. Hörl, J. Stepinski, C. Gantert, M. Hörl and A. Heidland, Evidence for the participation of proteases in protein catabolism during hypercatabolic renal failure, Klin. Wschr. 59: 751 (1981)

5. S. Neumann, G. Gunzer, N. Hennrich and H. Lang, PMN-elastase assay. Enzyme Immunoassay for human polymorphonuclear elastase complexed with α_1-proteinase inhibitor, J. Clin. Chem. Clin. Biochem. 22: 693 (1984)

6. A. Heidland, W.H. Hörl, N. Heller, H. Heine, S. Neumann, R.M. Schaefer and E. Heidbreder, Granulocyte lysosomal factors and plasma elastase in uremia: a potential effect of catabolism. Klin. Wschr. 62: 218 (1984)

7. M. Jochum, K.H. Duswald, S. Neumann, J. Witte and H. Fritz, Proteinases and their inhibitors in septicemia - basic concepts and clinical implications. Adv. Exp. Med. Biol. 167: 391 (1984)

PROTEASE HISTOCHEMISTRY IN NORMAL AND UREMIC RATS

[1]Reinhart Gossrau, [2]August Heidland, and [2]Jürgen Haunschild

[1]Department of Anatomy, Free University and
[2]Department of Internal Medicine
[1]1000 Berlin (West) 33, FRG and [2]8700 Würzburg, FRG

INTRODUCTION

In rats, acute renal failure (ARF) is accompanied by metabolic alterations in various organs such as liver, skeletal muscle and lung and increased protease activities in plasma and bronchial lavage fluid. Little or nothing is known, however, for methodological reasons about possible changes of protease activities at the cellular and intracellular level in these organs (for references see Flügel-Link et al., 1983; Heidland et al., 1984, 1987). Furthermore, investigations are lacking which deal with the possible response of proteases to ARF in other rat tissues. This lack of knowledge initiated the present study in which the cellular and intracellular effects of ARF on proteases are described for several rat tissues using qualitative histochemical means. In addition, we report on the most important findings for the normal protease pattern in these tissues for reasons of comparison but also due to the fact that proteases in particular may show not only striking interspecies but also intraspecies differences (Gossrau, 1985; Gossrau and Graf, 1987 a, b).

MATERIALS AND METHODS

Experiments were performed with Wistar rats from a closed colony of the animal house of the Department of Internal Medicine (University of Würzburg) as described by Gossrau and coworkers (1987 a). 10 μm thick cryostat sections of the parotid, submandibular, sublingual and extraorbital glands, thymus, lymph nodes, spleen, lung, heart, liver, stomach, duodenum, jejunum, colon, pancreas, skeletal muscle or smears of bronchoalveolar lavage (BAL) cells were incubated for microsomal alanyl aminopeptidase (mAAP), glutamyl aminopeptidase (EAP), γ-glutamyl transpeptidase (γ-GTP), dipeptidyl peptidases (DPP) I, II and IV and endopeptidases (EP) I (Ala-endopeptidase) and II (Arg-endopeptidase) for 30 to 120 min at 37°C and post-treated with the methods of choice as given by Lojda et al. (1979) and Gossrau (1981, 1985). The designation of the exopeptidases mAAP, EAP, DPP I, II and IV and γ-GTP bases on the recommendations as given by McDonals and Barrett (1986).

RESULTS

Normal distribution and activity pattern of proteases

Except for the duodenal and jejunal brush border of mature enterocytes and microvillous zone of renal proximal tubular cells which stained for EP I and II and EP II which reacted in the intercalated duct cells of the parotid gland, these endopeptidases could not be detected in any of the investigated organs. DPP IV was present in the mesothelial cells of all serous linings, i.e. the parietal and visceral pleura, pericardium and peritoneum.

In the following proteases will be described for parenchymal and special stromal cells; the elements of the ordinary stroma, e.g. connective tissue fibers or fibrocytes were neglected.

Parotid gland. mAAP was found in the epithelial cells of intercalated ducts. DPP I and II were present in the lysosomes of the basal cytoplasm of the acinar cells and in that of the epithelial cells of the striated ducts. Submandibular gland. DPP IV occurred in the luminal plasma membrane of the acinar cells and lysosomal DPP I and II in the basal cytoplasm of the acinar, tubular and striated duct cells. Sublingual gland. γ-GTP was present primarily in the luminal surface membrane of the epithelial cells of the striated ducts. DPP I and II reacted in a few lysosomes of the tubular and striated duct cells.

Extraorbital gland. In contrast to the comparatively weak aminopeptidase activities of the salivary glands (parotid, submandibular and sublingual gland) at least DPP IV (Fig.1) was very active in the extraorbital gland which belongs to the lacrimal glands (for literature see Gossrau et al., 1987 a). This protease was primarily localized in the luminal plasma membrane of the acinar cells but also in that of the intercalated duct and striated duct cells but not in that of the epithelial cells of the interlobular (excretory) ducts. mAAP (Fig.2) and EAP were found in the same cells of the intercalated and striated ducts and lysosomal DPP I and II in the acinar, intercalated, striated and interlobular duct cells.

Pancreas. DPP IV (Fig.3) occurred in the plasma membrane of the intercalated duct cells including their centroacinar portions and in the epithelium of the excretory ducts. γ-GTP was present in the luminal cytoplasm of the acinar cells and possibly also in their luminal and lateral plasma membrane.

Liver. The most active protease was DPP IV (Fig.4) which could be detected without regional differences in the plasma membrane at the biliary pole of hepatocytes, in the surface membrane of sinus endothelial cells and in that of epithelial cells of the terminal ductules (canals of Hering) and interlobular bile ducts. mAAP predominated but with far lower activity than DPP IV at the biliary pole of periportal hepatocytes. Very few of these structures reacted weakly and the Hering cells and epithelial cells of the interlobular bile ducts strongly for γ-GTP.

Lung. The most active and ubiquitous protease was DPP IV (Fig. 5) associated with the plasma membrane of the endothelium of all segments of the capillary bed. γ-GTP could be visualized in the luminal and lateral surface membrane of the epithelium of the middle and terminal part of the bronchiolar tree, EAP in some capillary endothelial cells and mAAP (Fig.6) in ovally-shaped cells either located in the alveolar epithelium or in the connective tissue. DPP I (Fig.7) and II were present with high activity in stromal and intraepithelial macrophages.

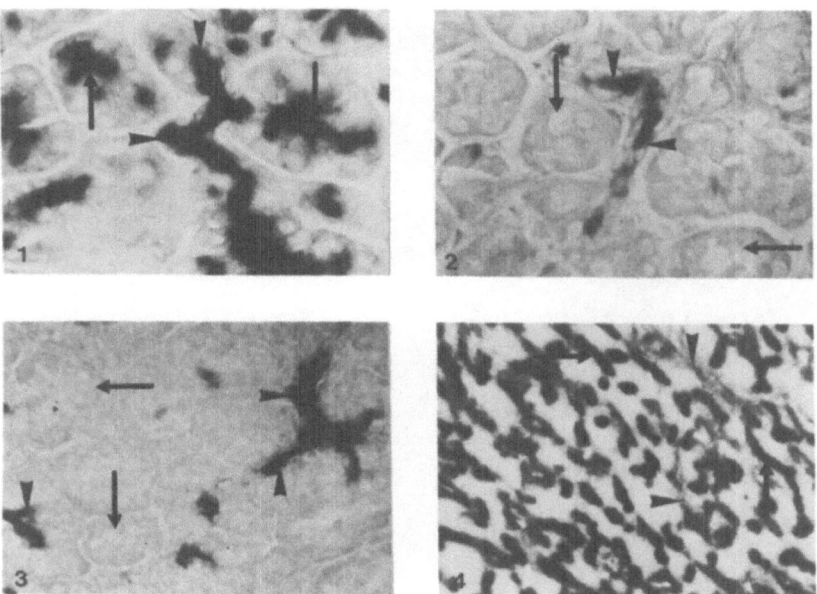

Fig. 1,2. Extraorbital gland. Fig. 1. DPP IV in acinar (arrows) and duct
cells (arrow heads). Fig. 2. mAAP in duct cells (arrow heads). Arrows
acinar cells. Fig. 3. Pancreas. DPP IV in duct cells (arrow heads). Arrows
acinar cells. Fig. 4. Liver. DPP IV in bile canaliculi (arrows) and sinus
endothelial cells (arrow heads). Magnification x450

Thymus. EAP (Fig.8) stained the plasma membrane of cortical reticular
cells and DPP IV that of cortical and medullary thymocytes.
DPP I (Fig.9) and II were present with low activity in the lysosomes of
thymocytes and reticular cells and with high activity in those of macrophages at the corticomedullary border. - Spleen. DPP IV (Fig.10) occurred
in the endothelium of the pulp veins and in the plasma membrane of T lymphocytes of the Malpighian corpuscles (white pulp) and lysosomal DPP I and II
in macrophages primarily at the border between the red and white pulp. -
In lymph nodes, DPP IV was associated with the surface membrane of T lymphocytes.

Heart. DPP IV reacted in the surface membrane of the endothelium of
special (not all) segments of the capillary bed and in that of endocardial
endothelial cells. Similar data were obtained for EAP primarily for the endocardial endothelium. - Skeletal muscles. Striated muscle fibers did not
stain for any of the investigated proteases.

Stomach. mAAP and DPP II were present in endocrine cells in the basal
region of the gastric glands proper and DPP IV in the capillary endothelium
of the middle part of the mucosa. DPP I and II occurred in lysosomes and
DPP I perhaps also in the secretion granules of chief cells. - Duodenum, Jejunum, Ileum. The brush border of enterocytes in the middle segments of the
villi showed the most intense staining by mAAP, DPP IV and to a lesser degree by EAP. The weakest staining was found with ɣ-GTP. Additional sites
with protease activity were the DPP IV-positive duodenal and jejunal entero-

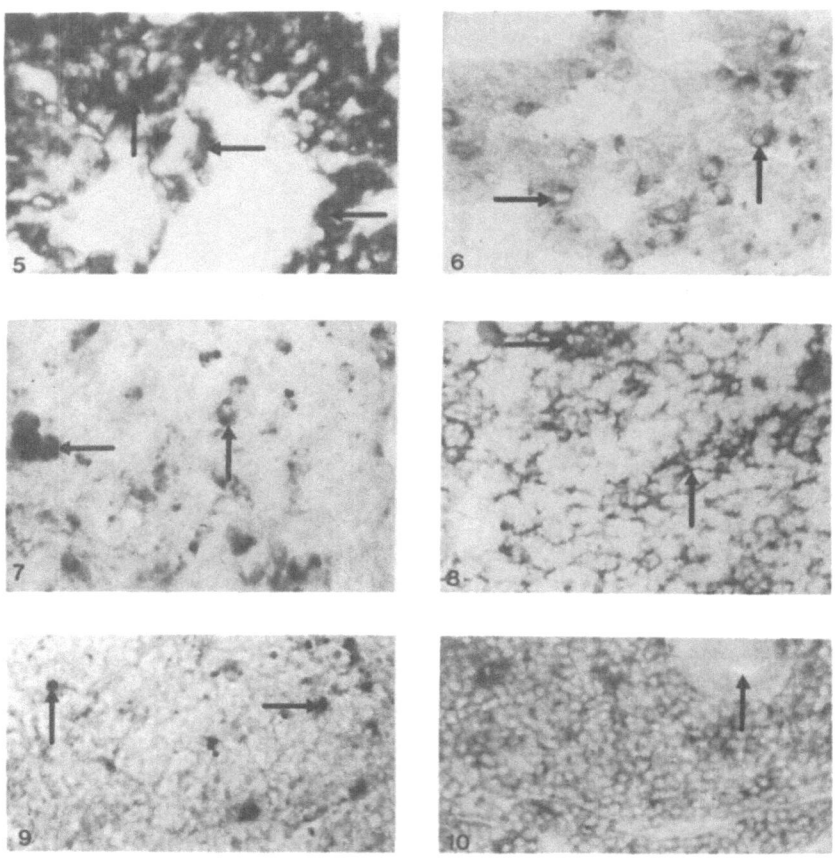

Fig. 5-7. Lung. Fig. 5. DPP IV in capillary endothelium (arrows). Fig. 6. mAAP in ovally-shaped cells (arrows). Fig. 7. DPP I in macrophages (arrows). Fig. 8,9. Thymus. Fig. 8. EAP in reticular cells (arrows). Fig. 9. DPP I in macrophages (arrows). Fig. 10. Spleen. DPP IV in T lymphocytes. Arrow follicular artery. Magnification x450

blasts, Golgi apparatus of jejunal enterocytes, γ-GTP-positive reticular fibers of the villous stroma and DPP IV- and GGT-positive epithelial and stromal lymphocytes. - Colon. DPP IV was found in the microvillous zone of crypt epithelial cells and lysosomal DPP I and II in the crypt and surface epithelial cells.

BAL cells. Moderate to high staining intensity was noticed for DPP II (Fig.11) but not for DPP I (Fig.12). DPP IV (Fig.13) and γ-GTP (Fig.14) were also not detected.

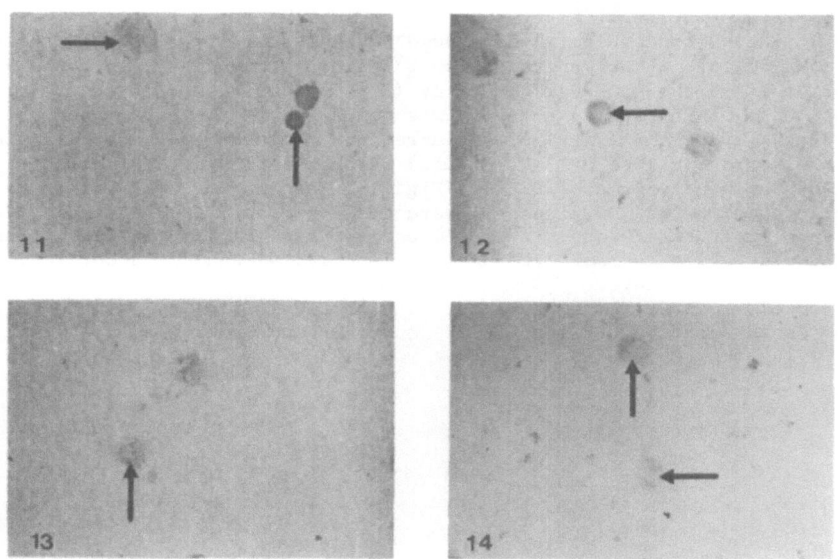

Fig. 11-14. BAL cells (arrows). Fig. 11. DPP II. Fig. 12. DPP I. Fig. 13. DPP IV. Fig. 14. γ-GTP. Magnification x450

Protease distribution and activity after ARF

Similar or identical observations were made after 24 or 72 hours.

The proteases of the parotid and sublingual gland, heart, liver (Fig. 15), stomach, small and large intestine, lymph nodes, spleen (Fig.16), and skeletal muscle did not show any clear-cut response following ARF. A slight response was found in the submandibular gland, pancreas and lung and a clear-cut response for the proteases in the extraorbital gland, thymus and BAL cells.

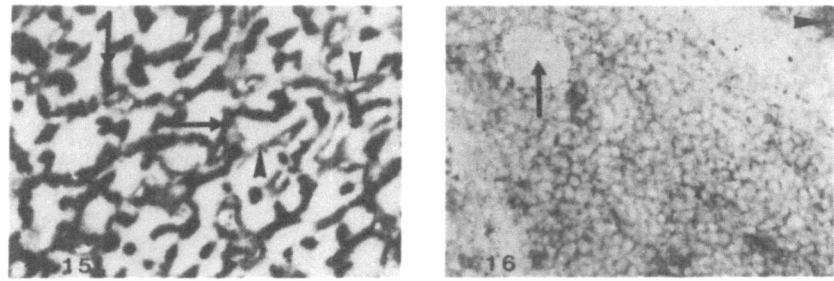

Fig. 15. Liver. DPP IV in bile canaliculi (arrows) and sinus endothelium
(arrow heads). Fig. 16. Spleen. DPP IV in T lymphocytes. Arrow central
artery, arrow head pulp vein. Magnification x450

    The submandibular gland showed reduced activity of DPP IV in the
acinar cells. In the pancreas, lower staining intensities were observed
for γ-GTP in the same cells; DPP IV (Fig.17) behaved as in controls. In
the lung, mAAP (Fig.18) activity was decreased in the ovally-shaped cells
and that of DPP II (Fig.19) was increased in macrophages. Capillary en-
dothelium-associated DPP IV (Fig.20) was not affected. The extraorbital
gland was nearly free of DPP IV (Fig.21), EAP and mAAP (Fig.22) in all
cell types whereas DPP I and II were still present. In the thymus, increase of staining intensity was found for EAP (Fig.23) in the cortical
reticular cells and for lysosomal DPP I (Fig.24) and II in cortical and
medullary macrophages. - Compared with controls BAL cells stained more
intensely for DPP II (Fig.25) but not for DPP I (Fig.26); single cells
showed in addition DPP IV (Fig.27) and mAAP and sometimes also γ-GTP
activity (Fig.28).

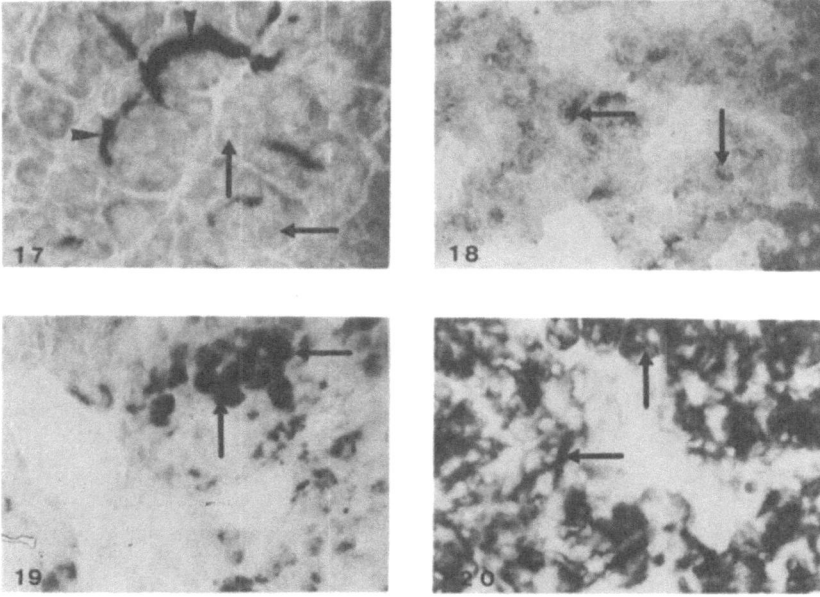

Fig. 17. Pancreas. DPP IV in duct cells (arrow heads). Arrows acinar cells.
Fig. 18-20. Lung. Fig. 18. mAAP in ovally-shaped cells (arrows). Fig. 19.
DPP II in macrophages (arrows). Fig. 20. DPP IV in capillary endothelial
cells (arrows).

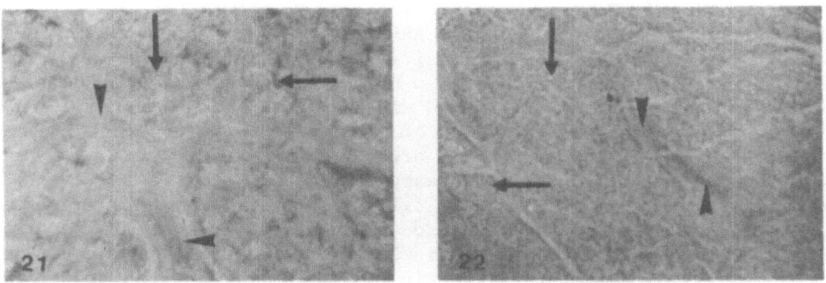

Fig. 21,22. Extraorbital gland. Fig.21. DPP IV. Arrows acinar cells, arrow heads duct cells. Fig. 22. mAAP. Symbols see Fig. 21. Magnification x450

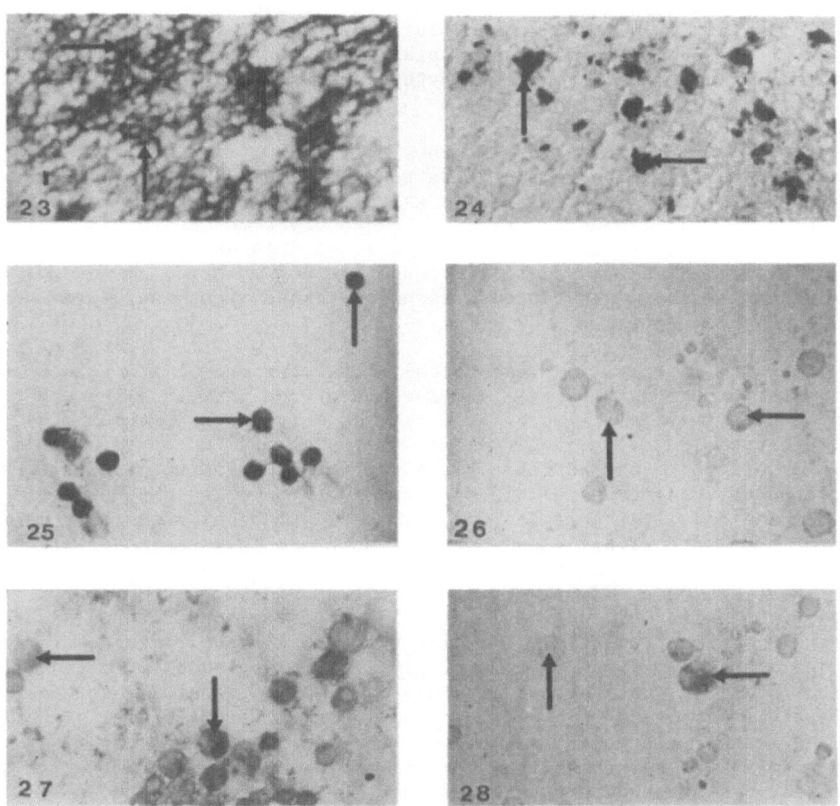

Fig. 23,24. Thymus. Fig. 23. EAP in reticular cells (arrows). Fig. 24. DPP I in macrophages (arrows). Fig. 25-28. BAL cells arrows. Fig. 25. DPP II. Fig. 26. DPP I. Fig. 27. DPP IV. Fig. 28. γ-GTP. Magnification x450

DISCUSSION

From the presented observations evidence occurs that we can differentiate between ARF-sensitive and ARF-insensitive protease activities and
organs in rats. This different sensitivity to ARF is independent, however,
of the respective protease and also cell type since the same protease is
affected in one organ and cell type whereas in the other it is not; e.g.
DPP IV responded in the duct cells of the extraorbital but not in those of
the pancreas to ARF. ARF-sensitive organs were primarily the extraorbital
gland, thymus and lung (when proteases are investigated). The reasons for
this different reaction to ARF are difficult to explain because parallel
changes of urea, proteases, protease inhibitors and other parameters in
the blood plasma (for references see Heidland et al., 1987) should be present everywhere in the body (systemic effect). However, the protease activities of all organs were not altered at all intracellular localizations.
From this one may conclude that primarily there is no strict relationship
between the altered plasma state and changes of protease activities.

At first side, plasma membrane-associated proteases seem to be more
affected than lysosomal proteases if one takes e.g. the extraorbital gland
as an example. However, the thymus shows that beside surface membrane-
linked proteases the lysosomal counterparts can also be affected. Here,
the increase of EAP is presumably a relative phenomenon due to the degeneration of thymocytes which results in a closer package of EAP-positive reticular cells. By contrast, an absolute increase of lysosomal protease activity was found for thymic macrophages which are needed for the degradation
of the dying and decomposing thymocytes (for references see Gossrau et al.,
1987 b).

The thymic response to ARF shows that it is a non-specific reaction
since this immune organ is sensitive to all kinds of noxae which always
lead to nearly identical changes (Gossrau et al., 1987 b). A comparable
situation should be true for alveolar macrophages which are also affected
by quite different events (for references see Spencer, 1977). In contrast,
to our knowledge such a general response was not observed so far for the
extraorbital gland whose protease reaction to ARF might therefore considered to be specific.

Despite the similar or identical ultrastructure of the acinar and
duct cells of the extraorbital, submandibular and parotid gland and pancreas of the rat these cells clearly reacted only in the extraorbital gland
to ARF which reduces or even abolishes the functional capacity of this organ. Normally, the extraorbital gland participates in the production and
secretion of the lacrimal fluid and thus contributes to the nourishing of
the cornea (for literature see Gossrau et al., 1987 a). Possibly, capillaries of the extraorbital gland are more numerous (Gossrau, unpublished observations) or differ in their fine structure which lead to higher local
concentrations, e.g. of urea or other waste products in the microenvironment of the acinar or duct cells of this special gland. Perhaps also ARF
affects exocrine gland cells not directly but indirectly via alterations
of vegatative nerve fibers which supply all these glands but may differ
for the single gland. Another unresolved question is how far corneal
function is disturbed by the damaged cells of the extraorbital gland and
whether it can be compensated, e.g. by the lacrimal gland proper. The
acinar cells of the submandibular gland and pancreas were only slightly
affected as shown by their reduced dipeptidyl peptidase IV and γ-glutamyl
transpeptidase activity; their duct cell proteases revealed normal activity, however, following ARF.

In the lung, uremic pneumonia which occurred after ARF in rats as described by Heidland et al. (1984) with different methods including electron microscopy can also be traced using protease histochemistry. The most conspicuous event is the increase of lysosomal protease activity in stromal, intraepithelial and bronchial lavage macrophages. Furthermore, dipeptidyl peptidase IV activity is increased and possibly even induced by ARF. Dipeptidyl peptidase IV activity is further enhanced in these cells after experimentally induced chronic uremia or renal failure respectively together with that for microsomal alanyl aminopeptidase and γ-glutamyl transpeptidase. All other effects observed after chronic uremia are less severe when compared with those found after ARF (Gossrau et al., 1987 b). Whether the macrophage proteases are secreted and occur also in the bronchial lavage fluid as is presumed for some serine and arginine endopeptidases of the kallikrein- and trypsin-type (Heidland et al., 1984, 1987) and what the functional meaning of this increase in exopeptidase activity is will hopefully be addressed by future investigations.

As far as the human situtation is concerned one has to be cautious with the transposition or extrapolation of these findings to man. For instance, the extraorbital gland and thymus which show the most severe damages in rats behave differently in man. In humans, an extraorbital gland does not exist, and the thymus of adults is involuted and therefore presumably of secondary functional importance. However, it is also possible that the extraorbital gland reflects well a corresponding decrease of protease activities in the lacrimal gland which exists in man, too. Then, and provided the protease response of the lacrimal gland would be species-independent, ARF might be of practical importance in ophthalmology.

The situation with ARF becomes even more complex if one takes into account that protease studies alone do not reflect the whole potential of an organ and its cells to respond to ARF; this, we were able to show, e.g. for the rat liver where protease activities were not altered but where the activities of non-specific alkaline phosphatase in the plasma membrane at the biliary pole of the hepatocytes was drastically increased (Gossrau et al., 1987 a). From this is also follows that ARF leads to multienzymatic changes to which also those of proteases belong. This does not exclude that certain proteases may play a key role in the pathogenesis of ARF (for references see Heidland et al., 1987).

SUMMARY

In rats, ARF caused protease changes which could be detected histochemically. The organs which clearly responded were the extraorbital gland, thymus and lung (ARF-sensitive organs); the others did not show a clear-cut response (ARF-insensitive organs). In the affected organs activities of plasma membrane-associated and lysosomal proteases were either increased or decreased. The response of the proteases in the extraorbital gland is considered to be specific whereas this in the thymus and lung is not. However, ARF has not a protease-specific effect which can be shown when further groups of enzymes are investigated using histochemical means.

ACKNOWLEDGEMENTS

The authors are indebted to Mrs. U. Sauerbier for photographic work and especially to Mrs. A. Hechel for her patience with the typing of the manuscript.

REFERENCES

Flügel-Link, R.M., Salusky, J.B., Jones, M.R., and Kopple, J.D., 1983, Protein and amino acid metabolism in posterior hemicorpus of acutely uremic rats, Am. J. Physiol. (Endocrinol. Metab. 7):E 615.

Gossrau, R., 1981, Investigation of proteinases in the digestive tract using 4-methoxy-2-naphthylamine (MNA) substrates, J. Histochem. Cytochem. 29:464.

Gossrau, R., and Graf, R., 1987 a, in press, Cytochemistry of dipeptidyl peptidase (DPP) IV, Akademie-Verlag, Berlin.

Gossrau, R., and Graf, R., 1987 b, in press, Species differences and the general biological significance of proteases, Histochem. J. 19.

Gossrau, R., Heidland, R., and Haunschild, J., 1987 a, in press, Enzyme cytochemistry of rat organs after uremia with special reference to proteases, Histochemistry.

Gossrau, R., Merker, J.-J., Günther, T., Graf, R., and Vormann, J., 1987 b, Enzymatic and morphological response of the thymus to drugs in normal and zinc-deficient pregnant rats, Histochemistry 86:321.

Heidland, A., Heine, H., Heidbreder, E., Haunschild, J., Weipert, J., Gilge, U., Kluger, G., and Hörl, W.H., 1984, Uremic pneumonitis, Evidence for participation of proteolytic enzymes, Contr. Nephrol. 41:352.

Heidland, A., Weipert, J., Schaefer, R., Heidbreder, E., Peter, G. and Hörl, W.H., 1987, in press, Proteases and other catabolic factors in renal failure, Kidney Int..

Lojda, Z., Gossrau, R., and Schiebler, T.H., 1979, Enzyme histochemistry, A laboratory manual, Springer, Berlin, Heidelberg, New York.

McDonald, J.K., and A.J. Barrett, 1986, Mammalian proteases, A glossary and bibliography, vol. 2, exopeptidases, Academic Press, London, Orlando, San Diego, New York, Montreal, Sydney, Tokyo, Toronto.

Spencer, H., 1977, 'Pathology of the lung', Pergamon Press, Oxford.

TOTAL KININOGEN LEVELS, PLASMA RENIN ACTIVITY, DOPAMINE-BETA -HYDROXYLASE AND PLASMA CATECHOLAMINES IN CHRONIC AND ACUTE RENAL FAILURE

Krzysztof Marczewski, Andrzej Książek,
Janusz Solski, and Zbigniew Pachucki

Clinic of Nephrology , Academy of Medicine
Lublin , Poland

INTRODUCTION

It has been shown that deranged sympathetic functions are common in chronic renal failure (CRF) (1,2,3,4), similarly as increased plasma renin activity (PRA) in acute renal failure (ARF) (5,6). Recently it was reported of the adrenergic disorders in oliguric and polyuric periods of ARF (7). It is well knownthat in healthly persons the renin-angiotensin –system is connected with the adrenergic system and with renal kinins (8). In spite of these data there are very few studies on adrenergic activity in ARF and CRF connected with investigations of PRA and kinins. The present study was undertaken to determine the activity of thesesystems simultaneously in CRF and ARF.

PATIENTS AND METHODS

Patients and controls

A total of 24 patients with predialysis CRF (creatinine clearance <10 ml/min ; 8 women, 16 men ; mean age 36,1±3,2 ; age range 22 to 61 years), 15 patients with ARF (5 women and 10 men ; mean age 39,1±11,6 ;age range 14 to 61 years; investigated in all periods of ARF) and 15 healthly volunters (8 women, 7 men; mean age 28.7±10,1 ; age range 22 to 60 years) were examined.

Investigetion procedures

The experimental procedure was performed in the morning between 7a.m. and 10 a.m.. The CRF patients and healthy volunters were investigated once and the ARF patients three times: in oliguric phase, before first dialysis ; the second time in polyuric phase, two days after last dialysis and last time in convalescence period, one day before the hospital leaving. The arterial pressure was measured with a sphygmomanomether and heart rate (R-R interval) was calculated from the ECG curve. Blood samples for examination were collected 30 min.after cannulisation of the vein in a supine position.

Biochemical analysis

Plasma epinephrine (E), norepinephrine (NE) and dopamine (DA) determined by a sensitive radioenzymatic method of Peuler and Johnson (9), PRA was determined by radio-immunoassay of generated angiotensin I using a gamma coat 125 I-PRA radio-immunoasay kit (Ca 553)(10) ,total kininogen level by the method of Briseid et al (11). DBH activity was examined using the photometric method of Nagatsu and Udenfriend (12).

Statistical analysis

All results are calculated as means,standard deviations and errors of the means (SEM) . Statistical analyses were performed with the Student t test and Wilcoxon's test.The Spearman rank order correlation and linear regresion analysis was used to demonstrate a correlation between values .

RESULTS

CRF patients versus controls

As shown in table 1, the CRF patients had mean values of kininogen, PRA, NE and MAP significantly higher than the controls.

Table 1. Resting values of several parameters in patients with CRF and ARF compared with healthly persons. The data are shown as mean ± error of means.

VALUES		C R F	A R F Oliguria	Polyuria	Convalescence	CONTROL
KININOGEN	ug/ml	5,41*** ±0,31	5,67*** ±0,36	6,72*** ±0,28	7,27*** ±0,37	4,02 ±0,20
P R A	ug/ml/hr	15,46*** ±1,9	14,05* ±3,5	7,9 ±3,2	7,6 ±2,4	5,07 ±0,95
EPINEPHRINE	ng/l	119,3 ±24,7	215,9? ±79,1	168,3 ±43,7	114,4 ±30,3	149,4 ±29,3
NOREPINEPHRINE	ng/l	426,2* ±71,1	485,0* ±96,0	455,8* ±86,6	427,8 ±99,1	270,6 ±26,3
DOPAMINE	ng/L	96,8 ±18,3	177,7** ±62,0	122,6 ±20,8	71,8 ±20,1	64,7 ±20,0
D B H	i.u.	24,2 ±0,7	15,8* ±2,5	21,3* ±6,6	17,1* ±3,6	26,0 ±3,0
M A P	mmHg	118.5*** ±3.8	105.0 ±6.0	103.5 ±3.8	92.3 ±3.0	95.3 ±3.4
R - R interval	msek	728 ±27	793 ±48	742 ±41	705 ±29	789 ±24

?-p<0,1;*-p<0,05 ;**-p<0,01 ;***-p<0,001 compared with controls

ARF patients versus controls

Mean values of kininogen, PRA, NE and DA were significantly higher and DBH activity significantly lower in the ARF patients in oliguric period versus controls. The patients had higher E levels, but the difference was not significant (p<0.1), owing to a large overlap of individual values. In polyuria the mean values of kininogen and NE were elevated and DBH activity was reduced significantly versus controls. Also in convalescence the mean values of kininogen were significantly increased and mean values of DBH activity were decreased in comparison with controls.

Correlations

In the 24 patients with CRF, there was a significantly negative correlation between E and kininogen (Sp=-0,735, r=-0,618, p<0,01) and between DA and kininogen (Sp=-0,448, p<0,05), whereas no correlations were ascertained in the group of patients with ARF.

DISCUSSION

In accordance with other authors (2,3,4), we observed profound disorders in adrenergic activity in CRF, but also the increase of PRA and total kininogen in this group of patients.

Similarly discorders performed in ARF patients in oliguric phase. The deramges of adrenergic function and PRA decreased in the next periods of ARF. Although we observed reduced DBH activity even in convalescence period. However total kininogen increased during of ARF and was the highest in convalescence.

The possible causes of increasing of free plasma catecholamines may be: enhancement of their synthesis, deminished reuptake ,decrease of katabolism,reduced urinary excretion, change of sensivity of adrenoceptors or/and other cases as yet unknown. It is difficult to state if a rise of PRA in CRF is due to the same reasons such as in ARF. It seemed likely that an increase of total kininogen levels indicating fall of kinin activity, was associated with an inhibition of proteases involved in kinin formation. Reduction of proteases activity may be depending on the presence of their plasma inhibitors, analogically to DBH plasma inhibitors in patients with uremia (3).

In present study we investigated only "rest" activity of the adrenergic system . We assumed plasma levels of free E, NE and DA and DBH activity as index of adrenergic activity, as also only total kininogen level as index of kinins. In accordance with this reduction we took care to draw a conclusion .

We conclude that the disturbances of adrenergic ,renin and kinin systems mutually connected is a stable symptom in acute and chronic renal failure. It is possible that important connecting factor of this disorders is the changed proteolytic activity of proteases in renal insufficiency.

REFERENCES

1. W. J. Hennesy, A. W. Siemsen , Autonomic neuropathy in chronic renal failure, <u>Clin. Res.</u>, 16:385 (1968).
2. N. O. Atuk, C. J. Bailey, S. Turner, M. J. Peach, F. B. Westervelt , Red blood cell catechol-O-transferase, plasma catecholamines and renin in renal failure, <u>Trans. Am. Soc. Artif. Intern. Organs,</u> 22:195 (1976)
3. A. Książek , Dopamine-beta-hydrokxylase activity and plasma catecholamine levels in the plasma of patients with renal failure, <u>Nephron</u> 24:170 (1979).
4. E. Heidbreder, K. Schafferhans, A. Heidland, Autonomic neuropathy in chronic renal insufficiency, Nephron, 41:50 (1985) Kokot, J. Kuska, Plasma renin activity in acute renal insufficiency, <u>Nephron,</u> 6:115 (1969).
6. B. Kehoe, G. R. Keeton, C. Hill, Elevated plasma renin activity associated with renal dysfunction, <u>Nephron,</u> 44:51 (1986).
7. E. Heidbreder, K. Schafferhans, A. Heidland, Sympathicus Funktion bei terminaler Niereninsuffizienz , <u>Dialyse Journal,</u> 14:2 (1986).
8. K. Abe, M.Sato, Y. Imai, T. Haruyama, K. Sato, M. Hiwatari, Y. Kasai, K. Yoshinaga, Renal kallikrein-kinin:Its relation to renal prostaglandins and renin-angiotensin-aldosterone in man, <u>Kidney Int.</u>,19:869 (1981).
9. J. D. Peuler, G. A. Johnson, Simultaneous single isotope radioenzymatic assay of plasma norepinephrine, epinephrine and dopamine, <u>Life Sci.</u>, 21:625 (1977).
10. H. L. Vader, L. M. Geuskens, C. L. J. Vink , On standardizing the determination of the plasma renin activity .<u>Clinica Chim. Acta</u> 84:93 (1978).
11. K. Briseid,O. K. Dyrud, Oie Svein, In vitro estimation of rate of release of kinin in human plasma , <u>Acta Phamacol, Toxicol.</u>, 25:201 (1967).
12. T. Nagatsu, S. Udenfriend, Photometric assay of dopamine-beta -hydroxylase activity in human blood, <u>Clin. Chem.</u>, 18:980 (1972).

BIOINCOMPATIBILITY OF DIALYSIS MEMBRANES: FACTOR H BINDING CORRELATES INVERSELY WITH COMPLEMENT ACTIVATION INDICATING A LOCAL IMBALANCE OF INVOLVED PROTEASES/ANTI-PROTEASES

Ernst Wilhard Rauterberg* and Eberhard Ritz^

Institute for Immunology* and
Department of Internal Medicine^
Ruperto-Carola University of Heidelberg, FRG

Introduction

The problem of bioincompatibility reactions has plagued those who applied hemodialysis from its very beginning. Although other mechanisms for triggering of adverse effects during dialysis, i.e. anaphylactic reactions against the sterilizer ethylenoxyde or against substances extracted from the membranes, from the potting or the tubings have recently come into focus, complement (C) activation is one potentially important factor in bioincompatibility. In the present report, we try to summarize evidence for an involvement of the C system in bioincompatibility of dialysis membranes. We add some recent findings of our laboratory relating to the roles of factor H, the regulatory antiprotease and inhibitor of the alternative pathway C3 convertase.

C-activation by dialysis membranes

Craddock and coworkers (1977a) were the first to show by ex-vivo and in-vitro studies that dialysis membranes can activate the C-system, i.e. cause cleavage of both C3 and factor B. Reinfusion of autologous, cellophane-incubated plasma into rabbits produced selective granulocytopenia and monocytopenia (similar to that seen in dialyzed patients) and sequestration of leukocytes in the lungs of the experimental animals. Acute leukopenia immediately after start of hemodialysis had been noted previously in 1968 by Kaplow and Goffinet but its mechanism had remained unknown. In further studies Craddock and collaborators (1977b) demonstrated that complement induced granulocyte aggregation in vitro. The possibility was suggested that this mechanism caused leukopenia and lung sequestration of leukocytes in hemodialysis patients.

With the availability of radioimmunologic tests for the measurement of C3a and C5a it became feasible to monitor time course and extent of C-activation by dialyzer membranes with greater sensitivity than achieved with immune precipitation tests for C3 or factor B conversion. Significantly elevated levels of C3a antigen in the venous effluent of the dialyzer were detected within 2 minutes after start of dialysis (Chenoweth et al., 1983). C3a levels increased up to the 15th minute of dialysis and gradually decreased thereafter. While the time course of changes in C3a concentration was correlated with the evolution of leukopenia, the C-factor assumed to be responsible for this effect, i.e. C5a, was not strikingly elevated until late in the course of dialysis when leukocyte counts had returned to normal values. Similar findings were reported by other investigators. Although their observations apparently argue against a role of C5a in-vivo, it should be considered that no measurable free C5a in the face of elevated C3a in plasma during the early phase of dialysis might be the result of (i) rapid adsorption of C5a to its respective receptors on cells and (ii) marked differences in the levels of the two respective anaphylatoxins generated by C-activation; such difference may result from the large difference of the plasma concentrations of the precursor proteins, i.e. C3 (about 1.2 mg/ml) and C5 (about 0.05 mg/ml). The differences in the levels of the two anaphylatoxins seem not to result from their individual binding behaviour to dialyzer membranes (Cheung et al., 1986).

Other studies have provided evidence that the risk of acute neutropenia after start of dialysis might differ between new and reused dialyzer membranes. Reused membranes induced no significant neutropenia unless bleached by 4.3% sodium hypochlorite before reuse (Gagnon and Kaye, 1984). Similar observations in a larger series of patients with more dialyses per patient were made by Kant and coworkers (1981). Recently, Cheung and collaborators (1987) observed that sodium hypochlorite treatment removes C3-related antigens from the dialyzer membrane and at the same time restores its C-activating capacity. These findings suggest that accumulation of bound C3-activation products inhibit further C-activation. But this effect seems not to be due to complete blocking of binding sites by covalently attached C3b, iC3b, C3d or C3d-g, since Parker and associates observed that the major proportion of C3-antigens eluted from membranes and analyzed by Western blotting consisted of C3c (1987).

A number of recent investigations addressed the issue whether complement activation might be involved in those patients developing severe adverse reactions. The spectrum of reactions ranges from generalized urticaria or angioedema to life-threatening hypotension or cardiopulmonary collapse immediately after initiation of dialysis. Interestingly, in long-term dialysis patients with a history of recurrent adverse reactions to new dialyzer membranes Hakim and coworkers (1984) observed significantly higher levels of C3a and an earlier peak of C3a in the venous effluent than in asymptomatic patients. In parallel, in-vitro alternate pathway of complement (APC) activation by small amounts of zymosan yielded significantly higher release of C3a-antigens in sera of patients with side-effects. The exact mechanism responsible for this observation is unclear. It is of interest that the plasma concentration of the rate limiting factor of the APC, i.e. factor D was reported to be significantly elevated in patients with renal failure and in patients with long-term dialysis (Volanakis et al., 1985). Even this finding cannot account for the occurrence of adverse reactions in a small subset of dialysis patients. The existence of additional factors has to be postulated. Recent findings point to leukocyte C5a receptor modulation or uncoupling as an attractive possibility: Neutrophils taken from patients during dialysis reveal a number of altered functions such as a diminished chemoluminiscence and superoxide response, abnormal chemotactic behaviour, decreased number of surface receptors for fMLP or of Fc-receptors, and defective adherence and aggregation response. To exclude the possibility that unresponsiveness of neutrophils resulted from chronic exposure to C5a, Lewis and coworkers (1987) measured binding of labeled C5a and fMLP to neutrophils and monocytes. These studies indicated that C5a receptors on both cell types obtained from normal donors and exposed to dialysis membranes are specifically blocked or internalized, while such receptor modulation was absent with leukocytes from chronically hemodialyzed patients. It is still unsolved whether this observation reflects adaptive uncoupling of C5a recep-

tor mediated functions and whether patients with adverse reactions differ
in this respect from the majority of hemodialyzed patients.

To summarize the information on adverse reactions and C-activation by
dialyzer membranes, one can state that the C-system may be part of the
effector mechanisms. Consequently, a number of investigators compared
different dialyzer membranes with respect to their potential to cause C-
activation. It is a logical approach to search for membranes with minimal
C-activating properties. Such search must take into account, however, some
new insights concerning the mode how the C-system is activated by dialysis
membranes.

Activation of the complement system

More than 20 serum and cell surface proteins comprise the mammalian
complement system with its two modes of activation, i.e. the classical and
the alternative pathway. The activation sequence is held in check by
control mechanisms which prevent unrestricted activation by the intervention of inhibitors and inactivators which control different steps or
activated components of the respective activation cascade (Fig.1).

Because of its ability to operate in the absence of specific antibodies, the alternative pathway of complement (APC) is generally viewed as
representing a non-specific system of natural defense which appears to be
phylogenetically older than the classical pathway of complement (CPC). The
APC comprises six plasma proteins (Table 1) and can be described as a
defense system based on a primitive mode of detection of foreign surfaces.
A variety of microorganisms, of virus-infected or -transformed cells and
of parasites activate the APC and are sensitive to neutralization or
killing induced by this pathway.

Table 1. Proteins of the APC

protein	abbreviation	number of peptides	molecular weight (kD)	serum concentration (µg/ml)
third component of complement	C3	2	180	1200
factor B	B	1	94	200
factor D	D	1	25	2
properdin	P	4	220*	25
factor H	H	1	155	400
factor I	I	2	88	35

*molecular weight of the tetrameric species of properdin.

In the CPC activation is directed towards the target via antibodies,
thereby initiating binding and activation of C1. In contrast, such initiating factor(s) is (are) lacking in APC. C-activation by dialysis membranes
is assumed to occur predominantly or exclusively via APC.

It is now generally accepted that activation of the APC by "foreign
surfaces" is the result of controlled continuous "ticking over" in the
fluid phase. Such "ticking over" is thought to result from spontaneous but

THE COMPLEMENT SYSTEM

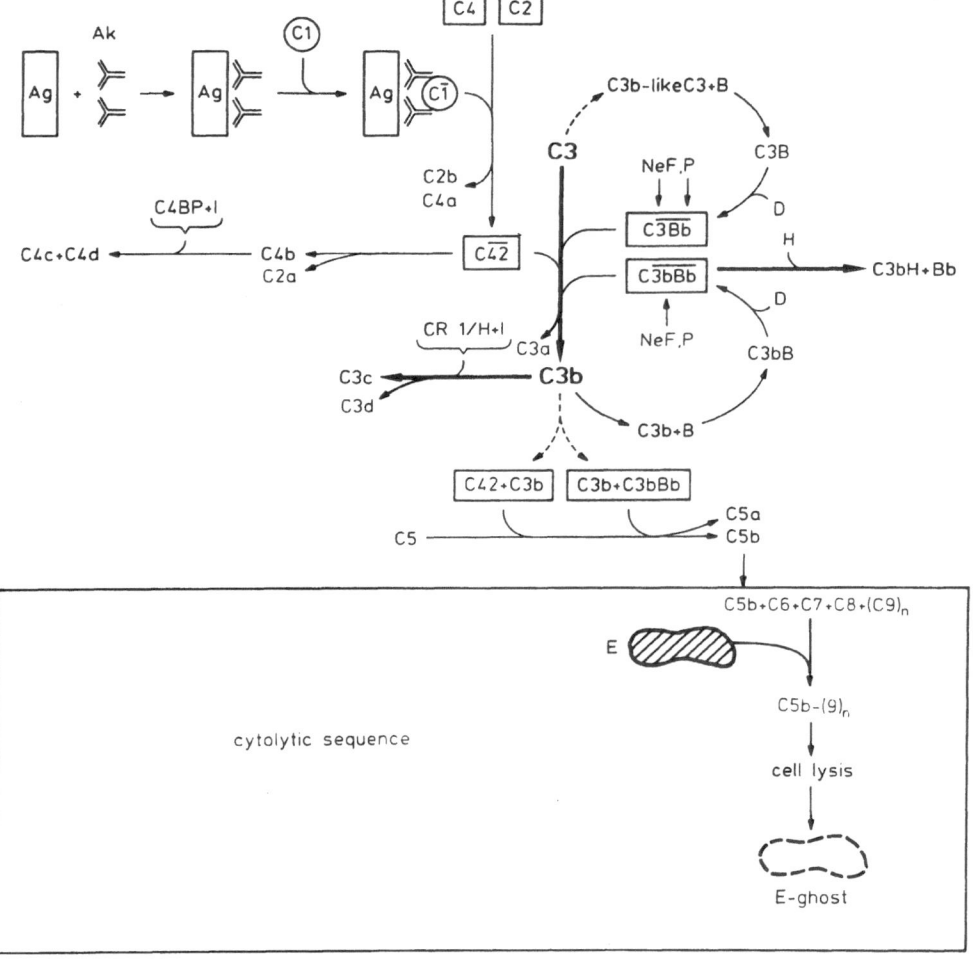

Fig.1   Schematic diagram of the complement (C) system.
(E: erythrocytes; all other abbreviations
according to the recommendation of the WHO).

slow hydrolysis of an internal thioester bond in the C3 molecule. The resulting hydrolyzed C3 product, i.e. C3(H_2O), aquires the ability to act as C3b, i.e. to form a factor B dependent C3-convertase similar to $\overline{C3bBb}$. Therefore, C3(H_2O) has been termed "C3-like C3". Further progression of the activation process depends on the fate of the small number of nascent C3b-molecules produced by the C3-cleaving activity of the continuously formed $\overline{C3(H_2O)Bb}$: Upon proteolytic cleavage of C3 to C3b the latter aquires a metastable and highly reactive conformation which enables it to covalently bind to OH- or NH_2- or SH-groups on "foreign surfaces" via the internal thioester bond in the alpha chain. The majority of nascent C3b molecules emerging from this "ticking over" process remain in the fluid phase. The thioester bond undergoes spontaneous hydrolysis in water with rapid inactivation of C3b by factor H and I. Yet, single C3b molecules can hit appropriate chemical groups on the "foreign surface", but will be inactivated almost instantly unless they attach in a protected micro-environment, restricting access of the regulating factors H and I.

Interestingly, in both fluid phase C3b or C3(H_2O), complex formation with factor B and cleavage of factor B in the C3bB-complex by factor D proceed more quickly than dissociation of newly formed C3-convertases of the APC. Dissociation by factor H and subsequent cleavage of C3b, occuring by interaction of factor H and I, will generate iC3b and finally C3d-g or C3c respectively. Therefore, factors H and I play different roles in maintaining the APC:
- They allow continuous formation of a small number of C3-convertase mole-cules $\overline{C3(H_2O)Bb}$ and, thereby, of a small number of nascent C3b-mole-cules.
- They effectively prevent massive fluid phase activation with consecutive consumption of C3 and factor B.
- They direct C-activation via APC to those surfaces on which efficient regulation by factors H and I does not occur.

The next step in the C-activation, essential for both CPC and APC is the formation of amplification C3 convertase. It is identical with the C3-convertases of the APC, i.e. $\overline{C3bBb}$ or its stabilized form $\overline{C3bBbP}$, regardless of the mode of primary C-activation. However, if we focus exclusively on the mode of C-activation in APC, the term `amplification C3 convertase` describes predominantly quantitative and kinetic aspects. Local unrestricted activity of $\overline{C3bBb}$ on protected sites will produce numerous nascent C3b molecules each of which has the potential to form new C3-convertases or to participate in the formation of C5-convertases. If the generation of C3b proceeds with sufficient rapidity and intensity, a local imbalance will build up concerning the controlling factors H and I which are absorbed during their reaction with a proportion of newly gene-rated C3-convertases. Local consumption of the inhibitory proteins in the fluid phase is assumed to permit uncontrolled C-activation and C3-cleavage even in the fluid phase. This is the C-activation observed in the superna-tant after incubation of serum with "APC-activators" like zymosan (an insoluble boiled residue of yeast) or bacterial walls or other "foreign surfaces" presumably including dialysis membranes.

The time course of interaction of APC proteins with "activator sur-faces" can be divided conveniently into two phases. In a first phase the C-attack is focused on the "foreign surface" and in a second phase biolo-gical activities of the complement system spill over into the systemic phase by recruitment of the peptides C3a, C5a and C4a. Apart from its anaphylatoxic action C5a exhibits potent chemotactic and leukocyte activa-ting properties.

As a final step of C-activation by the APC, C5 convertase can be formed. This functional activity, i.e. cleavage of C5 to C5a and C5b, is generated when additional C3b molecules bind in the vicinity of APC C3-convertases. By very elegant studies, it has been shown that such additional C3b serves as a binding site for C5, altering the conformation of the ligand C5 in such a way that it is susceptible to cleavage by $\overline{C3bBb}$. The larger cleavage product C5b is able to bind C6 via a metastable binding site and to initiate thus the assembly of the membrane attack complex (MAC) of complement, i.e. C5b678(9)$_n$. During its stepwise formation this complex aquires more and more hydrophobic binding sites on its surface. It is generally assumed that the increasing number of hydrophobic sites permits plugging of MAC into the lipid bilayer of target cells. Only those MAC complexes formed in the vicinity of cell membranes seem to exhibit biological activities. MAC generated in the fluid phase are inactivated by serum inactivators, e.g. S-protein (vitronectin) or a heterogeneous group of lipoproteins. Insertion of MAC has mainly been studies in artificial lipid membranes and in the lipid bilayer of erythrocytes. In these systems it causes formation of functional pores, as originally postulated by the "dough-nut" hypothesis of Mayer (1973), and results in osmotic lysis of erythrocytes. Recent studies have provided evidence that MAC causes lysis less readily when nucleated cells are attacked. In a number of in-vitro studies with isolated tissue-derived cells or blood leukocytes, MAC was shown to induce dose-dependent release of potent mediators of inflammation, i.e. of cytokines like Il-1; of prostanoids like PGE$_1$, thromboxan and prostacyclin; or of leukotrienes. One can even envisage a new positive feedback mechanism participating in inflammatory reactions: MAC might (directly or indirectly) induce release of proteinases from neutrophil granulocytes or from macrophages. This in turn could result in cleavage of C3, generation of nascent C3b and by this taken to activation of APC even on "foreign surfaces" which under cell-free conditions do not provide enough protected sites to efficiently activate the C-system. In-vitro studies at least suggest the possibility that functionally active C3b can be generated by leukocyte derived proteinases. The implications for interaction of C with dialysis membranes, i.e. agents known to cause leukocyte release reaction, are obvious. It is imperative to clearly differentiate between cell-free in-vitro bioincompatibility studies with serum on the one hand and studies using heparinized blood on the other hand. (Additional effects of heparin on the C-system must also be taken into account.)

Factors regulating APC-activation on "foreign surfaces"

At the moment it seems plausible to postulate that microenvironments, e.g. on cells, on microorganisms or on "foreign surfaces", permissive for APC operate via one or more of the following mechanisms (which are not mutually exclusive):

(i)   induction of a C3b conformation favouring accelerated interaction with factor B;

(ii)   modulation of spontaneous decay of $\overline{C3bBb}$ via interaction with the C3b- or the Bb-part of the enzymatic complex;

(iii)   restriction of access of factor H to bound C3b via alteration of the C3b-conformation;

(iv)   modulation of factor H/C3b interaction by chemical residues primarily affecting factor H conformation independent of or additional to factor H binding to C3b;

(v)   modulation of factor B binding to C3b by chemical residues favouring such an interaction;

(vi)   modulation of factor I binding and its enzymatic acitivy on C3bH-
        complexes.

        As one example of modulation of factor H binding to C3b, it has
initially been shown that the density of sialic acid residues on erythrocyte surfaces determines the extent of APC activation. Unmodified sheep
red blood cells (RBC) are poor activators; after neuraminidase treatment
they are strong APC activators similar to rabbit RBC. To achieve on normal
sheep RBC a decay rate of $\overline{C3bBbP}$ similar to that on rabbit RBC, a tenfold
higher concentration of factor H is required. Treatment of sheep RBC with
NaIO₄ which is known to remove the C8 and C9 carbon atoms from sialic acid
had the same effect as neuraminidase. In RBC of 21 inbred mice an inverse
correlation was found between sialic acid content and C-activating capacity. However, the concept of sialic acid as the exclusive regulator of
APC-activation has become questionable. Neuraminidase treatment of human
RBC was without effect on activation of APC and incorporation of LPS
changed sheep RBC into strong activators without modifying their saliac
acid content. With respect to factor I activity, it has been observed that
extensive cleavage of C3b by the regulatory activity of both factor I and
factor H is a prerequisite for cessation of APC.

        One can, therefore, conclude that factor H has a dual role in con-
trolling the C-activation: First, it serves as a direct inhibitor of $\overline{C3bBb}$
and $\overline{C3bBbP}$, both representing complexes with serine esterase activity.
Second, factor H serves as a cofactor for the proteolytic activity of
factor I. In the latter case, factor H behaves as an autonomous independent binding domain of a trypsin-type serine protease complex in which the
catalytic region is located on the factor H moiety. A similar organization
is found in other enzymatic complexes of the C-cascade, e.g. $\overline{C4b2a}$ and
$\overline{C3bBb}$. However, to the best of our knowledge, the factors H and I do not
form complexes in the fluid phase unless they interact with C3b.

Own studies

        Since factor H is obviously one of the most important regulators of
APC, we analyzed spontaneous binding of factor H to dialysis membranes.
The following parameters were evaluated:

(1)   binding of purified human factor H,

(2)   binding of factor H during incubation of serum or serum containing 10
      mM EDTA at 4°C and - kinetically - at 37°C,

(3)   cleavage of C3,

(4)   generation of C3d.

        All dialysis membrane materials were used in bead configuration
(Fig.2) to increase the surface. All incubations were made batchwise.
Binding of H was calculated measuring factor H concentrations in the
supernatants.

Fig.2 Cuprammonium "beads" in the microscope (x200).
Cuprammonium dialysis membrane material had been
crushed to a fine powder. The "beads" had a mean
diameter of 0.1 - 0.2 mm

It was the aim of the study to analyze the effect of limited chemical
modifications of dialysis membrane material on factor H binding and C-
activation. Unmodified cuprammonium beads (CU), and cuprammonium beads
derivatized by introduction of diethylaminoethyl-(DEAE-CU) or carboxy-
methyl-groups (CM-CU) were kindly provided by the ENKA company, Wuppertal,
FRG. Purified factor H was a gift from Dr. Karges, Behringwerke, Marburg,
FRG.

Table 2 gives the amount of factor H bound after incubating purified
factor H with modified and unmodified CU-beads; this is compared with C3d
generation after incubating the beads with serum for 120 minutes. There is
an obvious inverse relation between binding of purified factor H and C3d
generation. We failed to measure significant adsorption of factor H to
unmodified CU beads; correspondingly, with this material the highest
release of C3d was noted. Vice versa, substantial amounts of factor H were
bound to CM- and DEAE-CU beads; correspondingly C-activation and C3d
production were low.

Table 2

Binding of purified factor H to CU, DEAE-CU and CM-CU beads*

	µg H per ml swollen beads	C3d units in the supernatant
unmodified CU	0	16
DEAE-CU	340	14
CM-CU	360	6
serum without beads	---	3

* Data of two different experiments, i.e. incubation with purified H (H
adsorption) and with serum (C3d production) respectively

The inverse relation between binding (or adsorption) of factor H at 4°C and C-activation, still was noted when we compared differently modified cuprammonium materials, could also be confirmed when beads were incubated with serum. The results are demonstrated in fig.3. At low temperature we observed adsorption of 80 micrograms of factor H to unmodified CU beads while DEAE-CU and CM-CU beads bound 140 and 380 microgram respectively (see dotted bars in fig.3); all values refer to 1 ml pre-swollen beads.

FACTOR H BOUND TO CM-, DEAE- AND UNDERIVATIZED CUPROPHANE-BEADS

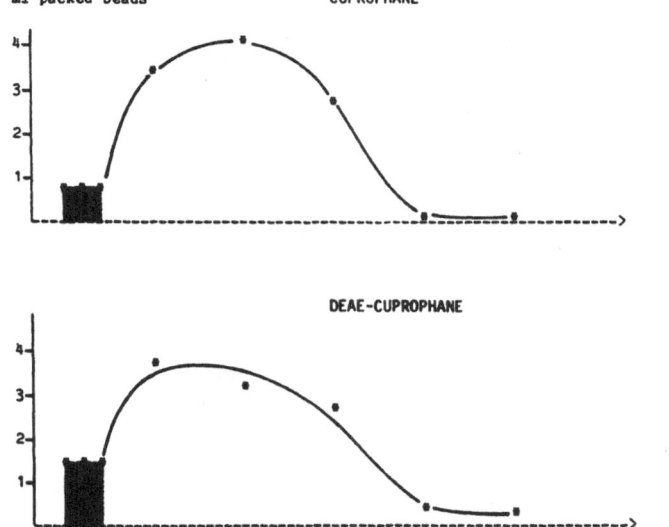

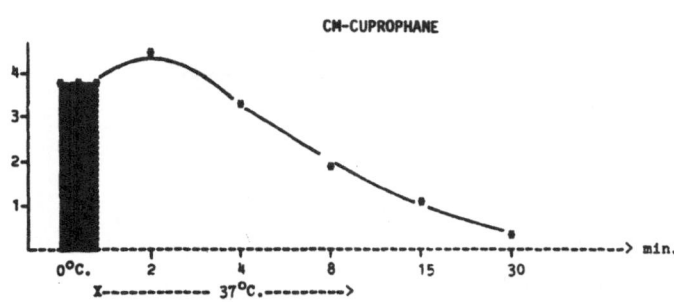

Fig.3 Kinetic of factor H binding to Cuprophane, DEAE-Cuprophane or CM-Cuprophane during incubation of serum at 37^0C. Solid bars represent spontaneous factor H binding at 0^0C at the start of the experiment. Note that spontaneous factor H binding varies more between the materials than factor H binding at 37^0C.

Fast binding of factor H was observed when the three different types of beads were incubated at 37°C; binding maximum was reached after 2 to 4 minutes. The maximal amount of factor H bound was similar with the three different types of beads. Kinetics of C3 cleavage in the supernatant

373

paralleled kinetics of factor H binding, i.e. maximal activity was seen after 2-4 min. at 37°C. Interestingly, for the three different types of beads the quantity of C3 cleaved was inversely related to the binding capacity for factor H at 4°C.

Fig. 4 shows the time course of the release of C3d into the supernatant with CU (a), DEAE-CU (c), CM-CU (d) and additionally washed CU beads (b). The kinetics of C3d release were different from those of factor H binding or C3 cleavage. The concentration of C3d in the supernatant increased at a constant rate during the 120 min period of incubation with all types of beads. Again, the quantitiy of C3d released was minimal with CM-CU.

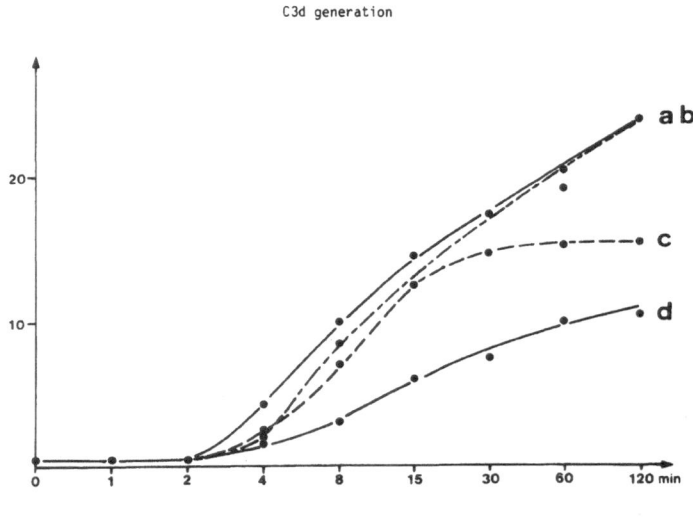

Figure 4

Our findings on binding of factor H to cuprammonium beads suggest that spontaneous, i.e. C3b independent, binding of factor H to "foreign surfaces", i.e. dialysis membranes, regulates C-activation via APC. Furthermore, evidence was provided that the quantity of C3d generated goes in parallel with C3 cleavage, but follows a different time course. The constant increase of C3d concentration in the supernatant with time can be interpreted as evidence for slow but constant cleavage of C3b on the surface of the beads. Further studies are needed to analyze whether the sites of C-activation on dialysis membranes are randomly spread or restricted to protected environments.

References

Chenoweth, D. E., Cheung, A. K., and Henderson, L. W., 1983, Anaphylatoxin formation during hemodialysis: Effects of different dialyser membranes, Kidney Int., 24:764

Cheung, A. K., Chenoweth, D. E., Otsuka, D., and Henderson, L. W., 1986, Compartmental distribution of complement activation products in artificial kidneys, Kidney Int., 30:74

Cheung, A. K., Parker, C. J., Wayman, A. L., 1987, Sodium hypochlorite (NaOCl) regenerates the complement(C)-activating potential of cuprophan membranes (CuM), Kidney Int., 31:230

Craddock, P. R., Fehr, J., Dalmasso, A. P., Brigham, K. L., and Jacob, H. S., 1977a, hemodialysis Leukopenia. Pulmonary vascular leukostasis resulting from complement activation by dialyser cellephane membranes, J. Clin. Invest., 59:879

Craddock, P. R., Hammerschmidt, D., White, J. G., Dalmasso, A. P., and Jacob, H. S., 1977b, Complement (C5a)-induced granulocyte aggregation in vitro, J. Clin. Invest., 60:260

Gagnon, R. F., and Kaye, M., 1984, Hemodialysis neutropenia and dialyzer reuse: Role of the cleansing agent, Uremia Invest., 8:17

Hakim, R. M., Breillatt, J., Lazarus, J. M., and Port, F. K., 1984, Complement activation and hypersensitivity reactions to dialysis membranes, New Engl. J. Med., 311:878

Kant, K. S., Pollak, V. E., Cathey, M., Goetz, D., and Berlin, R., 1981, Multiple use of dialyzers: Safety and efficacy, Kidney Int., 19:728

Kaplow, L. S., and Goffinet, J. A., 1968, Profound neutropenia during the early phase of hemodialysis, J. Am. med. Ass., 203:133

Lewis, S. L., Van Epps, S. E., Chenoweth, D. E., 1987, Leukocyte C5a receptor modulation during hemodialysis, Kidney Int., 31:112

Parker, C. J., Cheung, A. K., and Wayman, A. L., 1987, Characterization of the binding of complement (C) C3 fragments to cuprophan (CuM) and cellulose acetate membranes (CAM), Kidney Int., 31:241

Volanakis, J. F., Barnum, S. R., Giddens, M. and Galla, J., 1985, Renal filtration and catabolism of complement protein D, New Engl. J. Med., 312:395

HEMODIALYSIS WITH CUPROPHANE MEMBRANES LEADS TO ALTERATION OF GRANULO-CYTE OXIDATIVE METABOLISM AND LEUKOCYTE SEQUESTION IN THE LUNG

Gerald Kolb*, H. Schönemann*, W. Fischer*, K. Bittner°,
H. Lange°, H. Höffken", V. Damann", K. Joseph", and
K. Havemann*

*Div. Hematology/Oncology; °Div. Nephrology; "Div. Nuclear Medicine, Baldingerstrasse, D-3550 Marburg, FRG

INTRODUCTION

Transient leukopenia[1] and change in pulmonary function[2] are known to occur regular during the initial phase of hemodialysis (HD) with cuprophane membranes, suggesting that an intrapulmonary sequestration of granulocytes may happen[3]. Besides this, hypoxemia, complement-activation, increase of granulocyte adherence and enhanced release of granular enzymes have been reported[4,5,6]. In addition, alteration of granulocyte behavior, such as activation of phagocytosis and impaired oxidative metabolism[7,8] has been described.

Recent studies in granulocyte function were restricted to comparison of pre- and postdialytic samples or were based on very few intradialytic values. Therefore only poor data are existing elucidating the kinetics of granulocyte function during HD. For that reason we studied the granulocyte metabolism of 26 patients during 3-4 h of HD by a method of Cytochrome C reduction and a chemoluminescence assay. Further the interaction of isolated granulocytes with the cuprophane membrane was examined in vitro using a plasma free model.

The distribution and pulmonary sequestration of granulocytes during HD was investigated employing radiolabeled cells.

MATERIAL AND METHODS
Patients

Twenty-six patients, 15 male, 11 female, aged 61.4, $\pm$ 8.5 years gave their informed consent to participate at the study. There were 7 cases of acute renal failure and 19 of chronical renal failure, being on regulary dialysis treatment for 3.7 $\pm$ 2.1 years. HD was performed by an A 2008 C (Fresenius AG, Oberursel, FRG) or an AK-10 (Gambro, Hechingen, FRG) equipment. As dialyzer we used SMAD 125/140 (SMAD, Lyon, France) or GFS MC120 H (Hämophan^R, Gambro).

Blood sampling

Sodium citrate blood was taken from the arterial line before systemic heparinisation prior to dialysis and 10, 20, 30, 60, 120, 180, 240 min after the beginning of dialysis. All samples were stored at -4°C and assayed within 30 min.

Preparation of PMN (Polymorphnuclear neutrophils)

For the separation of PMN we used a two-step-method reported by BÖYUM [9], replacing dextrane with hemaceel 35 (Behringwerke, Marburg, FRG). Remaining erythrocytes were lyzed by hyptonic shock. The isolated cells were resuspended in HBSS without Ca^{++} and Mg^{++} (Gibco, Parisley, Scotland) to $2x10^7$ cells/ml.

PMN elastase assay

For the measurement of PMN elastase levels in plasma we used the enzyme-linked immunoassay described by Neumann et al.[10] (Merck, Darmstadt, FRG).

Statistics

Data are expressed as means $\pm$ SE. For statistical analysis the paired t-test was performed.

Cytochrome C reduction by PMN

Cytochrome C reduction was measured according to a modified method of CURNUTTE and BABIOUR[11]. $2x10^6$ PMN were incubated for 15 min at 37°C in 1 ml of HBSS without Ca^{++} and Mg^{++} containing 80 µM of Cytochrome C (Sigma Chemicals, St.Louis, USA) and for 10 min in presence or absence of 1 µg of PMA (Phorbol Myristate Acetate, Sigma), as a nonspecific soluble membrane stimulant. The amount of reduced Cytochrome C was calculated from the decrease of absorbance at 550 nm between the absence and presence of PMA, using the extinction coefficient $E_{550} = 2.1x10^4$ M^{-1} cm^{-1}. The results were expressed in nanomoles of reduced Cytochrome C x $2x10^6$ PMN x 10 min^{-1}.

Chemiluminescence (CL)

The chemiluminescence assay was performed according to DeCHATELET[12]. Every sample contained $5x10^5$ PMN in a final volume including 650 µl HBSS pH 7.2, 50 µl phorbole myrestate acetate (PMA) and 100 µl luminol. Chemiluminescence was determined in a luminometer at 37°C. Maximal values of CL were registrated.

In vitro HD of granulocytes

We used the small cuprophane dialyzer P 100 (Fresenius, Oberursel, FRG) with a membrane surface of 285 cm² an intracapillary volume of 1.7 ml. Before use the dialyzer was washed with 200 ml HBSS. In our experiments 1.7 ml HBSS containing $1x10^7$ PMN of a healty donor were applicated. CL was determined immediately after washing the cell suspension through the dialyzer or after 5 and 10 min incubation. Cell count was performed to determine the ratio of adherend to nonadherend PMN.

Granulocyte-radiolabeling and lung szintigraphy

PMN preparation an radiolabeling with 99m-Tc-hexamethyleprophylene-amino-oxime (99m-Tc-HM-PAO) was performed according to JOSEPH et al.[13], labeled cells were reinjected intravenously. Starting with the time of reinjection the activity in the lung was continously registrated during 75 min using a computer linked gamma camera. Dialysis was started either simultaneously or 10 min after cell-reinjection.

RESULTS

Figure 1 shows the effect of hemodialysis on WBC, Cytochrome C reduction by PMN and plasma E-a$_1$PI levels. Dialysis with cuprophane membrane cause the well-known profound leukopenia 10 min after starting HD (7.4 ± 2.6 to 4.4 ± 2.5 G/l). Coincidating with that a significant decrease of Cytochrome C-reduction by PMN occurs (27.5 ± 7.9 to 20.3 ± 7.7 nanomoles ferrocytochrome C x 10 min^{-1}x2x10^{-6}PMN^{-1}). 30 min after the beginning of HD this defect in oxidative metabolism begins to disappear (24.7 ± 7.8 nm). After that, Cytochrome C-reduction returns to predialytic values (28.2 ± 7.6 nm) at the end of HD. E-a$_1$PI levels raise from 163 ± 108 to 440 ± 178 µg/ml having its strongest increase during the first 10 min (55 µg/ml/10min).

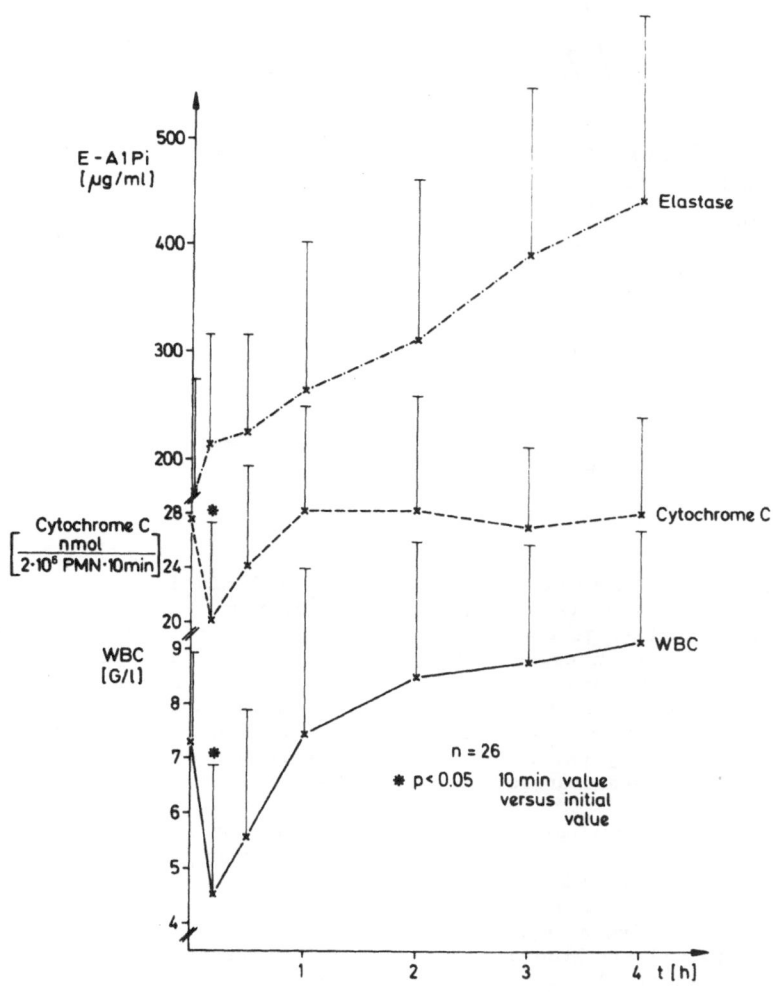

Figure 1 White blood cell count (WBC), granulocyte related Cytochrome C reduction and plasma levels of PMN elastase (as Elastase-alpha-1-proteinase-inhibitor-complex) during cuprophane HD.

Figure 2 shows the effect of HD on the granulocyte chemiluminescence (CL). It could be demonstrated that cuprophane HD leads to an inhibition of the PMA stimulated CL 10 min to 30 min and 3 h after starting HD. This cuprophane related depression of CL could not be registrated using the modified cuprophane membrane: Hemophane[R]. WBC showed
the well known course mentioned above.

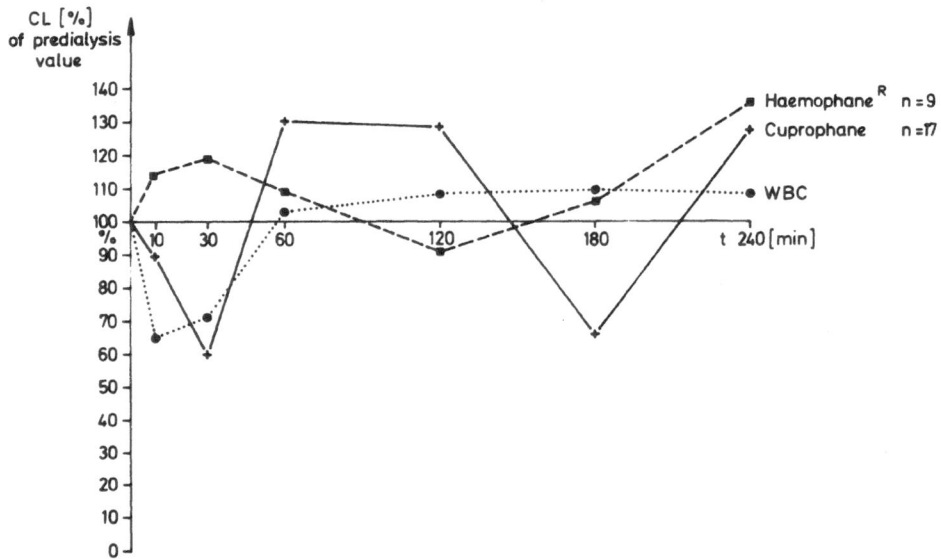

Figure 2 Effect of HD with cuprophane and Haemophane membrane on the
granulocyte chemiluminescence.

Isolated granulocytes were incubated 0-10 min in a small cuprophane fiber-dialyzer under plasma free conditions. As shown in Fig. 3
by CL the oxidative metabolism of the non adherend cells was stimulated
already after short cell-membrane contact during fast perfusion. It
could be also demonstrated, that the cells became adherend to the cuprophane fibers depending on incubation time. Whereas 98% of the cells
could be refound after perfusion, after 5 min 30% and after 10 min of
incubation only approx. 20% of the cells were not adherend. Viability
of the non adherend granulocytes was rather constant (96-88%).

In order to evaluate the accumulation of granulocytes in the lung
during hemodialysis we used radiolabeled autologous granulocytes. After
preparation and labeling the granulocytes were reinjected intravenously. Beginning with reinjection the activity in the lung was registrated
during 75 min using a computer linked gammacamera. In one study dialysis was started simultaneously, in two other cases 10 min and 20 min
after reinjection. The time-activity-curves over the lungs show an increasing activity 10-20 min after beginning with hemodialysis (Fig. 4)
caused by redistribution of labeled granulocytes. Redistribution was
not registrated in studies performed without dialysis (Fig. 4).

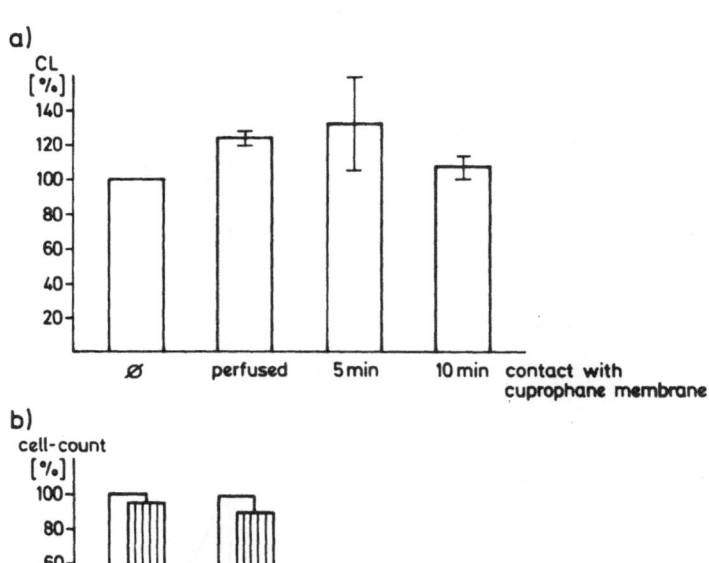

<u>Figure 3</u> a) Chemiluminescence of granulocytes exposed to cuprophane
fibers depending on incubation time.

b) Decrease of granulocyte cell count depending on incubation
time and related granulocyte adherence to cuprophane fibers
to which they were exposed. Filled collums = ratio of
viable cells.

DISCUSSION

The so-called oxidative metabolism of the granulocyte plays a central role in cell function. In any case of its alteration the oxidative metabolism is involved. For this reason methods like the Cytochrome C-reduction test and the chemiluminescence assay have become important for estimation of granulocyte activity in general without restriction to a single function.

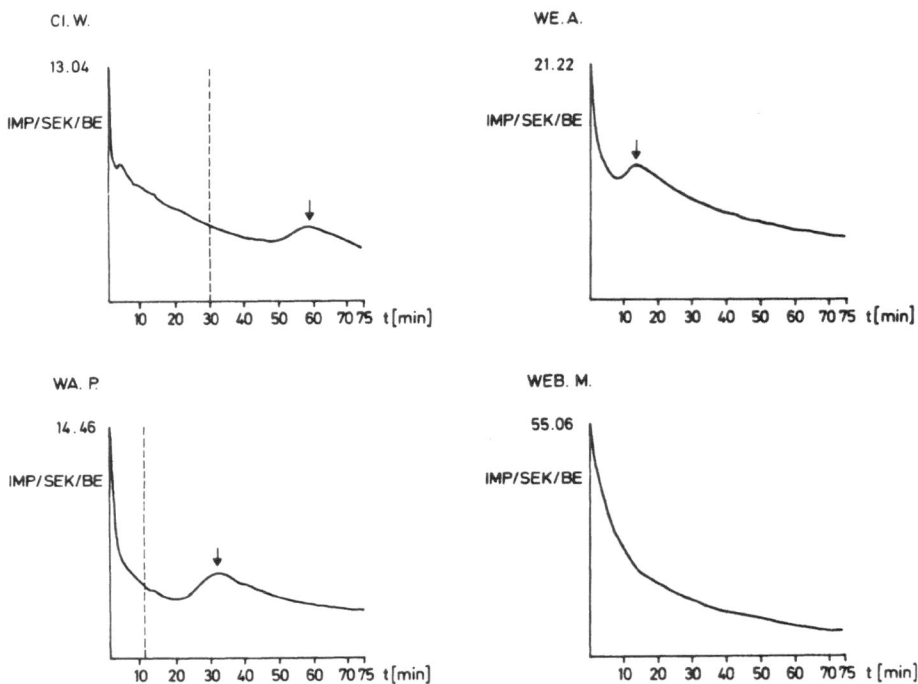

Figure 4 Time to activity curves of labeled granulocytes over the lung (lung scintigraphy) during cuprophane HD (n=3); control: without HD.

In this study we investigated the kinetics of granulocyte oxidative metabolism of patients undergoing HD treatment with cuprophane membranes. It could be demonstrated that granulocytes were less stimulable for chemiluminescence and H_2O_2 release (Cytochrom C-reduction) already after 10 min of contact to the cuprophane dialyzer; but in contrast to the Cytochrome C-reduction strongest inhibition of CL was found after 30 min and after 3 h HD. a second inhibition similar to the initial one occurs regulary in CL. Hemophane[R] which is a modified cuprophane membrane material failed the distinct inhibition phases in CL suggesting that these phenomenon will be strongly depend on the membrane material. We suggest that an impaired oxidative metabolism may be a function of an exhausting membrane induced prestimulation of the granulocyte. This could be due to the cell-membrane interaction[14] resulting in the so-called "frustrated phagocytosis" which is accompanied by release of proteolytic enzymes[15].

Recent studies suggested that alteration of granulocyte metabolism and function may be only depending on activated plasma factors especialy on the fifth complement C5a component which will be generated during cuprophane HD[14]. In contrast to this hypothesis and in according to new findings of Hörl[17] we could demonstrate that membrane contact allone in the absence of plasma leads to a distinct increase in granulocyte adherence and enhanced CL of the non adherent cells. However, both mechanisms - C5a and cell-membrane contact have the same effect on granulocyte function so that they may support and intensify each other.

Until now the physiological consequences of this alteration of granulocyte oxidative metabolism will not be clear in detail, but the strong temporal correlation to well known alterations in granulocyte function like enhanced granulocyte adherence and aggregation on the one site and accumulation of granulocytes in the lung on the other leads to some interesting speculations: It can be frequently observed that patients suffering from chronic renal failure develop lung fibrosis during a long terme HD career. As we know from studies on the early shock lung (acute respiratory distress syndrome = ARDS) the activation of granulocyte oxidative metabolism plays an important role in the development of lung fibrosis[18]. It has been confirmed that the so-called oxidative burst with an explosive release of oxygen radicals leads to an functional inbalance between proteolytic enzymes and their oxidized inhibitors. So one may speculate that HD membrane related alteration of granulocyte metabolism, increase of enzyme release and leukocyte sequestion in the lung could work together in the sense of the discussed model of the early shock lung.

REFERENCES

1. L.S. Kaplow, and L.S. Goffinet, Profound neutropenia during the early phase of hemodialysis, JAMA 203: 1135 (1968).
2. P.R. Craddock, J.Fehr, K.L. Brigham, and H.S. Jacobs, Complement and leukocyte mediated pulmonary dysfunction in hemodialysis, New Engl. J. Med. 296: 769 (1977).
3. P.R. Craddock, J.Fehr, A.P. Dalmasso, K.L. Brigham, and H.S. Jacob, Hemodialysis leukopenia: pulmonary vascular leukostasis resulting from complement activation by dialyzer cellophane membranes, J. Clin. Invest. 59: 879 (1977).
4. R.M. Hakim, and G.H. Lowrie, Effect of dialyzer raise on leukopenia, hypoxemia and total hemolytic complement system, Thromb. Am. Soc. Artif. Intern. Organs 26: 159 (1980).
5. P. Aljama, A. Martin-Malo, R. Pérez, D. Castillo, A. Torres, and F. Velasco, Granulocyte adherence changes during hemodialysis, Contr. Nephrol. 46: 75 (1985).
6. R.M. Schaefer, A. Heidland, and W.H. Hörl, Release of leukocyte elastase during hemodialysis, Contr. Nephrol. 46: 109 (1985).
7. A.T. Nguyen, C. Lethias, J. Zingraff, A. Herbelin, C. Naret, and B. Descamps-Latscha, Hemodialysis membrane-induced activation phagocyte oxidative metabolism detected in vivo and in vitro within microamounts of whole blood, Kidney Int. 28: 158 (1985).
8. M.S. Cohen, D.M. Elliott, T. Chaplinski, M.M. Pike, and J.E. Niedel, A defect in the oxidative metabolism of human polymorphonuclear leukocytes that remain in circulation early in hemodialysis, Blood 60: 1283 (1982).
9. A. Böyum, IV. Isolation of mononuclear cells and granulocytes from human blood-isolation of mononuclear cells by one centrifugation, and of granulocytes by combining centrifugation and sedimentation of 1 g, Scand. J. Clin. Lab. Invest. 21: 77 (1968).

10. S. Neumann, N. Hennrich, G. Gunzer, and H. Lang, Enzyme-linked immuno-assay for human granulocyte elastase in complex with α_1-proteinase inhibitor, in: Proteases: Potential role in health and disease, W.H. Hörl, A. Heidland, eds., Plenum Press, New York, London, p. 379 (1984).
11. J.T. Curnutte, and B.M. Babior, The effect of bacteria and serum on superoxide production by granulocytes, J. Clin. Invest. 53: 1662 (1974).
12. L.R. DeChatelet, G.D. Long, P.S. Shirly, D.A. Bass, M.I. Thomas, F.W. Hendergon, and M.S. Cohen, Mechanism of luminal-descendent chemi-luminescence of human neutrophils, J. Immunol. 129: 1589 (1982).
13. K. Joseph, V. Damann, G. Engerhoff, and K.R. Gruner, Markierung von Leukozyten mit 99m-Technetium-HM-PAO: Erste klinische Ergebnisse, Nuc-Compact 5: 277 (1986).
14. C. Tetta, G. Camussi, G. Segolini, and A. Verlellone, Direct inter-action between polymorphonuclear neutrophils (PMN) and dialysis membranes: role of the electrical charges, Kidney Int. 29: 609 (1986).
15. G. Weissmann, R.B. Zurier, and S. Hoffstein, Leukocytic proteases and their immunologic release of lysosomal enzymes. Am. J. Path. 68: 539 (1972).
16. S. Stannnat, J. Bahlmann, D. Kiessling, K.M. Koch, H. Deicher, and H.H. Peter, Complement activation during hemodialysis. Comparison of polysulfone and cuprophan membranes, Contr. Nephrol. 46: 102 (1985).
17. W.H. Hörl, W. Riegel, and P. Schollmeyer, Plasma levels of main granulocyte components in patients dialyzed with polycarbonate and cuprophane membranes, Nephron 45: 272 (1987).
18. A. Janoff, Proteases and lung injury. A state of - the art minireview, Chest 54: 835 (1983).

EFFECT OF IMMUNOSUPPRESSION ON THE RELEASE OF MAIN GRANULOCYTE

COMPONENTS: IN VIVO AND IN VITRO STUDIES

Christoph Wanner, Barbara Simon, Andreas Gösele, Werner Riegel, Peter Schollmeyer, and Walter H. Hörl

Department of Medicine, Division of Nephrology, University of Freiburg Hugstetterstr. 55, 7800 Freiburg/FRG

INTRODUCTION

In operative trauma or severe inflammatory responses, polymorphonuclear leukocytes release lysosomal proteinases extracellularly and into the circulation (Duswald et al., 1985). Immediately after surgery a decrease of elastase in complex with α_1-proteinase inhibitor (E-α_1PI) is observed due to the elimination of the infection focus (Jochum et al., 1984). During hemodialysis treatment there is also an increase of E-α_1PI (Hörl et al., 1983; Hörl and Heidland 1984) depending on the membrane material of the dialyzer used (Hörl et al., 1985a and b). There was also demonstrated that polymorphonuclear leukocytes release myeloperoxidase and lactoferrin during hemodialysis treatment (Hörl et al., 1986). Corticosteroids can stabilize membranes, inhibit locomotion, phagocytosis and degranulation of polymorphonuclear leukocytes (Smolen and Weissmann 1978; Dunham et al., 1977). In the present study we evaluated plasma levels of main granulocyte components of patients following cadaveric renal transplantation. The effect of different immune suppressive drugs on the release of main granulocyte constituents was investigated in vitro and also during hemodialysis.

PATIENTS AND METHODS

Six groups of patients were studied:

Group I consisted of 10 patients (3 male, 7 female), aged 64.5 ± 3.4 years, on regular hemodialysis treatment for 9.8 ± 2.2 months. Patients were dialzyed with hollow fiber dialyzer made from cuprophan (E3, Fresenius, Oberursel, FRG).

Group II consisted of 9 patients (3 male, 6 female), aged 43.8 ± 3.4 years had been subjected to cadaveric renal transplantation. Before transplantation patients had been on regular dialysis treatment for 51.0 ± 13.0 months. Because of oligo-anuric renal failure patients were dialyzed postoperatively with cuprophan hollow-fiber dialyzers for 12.4 ± 5.7 days.

Group III consisted of 10 patients (8 male, 2 female) aged 42.9 ± 4.5 years, who needed laparatomy following trauma. These patients were studied postoperatively during 9 days.

Group IV consisted of 15 patients (6 male, 9 female) aged 35.4 ± 2.8 years, following cadaveric renal transplantation. This group of patients was studied postoperatively during 9 days.

Group V consisted of 10 patients (6 male, 4 female), aged 47.2 ± 2.5 years, subjected to cadaveric renal transplantation 6.2 ± 2.5 months ago. Mean creatinine was 1.6 ± 0.3 mg/dl. Immunosuppression was maintained with ciclosporin and prednisolone.

Group VI consisted of 10 patients (5 female, 5 male), aged 39.3 ± 3.1 years, subjected to cadaveric renal transplantation 5.3 ± 1.1 years ago. Mean creatinine was 1.6 ± 0.2 mg/dl. Immunosuppression was maintained with azathioprine and prednisolone.

Venous blood samples were anticoagulated with sodium citrate and the plasma was separated immediately after centrifugation to prevent leakage of leukocyte constituents. The samples were stored at - 30° C and were thawed only once for processing.

Plasma E- α_1 PI, lactoferrin and myeloperoxidase were performed with highly sensitive enzyme-linked immunoassays as described by Neumann et al. (1984) and Hörl et al. (1986).

For in vitro experiments blood samples of 10 healthy controls were incubated with prednisolone (final concentration 100 ng/ml), ciclosporin (final concentration 400 ng/ml) or azathioprine (final concentration 3 µg/ml). The results of plasma elastase, lactoferrin and myeloperoxidase were compared with blood samples of 10 patients under immunosupression with ciclosporin and prednisolone and of 10 patients under immunosupppression with azathioprine and prednisolone.

Statistics

Data are given as mean ± SEM. For analysis of significance the Student's t test for paired and unpaired samples and the Mann-Whithney U test was used.

RESULTS

Figure 1 shows plasma levels of main granulocyte constituents at the start and the end of hemodialysis treatment. As compared to patients on regular hemodialysis treatment, patients after kidney transplantation under immunosupppression with ciclosporin and prednisolone showed significantly lower granulocyte lactoferrin levels at the end of hemodialysis treatment.

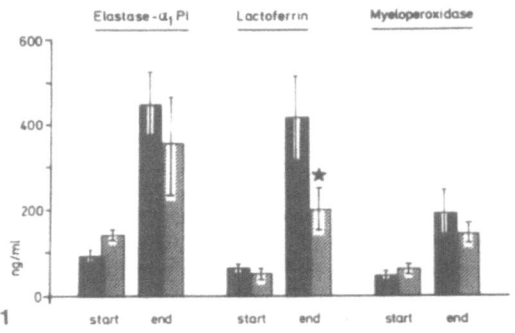

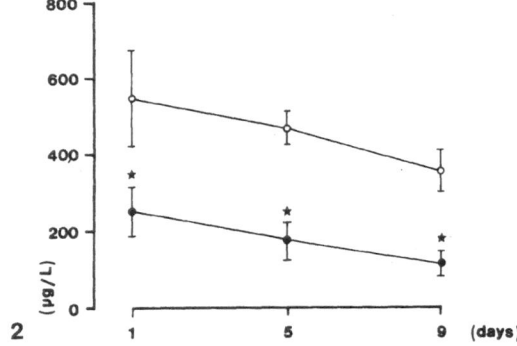

Fig. 1. Plasma E-α_1 PI, lactoferrin and myeloperoxidase in patients on regular hemodialysis treatment ■ and after kidney transplantation ▨ at the start and the end of hemodialysis. Data are given as mean ± SEM from 10 patients.

Fig. 2. Plasma E-α_1 PI of patients after cadaveric renal transplantation •——• and after abdominal surgery ○——○. Data are given as mean ± SEM from 15 and 10 patients respectively.

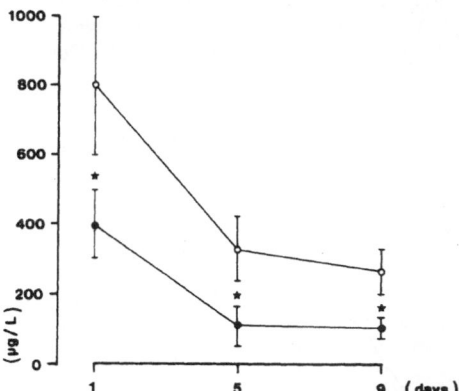

Fig. 3. Plasma lactoferrin of patients after cadaveric renal transplantation ●——● and after abdominal surgery o———o .

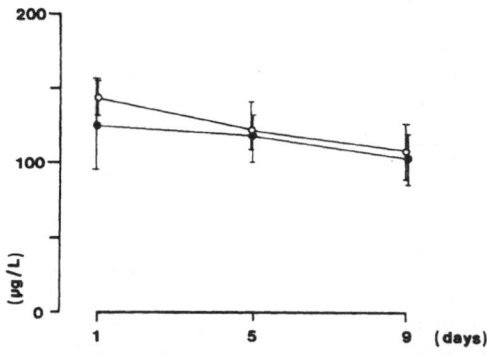

Fig. 4. Plasma myeloperoxidase of patients after cadaveric renal transplantation ●——● and after abdominal surgery o———o .

Figure 2 shows plasma levels of granulocyte lactoferrin of patients after CRT and after abdominal surgery. Significant lower lactoferrin values were observed in patients under immunosuppression. When plasma E- α_1PI values were determined, similar effects of immunosuppressive drugs were obtained (Fig. 3). On the other hand, plasma levels of granulocyte myeloperoxidase were comparable in patients after kidney transplantation and abdominal surgery (Fig. 4). The influence of different immunosuppressive agents on degranulation of polymorphonuclear leukocytes (PMNLs) was investigated in in vitro studies. Figure 5 demonstrates reduced degranulation of PMNLs regarding both lactoferrin and elastase values. No effect of azathioprine and prednisolone could be detected. However, in vivo administered ciclosporin did not influence spontaneous degranulation of human PMNLs in vitro (Fig. 6)

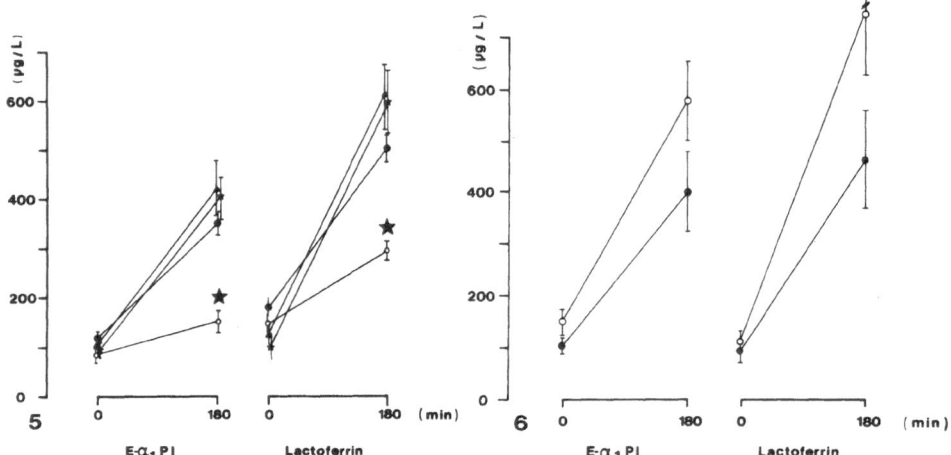

Fig. 5. In vitro degranulation of PMNLs in heparinized blood samples of healthy control subjects with and without addition of different immunosuppressive drugs. o———o reveals addition of ciclosporine, ▲———▲ azathioprine and ✻———✻ prednisolone.

Fig. 6. In vitro degranulation of PMNLs in heparinized blood samples of transplant patients under immunosuppression with ciclosporin and prednisolone (o———o) as well as azathioprine and prednisolone (•———•).

DISCUSSION

It has been shown that ciclosporin effectively inhibits in vitro superoxide release from pulmonary alveolar macrophages (Drath and Kahan 1983). On the other hand, the 30-day in vivo administration of immune suppressive drugs was without effect on superoxide release from both polymorphonuclear neutrophils and pulmonary alveolar macrophages. It was suggested that either ciclosporin was not achieving high enough concentration in the appropriate environment of these phagocytes, or the cells were not sensitive to the circulating from of the drug (Draht and Kahan 1983). In the present study comparable results have been made. Heparinized blood samples obtained from healthy controls were incubated in vitro with prednisolone, azathioprine and ciclosporin. A significant inhibition of the release of granulocyte elastase and lactoferrin was only observed in the presence of ciclosporin (Figure 5). However, in vivo administration of ciclosporin had no effect on the release of main granulocyte components (Figure 6). High doses of corticosteroids inhibited lactoferrin release but did not affect plasma levels of myeloperoxidase and elastase during hemodialysis therapy in patients with acute renal failure (Catalano et al., 1987). In our patients on hemodialysis after kidney transplantation, corticosteroids combined with ciclosproin prevented degranulation of specific granules of PMNLs as well. Compared to patients following abdominal surgery, however, plasma levels of granulocyte elastase and lactoferrin rapidly decreased under immunosuppression following cadaveric renal transplantation (Figures 2 and 3). These data suggest a direct effect of ciclosporin on human PMNLs. Comparable levels of plasma myeloperoxidase have been found in patients following renal transplantation and abdominal surgery (Figure 4). Studies of Schmidt (1978) demonstrate that there is not only differential release of enzymes from azurophilic and specific granules of PMNLs, but also a differential release of elastase and chymotrypsin (cathepsin B), both of which are localized in the primary granules. In conclusion, our results demonstrate in vitro inhibition of spontaneous degranulation of human PMNLs by ciclosporin, whereas in vivo administration of ciclosporin shows no effect under these conditions. In vivo administration of cicolosporin affects both degranulation of human PMNLs during hemodialysis and after operative trauma.

ACKNOWLEDGEMENTS

This work was supported by the "Deutsche Forschungsgemeinschaft", Ho 781/3-4. We thank Mrs. B. Meier for helpful assistance.

REFERENCES

Catalano, C., Maggiore, Q., Enia, G., and Hörl, W.H., 1987, Methylprednisolone reduces dialysis-induced degranulation of specific granules of polymorphonuclear leukocytes in patients with acute renal failure, Clin Nephrol., 28: 48.

Draht, D.B., and Kahan, B.D., 1983, Pulmonary macrophage and polymorphonuclear leukocyte function in response to immunosuppressive therapy, Transpl Proc., 15: (Suppl 1), 2367.

Dunham, P., Babiarz, P., Israel, A., Zerial, A., and Weissmann, G., 1977, Membrane fusion: studies with a calcium-sensitive dye, arzenazo III, in: liposomes, Proc Natl Acad Sci USA, 74:1580

Duswald, K.H., Jochum, M., Schramm, W., and Fritz, H., 1985, Released granulocyte elastase: an indicator of pathobiochemical alterations in septicemia after abdominal surgery, Surgery, 98:892.

Hörl, W.H., Jochum, M., Heidland, A., and Fritz, H., 1983, Release of granulocyte proteinases during hemodialysis, Am J Nephrol., 3:213.

Hörl, W.H., and Heidland, A., 1984, Evidence for the participation of granulocyte proteinases on intradialytic catabolism, Clin Nephrol., 21:314.

Hörl, W.H., Schaefer, R.M., and Heidland, A., 1985, Effect of different dialyzers on proteinases and proteinase inhibitors during hemodialysis, Am J Nephrol., 5:320.

Hörl, W.H., Steinhauer, H.B., and Schollmeyer, P., 1985, Plasma levels of granulocyte elastase during hemodialysis: effects of different dialyzers membranes, Kidney Int., 28:791.

Hörl, W.H., Riegel, W., Schollmeyer, P., Rautenberg, W., and Neumann, S., 1986, Different complement and granulocyte activation in patients dialyzed with PMMA dialyzers. Clin Nephrol., 25:304.

Jochum, M., Duswald, K.H., Neumann, S., Witte, J., and Fritz, H., 1984, Proteinases and their inhibitors in septicemia - basic concepts and clinical applications, in: Proteases: Potential Role in Health and Disease, W.H. Hörl., and A. Heidland, eds., Plenum Press, New York, London, 391.

Neumann, S., Hennrich, N., Gunzer, G., and Lang, H., 1984, Enzyme-linked immunoassay for human granulocyte elastase in complex with alpha $_1$-proteinase inhibior, in: Proteases: Potential Role in Health and Disease, W.H. Hörl, and A. Heidland, eds., Plenum Press, New York, London, 379.

Schmidt, W., 1978, Differential release of elastase and chymotrypsin from polymorphonuclear leukocytes, in: Neutral proteases of human polymorphonuclear leukocytes, Havemann, K., and Janoff, A., (eds), Baltimore - Munich, Urban & Schwarzenberg, 77.

Smolen, J.E., and Weissmann, G., 1978, The granulocyte: Metabolic properties and mechanisms of lysosomal enzyme release, in: Neutral protease of human polymorphonuclear leukocytes, K. Havemann., and A. Janoff, eds., Urban & Schwarzenberg, Baltimore - Munich, 56.

RELEASE OF GRANULOCYTE PROTEINS DURING CARDIOPULMONARY BYPASS:

EFFECT OF DIFFERENT PHARMACOLOGICAL INTERVENTIONS

Werner Riegel, Gerhard Spillner, Volker Schlosser
Klaus Lang, and Walter H. Hörl

Department of Medicine, Division of Nephrology and
Department of Surgery, Division of Cardiovascular Surgery
University of Freiburg, 7800 Freiburg, FRG

INTRODUCTION

Leukopenia and anaphylatoxin formation occur during cardiopulmonary bypass[1,2]. Both C3a and C5a are converted to their respective desArg forms by plasma carboxypeptidase, but C5a-desArg still retains its capacity to activate leukocytes[2-6]. After C5 cleavage C5b-9 complexes are generated on target membranes or SC5b-9 complexes in the fluid phase. Recent studies demonstrated deposition of C5b-9 on blood cells during cardiopulmonary bypass. It was suggested that C5b-9 complexes may be partly responsible for the hemolysis and may augment granulocyte activation by the stimulation of arachidonate metabolism in those cells[7].

Complement activation is also temporarily correlated with hemodialysis leukopenia in patients dialyzed with cuprophane hollow-fiber dialyzers. The complement-derived fragment C5a mediates the expression of a granulo-cyte-adhesion-promoting surface glycoprotein on granulocytes providing leukoaggregation, sequestration of granulocytes, and neutropenia during hemodialysis[8]. Granulocyte activation during hemodialysis, however, does not necessarily need complement activation. Recent studies from our laboratory demonstrated that granulocyte activation during hemodialysis in the absence of complement activation could be inhibited by calcium channel blockers[9].

Therefore, the present study was designed to investigate different pharmacological interventions on granulocyte activation during cardio-pulmonary bypass.

PATIENTS AND METHODS

A total number of 34 patients undergoing cardiac surgery were included in this study. The patients were selected to 5 different groups.

Cardiopulmonary bypass equipment consisting of a roller pump, a Shiley S-100 A oxygenator (Shiley Inc., California, USA) or a Capiox oxy-genator (Terumo Inc., Tokyo, Japan) was uniformly employed. The membrane material of both oxygenators consisted of polypropylene. The pump-oxygenator system was primed with 500 ml Gelifundol[R] (Biotest, FRG), Ringer-solution, 5% Fructose- and 10% Mannitol-solution to a final volume of 1,750 ml for

Shiley S-100 A oxygenator and 2,200 ml for Capiox II oxygenator. Immediately before institution of bypass, patient's blood was anticoagulated with heparin administered intravenously in a dosage of 3 mg/kg body weight. After canulation of vena cava superior and inferior as well as ascending aorta, bypass was instituted with a pump flow rate of 2.2 liters per minute per square meter of body-surface area. Body temperature was regulated both 32° C rectal and 29° C esophageal.

During completion of aortovenous anastomosis, while the perfusate temperature was gradually beeing decreased, an aortic cross clamp was applied, asanguinous, cold potassium cardioplegia solution was administered, and the coronary arteriovenous anastomosis was performed. With completion of the final anastomosis, the body temperature was gradually increased to 37° C, the cross clamp was then removed, and after resumption of cardiac activity the partial cardiopulmonary bypass was gradually discontinued.

Group I consisted of 10 patients (7 male, 3 female) with a mean age of 56.6 + 2.6 (mean + S.E.M.) years undergoing cardiac surgery by mitral valve replacement (n=2). Eight patients underwent elective saphenous-vein coronary-artery bypass. In this group of patients we compared the plasma levels of main granulocyte components (lactoferrin, myeloperoxidase, elastase) with those of the anaphylatoxins C3a and C5a.

In order to investigate the effects of different drugs, the calcium channel blocker nifedipine and the antiplatelet drug dipyridamol were used. Patients were randomly selected to the four following groups.

Group II consisted of 6 patients (2 male, 4 female) with a mean age of 36.8 + 9.1 years undergoing cardiac surgery by mitral valve (n=3) or aortic valve (n=3) replacement. Patients of this group received no drugs and acted as controls for groups III-V.

Group III consisted of 6 male patients with a mean age of 53.0 + 3.9 years. Five patients underwent elective saphenous-vein coronary-artery bypass and one patient operation of aortic valve stenosis. These patients received 300 mg dipyridamol 12 hours before surgery (3.9 + 0.2 mg/kg body weight).

Group IV consisted of seven male patients with a mean age of 56.3 + 1.7 years. All patients underwent elective saphenous-vein coronary-artery bypass. These patients received 300 mg dipyridamol 12 hours before surgery (3.7 + 0.1 mg/kg body weight). Furthermore, nifedipine (5.91 + 0.53 µg/kg body weight/hour) was given intravenously about 15 minutes before till the end of the extracorporeal circulation. All patients received calcium channel blockers orally days before surgery.

Group V consisted of five patients (4 male, 1 female) with a mean age of 57.2 + 3.9 years. All patients underwent elective saphenous-vein coronary-artery bypass. These patients received 300 mg dipyridamol (4.1 + 0.3 mg/kg. body weight) and nifedipine (n=3) or diltiazem (n=2) 12 hours before surgery. These patients were under steady oral treatment with nifedipine or diltiazem.

Blood samples required for determination of plasma C3a and C5a levels, granulocyte myeloperoxidase, lactoferrin, and elastase were obtained at the following times: on arrival in the operating suite but before administration of anesthetic (the base-line period); after institution of bypass at intervals of 15, 30 and 60 minutes; at the end of the operation; 2 hours postoperatively; and 24 hours postoperatively. Blood specimens were obtained from central vein catheter before and after operation or cardiopulmonary-bypass venous-return line during operation. All samples were anticoagulated immediately with sodium citrate for determination of the main granulocyte compo-

nents or disodium EDTA for C3a and C5a. Plasma was separated from the
sample within 30 min after its collection to prevent leakage of leukocyte
constituents by centrifugation for 10 minutes at 4000 x g/min. The plasma
specimens were stored at - 30°C until assayed.

The measurement of plasma levels of the granulocyte elastase in complex
with α_1-proteinase inhibitor (E-α_1PI) was performed with a highly sensitive
enzyme-linked immunoassay as described by Neumann et al. [10]. The determination of human lactoferrin and myeloperoxidase was performed as previously
described[11]. C3a and C5a were measured in sodium-EDTA plasma specimens after
removal of arginine by anaphylatoxin inactivator (C3a desArg, C5a desArg).
Radioimmunoassay kits for C3a desArg and C5a desArg were obtained from
Upjohn Diagnostics (Kalamazoo, Michigan, USA). The assay for C3a and C5a
follow those described by Hugli et al. [12].

Data are given as mean values + S.E.M. Wilcoxon's nonparametric rank
test for paired data samples was used to evaluate statistically significant
differences between the study period and base-line values. To compare one
effect of different drugs on granulocyte and complement activation an analysis of variance (ANOVA) was used.

RESULTS

In patients of group I plasma level of granulocyte lactoferrin was
97.0 + 22.8 ng/ml in the base-line period. Within 15 minutes after the
institution of bypass there was a mean plasma lactoferrin level of 696.1
+ 112.5 ng/ml. Maximal lactoferrin levels were observed after 60 minutes
(1,418.9 + 190.8 ng/ml). 24 hours postoperatively plasma lactoferrin values
returned to the range of the base-line period (101.4 + 17.7 ng/ml). Plasma
values of myeloperoxidase increased by 460% (from 37.1 + 4.3 to 170.9 +
34.9 ng/ml) at the end of operation. 24 hours postoperatively, plasma
myeloperoxidase level was still elevated (87.5 + 8.2 ng/ml).

E-α_1PI levels increased from 89.4 + 7.4 ng/ml in the base-line period
to 617.0 + 106.6 ng/ml 2 hours postoperatively. Even 24 hours postoperatively E-α_1PI values remained elevated (321.3 + 24.1 ng/ml).

Plasma C3a levels rose from 131.8 + 12.6 to 401.1 + 56.4 ng/ml and C5a
levels from 40.7 + 1.6 to 77.1 + 25.9 ng/ml within the first 15 minutes
after institution of bypass. Both parameters remained elevated until the
end of operation.

White blood cell count of group III, IV and V during cardiopulmonary
bypass is shown in Fig.1. A significant decrease was found after 15 minutes
in all three groups. As you can see there is no difference comparing the
effect of different pharmacologic interventions on WBC count (ANOVA). There
was also no difference when we compared the number of white blood cells of
these three groups with the results obtained in groups I and II.

Fig. 2 shows the effect of dipyridamol alone and in combination with
intravenously or orally administered nifedipine on plasma E-α_1PI values.
Dipyridamol alone and in combination with orally administered nifedipine
or diltiazem was without effect on E-α_1PI. However, continuous infusion
of nifedipine resulted in markedly lower elastase release. On the other
hand, plasma C3a levels were even higher in patients infused with nifedipine (Fig.3).

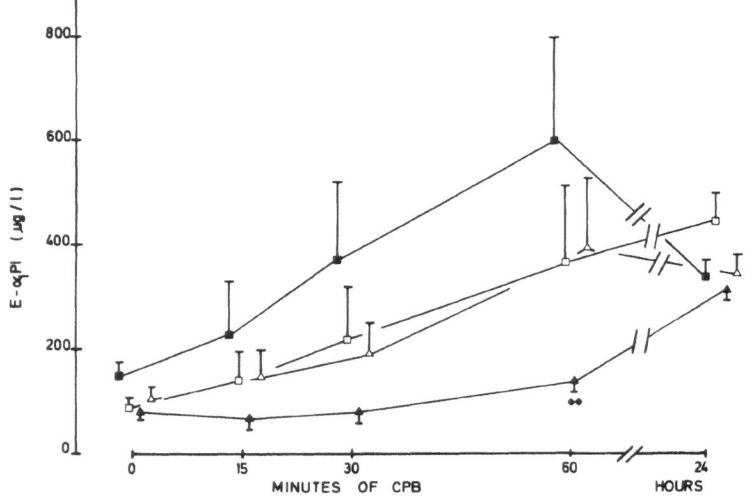

Fig.1 Effect of different pharmacological interventions during cardiopulmonary bypass (CPB) on white blood cell (WBC) count and the
quotient of WBC / PCV (hematocrit). Mean values ± S.E.M. from 5-6
patients. Symbols are □ for group III, ▲ for group IV, and
△ for group V.

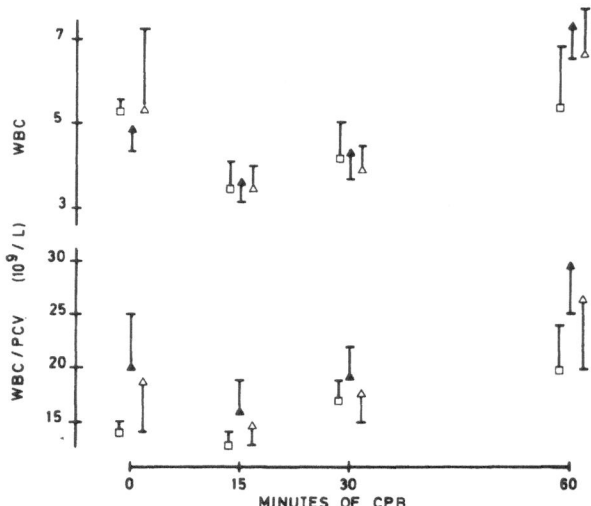

Fig.2 Effect of different pharmacological interventions during cardiopulmonary bypass (CPB) on plasma levels of granulocyte elastase in
complex with α_1-proteinase inhibitor (E-α_1PI). Mean values ± S.E.M.
from 5-6 patients. Symbols are ■ for group II, □ for group III,
▲ for group IV, and △ for group V.
**p < 0.01 group IV versus the three other groups.

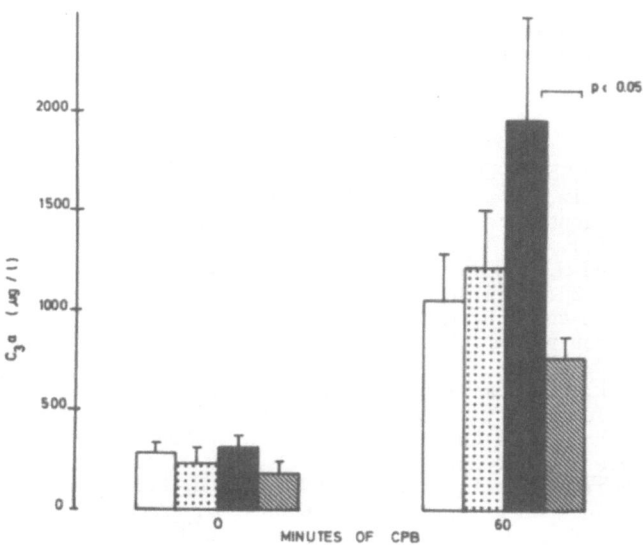

Fig.3 Effect of different pharmacological interventions on plasma C3a
levels before (0 min) and 60 min of cardiopulmonary bypass (CPB).
Mean values ± S.E.M. from 5-6 patients. Symbols are ☐ for group II,
⊡ for group III, ■ for group IV, and ▧ for group V.

DISCUSSION

Recent studies from our laboratory showed that granulocyte activation
during hemodialysis does not necessarily need complement activation[9]. The
same experimental observations need to be specifically reported for cardiopulmonary bypass, since there are many differences in the physiological
response to hemodialysis versus cardiopulmonary bypass. The present study
demonstrates inhibition of granulocyte activation by nifedipine infusion
during cardiopulmonary bypass, whereas plasma C3a levels were even higher
under these experimental conditions (Figures 2 and 3). Oral administration
of calcium channel blockers also has an effect on the release of lactoferrin
from the specific granules of polymorphonuclear leukocytes[13].

Plasma levels of granulocyte elastase with α_1PI complex were significantly elevated throughout the study. In contrast, plasma levels of lactoferrin were in the normal range 24 hours after operation. The half-live of
the elastase α_1PI complex is about 1 hour[14]. Increased plasma levels of
elastase α_1PI complex may be caused by the operative trauma. It has been
shown that in patients without preoperative infection, the operative trauma
was followed by an increase of the elastase α_1PI level up to threefold of
the normal value[15]. Plasma levels of elastase α_1PI complex remained significantly elevated for 48 to 72 hours in patients undergoing aortofemoral bypass operation[16].

Kokot and coworkers[17] could show in in-vitro experiments that the
calcium ionophore A 23187 induced a release of elastase from neutrophils in
a dose- and time-dependent manner. In vitro, verapamil caused only inhibition
of elastase release in unphysiological high doses of 10^{-4} and 10^{-3} M[17]
suggesting a direct effect of the drug. In agreement with our observations
on patients undergoing regular hemodialysis treatment , we found inhibition
of the release of granulocyte components during cardiopulmonary bypass with
relative low doses of intravenous administered calcium channel blockers[18].

395

Therefore, our data suggest a new mechanism of granulocyte activation during cardiopulmonary bypass via calcium ions.

REFERENCES

1. D.E. Chenoweth, S.W. Cooper, T.E. Hugli, R.W. Stewart, E.H. Blackstone, and J.W. Kirklin, Complement activation during cardiopulmonary bypass: evidence for generation of C3a and C5a anaphylatoxins. N. Engl. J. Med. 304: 497 (1981).
2. T.E. Hugli and H.J. Müller-Eberhard, Anaphylatoxins: C3a and C5a. Adv. Immunol. 26: 1 (1978).
3. D.E. Chenoweth and T.E. Hugli, Demonstration of specific C5a receptor on intact human polymorphonuclear leukocytes. Proc. Natl. Acad. Sci. USA 75: 3943 (1978).
4. P.R. Craddock, D. Hammerschmidt, J.G. White, A.P. Dalmasso, and H.S. Jacobs, Complement (C5a)-induced granulocyte aggregation in vitro: a possible mechanism of complement-mediated leukostasis and leukopenia. J. Clin. Invest. 60: 260 (1977).
5. J.T. O'Flaherty, D.L. Kreutzer, and P.A. Ward, Neutrophil aggregation and swellin induced by chemotactic agents. J. Immunol. 119: 232 (1977).
6. R.O. Webster, S.R. Hong, R.B. jr Johnston, and P.M. Henderson, Biological effects of the human complement fragments C5a and C5ades Arg on neutrophil function. Immunopharmacology 2: 201 (1980).
7. A. Salama, F. Hugo, D. Heinrich, R. Höge, R. Müller, V. Kiefel, C. Müller-Eckhardt, and S. Bhakdi, Deposition of terminal C5b-9 complement complexes on erythrocytes and leukocytes during cardiopulmonary bypass. N. Engl. J. Med. 318: 408 (1988).
8. M.A. Arnaout, R.M. Hakim, R.F. Todd, N. Dana, and H.R. Colten, Increased expression of an adhesion-promoting surface glycoprotein in the granulocytopenia of hemodialysis. N. Engl. J. Med. 312: 457 (1985).
9. M. Haag-Weber, P. Schollmeyer, and W.H. Hörl, Granulocyte activation during hemodialysis in the absence of complement activation: inhibition by calcium channel blockers. Europ. J. Clin. Invest. (in press).
10. S. Neumann, N. Hennrich, G. Gunzer, and H. Lang, Enzyme-linked immunoassay for human granulocyte elastase in complex with α_1-proteinase inhibitor, in: Proteases: Potential Role in Health and Disease, W.H. Hörl and A. Heidland, eds., New York, London, 1984, Plenum Press, p. 379
11. W.H. Hörl, W. Riegel, P. Schollmeyer, W. Rautenberg, and S. Neumann, Different complement and granulocyte activation in patients dialyzed with PMMA dialyzers. Clin. Nephrol. 25: 304 (1986).
12. T.E. Hugli and D.E. Chenoweth, Biologically active peptides of complement: techniques and significance of C3a and C5a measurements, in: Immunoassays: Clinical Laboratory Techniques for the 1980's, R.M. Nakamura, W.R. Dito, and E.S. Tucker, eds., New York, 1980, Alan R. Liss. Inc., p. 443
13. W. Riegel, G. Spillner, V. Schlosser, and W.H. Hörl, Plasma levels of main granulocyte components during cardiopulmonary bypass. J. Thorac. Cardiovasc. Surg. (in press).
14. K. Ohlsson and C.B. Laurell, The disappearance of enzyme-inhibitor complexes from the circulation of man. Clin. Sci. Mol. Med. 51: 87 (1976).
15. H. Fritz, M. Jochum, K.-H. Duswald, H. Dittmer, H. Kortmann, S. Neumann, and H. Lang, Granulocyte proteinases as mediators of unspecific proteolysis in inflammation: A review, Selected Topics in Clinical Enzymology, D.M. Goldberg and M. Werner, eds., Berlin, New York, 1984, Walter de Gruyter, p. 305
16. M. Hörl, M. Sperling, J. Mayer, and W.H. Hörl, Plasma proteinases, proteinase inhibitors and other selective plasma proteins following aortofemoral bypass operation. Eur. Surg. Res. 18: 69 (1986).
17. K. Kokot, M. Teschner, R.M. Schaefer, and A. Heidland, Stimulation and inhibition of elastase release from human neutrophils - dependence on the

calcium messenger system. <u>Mineral Electrolyte Metab.</u> 13: 189 (1987).

18. W. Riegel, G. Spillner, V.Schlosser, and W.H. Hörl, Nifedipine inhibits granulocyte activation during cardiopulmonary bypass. <u>Klin. Wochenschr.</u> 65: 581 (1987).

SIGNIFICANT ROLE OF PROTEASE INHIBITION BY APROTININ IN MYOCARDIAL

PROTECTION FROM PROLONGED CARDIOPLEGIA WITH HYPOTHERMIA

Makoto Sunamori, Ryuichi Innami, Hitoshi Fujiwara
Motoki Yokoyama, and Akio Suzuki

Dept. of Thoracic Surgery, Tokyo Medical and Dental
University, School of Medicine
1-5-45, Yushima, Bunkyo-ku, Tokyo, 113, Japan

Biochemical and morphological damage in the ischemic myocardium result from increased activity of proteases, particularly lysosomal enzymes. The concentration of lysosomal enzymes increases in the both myocardium in response to ischemia and the systemic circulation in response to extracorporeal circulation. Cold cardioplegia minimizes the effects of ischemia during open heart surgery. It has been reported that aprotinin protects the myocardium from ischemia followed by reperfusion by suppressing the release of lysosomal enzymes[2,3].

This investigation was aimed to characterize the changes in lysosomal enzymes and in indeces of myocardial damage by aprotinin during and after prolonged (3 - 4 hours) of cardioplegia associated with hypothermia followed by reperfusion.

Materials and Methods

Twenty-four mongrel dogs were studied to characterize the changes in lysosomal enzymes; beta-glucuronidase (Beta-G), N-acetyl-glucosaminidase (NAG) and in indices of myocardial damage by aprotinin during and after 3 - 4 hours of cardioplegia followed by reperfusion. Dogs were anesthetized with pentobarbital, intravenously and thoracotomized. Cardiopulmonary bypass (CPB) was instituted with moderate hemodilution (approximate hematocrit between 20 and 25 %) and systemic hypothermia around 20°C. Cardioplegia was induced by the lidocaine-Mg^{2+} cardioplegic solution (lidocaine 11 mM, Mg^{2+}-1-aspartate 8 mM, glucose 242 mM/L, pH 7.50, Osmolarity 450 mOsm/L) with initial dose of 10 ml/kg, and additional dose of 3 ml/kg every hour.
Myocardial temperature was strictly kept between 15 and 20°C. The aorta was cross-clamped for 3 - 4 hours. Reperfusion was observed for 60 minutes ; initial reperfusion was supported by CPB until systemic temperature reached 37°C, it took approximately 20 - 25 minutes, then CPB was terminated, and reperfusion was continued without either any mechanical circulatory support or cardiotonic drugs. Left ventricle was vented during CPB.

Hemodynamic changes (heart rate, arterial pressure, cardiac output, LV pressure and LVEDP) were monitored and measured before CPB, at 30 and 60 minutes of reperfusion. Lysosomal enzymes and MB-CK, mitochondrial aspartate aminotransferase (m-AAT) were studied in the samples withdrawn from the coronary sinus before pretreatment, before CPB, at 5, 30, and 60 minutes of reperfusion. Myocardial water content, adenosine triphosphate (ATP), creatine phosphate (CP) and ultrastructure were studied on the myocardial samples from the endocardium, epicardium and the ventricular septum of the left ventricle taken at the end of experiment. This study consists of following groups by dose of aprotinin and duration of cardioplegia :

Group I   : non-treated group, n = 5, cardioplegia for 3 hours.
Group II : aprotinin 1×10^4 KIU/kg, n = 7, cardioplegia for 3 hours.
Group III: aprotinin 1×10^4 KIU/kg, n = 6, cardioplegia for 4 hours.
Group IV : aprotinin 2×10^4 KIU/kg, n = 6, cardioplegia for 4 hours.
All dogs were pretreated with betamethasone, nifedipine and coenzyme Q_{10}.

Results

    All dogs survived in the group I, II and IV, however, 2 of 6 dogs (33 %) survived in the Group III.
Beta-G changes as shown in Fig 1; The groups treated with aprotinin had tendency to reduce beta-G activity as compared to the non-treated group. Beta-G at 5 minutes of reperfusion was significantly lower in the group II, III and IV than that of the Group I. NAG is shown in Fig 2. Aprotinin significantly suppressed NAG in the preischemic period and aprotinin, 1×10^4 KIU/kg tended to suppress NAG throughout experiment as compared to the non-treated group following 3 hours of cardioplegia. Aprotinin 2×10^4 KIU/kg is much effective to suppress NAG following 4 hours of cardioplegia as compared to aprotinin 1×10^4 KIU/kg.

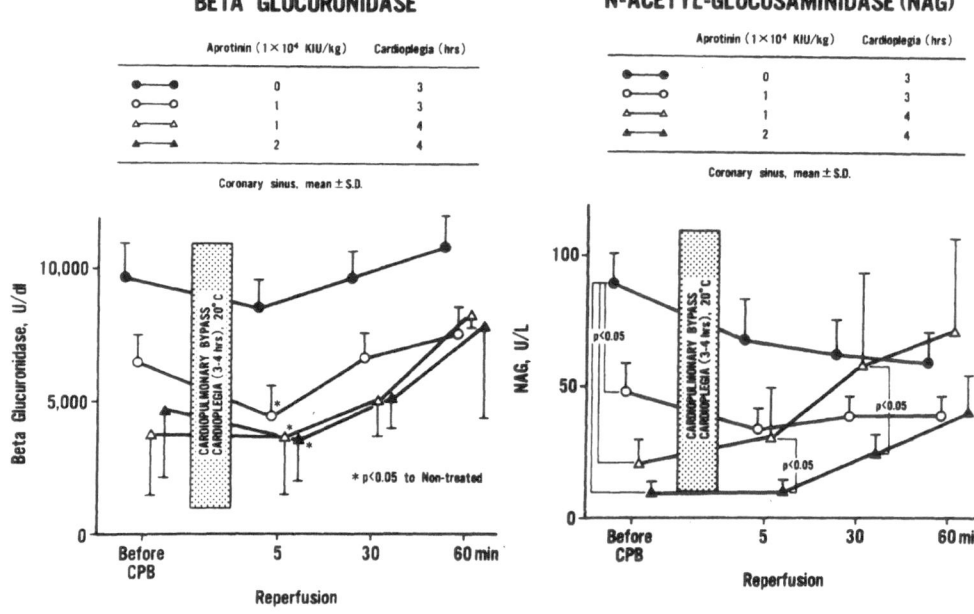

Fig. 1   Fig. 2

Serum MB-CK in the coronary sinus is shown in Fig 3. No significant difference was demonstrated among four groups. Serum MB-CK in groups rendered ischemic for 4 hours with either 1×10^4 KIU/kg or 2×10^4 KIU /kg of aprotinin is not significantly elevated as compared to those of cardioplegia for 3 hours. Serum m-AAT in the coronary sinus is shown in Fig 4. Animals treated with aprotinin showed significantly lower m-AAT values as compared to the non-treated animals. Aprotinin 1×10^4 KIU/kg is effective to suppress release of m-AAT during reperfusion following 3 hours of cardioplegia, and aprotinin 2×10^4 KIU/kg is effective to suppress release of m-AAT following 4 hours of cardioplegia.

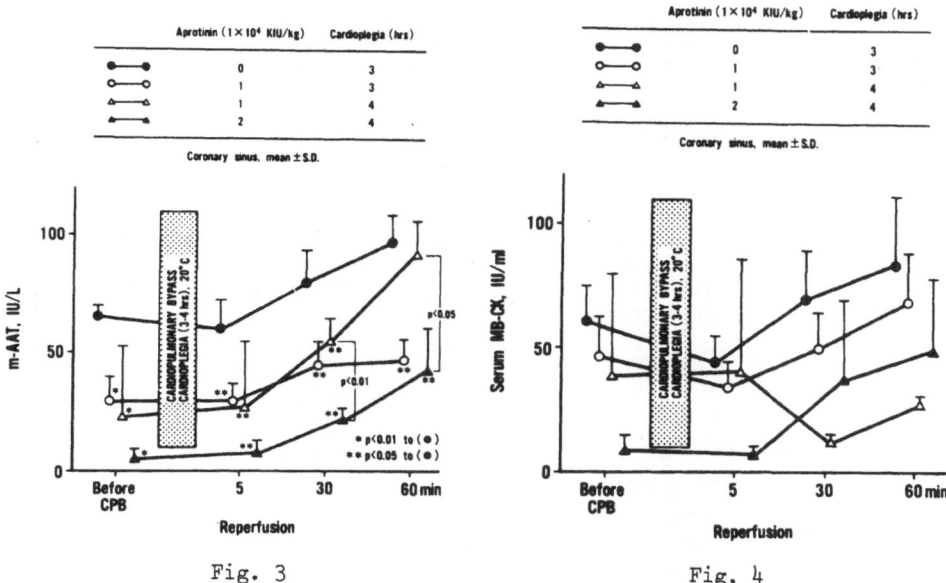

Fig. 3   Fig. 4

Myocardial ATP and CP were shown in Table 1. Myocardial ATP content was preserved in the half level of the normal tissue (normal range ; 30 - 32 ug/mg protein) in all groups without significant difference among them. Myocardial CP content was lower in the group III (1×10^4 KIU/kg of aprotinin, cardioplegia for 3 hours) as compared to the group IV (2×10^4 KIU/kg of aprotinin, cardioplegia for 4 hours). Normal range ; 30 - 32 ug/mg protein.

Table 1

Left Ventricular Endocardium

mean ± S.D.

Aprotinin (1×10^4 KIU/kg)	0	1	1	2
Cardioplegia (hrs)	3	3	4	4
Reperfusion (min)	60	60	60	60
ATP (μg/mg protein)	14.00 ± 3.57	15.96 ± 2.90	16.75 ± 2.35	15.63 ± 4.91
CP (μg/mg protein)	12.51 ± 6.54	16.50 ± 4.39	4.27 ± 1.95	14.44 ± 8.80
			└─ $0.05 < p < 0.1$ ─┘	
n	5	7	3	6

Myocardial water content was summarized in Table 2. No significant difference in myocardial water content was demonstrated among four groups. They were significantly increased as compared to the normal tissue.

Table 2

				mean ± S.D.
Aprotinin (1×10^4 KIU/kg)	0	1	1	2
Cardioplegia (hrs)	3	3	4	4
Reperfusion (min)	60	60	60	60
Water content (%)	81.1±0.9	81.2±0.7	80.5±1.7	80.8±1.3
n	5	7	3	6

Representative ultrastructural findings are demonstrated in Fig 5. Aprotinin 2×10^4 KIU/kg preserved mitochondrial ultrastructure better as compared to aprotinin 1×10^4 KIU/kg in cardioplegia for 4 hours followed by reperfusion.

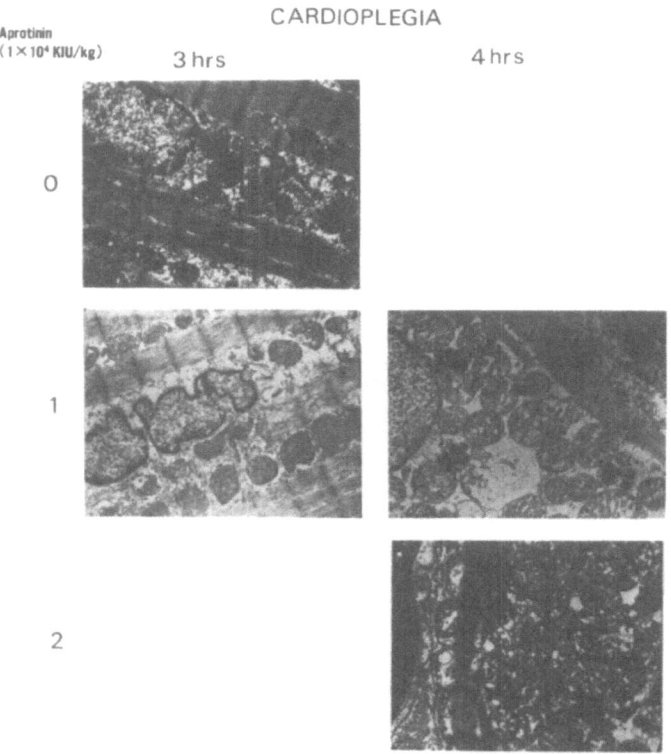

Fig. 5. Ultrastructure of LV Endocardium.

Discussion

Aprotinin protects the myocardium from ischemia followed by reperfusion by suppressing the release of lysosomal enzymes. Release of beta-G and NAG into the coronary sinus was also suppressed in prolonged ischemic condition in this investigation, particularly, NAG was remarkably suppressed by aprotinin 2×10^4 KIU/kg in reperfusion following 4 hours of cardioplegia. Various concentration of aprotinin for preservation of tissue viability or cellular function has been reported ; 500 - 2500 KIU/kg of aprotinin for kidney preservation[4],[5] higher concentration of aprotinin has been proposed in the model of tissue culture for preservation of kidney[6]. Further investigation is mandatory to determine optimal concentration of aprotinin in global ischemia followed by reperfusion under extracorporeal circulation.

It has been reported that aprotinin is effective to preserve adenosine pool in the myocardium[3],[7]. In our current study, myocardial ATP content was relatively well preserved although no significant difference was observed among four groups, and myocardial creatine phosphate content was well preserved in the dogs rendered ischemic for 3 hours treated with aprotinin 1×10^4 KIU/kg and animals rendered ischemic for 4 hours treated with aprotinin 2×10^4 KIU/kg.

It is noted that m-AAT in the coronary sinus was significantly suppressed in the group of cardioplegia for 4 hours treated with aprotinin 2×10^4 KIU/kg as compared to the group treated with aprotinin 1×10^4 KIU/kg. Since coronary sinus m-AAT concentration correlates to myocardial mitochondrial morphological damage[8], our result suggests that aprotinin 2×10^4 KIU/kg seems to be effective to preserve myocardial mitochondrial structural integrity. This is apparently supported by the ultrastructural findings in this study.

Eisenbach et al[9] reported that aprotinin prevents edema in the capillary endothelium. Myocardial water content in our study increased as compared to the normal myocardium, and also, no difference in myocardial water content was observed independent to dose of aprotinin. This result is most likely due to prolonged hemodilution of extracorporeal circulation.

In conclusion, aprotinin is effective to preserve the myocardium from ischemia followed by reperfusion, and pretreatment of aprotinin 2×10^4 KIU/kg protects the myocardium from 4 hours of cardioplegia associated with hypothermia followed by reperfusion. Further study needs to determine optimal dose of aprotinin for prolonged myocardial preservation.

Reference

1. Decker RS, Wildenthal K : Sequential lysosomal alterations during cardiac ischemia and cytochemical changes.
   Lab Invest 38: 662-673, 1978
2. Sunamori M, Amano J, Kameda T et al : Additive protection of aprotinin, protease inhibitor to cold cardioplegia from ischemic myocardium. Jpn Cir J 44: 771-776, 1980
3. Glenn TM, Tauber PF, Miller AG et al : Alterations of the hemodynamic and biochemical sequalae associated with acute coronary ligation by aprotinin. New Aspects of Trasylol Therapy Vol 8: Schattauer, pp 329-345, 1975
4. Godfrey AM, Salaman JR : Trasylol (aprotinin) and kidney preservation Transplantation 25 : 167-168, 1978
5. Hoyer J, Garbe L, Delpierre S et al : Functional evaluation of the transplanted lung after long-term preservation.
   Respiration 39: 323-332, 1980

6. Davis H, Gasho C, Kierman JA : Effects of aprotinin on organ culture of the rat's kidney. <u>In Vitro</u> 12: 192-197, 1976
7. Mustafa SJ : Effect of aprotinin on the metabolism of adenosine in cultured heart cells. <u>Biochem Pharm</u> 28: 340-343, 1979
8. Amano J, Sunamori M, Okamura T et al : Studies on the significance of serum mitochondrial aspartate aminotransferase activity following ischemic cardiac arrest.
   <u>Jpn Circ J</u> 46: 1345-1352, 1982
9. Eisenbach J, Heine H : Zur Wirkung des Proteaseninhibitors Trasylol auf ischemisches Lungengewebe. <u>Med Welt</u> 24: 1697-1700, 1973

FIBRINOLYSIS CAUSED BY CARDIO-PULMONARY BYPASS AND SHED

MEDIASTINAL BLOOD RETRANSFUSION - IS IT OF CLINICAL RELEVANCE?

W. Dietrich[1], A. Barankay[1], P. Wendt[2], A. stemberger[2],
G. Blumel[2], M. Spannagl[3], M. Jochum[3], and J.A. Richter[1]

[1]Institute for Anesthesiology, German Heart Center, Munich
[2]Technical University, Munich
[3]Ludwig Maximilian University, Munich

INTRODUCTION

The importance of blood saving methods during cardiac surgery is well accepted. It is commonly known that, in spite of heparin treatment, systemic coagulation and fibrinolysis will be activated by cardio-pulmonary bypass (CPB). Whether this activation is of clinical relevance was many times a point of discussion (1). Similar changes were found in the shed mediastinal blood postoperatively (2), which was retransfused to reduce homologous blood requirement (3). It is generally accepted now, that this method leads to a reduction of homologous blood requirement in cardiac surgery (4), but some reservations were made because of the low quality of the chest tube blood (5).

The aim of our study was to investigate the degree of fibrinolysis in the shed blood and in patients' circulation, the influence of lysosomal enzymes of desintegrated or activated granulocytes upon fibrinolyses, and to answer the question, whether retransfusion of shed mediastinal blood causes clinically relevant changes in patients' hemostasis.

METHODS

In twenty adult patients undergoing cardiac surgery parameters of fibrinolysis were analysed at the following instances: 1. before operation, 2. after CPB (comparing the data obtained at these two instances the effect of CPB on fibrinolysis can be estimated), 3. in the intensive care unit before and 4. 20 minutes after retransfusion. Samples were taken from patients' circulation as well as from the chest tube drainage. The applied analytical methods for measurements of plasminogen, antiplasmin, D-dimer, early fibrin(ogen) degradation products (N-terminal Bß-fragments 1-42 and 15-42) and elastase from PMN granulocytes complexed with a_1-proteinase-inhibitor (E-a_1PI) are shown in Table 1. Retransfusion was only performed when at least 250 cc blood were collected and when it was indicated by the patient's circulatory state.

Table 1.   Methods

plasminogen	–	chromogenic substrate S 2251 (Kabi)
antiplasmin	–	chromogenic substrate S 2251 (Kabi)
D–dimer	–	monoclonal antibodies (ELISA/MabCo)
N–terminal Bß–fragments	–	monoclonal antibodies (ELISA/NIH)
E–a$_1$PI	–	PMNE (ELISA/Merck)

For statistical analysis the paired t–test comparing the measurements before and after retransfusion was applied. A p–value < 0,05 was considered to be significant. Results are given as mean ± SD.

RESULTS

The mean amount of the first chest blood retransfusion was 325 ± 77 cc (ranging from 250 – 520 cc). The mean hemoglobin content of the chest drainage was 69 g/l, the mean hematocrit 21 % (ranging from 13 – 29 %), the protein content 40.1 ± 5.4 g/l.

Figure 1 shows the course of plasminogen and antiplasmin activity, the dotted line representing the value corrected for preoperative protein. When using this correction for hemodilution no statistically significant influence of CPB on these parameters was found. However, antiplasmin activity in the drainage fluid was significantly reduced. No difference of activity in the patients' blood could be demonstrated before and after retransfusion. The surprising fact that no reduction of plasminogen activity in the chest drainage was measured might be explained by the influence of fibrin and fibrin(ogen) split products on the measurement of plasminogen (6).

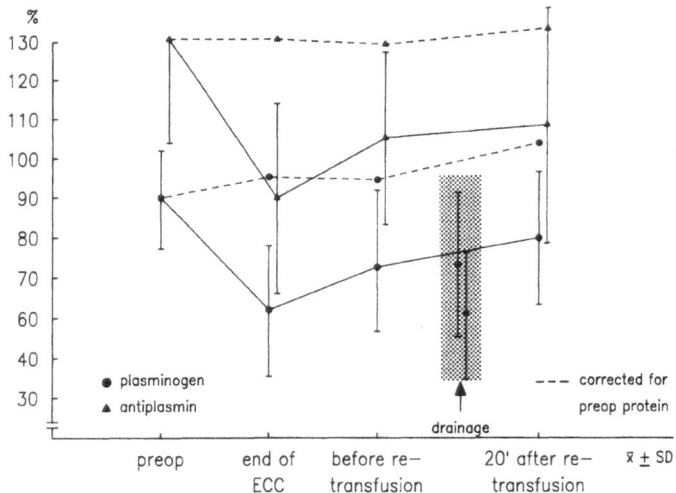

Fig. 1   Course of plasminogen and antiplasmin. The dotted line represents the value corrected for preoperative protein.

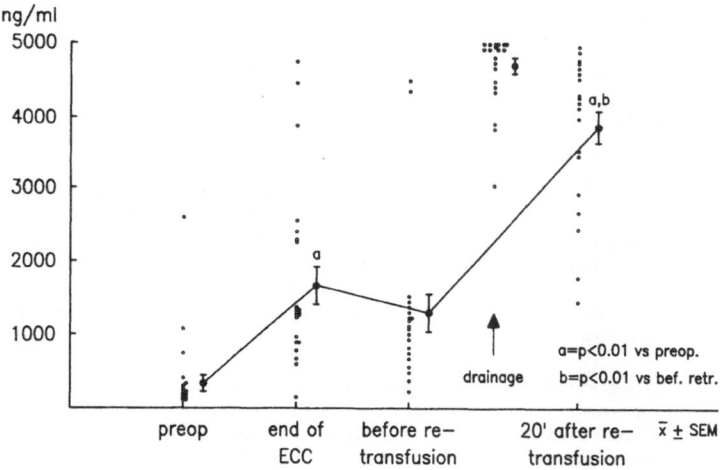

Fig. 2 Cross-linked fibrin fragments. Retransfusion of of shed blood caused a significant (p < 0,01) increase of the concentration in patients' circulation compared
to values before retransfusion.

The cross-linked fibrin fragments (Figure 2) showed a wide variation after CPB. The concentration decreased after operation, but
developed a tremendous increase up to 5,000 ng/cc in the drainage
fluid. The retransfusion caused a significant increase of cross-linked
fibrin fragments in patients' blood.

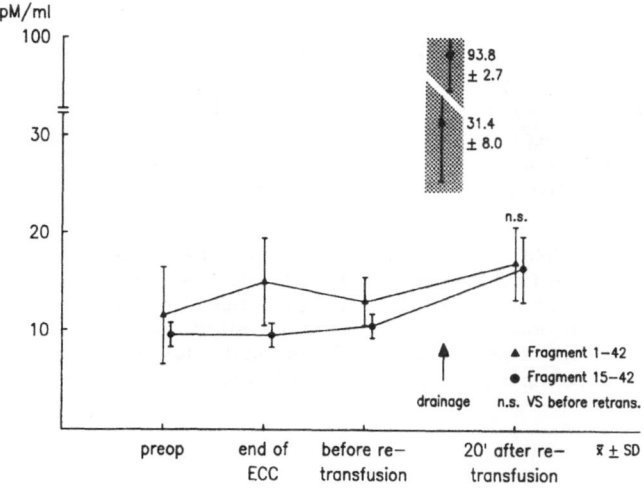

Fig. 3 Bß-related peptides. Note the interruption
of the Y-axis. Despite the high concentration in the drainage blood there was
- in contrast to the cross-linked fibrin
fragments - no significant increase in
patients' circulation after retransfusion.

The Bß-related peptides (Figure 3) showed a comparable course during CPB with an increase of the 1-42 fragments which was not significant. In the chest tube drainage a more pronounced increase of Bß 15-42 peptides was found. However, there were no differences of Bß-related peptides in patients' blood before and after retransfusion.

The PMN E-a$_1$PI complex (Figure 4) showed a tenfold increase in patients' blood after CPB, but there was an extraordinary increase in the drainage fluid with a mean concentration of 13,000 ng/cc - more than hundredfold the preoperative value. Retransfusion also caused a significant increase of the E-a$_1$PI complex in patients' blood. These values were significantly higher than those obtained prior to retransfusion.

No significant correlation between either blood loss or time lag between operation and retransfusion and fibrinolysis, represented by the antiplasmin activity, could be found in the shed blood. After retransfusion blood loss did not exceed the average of normal range.

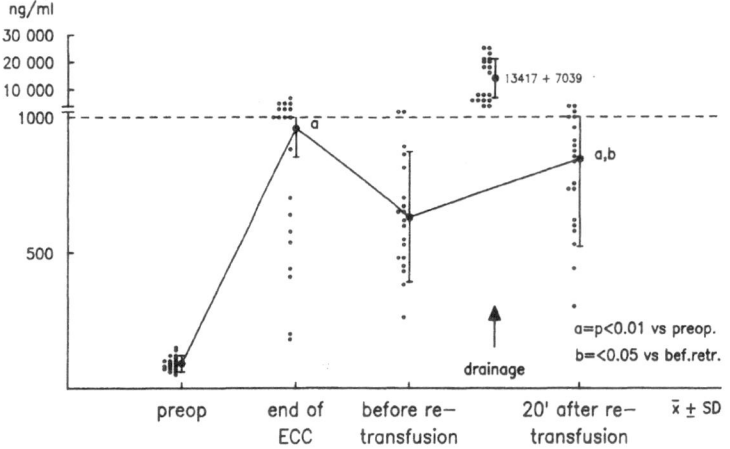

Fig. 4   PMN E-a$_1$PI complex. Note the interruption and different scale of the Y-axis. There was a tremendous increase in the shed blood. Retransfusion of shed blood caused a significantly increased level of the complex in patients' blood.

DISCUSSION

Shed mediastinal blood is not comparable to autologous or homologous (7) blood in regard to hemoglobin content, hematocrit or coagulation properties. However, in former investigations (8), comparing two groups of patients with and without retransfusion of shed mediastinal blood, no difference in the postoperative blood loss could be observed.

In the present study an activation of the fibrinolytic system in shed mediastinal blood was found. The magnitude of fibrinolytic activation in shed blood was comparable to that measured during CPB. This activation of the fibrinolytic system was indicated by comparable decreases of antiplasmin activity in patients' blood after CPB as well as in the shed blood. However, retransfusion of shed blood did not cause an additional decrease of antiplasmin in the patients' circulation.

Bß-related peptides are a most sensitive marker of the endogenous activation of the fibrinolytic system, which can reveal subclinical activation of fibrinolysis (9). A pronounced increase of fibrin(ogen) split products was found in the shed blood. After retransfusion only the D-dimers showed a significant increase in patients' blood, whereas the concentration of Bß-related peptides showed no increase. This can be explained by the lower molecular weight and therefore faster clearing rate of these peptides.

The constellation of increased fibrin(ogen) split products and stable antiplasmin activity after retransfusion can be interpreted as a waste effect of retransfusion. That means, the enhanced fibrinolytic or proteolytic activity within the shed blood is inhibited by the physiologic inhibitors. Some of these complexes and the split products obviously remain in the circulation after retransfusion without inducing fibrinolytic activity within the patient.

In this study the elevated levels of complexed elastase after CPB correspond to the results of other authors (10) indicating an activation or desintegration of PMN granulocytes due to CPB. The extraordinary increase of $E-a_1PI$ in the shed blood may be explained by desintegration of granulocytes caused by mechanical alteration due to pericardial and/or pleural movement in the thorax and in the chest tubing.

Which conclusions can be drawn from the results of this study?

The fibrinolytic and unspecific proteolytic activity in the drainage fluid does not induce fibrinolytic activation after retransfusion. The increased level of fibrin(ogen) split products seems not to be clinically relevant, at least not with the amounts having been retransfused. It can be concluded that retransfusion of shed mediastinal blood is a save method of autologous volume substitution in the early postoperative period. However, it has to be taken into account that retransfusion of greater volumes of shed blood may cause a considerable impairment of coagulation and/or the fibrinolytic system. This deterioration might be induced by the infusion of fibrin(ogen) split products and/or the release of the content of desintegrated granulocytes. In this respect the possible role of pharmacological intervention by protease inhibition should be object of further investigations.

REFERENCES

1. S. Popov-Cenic, H. Murday, P. G. Kirchhoff, G. Hack, J. Fenyes, Anlage und zusammenfassendes Ergebnis einer klinischen Doppelblindstudie bei aorto-koronaren Bypass-Operationen, in: "Proteolyse und Proteinaseninhibition in der Herz- und Gefäßchirurgie, R. Dudziak et al., eds., Schattauer Verlag Stuttgart/New York (1985).

2. W. Dietrich, E. Göb, P. Späth, M. Jochum, G. Heinemann, E. Gams, J. A. Richter, Qualitative Untersuchungen des nach herz-chirurgischen Eingriffen retransfundierten Drainageblutes, Anästhesist 34:93 (1985).

3. H. V. Schaff, J. Hauer, T. J. Gardner, J. S. Donahoo, L. Watkins, V. L. Gott, R. K. Brawley, Routine use of autotransfusion following cardiac surgery: experience in 700 patients, Ann Thorac Surg 27;493 (1979).

4. J. Weniger, R. Shanahan, Reduction of bank blood requirements in cardiac surgery, Thorac Cardiovasc Surgeon 30:142 (1982).

5. R. F. Carter, B. McArdle, G. M. Morritt, Autologous transfusion of mediastinal drainage blood, Anaesthesia 36:54 (1981).

6. J. Gram, J. Jespersen, A functional plasminogen assay utilizing the potentiating effect of fibrinogen to correct for the over-estimation of plasminogen in pathological plasma samples. Thromb Haemostas, 53:255 (1985).

7. J. Weniger, J. van der Emde, K. Th. Schricker, J. Blechschmidt, Retransfusion von Drainageblut nach herzchirurgischen Ein-griffen, Langenbecks Arch Chir 351:229 (1980).

8. W. Dietrich, E. Göb, P. Mitto, A. Barankay, J. A. Richter, Reduction of homologous blood requirement using postoperative shed mediastinal blood retransfusion in myocardial revascularisation, Thorac Cardiovasc Surgeon 33:93 (1985).

9. J. M. Walenga, J. Fareed, G. Mariani, H. L. Mossmore, R. L. Bick, R. M. Emanuele, Diagnostic efficacy of a simple radio-immunoassay test of fibrinogen/fibrin fragments containing the Bß 15-42 sequence, Semin Thromb Hemost 10:252 (1984).

10. M. Jochum, S. Popov-Cenic, A. Philapitsch, Release of granulo-cytic proteins during open heart surgery, 2nd International Con-gress on Proteasis: Potential role in health and disease, Rothen-burg, this issue (1987).

NUTRITION AND PROTEASE ACTIVITY

Joel D. Kopple

Divison of Nephrology and Hypertension, Harbor-UCLA Medical
Center and UCLA Schools of Medicine and Public Health
Los Angeles, California

The influence of nutritional intake and nutritional status on the activity of tissue proteases has been the subject of research for many years. In general, two different aspects of this subject have been investigated: 1) The effects of protein malnutrition, calorie malnutrition and protein-calorie malnutrition on protease activity and the contribution of proteases to wasting in starvation; and 2) The effect of specific nutrient deficiencies on specific proteases. This manuscript will briefly review these two areas of research.

The Relationship between Starvation, Protein-Calorie Malnutrition or Protein Malnutrition on Protease Activity

It is well known that when protein intake is discontinued, there is a rapid fall in net urea production. This indicates that either protein synthesis has increased or protein degradation has fallen (1,2). The fall in net urea production occurs rapidly, and in humans without renal failure urinary urea excretion falls to near plateau within seven to ten days after starting a high calorie, protein-free diet.

The response to short-term and long-term fasting has been compared in obese humans. The results indicate that there are major differences between the response to these two conditions (2,3). After one or two days of a fast, blood amino acid concentrations are well preserved (2-4); plasma branched chain amino acids began to rise. There is net protein breakdown and release of amino acids in skeletal muscle and uptake of amino acids by the liver (2-4). With prolonged fasting, for five to six weeks, a marked adaptation in protein conservation occurs (3). The net rate of protein degradation and release of amino acids in skeletal muscle, the major source of protein in the body, has fallen. Arterial blood amino acid levels are lower; and there is decreased uptake of amino acids by the liver.

A number of investigators have examined the metabolic response to starvation, protein calorie malnutrition, or protein malnutrition in greater detail in different organs of animals. Millward and Waterlow studied rats that were fed protein free diets for 30 days (5). Fractional protein synthesis in skeletal muscle fell quickly, within the first day, and continued to fall over the 30 days of study, to less than one-third of the values in well fed rats. Fractional muscle protein degradation rate fell less dramatically, but did not rise over the 30 day period. During four

rats. Hence, these two groups are not strictly comparable.

The response of the heart to malnutrition has been examined in several studies. Wildenthal and associates starved rats for one to six days and measured activities of three lysosomal enzymes in the myocardium (9). The authors found that in the starved animals in comparison to the control animals, cathepsin D activity rose by 20%, acid phosphatase activity increased slightly, by only 7%, and glucosaminidase activity decreased by 14%; there was no change in creatine phosphokinase activity. The increase in cathepsin D activity occured primarily in the myocytes. The changes often were time dependent. The alterations in cathepsin D and glucosaminidase activities were not significant after 1-2 days of starvation, but were present at three days, and the effect was twice as great at six days. Rats fed diets containing carbohydrate or fat but no protein did not develop a fall in heart weight or an increase in cathepsin D activity (9). This is consistent with the observations in skeletal muscle that energy deprivation rather than protein deficiency increases cathepsin D activity. The findings also indicate that starvation does not cause a generalized increase in all lysosomal enzymes in the myocardium.

The changes in the protease activity did not occur early enough to account for the initial reduction in heart weight or protein content, although the rise in activity of cathepsin D occured in association with most of the reduction in heart size and protein content (9). This lack of a close correlation between weight loss and increased protease activity is similar to the findings of some studies on the effect of malnutrition on weight loss and protease activity in skeletal muscle and liver (6-8). However, Desai reported that in rats starved for 120 hours, free and total hepatic activities (expressed per mg protein per minute) of five lysosomal hydrolases - acid phosphatase, beta-glucuronidase, beta-galactosidase, arylsulfatase and cathepsin D - rose as liver and body weights fell (10). The activities of these enzymes fell within 72 hours of refeeding. In contrast both kidney weight and kidney activities of these five lysosomal enzymes did not change much during starvation or refeeding, except for a small rise in total activity of cathepsin D.

One common and important weakness in the foregoing sudies is that the activity of enzymes are almost always expressed per gram of tissue or per weight of protein in tissue. Since both tissue mass and tissue protein content fall in protein-energy malnutrition, much of the reported increase in enzyme activity is only relative or even artifactual. Hence, it may be more appropriate to conclude from many of these studies that tissue protease activity is maintained relative to the loss in weight or protein content of the organ; but at least in the majority of the reports, there appears to be no clear increase in the absolute activity of these proteinases.

Also, most studies have not addressed the issue of endogenous enzyme activators or inhibitors. Changes or lack of changes in protease activity might reflect only alterations in protease activators or inhibitors. Moreover, it is theoretically possible that the observed changes in measured protease activity may occur in vitro and may bear no relationship to the actual physiology of the system.

In this regard, the research of Samarel and associates is pertinent (11). These authors showed that in the myocardium of starved rabbits, cathepsin D activity increased progressively, by 46 percent, over a 14 day period. When these authors measured cathepsin D by radioimmunoassay, the myocardial concentration of cathepsin D, expressed as micrograms/g wet weight, increased by 95 percent by 14 days. However, the specific activity of cathepsin D, expressed as units of enzyme activity per microgram of immuno-reactive enzyme, fell with starvation. Sephadex gel filtration of cardiac

days of starvation, when there was no energy intake, the fractional synthesis rate of muscle protein fell even faster. However, fractional protein muscle protein degradation increased dramatically by the third day of starvation.

Investigators also have examined the activity of proteolytic enzymes in animals fed diets deficient or absent in protein and/or calories. Millward reported that rats fed a protein-free diet did not show increased skeletal muscle cathepsin D or alkaline proteolytic activity until wasting had progressed to a near terminal state (6). Short term starvation for four days also was not associated with increased activity of these enzymes.

Rosochaki and Millward examined the effects of both protein malnutrition on cathepsin D and acid autolytic activity in skeletal muscle (7). Male rats were fed ad libitum a protein deficient diet providing 4% protein or a protein-free, calorie deficient diet providing 5 g/day of total diet for 21 days (7). The protein-free, low energy diet caused a 33% loss of muscle mass. In contrast, the low protein, normal energy diet almost abolished growth in these rats but did not lead to weight loss. After eating these diets for 21 days, all rats were fed a 20% protein repletion diet for 14 days.

Cathepsin D activity (measured by hemoglobin assay) and acid autolytic activity (measured at pH 3.5) were determined in these animals (7). In contrast to the previous study, the results showed that the marked protein wasting during feeding of the protein-free, low energy diet, was accompanied by an increase in muscle cathepsin D and acid autolytic activity. However, the rats fed the low protein diet that did not show muscle protein wasting, except for one observation, had no increase in proteinase activity. With refeeding of the rats fed either diet, the two proteolytic activities decreased transiently and then rose above control levels.

The authors concluded that marked protein wasting in severely protein or protein-energy deficient rats is accompanied by increased proteinase activity. Protease activity also increased during nutritional repletion. However, the increased protease activities were not always observed during the time when the rats began to waste (6,7). This latter finding is consistent with the observation that during the early phase of protein or protein-energy malnutrition, it is primarily a decrease in protein synthesis that accounts for the skeletal muscle protein wasting (5). Alternatively, the authors may not have chosen the appropriate proteases to assay or the available methods for assaying proteases may not have been sufficiently sensitive.

Roobol and Alleyne studied lysosomal hydrolase activity in young rats with protein malnutrition or protein-energy malnutrition (8). The authors found a rise in acid hydrolase activity in plasma and liver of the protein deficient and protein-energy deficient rats. In general, the enzyme activities were greater in the latter as compared to the fomer animals. In the protein deprived rats, acid hydrolase and cathepsin A activities were not altered in skeletal muscle; liver cathepsin A activity was unchanged in most groups of protein depleted rats. On the other hand, in the rats with protein-energy malnutrition, acid phosphatase activity was also increased in muscle, and cathepsin A activity was elevated in both liver and muscle.

One technical difficulty with these comparisons between the animals with protein-energy malnutrition and those with protein malnutrition is that the protein intake in the two groups was not identical and the total food intake in the two groups was not strictly controlled or monitored. In addition, in the study of Roobol and Alleyne (8), the rats with protein-energy malnutrition were studied at a younger age than the protein depleted

413

lysosomal extracts did not indicate the presence of high or low molecular weight forms of cathepsin D. These findings suggest that the increase in cathepsin D activity is related to an increase in synthesis or a reduction in degradation of this enzyme. Since the rise in enzyme activity and immunoreactive enzyme was greater than the fall in heart weight, it seems likely that reduced degradation of the enzyme can not account for the entire increase in enzyme levels. The fact that the specific activity of cathepsin D fell with starvation raises the possibility that an inhibitor of this enzyme was present.

Given the potential role of proteinases in the protein depletion that occurs during starvation, it would not be surprising if anatomical abnormalities in the heart also occured during protein-calorie malnutrition. Indeed, Smith observed many abnormalities in myocardial structure during starvation in rats (12). These changes include deformities of nuclei, loss of definition of nuclear granules, an increased diameter of vacuolar membranes, and an abnormal appearance of the golgi apparatus. In comparison to controls, Smith also found that starved rats had a marked increase in the myocardial activities of a number of proteases. This was particularly true for alkaline and neutral proteases.

Smith postulated two mechanisms for the pervasive myocardial injury with starvation: a) An increase in lysosomes, as shown by electron microscopy, and b) A rise in the activities of neutral and alkaline proteases that the author inferred was due to increased tissue calcium content. The calcium concentration of the myocardium was increased in the starved rabbits, although the author did not show whether this increase was in the free cytosolic or mitochondrial calcium. This question, of course, would have been very difficult to answer at the time that this study was carried out. Nonetheless, the contribution of calcium to the increased myocardial proteolytic activity during starvation cannot be considered to be established.

Interestingly, Decker et al. reported that in rabbits starved for six days the myocardium actually became more resistant to ischemic injury (13). This protective effect might be related to the accumulation of glycogen in the heart during starvation or possibly to a reduction in the metabolic rate and oxygen demands of myocytes.

It is not clear what activates proteases in starvation. Smith postulated that changes in intracellular pH or ion shifts may stimulate protease activity (12). On the other hand, Goldberg has shown that in skeletal muscle, a reduction in thyroid hormones decreases activities of cathepsins B and D, acid phosphatase, and N-acetylglucosaminidase (14). It is known that serum T_3 and T_4 fall in starvation. Hence, it is possible that the reduced activity of some proteases that has been observed in starvation may be due to the decrease in thyroid hormone levels.

The pancreas secretes a number of proteases, and these exocrine functions of the pancreas also could be affected by malnutrition. There is evidence that the pancreas is affected by both kwashiorkor, a condition caused primarily by protein malnutrition, and marasmus, a syndrome caused by both protein and energy malnutrition (15-17). Barbezat and Hansen reported that pancreatic secretory volume and the secretory rate for amylase, lipase, trypsin, chymotrypsin, and ribonuclease are decreased in both kwashiorkor and marasmus (17). This impaired secretion was observed in the basal state and in response to secretogogues. There was a positive correlation between the output of total enzyme activity from the pancreas and the serum albumin concentration. Dietary therapy often led to complete and early restoration of the pancreatic secretion of volume, pH and enzyme activities.

The mechanisms responsible for the reduction in pancreatic exocrine function in malnutrition are not known. However, pancreatic fibrosis and and atrophy of acinar tissue are well described in severe malnutrition (18). Children with a past history of kwashiorkor who have not completely re-covered from this condition may continue to have reduced pancreatic secretion of enzymes (17). Animal studies confirm that pancreatic atrophy may occur in severe malnutrition and that pancreatic enzyme secretion may not return to normal after nutritional repletion (19). The point has been made that in malnourished patients with acute pancreatitis, serum amylase levels may be inappropriately low because of the reduced ability of the pancreas to syn-thesize amylase.

The Effect of Specific Nutrient Deficiencies on Protease Activity

In addition to the general effect of protein or protein-calorie mal-nutrition on protease activities, there is evidence that deficiencies of specific nutrients as well as obesity may also influence protease function. Deficiencies of vitamin A, vitamin E, vitamin D with phosphorus, and zinc have been associated with specific disorders in protease activity.

In children with vitamin A deficiency, keratomalacia (i.e. the rapid ulceration or liquefaction of the cornea) may occur. It has been estimated that tens of thousands of new cases of this syndrome occur each year in malnourished children and cause blindness (20-22). Researchers have postu-lated that in vitamin A deficiency, there is increased activity of colla-genase and other proteases, particularly in response in trauma (20-23). The activation or release of these proteases may cause the corneal injury and ulceration.

Seng and coworkers found that mild trauma caused by thermal burns to the cornea of vitamin A deficient rats caused stromal edema, leukoma, and sometimes ulceration; corneal collagenase activity increased (20,23). In the rats that were not vitamin A deficient, there was rapid reepithelia-lization of the cornea with little rise in collagenase activity. Leonard et al. has suggested that the enhanced corneal collagenase activity in vitamin A deficient rats may be derived from leukocytes that migrate into the cornea (21).

Twining et al. studied New Zealand rabbits that were fed a vitamin A deficient diet (22). Early xerophthalmia was observed by 4.5 to 5.5 months. The investigators then induced corneal ulceration by scraping the corneal epithelium - this was the trauma that initiated the keratomalacia. In the vitamin A deficient rabbits, corneal ulceration with inflammatory cells, necrosis and stromal degradation were observed 8 to 14 days later. The authors found that in the rabbits with early keratomalacia, as well as with corneal ulceration, there was enhanced cathepsin D-like activity and cathepsin B-like activity. This was observed whether the cathepsin D or B activities were expressed per mg of corneal protein or per total corneal mass. Greater activities for cathepsin D and B were observed in the corneas that had more severe, ulcerating xerophthalmia. The scraped cornea from pair-fed rabbits had lower levels of protease activities that were not different form normal values.

Vitamin E deficienciy in experimental animals can cause protein wasting in skeletal muscle and a syndrome resembling muscular dystrophy (24-27). It is speculated that the antioxidant functions of vitamin E may help to stabilize lysosomes of may inhibit activation of lysosomal enzymes. Hence, in vitamin E deficiency, ther may be increased protease release or activation which may cause both enhanced protein degradation and a muscular dystrophy syndrome.

Activity of acid and alkaline proteases and cathepsin B has been reported to be increased in skeletal muscle of vitamin E deficient animals (25,26). Zalkin and coworkers reported an increase in free and total activities of ribonuclease, cathepsin, beta-galactosidase and arylsulfatase in leg muscle from vitamin E deficient rabbits (27). There also was increased urinary excretion of creatine, allantoin and amino acids. Histological signs of muscle degeneration were observed a short time after the activities of these acid hydrolases began to rise. These findings are consistent with the possibility that in animals with vitamin E deficiency, muscle proteases are released and/or activated and cause the skeletal muscle degeneration.

Dean and associates also reported that in rachitic rats deficient in vitamin D and phosphorus, there is an increase in the collagenase activity from tibial growth plates (28). Cell volume was increased sixfold in these animals. In rachitic rats fed vitamin D and phosphorus that were undergoing healing, the collagenase activity and the increased cell volume fell virtually to normal.

Angiotensin converting enzyme (ACE) from lung and other tissues contains zinc (29). The zinc is tightly bound to the ACE in tissues. ACE activity can be inhibited by agents that chelate zinc. Indeed, antihypertensive ACE inhibiting drugs may act, in part, by binding to the zinc moiety. Reeves and O'Dell examined whether rats deficient in zinc will have reduced serum ACE activity (30). Rats fed a zinc-free but otherwise adequate diet for four days had low serum zinc concentrations and significantly reduced serum ACE activity. Rats treated similarly but then repleted with zinc for 12 hours had normal serum ACE activity. As another control, the authors added zinc to the serum of rats fed a zinc-free diet for four days; the medium zinc concentration was raised to the same level as in the media used for the zinc replete control animals. The zinc supplemented assay of the serum from these zinc deprived rats also showed normal ACE activity. These findings indicate that zinc depletion can impair the serum activity of this physiologically important enzymes.

The Effect of Overnutrition on Protease Activity

Patients who are obese have a high incidence of hypertension (31). Weight loss is associated with a reduction in blood pressure in these individuals. Tuck and associates have shown that in markedly obese humans, severely restricted energy intake and weight loss my lower plasma renin activity (32). This effect appears to be independent of whether the patients are fed a low or moderate sodium intake. It has been suggested that since obese patients have increased plasma catecholamine concentrations and sympathetic nervous system activity, this may be the cause for the increased plasma renin activity in obesity (31).

Summary and conclusion

Clearly, both the metabolism of many proteins and the size of protein pools are affected or regulated by the nutrient intake. At least some of these effects are probably due to changes in the activities of proteolytic enzymes. Indeed, starvation, protein-calorie malnutrition and protein malnutrition probably increase activities of a number of proteases in plasma, liver, skeletal muscle and heart. Pancreatic secretion of proteases and other enzymes is impaired. However, the changes in proteolytic activities often follow rather than precede alterations in the net rate of protein degradation. Thus, the contribution of the proteolytic enzymes to changes in protein degradation in these conditions is not clear.

The effects of nutrition on proteases also are difficult to evaluate because the protease activities have been measured in vitro. These activities may not necessarily reflect the protease activity in vivo, where lysosomes, membrane barriers and transport systems are intact, and inhibitors and facilitators of proteinases may act differently from the in vitro state. Also, proteinase activities often are expressed per weight of tissue or per mg of protein. Tissue weight and protein content often fall during protein and/or calorie malnutrition. Hence, the previous reports of increased protease activities in malnutrition may be incorrect. The observed increase is only relative to the fall in weight and protein content of the tissue. When expressed in terms of total tissue or organ mass, there was no rise in protease activities in many of these studies. Thus, arguably, it might be difficult to ascribe enhanced protein degradation to increased protease activity, when the enzyme is only increased relative to a fall in protein or weight of the tissues.

Deficiency of specific nutrients also may increase or decrease the activities of specific proteoloytic enzymes and may thereby alter important physiological processes. The nutrient deficiencies for which this has been reported include vitamin A, vitamin E, vitamin D with phosphorus, and zinc. Some of these nutrients probably act as cofactors for proteases or may affect the stability of lysosomes. Hence, the changes in the intake or body's pools of these nutrients may influence the activities of specific proteases. Persons who are markedly obese may have increased plasma renin activity. Weight reduction and restriction of energy intake may reduce plasma renin activity.

The evidence cited in this brief review indicates that nutritional intake and nutritional status may have profound effects on proteases. These effects are just beginning to be understood, and more research will be needed to delineate more completely these interrelationships between nutrition and protease activity.

(Supported by the UCLA Clinical Nutrition Research Unit (P01 CA-42710) and NIH Grant R01 AM-33112.)

REFERENCES

1. N.H. Munro, and M.C. Crim, The proteins and amino acids. in: Modern Nutrition in Health and Disease, 7th Edition, M.E.Shils, V.R.Young (eds.), Lea & Febiger, Philadelphia, pp. 1 (1988).

2. L.J. Hoffer, Starvation, in: Modern Nutrition in Health and Disease, 7th Edition, M.E.Shils, V.R.Young, Lea & Febiger, Philadelphia, pp. 774 (1988).

3. G.F. Cahill Jr., Starvation in man, N. Engl. J. Med. 282: 668 (1970).

4. P. Felig, O.E. Owen, J. Wahren, and G.F. Cahill Jr., Amino acid metabolism during prolonged starvation, J. Clin. Invest. 48: 584 (1969).

5. D.J. Millward, and J.C. Waterlow, Effect of nutrition on protein turnover in skeletal muscle, Federation Proc. 37: 2283 (1978).

6. D.J. Millward, The effect of diet on proteolytic activity in rat skeletal muscle, Proc. Nutr. Soc. 31: 3A (1972).

7. S. Rosochacki, and D.J. Millward, Cathepsin D and acid autolytic activity in skeletal muscle of protein deficient, severely protein-energy restricted and refed rats. Proc. Nutr. Soc. 38(3): 137A (1979).

8. A. Roobol, and G.A.O. Alleyne, Changes in lysosomal hydrolase activity associated with malnutrition in young rats. Br. J. Nutr. 32: 189 (1974).

9. K. Wildenthal, A.R. Poole, A.M Glauert, and J.T. Dingle, Dietary Control of cardiac lysosomal enzyme activities. Recent Adv. Studies Cardiac Struct. Metab. 8: 519 (1975).

10. I.D. Desai, Regulation of lysosomal enzymes. I. Adaptive changes in enzyme activities during starvation and refeeding. Can. J. Biochem. 47: 785 (1969).

11. A.M. Samarel, E.A. Ogunro, A.G. Ferguson, P. Allenby, and M. Lesch, Rabbit cardiac immunoreactive cathepsin D content during starvation-induced atrophy. Am. J. Physiol. 240: H222 (1981).

12. A.L.N. Smith, Effects of starvation on vacuolar apparatus of cardiac muscle tissue determined by electron microscopy, marker-enzyme assays and electrolyte studies. Cytobios 18: 111 (1977).

13. R.S. Decker, J.S. Crie, A.R. Poole, J.T. Dingle, and K. Wildenthal, Resistance to ischemic damage in hearts of starved rabbits - Relation with lysosomal alterations and delayed release of cathepsin D. Lab. Invest. 43: 197 (1980).

14. A.L. Goldberg, The regulation of protein turnover by endocrine and nutritional factors, in: Plasticity of Muscle, D. Pette (ed.), Walter de Gruyter & Co., Berlin, p. 469 (1980).

15. M.D. Thompson, and H.C. Trowell, Pancreatic enzyme activity in duodenal contents of children with a type of kwashiorkor, The Lancet 1: 1031 (1952).

16. F. Gomez, R.R. Galvan, J. Cravioto, and S. Frenk, Studies on the under-nourished child - Enzymatic activity of the duodenal contents in children affected with third degree malnutrition. Pediatrics 13: 548 (1954).

17. G.O. Barbezat, and J.D.L. Hansen, The exocrine pancreas and protein-calorie malnutrition, Pediatrics 42: 77 (1968).

18. J.N.P. Davies, The essential pathology of kwashiorkor. The Lancet 1 : 317 (1948).

19. K. Gyr, R.H. Wolf, A.R. Imodi, and O. Felsenfeld, Exocrine pancreatic function in protein-deficient patas monkeys studied by means of a test meal and an indirect pancreatic function test. Gastroenterology 68: 488 (1975).

20. W.L. Seng, J.A. Glogowski, G. Wolf, M.B. Berman, K.R. Kenyon, and T.C. Kiorpes, The effect of thermal burns on the release of collagenase from corneas of vitamin A-deficient and control rats, Invest. Ophthalmol. Vis. Sci. 19: 1461 (1980).

21. M.C. Leonard, L.K. Maddison, and A. Pirie, A comparison between the enzymes in the cornea of the vitamin-A deficient rat and those of rat leukocytes, Exp. Eye Res. 33: 479 (1981).

22. S.S. Twining, D.L. Hatchell, R.A. Hyndiuk, and K.F. Nassif, Acid pro-teases and histologic correlations in experimental ulceration in vitamin a deficient rabbit corneas. Invest. Ophthalmol. Vis. Sci. 26: 31 (1985).

23. W.L. Seng, K.R. Kenyon, and G.Wolf, Studies on the source and release of collagenase in thermally burned corneas of vitamin A-deficient and control rats. Invest. Ophthalmol. Vis. Sci. 22: 62 (1982).

24. I.M. Weinstock, A.D. Goldrich, and A.T. Milhorat, Enzyme studies in muscular dystrophy. I. Muscle proteolytic activity and vitamin E-deficiency. Proc. Soc. Exptl. Biol. Med. 88: 257 (1955).

25. T.R. Koszalka, K.E. Mason, and G. Krol, Relation of vitamin E to proteolytic and autolytic activity of skeletal muscle. J. Nutr. 73: 78 (1961).

26. A.M Spanier, and J.W.C. Bird, Endogenous cathepsin B inhibitor activity in normal and myopathic red and white skeletal muscle, Muscle & Nerve 5: 313 (1982).

27. H. Zalkin, A.L. Tappel, K.A. Caldwell, S. Shibko, I.D. Desai, and T.A. Holliday, Increased lysosomal enzymes in muscular dystrophy of vitamin E-deficient rabbits. J. Biol. Chem. 237: 2678 (1962).

28. D.D. Dean, O.E. Muniz, I. Berman, J.C. Pita, M.R. Carreno, J.F. Woessner Jr., and D.S. Howell, Localization of collagenase in the growth plate of rachitic rats. J. Clin. Invest. 76: 716 (1985).

29. M. Das, and R.L. Soffer, Pulmonary angiotensin-converting enzyme. Structural and catalytic properties. J. Biol. Chem. 250: 6762 (1975).

30. P.G. Reeves, and B.L. O´Dell, An experimental study of the effect of zinc on the activity of angiotensin converting enzyme in serum. Clin. Chem. 31: 581 (1985).

31. L.P. Dornfeld, M.H. Maxwell, A. Waks, and M. Tuck, Mechnanisms of hypertension in obesity, Kidney Int. 32 (Suppl.22): S254 (1987).

32. M.L. Tuck, J. Sowers, L.P. Dornfeld, G. Kledzik and M.H. Maxwell, The effect of weight reduction on blood pressure, plasma renin activity, and plasma aldosterone levels in obese patients, N. Engl. J. Med. 304: 930 (1981).

INSULIN DEGRADATION AFTER INJURY IN MAN

S.M. Hoare [1*], K.N. Frayn [2] and R.E. Offord [3]

[1]Lab. of Molecular Biophysics, Oxford, UK. [2]MRC Trauma unit University Medical School, Manchester, UK. [3]Dépt. de Biochimie Médicale, Univ. de Genève, Genève, Switzerland

*Present address: Fisons Scientific, Bishop Meadow Road Loughborough, Leics., UK   LE11 0RG

Plasma insulin concentrations in some severely injured patients are low despite hyperglycaemia (Carey, Lowery & Cloutier, 1970; Jordan, Fischer & Lefrak, 1972; Meguid et al., 1974). Although the main cause of this has been shown in animal experiments to be *alpha*-adrenergic inhibition of insulin secretion (Cerchio, Persico & Jeffay, 1973; Vigas, Németh & Jurcovicova, 1973), insulin degradation may also be increased (Meguid et al., 1978). These low plasma insulin concentrations are observed within a few hours of injury, before major changes in enzyme synthesis are likely to have taken place. Since it seemed likely that enzymes and cofactors involved in insulin degradation would be released from damaged tissue, we have investigated the possibility that insulin degradation is increased in plasma from injured patients. This has been done by assessing the degradation of semisynthetic [^{3}H-Phe^{B1}]-insulin by plasma *in vitro*.

METHODS

Plasma samples were obtained between 0.5h and 3h following injury from 14 patients aged 17-86 years. Injuries, which were of various types, were graded on the Injury Severity Score (ISS) scale (Baker et al., 1974). There was only one case of burns, a patient with a small leg burn (ISSD=1). All other patients had at least one fracture. Follow-up samples were obtained from two patients (ISS of 14 and 43) on the 1st and 3rd or 2nd and 3rd days after injury respectively.

Plasma samples were frozen for up to 19 days before assessing their degradative capacity by incubation with semisynthetic [^{3}H-Phe^{B1}]-insulin (Halban & Offord, 1975; Halban et al., 1976). Preliminary experiments showed that such storage may have lessened the degradative capacity of the plasma, but did not increase it. The plasma was incubated with approx. 5×10^5 to 1×10^6 c.p.m. of [^{3}H-Phe^{B1}]-insulin, which normally entailed addition of 0.02-0.025ml of insulin solution (in 0.2M glycine buffer containing 0.25 % w/v BSA, pH 7.4) to 0.2ml of plasma. The final concentration of exogenous insulin did not exceed 40nM. The plasma was incubated for 2h at 37°C, at the end of which it was frozen and stored at -20°C. Degradation of the insulin was assessed by gel filtration on Sephadex G-50 as described by Offord et al. (1979), with the exception that the elution buffer was, instead of pH 8.8, as previously described, at pH 7.4. Degradation was expressed as the percentage of the initial amount of intact insulin that remained.

Plasma concentrations of lactate and cortisol were also measured, using methods

Table 1 - Insulin Degradation and other Characteristics of Plasma from Injured Patients and Controls

Measurement	Injured	Control	P value
No. of patients	15	6	
Degradation, %	17 (7-65)	7 (0-10)	<0.01
Lactate, mmol/l	3.5 (2.0-11.6)	0.9 (0.6-1.2)	<0.01
Cortisol, umol/l	0.86 (0.62-1.31)	0.39 (0.23-1.54)	NS
Haemoglobin, g/l	1.6 (0.4-6.5)	1.2 (0.4-1.3)	NS

Results are median with range in parentheses.

NS = not significant.

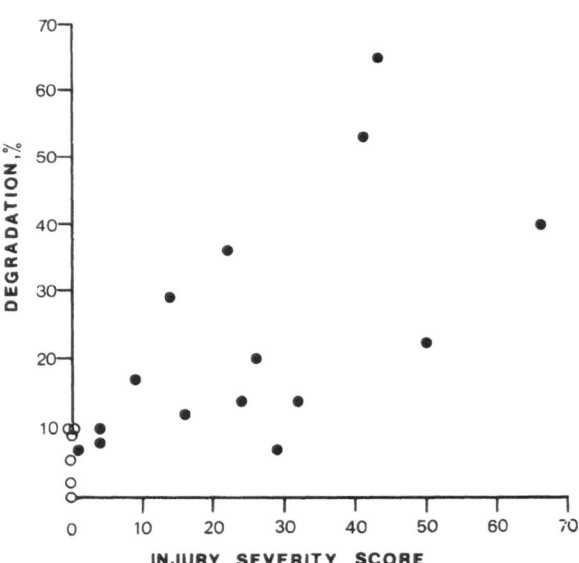

Fig. 1 - Relationship between degradation of insulin in plasma from injured patients (% in 2h) and ISS. Each point (closed circle) represents the result from one injured patient. Results from six control subjects (open circles) are also plotted at ISS=0. The correlation between degradation and ISS for the samples from the injured patients was significant at p < 0.01.

described previously (Stoner et al., 1979) and the plasma haemoglobin concentration was measured by the oxahaemoglobin method (Varley, 1969).

Control plasma samples were taken from laboratory personnel and from in-patients at least two weeks after minor injury. The significance of differences was assessed using Wilcoxons's 2-sample rank sum test, and of correlations using Kendall's rank correlation method (Snedecor & Cochran, 1967).

RESULTS

Degradation of insulin by plasma from injured subjects was significantly greater than by plasma from control subjects (Table I). The amount of degradation was positively related to the Injury Severity Score (Fig. 1). The plasma lactate concentration was also significantly correlated with ISS ($p < 0.05$), whilst cortisol and haemoglobin concentrations were not. Insulin degradation was unrelated to plasma concentrations of lactate, cortisol or haemoglobin. There was no obvious relationship between the nature of the injuries and the amount of insulin degradation.

In the four follow-up samples taken 1-3 days after injury, net insulin degradation was from 0-6% despite initial values for degradation of 29% and 65%.

DISCUSSION

Our results show that insulin degradation is increased in plasma from patients after injury and that this is correlated with ISS. This increased degradation was probably as a result of tissue damage. There seemed to be no obvious relationship between the type of injury and the degree of degradation. Most of the injuries were multiple fractures: abdominal injury did not seem to be associated with a higher degradative activity than would be expected purely on the grounds of ISS. The initial plasma lactate concentration, which is also used as a measure of the severity of injury (Peretz et al., 1965; Stoner et al., 1979), was not related to the degree of degradation. Taken together with the lack of relationship between degradation and plasma cortisol, this suggests that other aspects of the response to injury, such as fluid loss, were not so important. Degradation was not related to the plasma haemoglobin concentration, which would exclude haemolysis as a main cause of degradation: haemolysis enhances insulin degradation in plasma or whole blood (Brodal, 1971; Halban et al., 1978).

Injury Severity Score (ISS) is a measure of the amount of tissue damage caused by an injury (Baker et al., 1974; Stoner et al., 1980). As there is a correlation between the amount of insulin degradation and ISS, it seems likely that the degradation observed is caused by factors released into the circulation from damaged tissue. The rapid loss of degradative activity (returning to values indistinguishable from normal within 24h) suggests that these degrading factors are cleared rapidly from the circulation. It is possible that these factors could be enzymes or cofactors associated with insulin degradation. The enzymes might be relatively non-specific or with a degree of specificity for insulin (e.g. insulin proteinase, Duckworth & Kitabchi, 1981).

The quantitative contribution of this increased insulin degradation to the metabolic events following injury is not certain. Although insulin degradation in vitro, with a half-time around 80 minutes for the most active sample, was considerably slower than the insulin clearance in vivo the half-time of which is of the order of minutes (Samols & Marks, 1966; Sonksen et al., 1973), our method almost certainly underestimates the degradation which would occur in fresh plasma. In addition, plasma insulin in vivo may be exposed to higher local concentrations of degradative factors in the region of the damaged tissue.

ACKNOWLEDGEMENTS

We thank the U.K. Medical Research Council and the Fonds National Suisse pour la Recherche Scientifique for financial support.

REFERENCES

Baker, S.P., O'Neill, B., Haddon, W. and Long, W.B. The Injury Severity Score: a method for describing patients with multiple injuries and evaluating emergency care. J. Trauma 1974;14: 187-196.

Brodal, B.P. Evidence of an enzymatic degradation of insulin in blood in vitro. Eur. J. Biochem. 1971; 18: 201-206.

Carey, L.C., Lowery, B.D. and Cloutier, C.T. Blood sugar and insulin response of humans in shock. Ann. Surg. 1970; 172: 342-350.

Cerchio, G.M., Persico, P.A. and Jeffay, H. Inhibition of insulin release during hypovolemic shock. Metabolism 1973; 22: 1449-1458.

Duckworth, W.C. and Kitabchi, A.E. (1981) Insulin Proteinase. Endocrinol. Rev.; 2: 210-233.

Halban, P.A. and Offord, R.E. The preparation of a semisynthetic tritiated insulin with a specific radioactivity of up to 20 Curies per millimole. Biochem. J. 1975; 151: 219-225.

Halban, P.A., Karakash, C., Davies, J.G. and Offord, R.E. The degradation of semisynthetic tritiated insulin by perfused mouse livers. Biochem J. 1976; 160: 409-412.

Halban, P.A., Berger, M., Gjinovci, A., Renold, A., Vranic, M. and Offord, R.E. Pharmacokinetics of subcutaneously injected semisynthetic tritiated insulin in rats. In: Offord, R.E., Di Bello, D. eds. Semisynthetic peptides and proteins. London, New York Academic Press 1978; 237-247.

Jordan, G.L., Fischer, E.P., Lefrak, E.A. Glucose metabolism in traumatic shock in the human. Ann. Surg. 1972; 175: 685-692.

Meguid, M.M., Brennan, M.F., Aoki, T.T., Muller, W.A., Bal, M.R. and Moore, F.D. Hormone-substrate interrelationships following trauma. Arch. Surg. 1974; 109: 776-783.

Meguid, M.M., Aun, F., Soeldner, J.S. and Albertson, D.A. Biological half-life of endogenous insulin in man following trauma. Surg. Forum 1978; 29: 98-100.

Offord, R.E., Philippe, J., Davies, J.G., Halban, P.A. and Berger, M. Inhibition of degradation of insulin by ophthalmic acid and by a bovine pancreatic proteinase inhibitor. Biochem. J. 1979; 182: 249-251.

Peretz, D.I., Scott, H.M., Duf, J., Dossetor, J.B., MacLean, L.D. and McGregor, M. Metabolite levels after injury. (1965) Ann. N.Y. Acad. Sci.; 119: 1133-1141.

Samols, E. and Marks, V. Disappearance-rate of endogenous insulin in man. Lancet 1966; ii: 700.

Snedecor, G.W. and Cochran, W.G. Statistical methods 6th ed. Ames: Iowa State University Press, 1967.

Sönksen, P.H., Tompkins, C.V., Srivastava, M.C. and Nabarro J.D.N. A comparative study on the metabolism of human insulin and porcine proinsulin in man. Clin. Sci. Mol. Med. 1973; 45: 633-654.

Stoner, H.B., Frayn, K.N., Barton, R.N. Threlfall, C.J. and Little, R.A. The relationships between plasma substrates and hormones and the severity of injury in 277 recently injured patients. Clin. Sci. 1979; 56: 563-573.

Stoner, H.B., Heath, D.F. Yates, D.W. and Frayn, K.N. (1980) J. Roy. Soc. Med.; 73: 19-22.

Varley, H. Practical clinical biochemistry. 4th ed. London: William Heineman Medical Books, 1969; 586-587.

Vigas, M., Németh, S. and Jurcovicova, J. The mecanism of trauma-induced inhibition of insulin release. Horm. Metab. Res. 1973; 5: 322-324.

ENDOTOXIN ABOLISHES THE INDUCTION OF α2-MACROGLOBULIN SYNTHESIS IN CULTURED HUMAN MONOCYTES INDICATING INHIBITION OF THE TERMINAL MONOCYTE MATURATION INTO MACROPHAGES

Joachim Bauer, Ursula Ganter, and Wolfgang Gerok

Medizinische Universitätsklinik Freiburg

D-7800 Freiburg, West Germany

ABSTRACT

Monocytes were prepared from human blood and allowed to differentiate into macrophages in vitro. Synthesis and secretion of the potent proteinase inhibitor α2-macroglobulin is strongly induced during maturation of monocytes into macrophages establishing α2-macroglobulin as a differentiation marker protein. When the monocytes were incubated with endotoxin (S. typhi) in a concentration as low as 100 ng/ml, induction of α2-macroglobulin synthesis was completely abolished indicating impaired monocyte/macrophage differentiation by lipopolysaccharides (LPS). In contrast to α2-macroglobulin, synthesis of α1-proteinase inhibitor in mononuclear phagocytes was stimulated by endotoxin.

INTRODUCTION

The cells of the mononuclear phagocyte system derive from stem cells in the bone marrow, linking monoblasts, promonocytes, monocytes and the heterogenous tissue macrophages (for review see[1,2]). Cells of this lineage develop over several days in the marrow, circulate as monocytes in the peripheral blood for a few further days, followed by migration into different tissues, where they remain resident for up to several months as macrophages. Macrophages are present in the liver (as Kupffer cells), lung (as alveolar macrophages), bone (as osteoclasts) and several other tissues. The development of the the blood monocytes into macrophages is accompanied by marked morphological and biochemical changes and is designated as terminal maturation. Differentiation of monocytes into macrophages can be studied in vitro by cultivation of purified populations of human peripheral blood monocytes within hydrophobic teflon foils in the presence of human serum[3,4] (Fig. 1). In vitro maturation closely resembles differentiation in vivo[5].

Secretory products are involved in several macrophage functions, such as immunoregulation and the killing of bacteria (for review see[6]). Several proteinase inhibitors were found to belong to the proteins secreted by mononuclear phagocytes including α2-macroglobulin[7] and α1-proteinase inhibitor[8]. α2-Macroglobulin is a tetrameric protein consisting of four identical subunits each with a molecular weight of 182 kDa

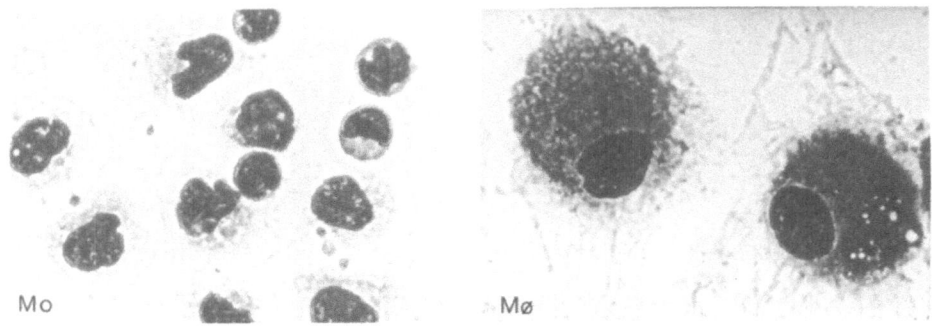

Figure 1. Human blood monocytes (Mo) and macrophages (MØ) derived from monocytes by in vitro maturation as described in Materials and Methods were sedimented on polylysine-coated plates and Pappenheim-stained. Cells were mounted with Entellan (Merck) and viewed by a Zeiss microscope (interference contrast, oil immersion, same magnification for Mo and MØ, around 1:760).

and is regarded as the most potent proteinase inhibitor in man[9-11]. Its broad specificity is directed against endopeptidases of all four major classes (serine, aspartic acid, cysteine, and metalloproteinases) and includes inhibition of kallikrein[12-14], plasmin[15-18], and thrombin[19-21]. This indicates its function helping to keep proteolytic activation of complement, kallikrein-kinine and the coagulation cascade under control.

Many efforts have been made to describe biochemical markers of the monocytes/macrophage differentiation. Mainly differentiation-specific membrane proteins were studied for an immunohistochemical monitoring of the terminal maturation[4]. Here we describe that synthesis and secretion of α2-macroglobulin is an excellent biochemical marker of the monocyte/macrophage differentiation. We found that endotoxin abolishes the induction of α2-macroglobulin synthesis in cultured human monocytes indicating inhibition of the terminal monocyte maturation into macrophages.

MATERIALS AND METHODS

RPMI 1640 medium was purchased from Seromed (Berlin, FRG) and supplemented according to Andreesen et al.[4] with L-glutamine, antibiotics, vitamins, which were all obtained from Gibco (Karlsruhe, FRG). PBS, Dulbecco was also from Gibco. Ficoll-paque and Protein A-Sepharose Cl-4B were obtained from Pharmacia (Freiburg, FRG). LPS from Salmonella typhi was a generous gift from Dr. C. Galanos Max-Planck-Institute of Immunobiology, Freiburg, FRG. Teflon foils (Biofolie 25) were from Heraeus (Hanau, FRG). |^{35}S|methionine (1330 Ci/mmol = 49,2 TBq/mmol) was from the Radiochemical Centre (Amersham, UK). Protosol was purchased from New England Nuclear (Boston, MA). X-Ray films (XAR5) were from Kodak. Alternatively, also hyperfilm-MP from Amersham was used. Human α2-macroglobulin and human α1-proteinase inhibitor were obtained from Sigma. Human serum was prepared from healthy donors under sterile conditions, heat-inactivated at 56°C for 40 min and pooled. Purified human interferon α and recombinant interferon β and was a generous gift from Bioferon (Laupheim, FRG).

Antibodies

Anti-human α1-proteinase inhibitor was from Sigma. A monospecific polyclonal antibody against human α2-macroglobulin was raised in rabbit with α2-macroglobulin purified according to[22]. Before subcutaneous injection into the rabbit, purified α2-macroglobulin was subjected to a reducing preparative SDS-PAGE. The Coomassie-stained α2-macroglobulin band (182 kDa) was excised from the gel, homogenized with Freund's complete adjuvans for the first injection and with incomplete adjuvans for two booster injections, thus administering three times 500 μg α2-macroglobulin. The antiserum immunoprecipitated a 182 kDa band from radiolabeled cell homogenates or media, which disappeared if an excess of commercially available unlabeled α2-macroglobulin was added before addition of the antiserum.

Human monocyte/macrophage cultures

The preparation of human blood monocytes and their further cultivation on hydrophobic teflon foils was carried out essentially as described by Andreesen et al.[3,4]. Briefly, monocytes were prepared from blood of normal donors by centrifugation over Ficoll-paque to separate mononuclear cells from granulocytes and red cells. Monocytes were then separated from lymphocytes by adherence to plastic at 37°C in RMPI 1640 medium containing 10% human AB-serum completely free of LPS. Complete absence of LPS was assayed according to Northoff et al.[23]. The nonadherent cells (lymphocytes) were removed by repeated gentle washing with warm RMPI 1640 medium. The adherent cells were then incubated overnight with RPMI 1640 medium containing 5% human serum at 37°C in an humidified atmosphere containing 5% CO_2. After a short exposure to 4°C the monocytes were recovered from the dishes. Purity of the monocyte cultures was more than 90%, as revealed by morphology and phenotype analysis. To allow terminal maturation in vitro, the monocytes (5×10^5 cells/ml) were cultivated within hydrophobic teflon foils in RPMI 1640 medium containing 5% human AB-serum. Fresh medium was added to the cells after 5 days in culture. Cells were taken from the teflon foils after varying periods of cultivation according to different states of maturation. Viability was tested by exclusion of Trypan blue and amounted to more than 85%.

Radiolabeling and immunoprecipitation

The cells were labeled with 30 μCi |[35]S|methionine for 3 h, except the pulse-chase experiments, where the pulse-labeling was done with 100 μCi for 20 min. In one dish 10^6 cells were incubated in 1 ml methionine-free RPMI 1640 medium at 37°C in the presence of |[35]S|methionine. After labeling, the culture media were separated from the cells. The cells from each dish were homogenized in 1 ml of 25 mM Tris/HCl, pH 7.5, 20 mM NaCl, 1% deoxycholate (Na^+) and 1% Triton X-100. The non-soluble material was removed by centrifugation at 12,000 rpm for 10 min. The supernatants of the cell homogenates and the culture media were used for immunoprecipitation.

Newly synthesized total proteins were determined by TCA-precipitation of the cell homogenates or the culture media. Briefly, 10 μl aliquots were spotted onto 1 cm^2 Whatman No. 3 paper. The pieces were air-dried, placed into ice-cold 10% TCA (w/v) for 5 min, transferred to 5% TCA (70°C) for 5 min (to remove non-incorporated methionine), rinsed 3 times with ice-cold 5% TCA and dried. The radioactivity was counted in vials with 7 ml Rotiscint scintillation fluid.

For immunoprecipitations, cell homogenates and culture media were diluted into 2 volumes of TNET buffer (20 mM Tris/HCl, pH 7.6, 0.14 M NaCl, 5 mM EDTA, 1% Triton X-100). Before specific immunoprecipitation, non-specifically binding material was extracted with a non-immune rabbit control serum. The specific immunoprecipitation was performed in the presence of an excess of antibody (anti-human α2-macroglobulin, anti-human α1-proteinase inhibitor). After addition of 10 μl antiserum and incubation overnight, the immune complexes were bound to 7 mg protein A-Sepharose, washed 4 times with TNET buffer and 3 times with 50 mM sodium-phosphate pH 7.5 buffer. The immunoprecipitates were dissociated in Laemmli buffer (0.1 M Tris/HCl, pH 6.8, 5% β-mercaptoethanol, 5% sodium dodecyl sulfate and 10% glycerol) for 4 min at 95°C and analyzed by SDS-PAGE[24] and fluorography according to Bonner and Laskey[25]. The specific radioactive bands (α2-macroglobulin, α1-proteinase inhibitor) were excised from the gels, solubilized with Protosol/water (9:1) at 45°C overnight and counted in a liquid scintillation spectrometer.

RESULTS

The course of terminal in vitro maturation of human blood monocytes into macrophages was followed by determination of the total protein synthesis. After different days of cultivation representing different states of terminal maturation, equivalent numbers of cells were metabolically labeled and the rate of total intracellular and secreted protein synthesis was determined. As shown in Fig. 2A, monocyte differentiation

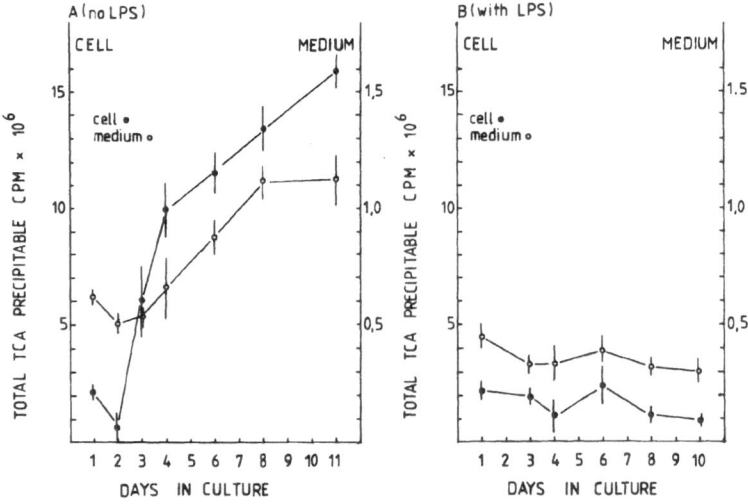

Figure 2. Total protein synthesis of human blood monocytes during in vitro maturation into macrophages. (A) Human monocytes prepared from normal blood were cultivated within hydrophobic teflon foils (5 x 10^5 cell/ml RPMI 1640 containing 5% human AB serum) as described in Materials and Methods. After different times of in vitro maturation, as indicated in the figure, 10^6 cells were incubated in 1 ml methionine-free RPMI 1640 medium and allowed to incorporate 30 μCi |35S|methionine for 3 h into de novo synthesized proteins. Radioactivity of total TCA-precipitable proteins was determined in cell homogenates (●——●) and culture media (o——o). (B) Cells were cultured as in (A), except that 100 ng/ml LPS (Salmonella typhi) were present throughout cultivation.

into macrophages which has been previously reported to occur during day 3-5 of in vitro cultivation[4] is accompanied by a drastic increase of total protein synthesis. The increase of the rate of intracellular protein synthesis (7-fold) as compared to that of secreted proteins (2-fold) was greater by 3-fold. This is consistent with the observation that the morphological transition from monocytes to macrophages is accompanied by an increase in cellular size and a maturation of cell organelles (Fig. 1).

The induction of α2-macroglobulin synthesis and secretion by monocytes during their terminal in vitro maturation into macrophages was analyzed (Fig. 3A). α2-Macroglobulin synthesis expressed as percentage

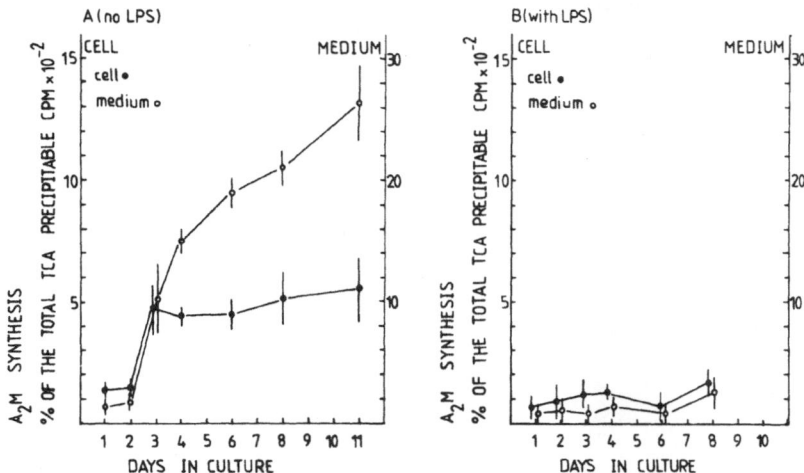

Figure 3.   α2-Macroglobulin synthesis in human blood monocytes during the course of terminal maturation into macrophages. (A) Human blood monocytes were prepared and cultivated (5 × 10^5 cells/ml RPMI 1640 containing 5% human AB serum) as described in Materials and Methods. Cells were harvested from the teflon foils after different days of cultivation as indicated in the figure. 10^6 cells were incubated for 3 h in 1 ml methionine-free RPMI 1640 medium in the presence of 30 μCi |^35S|methionine. The radioactivity incorporated into total TCA-precipitable proteins was determined in the cell homogenates and culture media. α2-Macroglobulin was immunoprecipitated with a monospecific antibody from equivalent cpm of cell homogenates (1.5 × 10^6 cpm) (●——●) and culture media (0.4 × 10^6 cpm) (o——o). The immune complexes were absorbed to Protein A-Sepharose and subjected to SDS-PAGE. After autoradiography, the α2-macroglobulin bands were excised from the gels, solubilized, counted and related to total TCA-precipitable radioactive proteins. (B) Synthesis of α2-macroglobulin when the monocytes were incubated in the presence of 100 ng/ml LPS (S. typhi) throughout cultivation.

of total protein synthesis showed a considerable increase concomitant with the monocyte-macrophage transition and closely paralleled the kinetics of total protein synthesis (Fig. 2A). Thus, both increase in total protein, as well as in specific α2-macroglobulin synthesis signal characteristic changes during monocyte-macrophage differentiation.

429

Specific induction of α2-macroglobulin synthesis during monocyte differentiation into macrophages was nearly completely abolished when the monocytes were maintained in culture in the presence of endotoxin at a concentration as low as 100 ng/ml (Fig. 3B). In addition to the observed lack of specific α2-macroglobulin induction during the course of in vitro cultivation, base level synthesis in 1 day-old monocytes was also decreased (to about 50% of control). Moreover, after endotoxin treatment throughout cultivation, the increase of total protein synthesis which could ordinarily be observed during the course of in vitro maturation (Fig. 2A) was now completely abolished (Fig. 2B). Morphologically, cells cultured in the presence of endotoxin retain their monocytic shape and do not develop a macrophage-specific morphology.

Endotoxin exerts its effect on α2-macroglobulin synthesis in a dose-dependent manner (Fig. 4). When given to the teflon-cultured monocytes for 7 days in a concentration as low as 1 ng/ml, LPS (S. typhi) was able to repress α2-macroglobulin synthesis markedly. A full repressory response was already seen with 10 ng/ml LPS (Fig. 4, lane 3).

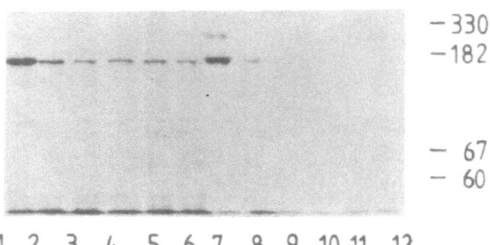

Figure 4. Dose-dependence of inhibition of α_2-macroglobulin in mononuclear phagocytes by long-term treatment with LPS. Monocytes were prepared as described and cultivated within hydrophobic teflon foils in the presence of 5% human serum for 7 days either in the absence or presence of different concentrations of LPS (S. typhi). After labeling 10^6 cells/dish with 30 μCi $|^{35}S|$methionine, α2-macroglobulin was immunoprecipitated from equivalent total TCA-precipitable cpm of cell homogenates (2 × 10^6 cpm) (lanes 1-6) and culture media (0.3 × 10^6 cpm) (lanes 7-12). No LPS present: lanes 1,7; 1 ng/ml LPS: lanes 2,8; 10 ng/ml: lanes 3,9; 100 ng/ml: lanes 4,10; 1 μg/ml: lanes 5,11; 10 μg/ml: lanes 6,12.

Endotoxin also exerted an inhibitory effect on α2-macroglobulin synthesis if given to the cells only for a short period of time. Newly synthesized α2-macroglobulin was measured in monocytes as well as maturated macrophages after only 24 h of LPS pretreatment (100 ng/ml). In this case endotoxin-treated cultures again exhibited diminished specific α2-macroglobulin synthesis (expressed as percentage of total protein synthesis). α2-Macroglobulin secretion in these experiments was decreased by LPS to less than 60% of control in monocytes and to less than 50% in macrophages (not shown).

Mononuclear phagocytes have been reported to synthesize α1-proteinase inhibitor[8]. The influence of endotoxin on the synthesis of this protein was compared with the effects seen on α2-macroglobulin synthesis. Treatment of monocytes with different concentrations of LPS during 7 days of cultivation resulted in a strong stimulation of α1-proteinase inhibitor synthesis (Fig. 5). An endotoxin concentration as low as 1 ng/ml led to a full stimulatory response (Fig. 5, lane 2) from 0.05% to more than 0.15% of total protein secretion.

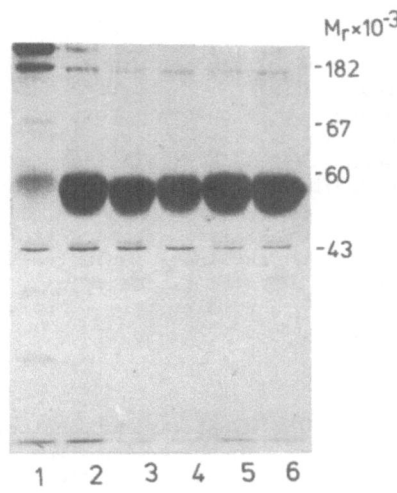

Figure 5. Dose-dependence of stimulation of α1-proteinase inhibitor synthesis in mononuclear phagocytes by long-term treatment with LPS. The experiment was the same as described in the legend to Figure 4. Instead of α2-macroglobulin, α1-proteinase inhibitor was immunoprecipitated from media of monocyte cultures treated without (lane 1) and with different concentrations of LPS (lane 2: 1 ng/ml LPS; lane 3: 10 ng/ml LPS; lane 4: 100 ng/ml LPS; lane 5: 1 µg/ml LPS; lane 6: 10 µg/ml LPS).

DISCUSSION

We applied cultivation of human blood monocytes within hydrophobic teflon foils to study α2-macroglobulin synthesis during terminal monocyte/macrophage differentiation. It was found that the increase in rate of total protein synthesis is a suitable monitor of the maturation process and that transition of monocytes into macrophages is accompanied by a strong induction of specific α2-macroglobulin synthesis and secretion (expressed as percentage of total protein synthesis and secretion in these cells). Increase of α2-macroglobulin synthesis was correlated by increased hybridizable α2-macroglobulin mRNA indicating pretranslational regulation (not shown). Thus, α2-macroglobulin expression is a biochemical marker of the monocyte/macrophage differentiation. Endotoxin was found to repress α2-macroglobulin synthesis. Our report for the first time describes a regulatory mechanism of α2-macroglobulin synthesis in man.

It is still unknown which biochemical events underly the LPS effect on α2-macroglobulin synthesis in mononuclear phagocytes. Unspecific toxic influences can be excluded since the strong repression of specific α2-macroglobulin synthesis was accompanied by a simultaneous increase of specific synthesis of α1-proteinase inhibitor. LPS has been shown to induce secretion of interferon α and ß by mononuclear phagocytes[26-29]. On the other hand, interferon α and ß[30,31] were reported to impair monocyte differentiation into macrophages. Therefore, LPS may exert the effects described in the long-term incubation experiments by induction of interferon α and ß and by the subsequent autoregulatory inhibition of monocyte differentiation into macrophages. In preliminary experiments with purified human interferon α and recombinant human interferon ß we found a repression of α2-macroglobulin synthesis.

The findings reported in this paper are of biological significance. Our data indicate that endotoxin, in addition to causing leukocyte degranulation and consecutive release of proteinases, may further influence the precarious balance between proteolytic enzymes and proteinase inhibitors by repressing the synthesis of α2-macroglobulin, the most potent proteinase inhibitor in many mammalian systems including man due to its broad target specificity. Not only in serum, but also in tissues, interactions between proteinases and their inhibitors play a crucial role. By synthesizing proteinase inhibitors, macrophages contribute to the demarcation of infected tissue areas, thereby restricting proteolysis to the site of injury. Also this function may be impaired by the presence of endotoxin.

ACKNOWLEDGEMENTS

We are grateful to Isolde Richter for technical assistance and to Helga Gottschalk for her help with the preparation of the manuscript.
This work was supported by the Deutsche Forschungsgemeinschaft, Bonn, F.R.G., and the Müller-Fahnenberg-Stiftung, Freiburg.
Ursula Ganter is a fellow of the Landesgraduiertenförderung Baden-Württemberg.

REFERENCES

1. R. Van Furth, (ed) (1980) Mononuclear Phagocytes: Functional Aspects, Martinus Nijhoff, The Hague.
2. R. T. Dean and W. Jessup (eds) (1985) Mononuclear Phagocytes: Physiology and Pathology, Elsevier Science Publishers, Amsterdam.
3. R. Andreesen, J. Picht, and G.W. Löhr. Primary cultures of human blood-born macrophages grown on hydrophobic teflon membranes, J. Immunol. Methods 56:295 (1983).
4. R. Andreesen, K.J. Bross, J. Osterholz, and F. Emmrich. Human macrophage maturation and heterogeneity, Blood 5:1257 (1986).
5. J. C. Waldrep, A.M. Kaplan, and T. Mohanakumar. Human mononuclear phagocyte-associated antigens J. Reticuloendothel. Soc. 34:323 (1986).
6. C. F. Nathan. Secretory products of human mononuclear phagocytes, J. Clin. Invest. 79:319 (1987).
7. T. Hovi, D. Mosher, and A. Vaheri. Cultures human monocytes synthesize and secrete α2-macroglobulin, J. Exp. Med. 145:1580 (1977).
8. A. B. Cohen. Interelationship between the human alveolar macrophage and α1-antitrypsin, J. Clin. Invest. 52:2793 (1973).
9. A. J. Barrett and P.M. Starkey. The interaction of α2-macroglobulin with proteinases, Biochem. J. 133:709 (1973).

LOCAL AND GENERAL DEFENCE MECHANISMS IN BACTERIAL

AND CHEMICAL PERITONITIS

Åke Lasson, Magnus Delshammar, and Kjell Ohlsson

Departments of Surgery and Surgical
Pathophysiology
University of Lund, Malmö General Hospital
214 01 Malmö, Sweden

INTRODUCTION

The inflammatory reaction in peritonitis often leads to profound systemic changes, indeed peritonitis accounts for half of all cases of fatal septic shock (1). Although there are numerous causes of peritonitis, the clinical picture is almost the same in most peritonitis patients. Thus, from a clinical point of view, a widespread inflammatory reaction in the peritoneum, whatever the cause, seems to form the basis for the systemic reactions which follow (2).

The aim of the present study was twofold. Firstly, to search for protease activation and protease-antiprotease interaction or imbalance in peritonitis in the peritoneal fluid, peritoneum or plasma. Secondly, to look for any differences in these reactions between patients with different aetiology, i.e. chemical and bacterial peritonitis.

MATERIAL AND METHODS

Sixteen patients undergoing urgent laparotomy because of peritonitis or intra-abdominal abscess were studied. Six of them (group A) had sterile (= chemical) peritonitis, while 10 (group B) had septic (= bacterial) peritonitis (table I). One patient in the chemical and one in the bacterial group had heart-disease pre-operatively, and another two patients in the bacterial group had respiratory disease. Although all patients were operated on within 12 hours after the diagnosis of peritonitis, the median pre-operative duration of symptoms was different, 1 day (range 1-5) in group A and 3 days (range 1-17) in group B.

Blood samples were taken before operation (day 1) and once daily thereafter. Peritoneal fluid and a peritoneal biopsy near the abscess or intestinal perforation were taken at the primary operation. Blood and peritoneal fluid were centrifuged and stored within one hour at -70°C. The peritoneal samples were fixed in buffered formalin and later embedded in paraffin.

Protease inhibitors, fibrinogen, fibronectin and C3d were quantified using electroimmunoassay (3,4). Antithrombin III (AT III), alpha-2-antiplasmin (alpha-2-AP), C1-inhibitor, prekallikrein, factor X and plasminogen were measured amidolytically (5). The amount of leucocyte elastase was determined using radioimmunoassay (6).

The peritoneal samples were examined with light microscopy after hematoxylin-eosin staining. The peroxidase-antiperoxidase (PAP) technique (7) was used for localizing leucocyte elastase, alpha-2-macroglobulin (alpha-2-M) and alpha-1-protease inhibitor (alpha-1-PI). The staining reaction was controlled using non-immune serum and also compared with normal peritoneal biopsies taken at normal non-infected abdominal operations.

Table I. Diagnosis and complications in 6 chemical and 10 bacterial peritonitis patients

Diagnosis	Preoperative	Postoperative
Chemical (aseptic; n=6)		
Perforated gastro-duodenal ulcer (5)	0	Assisted vent. (1)
Torsion of gall-bladder (1)	+	Assisted vent. (1)
Septic (baterical; n=10)		
Intra-abdominal abscess (6)	Septicaemia (5)	DIC (1)
		Septicaemia (3)
		Assisted vent. (4)
Perforated diverticulitis (3)	3+	Wound infection (1)
Perforated appendicitis (1)	0	Wound infection (1)

RESULTS

Clinical course

There were no post-operative complications in group A, except that two patients in this group needed two days of post-operative assisted ventilation (table I). The median hospital stay was 6.5 days (range 4-15). Four patients in group B needed assisted ventilation for several days post-operatively. Several post-operative complications were seen (table I), and the median hospital stay was 21.5 days (range 5-90). One patient developed a full-blown DIC syndrome. She survived, but had permanent renal insufficiency necessitating later renal transplantation. One patient with a malignancy in group B died 35 days after the primary operation.

Peritoneal fluid

Changes in the peritoneal fluid, i.e. consumption of proenzymes, high C3d levels, high levels of leucocyte elastase – alpha-1-PI complexes and protease inhibitory consumption, indicated proteolytic activity (table II). These quantitative results were in agreement with the results of crossed immunoelectrophoresis; showing protease – AT III complexes, protease-alpha-2-AP complexes, leucocyte elastase-alpha-1-PI complexes and cleavage of the native C3 (results not shown). All these changes were more pronounced in the bacterial group B (table II).

Table II. Levels of proenzymes and protease inhibitors in the peritoneal fluid at operation in 6 chemical and 10 bacterial peritonitis patients

	Chemical (6)	Bacterial (10)
Prekallikrein (%)	67.5 +/- 39.3	53.2 +/- 27.2
C3d (%)	47.0 +/- 8.9	57.7 +/- 34.3
Factor X (%)	24.0 +/- 29.0	30.6 +/- 57.3
Plasminogen (%)	74.5 +/- 32.2	35.3 +/- 28.1
Leucocyte elastase-alpha-1-PI complexes (microgr/l)	2560 +/- 1067	4141 +/- 2062
Alpha-2-M (%)	34.5 +/- 6.6	28.7 +/- 84.9
Alpha-1-PI (%)	95.0 +/- 26.6	84.9 +/- 34.9
AT III (%)	43.5 +/- 14.0	65.7 +/- 29.2
Alpha-2-AP (%)	132.5 +/- 26.5	75.7 +/- 40.0
C1-inhibitor (%)	0	1

Peritoneum

Immunohistological studies of the peritoneum showed leucocyte elastase (fig. 1), alpha-1-PI and alpha-2-M-reactive material to be present mainly in macrophage-like cells. Alpha-2-M-reactive material was also found in fibroblasts. There were no obvious differences between chemical and bacterial peritonitis. Control specimens, using both non-immune serum and normal peritoneum, gave no staining.

Plasma

The levels of leucocyte elastase – alpha-1-PI complexes were elevated, especially in group B (fig. 2).The levels of two of the main protease inhibitors, alpha-2-M and AT III, were low, especially in group B (fig. 3). Alpha-2-M did not reach normal levels during the 12-day

study period in group B. The levels of both C1-inhibitor and alpha-2-AP, the other two main protease inhibitors, were high throughout the disease, as for acute phase proteins (figs. 4

and 5). Levels of prekallikrein (fig. 4) and plasminogen (fig. 5), both proenzymes, were low during the first days of disease. In addition, six of the 10 patients in group B, but none of the patients in group A, had high levels of C3d for 2-7 days (results not shown).

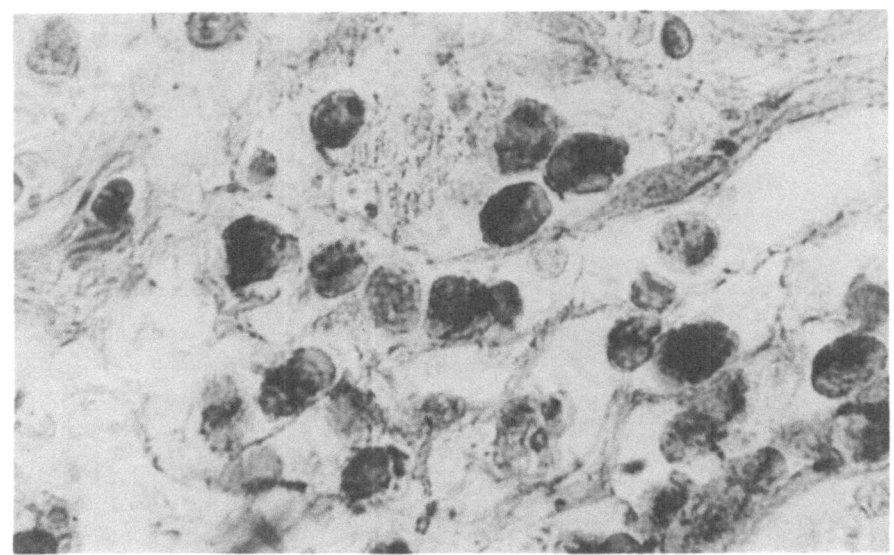

Fig.1 Peritoneum in bacterial peritonitis. Positive staining for human leucocyte elastase in macrophage-like cells (PAP-method). Rabbit anti-leucocyte elastase serum 1/5000.

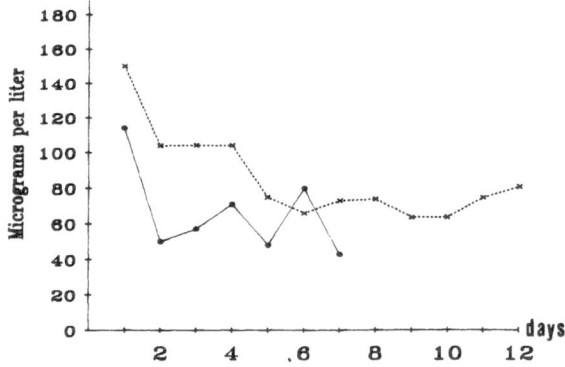

Fig.2 Levels of leucocyte elastase-alpha-1-PI complexes in plasma in 6 patients with chemical peritonitis(solid line) and 10 patients with bacterial peritonitis (broken line).

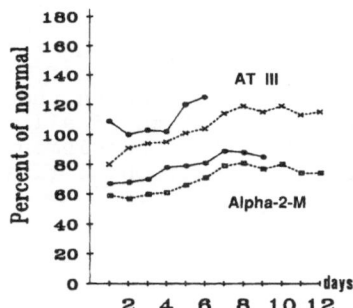

Fig.3 Levels of alpha-2-M and AT III in plasma. For symbols see fig 2.

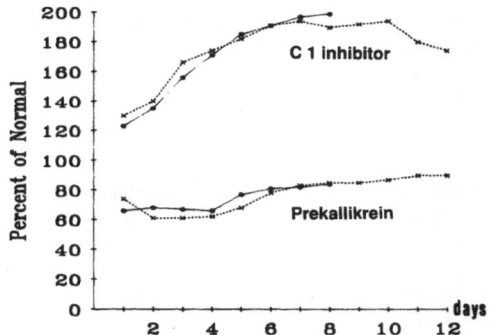

Fig.4 Levels of C1-inhibitor and prekallikrein in plasma. For symbols see fig 2.

DISCUSSION

The results of the present study show consumption of both proenzymes and protease inhibitors, indicating proteolysis and protease-antiprotease interactions in both chemical and bacterial peritonitis.

The biochemical changes were most pronounced intraperitoneally, that is at the site of the disease. The levels of the proenzymes prekallikrein, factor X and plasminogen were low, especially in the bacterial group (table II). C3d-levels were high, and over half of all the C3 present in the peritoneal fluid was cleaved in the bacterial group. These findings indicate activation of all the four main cascade systems, i.e. the kinin, complement, coagulation and fibrinolytic systems, in the peritoneal fluid. Low levels were

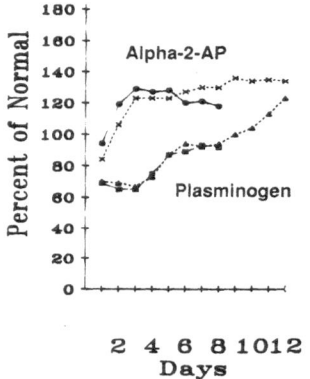

Fig.5 Levels of alpha-2-AP and plasminogen in plasma. For symbols see fig 2.

found for all protease inhibitors, except alpha-2-AP, in the peritoneal fluid. The level of C1 inhibitor, the main inhibitor of both the kinin and complement systems, was in fact 0. All these changes were similiar to, although not as pronounced as in another intra-abdominal inflammatory disease, acute pancreatitis (8,9,10).

The immunohistological studies of peritoneum samples showed the presence of proteases and protease inhibitors in macrophage-like cells. This indicates an active uptake of protease-antiprotease complexes, in agreement with our recent experimental studies in pigs(11). Thus, the peritoneum seems to play an important role in the defence against and clearance of activated proteases.

The peritoneal defence seemed in fact good, since the biochemical changes seen in plasma were far less pronounced than in the peritoneal fluid, also in agreement with the findings in acute pancreatitis(8,9). This could be seen both for leucocyte elastase-alpha-1-PI complexes (fig. 2), proenzymes (figs. 4 and 5) and protease inhibitors (figs. 3,4 and 5).

Furthermore, the levels in plasma were nearly the same both in chemical and bacterial peritonitis. The most probable reason for this similarity is that the leakage of highly acidic gastric juice together with pancratic juice in gastro-duodenal ulcer perforation and the bacterial and/or intestinal content in intestinal perforation or abscesses is equally chemotactic, and that most events thereafter are triggered by the leucocytes. Leucocyte elastase, among other leucocyte proteases, is known to cleave and degrade complement factors (12), clotting factors (13), alpha-2-AP and C1-inhibitor (14). Other explanations are however possible, i.e. a general scheme of reactions of the body to all sorts of trauma, mediated primarily through interleukin 1 (15). Or, different aetiologic agents may trigger one or several cascade systems. Once triggered, the reaction pattern is perhaps more or less self-perpetuating, irrespective of the causative agent. Whatever the cause, the intensity and not the type of the intra-abdominal challenge seems to determine the biochemical changes seen.

In conclusion, proteolysis and protease-antiprotease interactions were seen in both chemical and bacterial peritonitis. The changes were most pronounced intraperitoneally, and macrophage-like cells in the peritoneum showed an uptake of protease-antiprotease complexes. The general inflammatory reaction in peritonitis seemed to predominate, since the biochemical changes were almost the same in peritonitis of different aetiologies.

ACKNOWLEDGEMENTS

This investigation was supported by grants from the Swedish Medical Research Council (project no. B86-17X-03910-14A), the Swedish Society of Medicine, The Medical Faculty, University of Lund, the Påhlsson Foundation, Malmö General Hospital Foundation for Medical Research, the Tore Nilsson Foundation and Kungl. Fysiografiska Sällskapet i Lund.

REFERENCES

1. J.J. Schuler and L.M. Nyhus, 1980, Shock following abdominal operations, in: "Abdominal operations", R. Maingot, Ed., Appleton-Century-Crofts, New York.
2. R.L.Simmons and D.H.Ahrenholz, Pathobiology of peritonitis, a review, J.Antimicrob.Chemother.suppl A, 29-36 (1981).
3. C.-B.Laurell, Electroimmunoassay, Scand J Clin Lab Invest 29,suppl 124,21-37 (1972).
4. I.Brandslund,B.Teisner,P.Hyltoft-Petersen and S.-E.Svehag,Developement and clinical application of electroimmunoassays for the direct quantification of the complement C3 split products C3c and C3d, Scand J Clin Lab Invest 44,suppl 168,57-73 (1984).

5. P.Friberger,Synthetic pepetide substrate assays in caoagulation and fibrinolysis and their application on automates, Thromb.Haemostas.9,281-300 (1983).
6. K.Ohlsson and A.-S.Olsson, Immunoreactive granulocyte elastase in human serum, Hoppe-Seyler's Z.Physiol Chem,359,1531-1539 (1978).
7. L.A.Sternberger,P.H.Hardy,J.J.Cuculis Jr and H.G.Meyer, the unlabelled antibody enzyme method of immunochemistry, J.Histochem.Cytochem.18,315-333 (1970).
8. Å.Lasson and K.Ohlsson,Protease inhibitors in acute human pancreatitis.Correlation between biochemical changes and clinical course, Scand J Gastroenterol 19,779-786 (1984).
9. Å.Lasson and K.Ohlsson,Consumptive coagulopathy,fibrinolysis and protease-antiprotease interactions during acute human pancreatitis, Thromb Res 41,167-183 (1986).
10. P.Wendt,A.Fritsch,F.Schulz,G.Wunderlich and G.Blümel,Proteinases and inhibitors in plasma and peritoneal exudate in acute pancreatitis, Hepato-gastroenterol 31,277-281 (1984).
11. B.Sjölund,Å.Lasson and K.Ohlsson,The elimination of intraperitoneal trypisn-alpha-macroglobulin complexes in healthy and pancreatitis pigs (submitted 1987).
12. U.Johnson,K.Ohlsson and I. Olsson,Effects of granulocyte neutral proteases on complement components, Scand J Immunol 5,421-426 (1976).
13. W.Schmidt,R.Egbring and K.Havemann,Effects of elastase-like and chymotrypsin-like neutral proteases from human granulocytes on isolated clotting factors, Thromb Res 6,315-326 (1975).
14. M.S.Brower and P.C.Harpel,Proteolytic cleavage and inactivation of alpha-2-plasmin inhitor and C1 inactivator by human polymorphonuclear leukocyte elastase, J Biol Chem 252,9849-9854 (1982).
15. C.Dinarello,Interleukin-1 and the pathogenesis of the acute phase response, N.Engl.J Med.311,1413-1418 (1984).

DEFICIENT PHAGOCYTOSIS SECONDARY TO PROTEOLYTIC BREAKDOWN OF OPSONINS IN PERITONITIS EXUDATE

A. Billing[a], D. Fröhlich[a], M. Jochum[b] and
H. Kortmann[a]

a) Chirurgische Klinik und Poliklinik der Universität München, Klinikum Großhadern, Marchioninistr. 15, 8000 München 70, Germany
b) Institut für klinische Chemie und klinische Biochemie der Universität München

INTRODUCTION

In peritonitis, proper functioning of the intraabdominal local defence system is crucial for a favourable outcome and survival of the patient. Peritonitis exudate is characterized by the presence of a large number of viable bacteria despite a huge population of intact PMN-leukocytes. Although phagocyte function, the main factor of cellular defence, is intact or even stimulated in peritonitis exudate (1,2), there is no adequate explanation of how bacteria can persist in surroundings rich in PMN-leukocytes.

An adequate sufficient intraabdominal host defence results from a balanced cooperation between cellular and fluid phase components. The humoral immune process of recognizing and labelling a microbe as antigenically foreign is described as opsonization. This can proceed via non-specific or specific mechanisms (3). The latter is immunoglobulin G (IgG) dependent. Both pathways result in complement activation which leads to a liberation of opsonins, mediators of inflammation and microbicidal components. The main factors of opsonization are C3-derived complement components and IgG. Physiological C3 activation results in its breakdown into the fragments C3c and C3d. Unspecific proteolytic breakdown of opsonic factors in pleural empyema has also been described (4).

Phagocytosis leads to cell activation and also results in an extracellular release of lysosomal and oxidative granulocyte enzymes (5). Myeloperoxidase is known to impair opsonization (6). Proteolytic and oxidative destruction can destroy biological activity of protein components without altering their antigenicity. Thus, despite immunologically determined high concentrations, there may be a functional deficit in such factors. In parallel to enzyme release, particle attachment leads to a strong activation of oxygen metabolism in phagocytes, resulting in the generation of oxygen-derived free radicals. These microbicidal and cytotoxic substances are known to destroy α_1-proteinase inhibitor (α_1PI). Using a photometric amplification system, the release of oxygen-derived free radicals can be measured as chemiluminescence (CL) and is assumed to be a quantitative parameter for phagocytic activity. Using a constant number of phagocytes, CL-measurements can be used as a direct parameter for the quality of particle opsonization (7,8).

Little is known about intraabdominal opsonization. We developed a simple CL assay to evaluate opsonic activity (OA) in peritonitis exudates and serum samples of patients with acute and persisting peritonitis. The latter group was treated with Etappenlavage which means planned relaparotomy until clearance of the abdominal cavity. In addition, we investigated opsonin levels as well as released granulocytic proteins.

MATERIAL AND METHODS

Patients

50 abdominal exudates and corresponding blood samples were drawn intraoperatively from 27 patients with diffuse purulent peritonitis. Exudates were centrifugated, while blood was processed into serum and EDTA-plasma.

Chemiluminescence assay for opsonic activity

Zymosan was preopsonized with pooled normal serum, patients' serum or patients' exudate. The final chemiluminescence assay contained 0,05 ml diluted EDTA-blood (1/15) from healthy volunteers, 0,8 ml Veronal buffer and 0,1 ml Luminol solution (9). The reaction was started by adding 0,05 ml of the opsonized zymosan (20 mg/ml). The 30 min. integral of chemiluminescence was calculated. In each assay zymosan opsonized with normal serum, patients' serum and patients' exudate was tested simultaneously. As all other conditions (blood and buffer concentration) were identical, the resulting different chemiluminescence response is due to the quality of opsonization (10). Opsonic activity was expressed as a percentage of the normal serum value.

Opsonin studies

C3 and IgG levels were measured with a standard radial immunodiffusion assay (Behringwerke Marburg, normal values: IgG 1250 mg/dl and C3 82 mg/dl). C3 splitting was demonstrated by crossed immunoelectrophoresis according to Ganroth (11) employing a C3c antibody (Behringwerke Marburg).

Tests for PMN-enzymes

Elastase in complex with α_1-proteinase inhibitor (α_1PI) and myeloperoxidase concentrations were measured by ELISA: plasma reference for complexed elastase: 50-181 µg/l, and for myeloperoxidase: 25-47 µg/l (12, 13). Free elastase activity was measured with a chromogenic substrate (14) or by adding α_1PI and then re-assaying elastase-α_1PI complex.

442

RESULTS

Opsonization in serum

In acute peritonitis C3 and IgG serum levels were close to the lower limit of the normal range and were increased in patients with persisting peritonitis (Table 1). Opsonic activity in patients' serum was well correlated with a C3/IgG index which results from addition of C3 and IgG concentrations. Computerized correlation analysis resulted in a S-shaped curve, very similar to a dilution curve of normal serum (Fig. 1). In Fig. 1 serum samples of patients (n=23) with acute and persisting peritonitis at the time of sample collection were included.

Table 1. Opsonin levels (IgG, C3) and opsonic activity (OA) in patient serum (% of normal $\pm$ standard deviation)

	acute peritonitis (n=13)	persisting peritonitis (n=14)
IgG	62.8 $\pm$ 29.3	109.1 $\pm$ 30.8
C3	65.3 $\pm$ 23.1	83.0 $\pm$ 21.6
OA	85.8 $\pm$ 33.5	115.4 $\pm$ 20.8

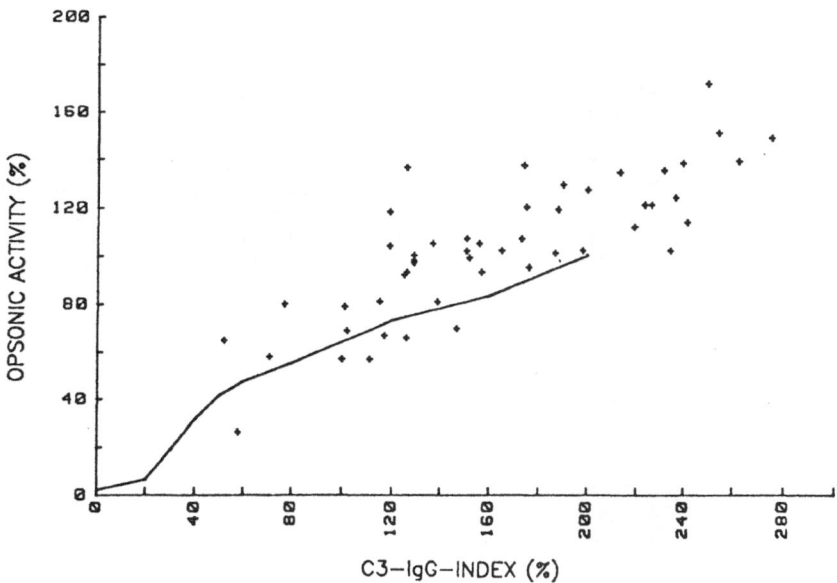

Fig. 1. Correlation of opsonic activity and opsonin concentration in patient serum. The C3-IgG-index results from addition of serum concentrations of both parameters. The curve demonstrates the correlation from serial dilutions of normal serum (1:2 to 1:10).

Opsonization in exudate

In peritonitis exudates, mean protein content was 66 % of

serum levels in acute peritonitis and 62 % of serum levels in
Etappenlavage. The electrophoretic protein distribution pattern
was similar to serum, indicating peritoneal permeability even
for large molecules. Opsonin concentrations are listed in
Table 2. According to the serum correlation of opsonin concentration and function, these opsonin levels should result in
an opsonic activity of 58 % of normal in acute peritonitis and
56 % of normal in Etappenlavage. The experimental determination
of opsonic function, however, showed a much lower activity
(8.4 % and 4.6 % of normal, respectively) indicating a pronounced deficit in particle opsonization in peritonitis
exudates.

Table 2. Opsonin levels (IgG, C3) and opsonic activity (OA) in
peritonitis exudates (% of normal serum value $\pm$ standard deviation). OA_{exp} is the expected OA according
to the correlation between IgG/C3 and OA in serum,
OA_{real} is the actual OA in exudate.

	acute peritonitis (n=13)	Etappenlavage (n=14)
IgG	43.9 $\pm$ 21.3	52.9 $\pm$ 18.1
C3	35.8 $\pm$ 27.8	25.5 $\pm$ 6.8
OA_{exp}	58	56
OA_{real}	8.4	4.6

To evaluate opsonin breakdown, crossed immunoelectrophoresis was carried out in 6 patients' serum and exudate samples
employing a C3c antibody. In peritonitis serum only a small
amount of C3 was fragmented (Fig. 2b). In exudate, however,
depending on the leukocyte concentration, a great part (Fig. 3c)
or almost all (Fig. 3d) C3 was split into fragments of lower
molecular weight. Thus, the opsonic deficit in purulent exudates
was accompanied by an extensive breakdown of the complement
factor C3.

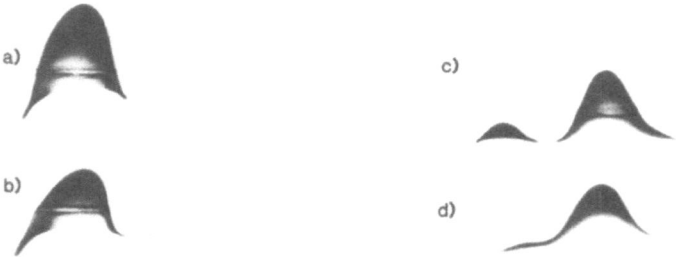

Fig. 2. Crossed immunoelectrophoresis for C3 in serum and
exudate.
a) Normal serum, no C3 splitting
b) Patient serum, only trace amounts of C3 breakdown
products
c) Exudate (22,000 leukocytes/mm³), pronounced C3
splitting into smaller components (right peak)
d) Purulent peritonitis exudate (110,000) leukocytes/
mm³), almost complete breakdown of C3

Unspecific proteolytic and oxidative activity in peritonitis
exudate

In 27 exudates, we quantified complexed PMN-elastase and
myeloperoxidase levels. Elastase concentrations were elevated
up to 250 mg/l, which is 2000 times higher than the normal
plasma range. We also found extremely high concentrations for
myeloperoxidase, reaching up to 160 mg/l (Table 3).

Table 3. Complexed PMN-elastase and myeloperoxidase levels in
peritonitis exudates (mean $\pm$ standard deviation,
μg/l).

	acute peritonitis (n=13)	Etappenlavage (n=14)
elastase (-α_1PI)	75,972 $\pm$ 52,366	89,853 $\pm$ 67,570
myeloperoxidase	34,458 $\pm$ 42,661	55,402 $\pm$ 39,005

In several exudates we could demonstrate free elastase
activity, both with a specific chromogenic substrate and an
α_1PI-binding assay. In some exudates up to 70 % of the total
elastase content was found to be uninhibited free elastase.
The concentration for α_1-proteinase inhibitor in these exudates
ranged from 99 to 341 mg/dl, which, calculated on the basis
of the molar ratio of inhibitor concentration versus proteinase,
should be sufficient for complete elastase inhibition. In
patients with gastrointestinal perforation, intraabdominal
elastase levels varied within a wide range (Table 4).

Table 4. Elastase (in complex or active) and α_1-proteinase
inhibitor (α_1PI) in peritonitis exudates.

	α_1PI-elastase complex(mg/l)	free elastase (mg/l)	total elastase (mg/l)	total α_1PI (mg/l)
Pat. 1	120.6	272.3	392.9	2250
Pat. 2	45.7	0.5	46.2	1610
Pat. 3	111.1	1.6	112.7	1060
Pat. 4	54.6	73.1	127.7	990

To investigate the possible influence of proteolytic lysosomal enzymes on opsonic activity we compared both parameters.
Correlation of the opsonic deficit with elastase levels in
exudates revealed that only exudates with a low concentration
of complexed elastase (<10 mg//l) reached an almost normal
opsonic activity, whereas exudates with high elastase concentrations were deficient in opsonic function (Fig. 3).

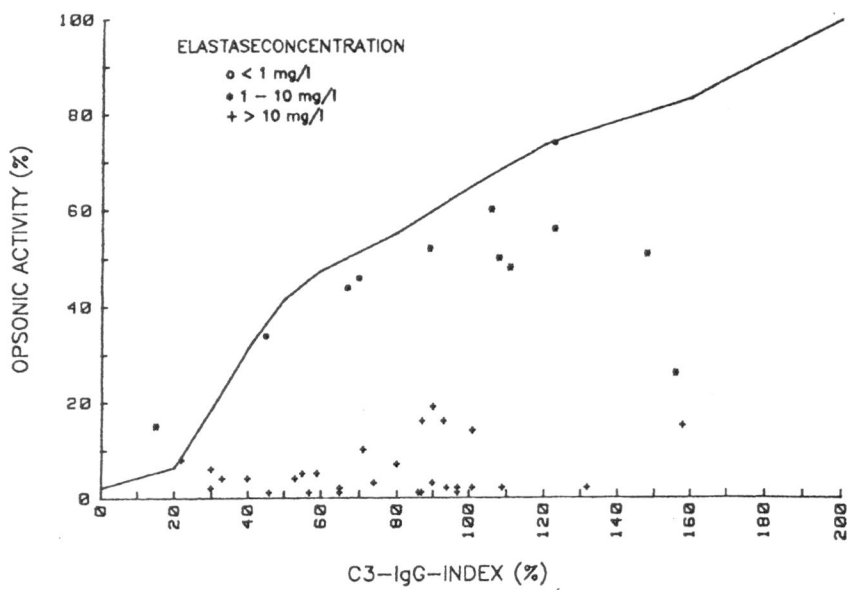

Fig. 3. Correlation of opsonic activity and elastase concentration in peritonitis exudates. The curve demonstrates
the relation between opsonin concentration and opsonic
activity in normal serum. Only exudates with elastase
concentrations <10 mg/l reveal adequate opsonic activity.

DISCUSSION

The CL-approach described here, provides a rapid,re-
liable non-destructive method for quantitative analysis of
opsonic capacity in serum and exudate. Thereby, in normal
as well as in patient serum the key role of IgG and C3 for
opsonization could be confirmed.

Little information is available about intraperitoneal
fluid-phase defence activity. In pleural empyema, deficient
phagocytosis due to breakdown of opsonins has been described
(4). We could demonstrate a high peritoneal permeability in
peritonitis giving way even for large proteins. Despite
sufficient immunologically measurable opsonin levels our results revealed an extended dysfunction of particle opsonization
in human peritonitis exudates. Most of the immunologically
found opsonins were functionally destroyed. The crossed immunoelectrophoresis gave evidence that purulent peritonitis exudates
contained hardly any intact physiologically active C3. The
identified C3 fragments seem to be degraded products without
opsonic function.

In exudates we found extremely high levels of PMN-elastase
(most of it in complex with α_1PI) and myeloperoxidase, indicating the release of a major part of the total phagocytic
enzyme content. Due to a slow peritoneal clearance of elastase-
α_1PI complexes, enzyme concentrations are further increased.
Despite an immunologically sufficient concentration of α_1PI we
could demonstrate free elastase activity in some exudate
samples. Oxidative impairment of α_1PI has been described (15)

and may be due to the release of myeloperoxidase and highly
reactive oxygen products during phagocytosis.

For the first time these data reveal clearly that the
dysfunction of the intraabdominal defence system in acute
peritonitis results from impaired opsonic capacity. One major
underlying pathomechanism may be oxidative inactivation of
the α_1-proteinase inhibitor thus allowing unspecific proteolytic opsonin degradation by free lysosomal enzymes. For
further improvement in therapy the effect of local proteinase-
inhibitor application has to be considered.

Acknowledgment

We thank Dipl.-Ing. B. Schmidt from the nephrology research
lab (Med. Klinik I der Universität München) for the performance of the crossed immunoelectrophoresis.

REFERENCES

1) J.Freischlag,B.Backstrom,D.Kelly,G.Keehn,B.a.R.Busuttil
   Comparison of blood and peritoneal neutrophil activity
   in rabbits with and without peritonitis
   J. of Surg. Res.; 40: 145-151, 1986

2) A. Billing, H. Kortmann
   Nachweis zellulärer und humoraler Abwehrdefekte bei der
   eitrigen Peritonitis mit einem modifizierten Chemilumines-
   zenzverfahren
   Acta chir. Austriacae 3: 340-341, 1986

3) H. Hahn
   Mechanismen der körpereigenen Infektabwehr
   FAC, Band 3-2, 139-150, 1984

4) F.A.Waldvogel,P.Vaudaux,P.D.Lew,A.Zwahlen,S.Suter,U.Nydegger
   Deficient phagocytosis secondary to breakdown of opsonic
   factors in infected exudates
   Adv.Exp.Med.and Biol. 141: 603-610, 1984

5) K. Ohlsson, I. Olsson
   The extracellular release of granulocyte collagenase and
   elastase during phagocytosis and inflammatory processes
   Scand.J.Haematol. 19: 145-152, 1977

6) B.I.Coble,C.Dahlgren,J.Hed,O.Stendhal
   Myeloperoxidase reduces the opsonizing activity of immun-
   globulin G and complement component C3b
   Biochem. et Biophys. Acta 802: 501-505, 1987

7) P.Bellavite,P.Dri,V.Della Bianca,M.C.Serra
   The measurement of superoxide anion production by immun-
   globulin G and complement component C3b
   Europ.J.Clin.Invest. 13: 363-368, 1983

8) R.C.Allen, M.Lieberman
   Kinetic analysis of microbe opsonification based on stimu-
   lated polymorphnuclear leukocyte oxygenation activity
   Inf.and Imm. 45: 475-482, 1984

9) D.Inthorn,Th.Szczeponik,B.Mühlbayer,M.Jochum,H.Redl
Studies of granulocyte function (chemiluminescence response) in postoperative infection. In: Schlag, G., Redl, H.
eds.; First vienna shock forum, Part B.
Progr.in Clin.and Biol.Res. 263B: 51-58, 1987

10) R.C. Allen, M.M. Liebermann
Kinetic analysis of microbe opsonification based on sti-
mulated polymorphonuclear leukocyte oxygenation activity
Inf. and Imm. 45: 475-482, 1984

11) PO Ganroth
Crossed immunoelectrophoresis
Scand.J.Clin.Lab.Invest. 29: 39-41, 1972

12) S.Neumann, G.Gunzer, N.Hennrich, H.Lang
PMN-elastase assay: enzyme immunoassay for human polymorphonuclear elastase complexed with α_1-proteinase inhibitor
J.Clin.Chem.Clin.Biochem. 22: 693-697, 1984

13) S.Neumann, G.Gunzer, H.Lang, M.Jochum, H.Fritz
Quantitation of myeloperoxidase from human granulocytes
as an inflammation marker by enzyme-linked immunosorbent
assay
Fresenius Z.Anal Chem. 324: 365, 1986

14) M. Jochum, A. Bittner
Inter-α-trypsin inhibitor of human serum: an inhibitor of
polymorphonuclear granulocyte elastase
Hoppe Seyler's Z.Physiol.Chem. 364: 1709-1715, 1983

15) N.R.Matheson, P.S.Wong, J.Travis
Enzymatic inactivation of human α_1PI by neutrophil myelo-
peroxidase
Biochem.Biophys.Res.Comm. 88: 402, 1979

PROTEOLYSIS AND LIPID PEROXIDATION -

TWO ASPECTS OF CELL INJURY IN EXPERIMENTAL HYPOVOLEMIC-TRAUMATIC SHOCK

Heinz Redl,Seth Hallström,Camille Lieners
Walter Fürst, and Gunther Schlang[*]

Ludwig Boltzmann Institute for Experimental
Traumatology
Vienna, Austria

INTRODUCTION

Polytrauma in patients leads to diffuse organ changes immediately after injury (Schlag et al., 1985).

These cell injuries are thought to result from the action of proteinases, oxygen radicals and other mediators (Schlag and Redl, 1987). The primary target organ of those mediators is the endothelial cell. We previously demonstrated endothelial changes in the microvascular system of the lung and skeletal musculature (Schlag et al., 1977).

The source of both proteinases and oxygen radicals are either activated phagocytes or injured tissue (via plasma cascade activation or ischemia/reperfusion).

To study these events we have set up a canine model of hypovolemic-traumatic shock (Schlag et al., 1980), in which we previously demonstrated the leukostasis of lung and liver and simultaneously, endothelial swelling in lung, liver and skeletal muscle (Pretorius et al., 1987; Redl et al., 1984, Schlag and Redl, 1987).

AIM OF THE STUDY

We used this model to evaluate the source and effect of shock-related proteolysis and oxygen radical release in tissue and plasma and collected data on (A) activation of humoral cascade systems (complement, kallikrein-kinin), (B) activation (accumulation) of phagocytes, (C) release of hydrolases (proteinases) and oxygen radicals (or reaction participant), (D) action of proteinases and oxygen radicals.

---

* with techn. support of A. Schiesser, E. Paul, C. Vogl, C. Wilfing, C. Kober, M. Thurnher,   W. Junger, R. Kneidinger and the Biotechnology Department

MATERIAL AND METHODS

The animals were instrumentated as described before (Redl et al., 1986). In a total of 10 anesthetized mongrel dogs both femoral bones were closed fractured and 100 blows were administered to one thigh. Subsequently, the animals were bled to a mean arterial blood pressure of 40 mmHg and kept for 2 (or 4) hours.

They were initially reinfused for 45 minutes with their own shed blood (2,000 IU heparin / 500 ml blood) plus Ringer's solution to a cardiac output of 120% of the control value, without exceeding a pulmonary artery pressure of 25 mmHg. This was followed by discontinuous fluid replacement until the end of the experiment (3 (or 1) hours) using the same criteria. The sham group (6 animals) was treated equally except for the shock procedure.

Plasma and tissue hypoxanthin and thiobarbituric acid reactive material (TBAR) (calculated as malondialdehyde) determinations were performed in modifications of previously described methods (Redl et al. 1986). Total complement level was determined as CH_{50} according to Kabat and Mayer (1971).

Dextransulfate released kallikrein (KK-DS) was determined (Philapitsch - personal communication) and protein (Cobas Bio based on the BCCA method (Pierce, USA)) with or without trichloracetic acid precipitation according to Hörl et al. (1981). Nitrogen in ultra-filtrates (MW < 1000) and inotropic activity was checked according to Hallström (1985). Superoxide and hydrogen peroxide in whole blood specimens were analyzed using the fluorimetric technique of Lieners (1986). N-Acetylglucosaminidase (NAG) was photometrically analyzed (Cobas-Bio) using p-Nitrophenyl-N-Acetyl-ß-glucosaminide.

The differences between groups were evaluated with the unpaired t-test. Differences between samples were analyzed by paired t-test. The significance levels were defined as at least 5% probability of error (p < 0.05). Data are presented as mean $\pm$ standard error.

RESULTS

A) An activation of the humoral cascade systems was seen early after trauma as demonstrated by the decrease of total complement (CH_{50}) (Schlag and Redl, 1986) and the decrease of liberable kallikrein (pre-kallikrein) (Fig. 1).

B) Activation of granulocytes in vivo and the resulting leukostasis were previously demonstrated (Redl et al., 1984; Schlag and Redl, 1987) in the same model. However ex vivo zymosan or phorbolmyristic acid (PMA) induced superoxide or hydrogen peroxide release (Table 1) was not enhanced.

C) Release of hydrolases (lysosomal enzymes) is demonstrated with the increase of NAG in plasma (Fig. 2).
The necessary substrate (hypoxanthine) (HX) accumulation for xanthin-oxidase-based radical formation could be demonstrated both in plasma and muscle tissue (Table 1).

D) The action of proteinases was demonstrated by elevated levels of TCA - non precipitable protein (Fig. 3) and increased nitrogen from plasma ultrafiltrates (MW $\leq$ 1000) (Table 2). These ultrafiltrates showed negative inotropic activity in a papillary muscle assay (Table 2), except for one non-responder.

Lipid peroxidation (TBAR) was seen in plasma samples and in some muscle samples of the shock group (Table 1).

TABLE 1

	hours	H_2O_2 - PMA (nmol/10^6 PMN)	H_2O_2 - Zym (nmol/10^6 PMN)		HX - Muscle (nmol/g)
	0	6.55 + 2.49	9.02 + 3.47		7.19 + 1.94
Sham	2	5.24 + 2.18	6.46 + 1.96		8.65 + 2.20
	4	7.47 + 1.09	6.43 + 0.69	(5)	9.66 + 3.68
	0	7.65 + 0.74	6.28 + 1.22		8.99 + 1.78
Shock	2	8.97 + 1.13	8.30 + 2.74		122.66 + 27.63
	4	12.31 + 3.30	4.38 + 0.89	(5)	191.32 + 74.05

	hours	HX - Plasma (nmol/ml)	TBAR - Muscle (nmol/g)	TBAR-Plasma (nmol/ml)
	0	0.70 + 0.46	46.17 + 3.69	1.24 + 0.07
Sham	2	0.39 + 0.11	47.90 + 2.45	1.31 + 0.16
	5	0.33 + 0.12	43.44 + 5.65	1.17 + 0.11
	0	0.32 + 0.06	43.61 + 3.05	1.36 + 0.17
Shock	2	3.14 + 0.90	44.89 + 4.46	2.40 + 0.25
	5	1.31 + 0.35	51.85 + 4.79	3.10 + 0.75

TABLE 2   (prolonged shock phase - 4^h, resuscitation 1^h)

	% Nitrogen	% Inotrop. Activity (of Tyrode control)
Co	1.09 + 0.14	97.4 + 2.92
Shock	3.67 + 0.25	44.2 + 11.32
Reinf.	4.03 + 1.09	31.3 + 2.64

451

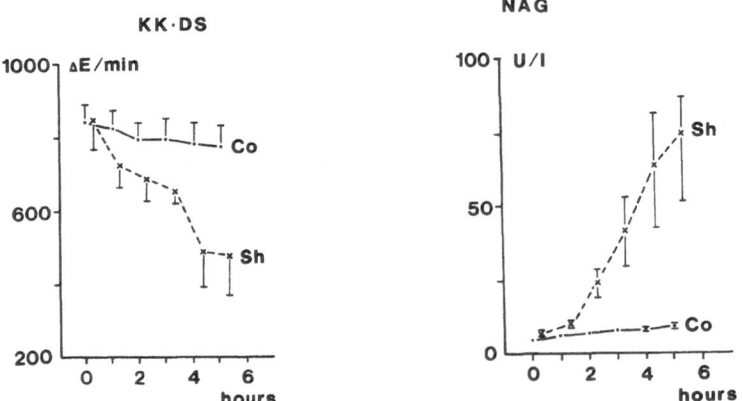

Fig. 1. Dextranesulfate liberable kallikrein activity of plasma (KK-DS) (Co-control, Sh - shock).

Fig. 2: N-acetylglucosaminidase (NAG) plasma levels (U/l).

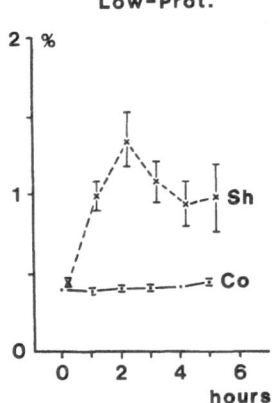

Fig. 3: Trichloracetic acid non-precipitable protein as percentage of total protein in plasma (% Low Prot).

DISCUSSION

PMN activation is assumed to occur after trauma, as demonstrated by leukostasis (Redl et al. 1984). Furthermore, we observed release of potential mediators, activation of complement (complement consumption = CH_{50}) and release of kallikrein (Wachtfogel et al. 1983). However, we were not able to prove PMN activation in this model ex vivo, although we used a detection system to avoid the pitfalls of cell separation. The reason may be the 2- to 3-fold increase in circulating PMN during the observation period, which also occurred in the control group (data not shown). Another reason could be the fact that circulating PMN might be different from the marginated ones. Nevertheless, it was corroborated by patient studies that an activation associated with a release reaction occurs, as demonstrated by the (human specific) granulocyte elastase assay after trauma (Dittmer et al., 1986; Nuytinck et al., 1986) and in association with ARDS (Redl et al. 1987).

In this study we could demonstrate the release or formation of potent proteases only indirectly by an increase in another lysosomal enzyme, i.e. N-acetylglucosaminidase, the decrease in prekallikrein and the higher levels of TCA non-precipitable polypeptides as well as increased nitrogen content of low molecular weight ultrafiltrates. These fractions develop a negative inotropic activity in a papillary muscle assay (Hallström, 1985), but are probably not the result of a newly formed "myocardial depressant factor" according to Lefer (1982), which we identified as a mere artefact (Hallström et al., 1987). The negative inotropic activity is rather the sum of several single activities that are elevated in shock and already present in normal plasma, of which some are positive inotropic and are currently being identified in our laboratory (Hallström et al., 1987).

Whether the underlying proteolytic events are related to phagocytes, to other cells or plasma borne proteases via tissue thromboplastin and activated F XII cannot be decided. The same applies to the question of the source of oxygen radical formation. Although in the recent series of experiments we found less radical-based formation of thiobarbituric acid reactive material in shock than in previous investigations (Schlag and Redl, 1986), the increased levels of hypoxanthin in plasma and tissue from degraded ATP in shock might be a good substrate for a xanthinoxidase based radical formation upon reperfusion in addition to the reaction of phagocytes. That both sources have to be taken into consideration can be concluded from experiments in which either allopurinol (Crowell et al., 1969) or the absence of neutrophils (Bishop et al., 1986) were proven to be protective.

REFERENCES

Bishop MJ, Cheney FW, Chi EY, et al. (1986) White cell depletion with hydroxyurea protects lungs from reperfusion injury. J Crit Care 1: 103.

Crowell JW, Jones CE, Smith EE (1969) Effect of allopurinol on hemorrhagic shock. Am J Physiol 216: 744.

Dittmer H, Jochum M, Fritz H (1986) Freisetzung von granulozytärer Elastase und Plasmaproteinveränderungen nach traumatischem hämorrhagischem Schock. Unfallchir 89: 160.

Hallström S (1985) Untersuchungen über das Auftreten von niedermolekularen herzwirksamen Plasmafaktoren im schweren hypovolämisch-traumatischen Schock. Dissertation, Technische Universität Wien.

Hallström S, Vogl C, Krösl P, et al (1987) Studies on low molecular weight inotropic plasma substances in prolonged hypovolemic traumatic shock. In: Schlag G, Redl H (eds) Progress in Clinical and Biological Research, subseries: Vienna Shock Forum, Part A: Pathophysiological Role of Mediators and Mediator Inhibitors in Shock, Alan R. Liss Inc, New York, 591.

Hörl WH, Stepinski J, Ganter C, et al (1981) Evidence for the participation of proteases in protein catabolism during hypercatabolic renal failure. Klin Wochenschr 59: 751.

Horpaczy G, Hügel W, Müller H, et al. (1987) Biochemical monitoring of the lung during and after extracorporeal circulation. In: Schlag G, Redl H (eds) Progress in Clinical and Biological Research, subseries: Vienna Shock Forum, Part A: Pathophysiological Role of Mediators and Mediator Inhibitors in Shock, Alan R. Liss Inc, New York, 95.

Junger W, Bahrami S, Spragg R, et al (1987) Simple chromatographic conditions for the HPLC separation of nucleosides in tissue samples. (submitted for publication)

Kabat EA, Mayer MM (1971) Experimental immunochemistry, 2nd ed., Thomas Springfield.

Lefer AM (1982) The pathophysiologic role of myocardial depressant factor as a mediator of circulatory shock. Klin Wochenschr 60: 713.

Lieners C (1986) Beeinflussung der Phagozytosetätigkeit von Granulozyten durch Inhalationsnarkotika. Diplomarbeit, Technische Universität Wien.

Nuytinck JKS, Goris RJA, Redl H, et al (1986) Posttraumatic complications and inflammatory mediators. Arch Surg 121: 886.

Pretorius JP, Schlag G, Redl H, et al (1987) The "lung in shock" as a result of hypovolemic-traumatic shock in baboons. J Trauma, in press.

Redl H, Pacher R, Woloszczuk W (1987) Acute pulmonary failure – Comparison of granulocyte elastase and neopterin in septic versus non-septic patients. In: Blair JA, Pfleiderer W, Wachter H (eds) Biochemical and clinical aspects of pteridines, Vol. 5, Walter de Gruyter, Berlin, New York, in press.

Redl H, Schlag G, Hammerschmidt DE (1984) Quantitative assessment of leukostasis in experimental hypovolemic-traumatic shock. Acta Chir Scand 150: 113.

Redl H, Schlag G, Schieser A, et al. (1986) Oxidant-induced alterations in lung adenine nucleotide precede edema formation. In: Novelli GP, Ursini F (eds) Oxygen free radicals in shock, Int. Workshop, Florence 1985, S. Karger, Basel, 180.

Schlag G, Redl H (1985) Morphology of the posttraumatic human lung after traumatic injury. In: Zapol WM, Falke KJ (eds) Pathophysiology and therapy of severe acute lung disease, Dekker, New York, 161.

Schlag G, Redl H (1986) Oxygen radicals in hypovolemic-traumatic shock. In: Novelli, Ursini (eds) Oxygen free radicals in shock, Int. Workshop, Florence 1985, S. Karger, Basel, 94.

Schlag G, Redl H (1987) The morphology of the microvascular system in shock: Lung, liver and skeletal muscles. Crit Care Med 13: 1045.

Schlag G, Redl H (1987) Mediators in sepsis. In: Vincent JL, Thijs LG (eds) Update in intensive care and emergency medicine – Septic shock: An European view, Springer, in press

Schlag G, Voigt WH, Redl H (1980) Vergleichende Morphologie des posttraumatischen Lungenversagens. Anästh Intensivther Notfallmed 15: 315.

Schlag G, Voigt HW, Schnells G (1977) Vergleichende Untersuchungen der Ultrastruktur von menschlicher Lunge und Skelettmuskulatur im Schock. Anaesthesist 26: 612.

Wachtfogel YT, Kucich U, James HJ, et al (1983) Human plasma kallikrein releases neutrophil elastase during blood coagulation. J Clin Invest 72: 1672.

PLASMA LEVELS OF ELASTASE 1 PROTEASE INHIBITOR COMPLEX IN THE MONITORING

OF ARDS AND MULTI-ORGAN FAILURE - A SUMMARY OF THREE CLINICAL TRIALS

H. Redl, E. Paul, R.J.A. Goris*, R. Pacher**,
W. Woloszczuk***, and G. Schlag

Ludwig Boltzmann Institute for Experimental Traumatology
Vienna, Austria
  * University Nijmegen, Netherland
 ** I. Department of Medicine, University Vienna, Austria
*** Ludwig Boltzmann Institute for Experimental Endocrinology
Vienna, Austria

INTRODUCTION

Sepsis and trauma are associated with the release of numerous mediators, which may contribute to cell injury in vital organs. The microcirculation is the primary target to undergo morphologic as well as functional changes. These alterations are caused by nonbacterial inflammation of the specific organ, which may represent a place of least resistance for additional bacterial infection. As a result, MOF (multiple organ failure syndrome) develops, and in many instances is fatal (Eiseman et al. 1977; Goris et al. 1985).

This sequence, among others, involves activation of granulocytes as well as their increased margination to the vascular endothelium with resulting leukostasis, e.g. in the lung (Redl et al. 1984; Redl et al. 1987) or liver (Schlag and Redl 1983), where it is partially induced by anaphylatoxins (Heideman et al. 1978). Subsequently, toxic oxygen species and proteolytic enzymes such as elastase are released and probably damage endothelial cells with resulting permeability edema. Endothelial swelling and leukostasis were first discovered in lung biopsy specimens of polytrauma patients (Schlag et al. 1978,Schlag and Redl 1985). These findings were corroborated by subsequent animal studies (Schlag and Redl 1983). Consequently, in 1984 we were rather keen to analyse plasma granulocyte elastase alpha$_1$ antiprotease complex, as described (Neumann et al. 1984) and successfully used in septic patients, and its correlation with severity of illness (Duswald et al. 1985). We then performed three studies to explore the relationship between severity of injury (in the trauma trial), incidence of ARDS, severity of MOF and the plasma levels of granulocyte elastase. In part of these studies we introduced the macrophage marker neopterin to monitor septic complications in intensive care patients (Strohmaier et al. 1987, Pacher et al. 1987, Redl et al. 1987). Before, neopterin was already used in several other indications (for review see Huber et al. 1984).

SUBJECT AND METHODS

PATIENTS - STUDY I

71 arbitrarily selected patients with polytrauma admitted to the University Hospital Nijmegen were included into the study (Nuytinck et al. 1986). The Injury Severity Score (ISS) was calculated with the Hospital Trauma Index (HTI) as previously described (Goris 1983). In this group of patients ARDS was defined as the necessity to administer ventilation therapy to the patient for more than two days if ISS was less than 50, and for more than four days if ISS was 50 or more (Goris et al. 1985). Patients with unstable thorax and/or head injuries were excluded from the ARDS definition.

PATIENTS STUDY II

Patients (cardiac, orthopedic and tumor surgery, polytrauma, pneumonia) admitted to the ICU of the Second Department of Internal Medicine, Vienna, with symptoms of "septicemia" as defined below (Group SEP) (n= 32) were divided into non-survivors (Group SEP/NS) (n = 14) and survivors (Group SEP/SURV) (n = 18). Survivors were defined as patients referred from the ICU to a general ward, while non-survivors died at the ICU during the observation period.

24 postoperative patients (cardiac, orthopedic and tumor surgery, polytrauma) without signs of septicemia served as control group (Group Co).

Definition of septicemia was similar to the criteria of Montgomery et al. (1985), except that if blood cultures were not positive, 3 out of 4 additional hemodynamic criteria had to be fulfilled (Pacher et al. 1987).

PATIENTS - STUDY III

27 patients with symptoms of lung failure admitted to the ICU of the Second Department of Internal Medicine, Vienna, were divided into septic patients (polytrauma, cardiac, orthopedic, tumor surgery or pneumonia) (n= 21) and non-septic patients (multiple transfusion, gas intoxication, near drowning, Akayama) (n = 6) according to the criteria described below. Patients with an uncomplicated postoperative course (polytrauma, cardiac, orthopedic and tumor surgery) (n= 21) in the ICU served as control group. Septicemia was defined as in study II. Acute lung failure (ARF) was defined according to the criteria of Pepe et al. (1982), and quantified (I) by the pulmonary failure score, which is part of the MOF score according to Goris et al. (1985) (O= no mechanical ventilation (MV), 1 = MV with PEEP $<$10 cm H_2O and/or $< FiO_2$ 0.4, 2= MV with PEEP $>$ 10 cm H_2O and/or $FiO_2 >$ 0.4) and (II) by the arterial alveolar oxygen tension ratio.

PLASMA ANALYSIS

Blood samples were taken in EDTA tubes and were centrifuged within one hour after sampling. Plasma samples were stored frozen. Granulocyte Elastase-alpha$_1$-Protease Inhibitor complex (E-alpha$_1$ PI) was measured by enzyme-immunoassay (Neumann et al. 1984) supplied in a kit by Merck (FRG), with a normal range of 60 - 110 ug/L.

Neopterin was determined using a radioimmunoassay distributed by Henning (FRG). The normal range in human plasma is 3-10 nmol/L.

Based on the author's personal experience (see also Strohmaier et al. 1987, Pacher et al. 1987), the limits for "highly pathological" plasma levels were set at 40 nmol/L neopterin and 400 ug/L Elastase complex. Many other plasma parameters were analysed and invasive hemodynamic measurements were performed, but are beyond the scope of this article.

Statistical analysis was done by E. Schuster (Department of Medical Computer Science, University of Vienna) and by T. de Boo, MSc, and W. Lemmens, MSc, Department of Statistics, Catholic University Nijmegen. Statistical analysis of septic patient data was performed with the Kruskal-Wallis and Wilcoxon tests. For changes between time points the paired t-test was used, and the Kendall-Tau statistic as well as the Spearman correlation coefficients were employed for the correlation matrix.

To predict a clinical event the definitions of sensitivity, specificity or accuracy were used.

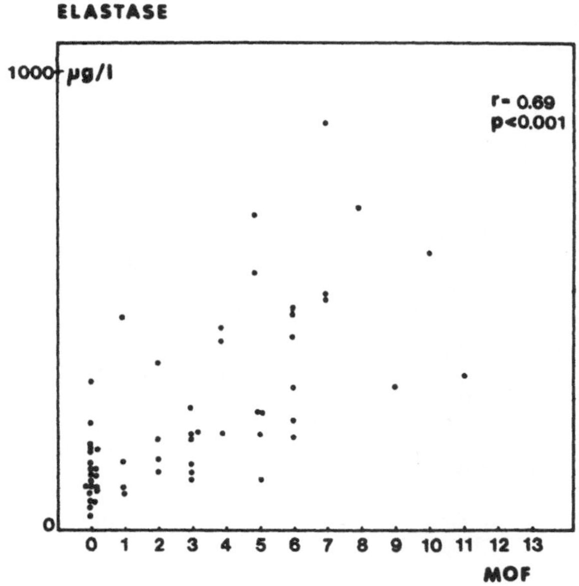

ELASTASE

Fig. 1. Relationship between severity of organ failures (quantified using the multi organ failure score according to Goris) and plasma levels of elastase alpha$_1$ antiprotease complex in polytrauma patients (Nuytinck et al. 1986).

RESULTS - STUDY I

Granulocyte elastase levels correlated only weakly (though significantly) with the severity of injury but were the best of all inflammatory parameters, e.g. complement components. There was a significant difference (p < 0.01) between elastase plasma levels in patients developing ARDS later on, resulting in a sensitivity of 87 %, a specificity of 77 % and an accuracy of 80 % for the prediction of ARDS. Upon admission, there was a significant difference in E-alpha$_1$ PI between later death and surviving patients.

Furthermore, E-alpha$_1$ PI was the only parameter that correlated with severity of organ failure (MOF-score) during the entire observation period, with correlation coefficients ranging between 0.69 – 0.79 (Fig. 1).

RESULTS – STUDY II

Granulocyte enzyme elastase could differentiate between groups SEP/NS and Co during the six-day observation period (Fig. 2). Group SEP/SURV vs. Group Co showed no significant difference, Group SEP/NS vs. SEP/SURV differed only on days 2, 3 and 5 of the six-day observation period, whereas neopterin could differentiate SEP/SURV and Co (post-operative patients without septic complications) (data not shown) (Pacher et al. 1987).

ELASTASE

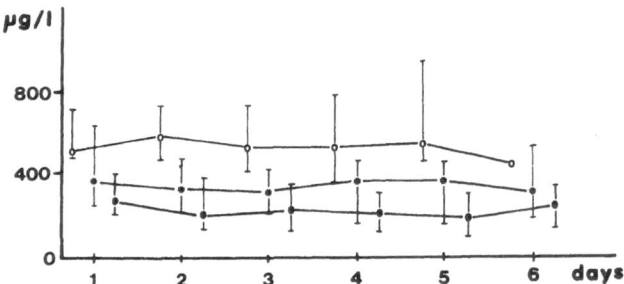

Fig. 2. Elastase alpha$_1$ antiprotease levels in 56 postoperative patients (Study II) (Pacher et al 1987), control group without complications (●), with so called septic complications (survivors *, non-survivors o). Data are shown as median with 25 and 75 percentile. There are significant differences between controls and non-survivors.

Septic patients (Groups SEP/SURV and SEP/NS) again demonstrated a significant correlation between E-alpha$_1$ PI and MOFS (r = .67, p >.0001, n_{Obs} = 110).

Acute pulmonary insufficiency was predicted by elevated E-alpha$_1$ PI values (> 400 ug/L) one day before mechanical ventilation had to be performed with a sensitivity of 81%, a specificity of 82 % and an accuracy of 82%. Furthermore the combined evaluation of E-alpha$_1$ PI (>400 ug/L) and neopterin (> 40 nmol/L) allowed prediction of a critical MOF score > 5 on the following day.

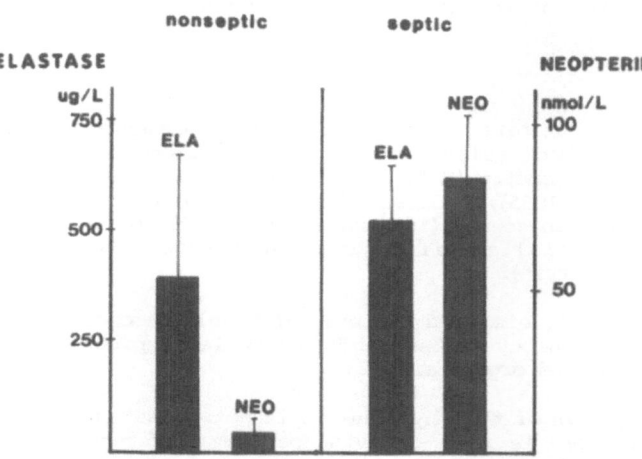

Fig. 3. Granulocyte marker elastase alpha$_1$ antiprotease (Neumann et al.
1984) and macrophage activation marker neopterin (Huber et al.
1984) in 27 postoperative and internal patients (Study III (Redl
et al. 1987)) during periods of acute respiratory failure.
Patients are divided in so called septic and non-septic patients
according to the criteria of Montgomery et al. (1985). Data are
shown as mean ± standard deviation. There is a significant
difference for neopterin plasma levels between septic and non-
septic patients.

RESULTS - STUDY III

In patients who were admitted to the ICU before acute lung failure E-alpha$_1$ PI continuously increased until full development of ARF. The increase one day before ARF (with plasma levels > 400 nmol/L as in study II) as compared to the first day of ARF (pulm. score = 2) was statistically significant (p < 0.002; n = 10). These plasma levels were significantly higher than in controls with no pulmonary failure (p < 0.001) (data not shown) (Redl et al. 1987). There was no significant difference between septic and non-septic patients for Elastase (Fig. 3). In contrast, neopterin plasma levels were significantly different in septic and non-septic patients, with "normal" values (< 20 nmol/L) encountered in the latter group (Fig. 3).

In the 9 survivors (4 septic, 5 non-septic) there was continuous improvement in lung function together with a significant decrease in E-alpha$_1$ PI (data not shown) (Redl et al. 1987).

DISCUSSION

There are several possibilities to demonstrate phagocyte stimulation in shock related situations, e.g. proof of cell margination (morphology, morphometry (Redl et al. 1987), scintigraphy-labelled cells (Schlag and Redl 1983), studies of cell function ex vivo (whole blood for e.g. chemiluminescence (Inthorn et al. 1987)), or analysis of plasma levels of potential specific marker substances (Elastase alpha$_1$ PI complex, lactoferrin).

In our report, elevated E-alpha$_1$ PI levels demonstrate the presence of activated granulocytes that might be at least partially responsible for lung injury and organ failure.

Direct proof of this hypothesis is not possible in the clinical setting, but there is indirect evidence from different patient studies. First, there is the weak but significant correlation of E-alpha$_1$ PI with injury severity in trauma patients (the best parameter among several others). This correlation would have been better if multiple sampling had been applied during the first hours after trauma, similar to the studies by Dittmer et al. (1986). Secondly, interesting correlations were found for organ failure, particularly of the lung, with an inverse correlation between P(a/A)O$_2$ and E-alpha$_1$ PI, as well as for prediction of necessary mechanical ventilation on the following day with high sensitivity, and eventually for prediction of ARDS several days later, based on the use of more or less arbitrarily chosen critical values (E-alpha$_1$ PI > 400). Similar studies were reported by other authors, too (Duswald et al. 1985; Zheutlin et al. 1986).

Furthermore we must consider that activated PMN are related to failure of several organs, leading to multi-organ failure. A correlation between occurrence of MOF and elevated E-alpha$_1$ PI was established both in trauma patients (Study I) as well as in postoperative patients (Study II). In both, similar correlation coefficients ranging between 0.6 - 0.7 were found. This correlation is not only valid for the PMN marker elastase but also for the macrophage activation marker neopterin (Pacher et al. 1987), which we have shown to be useful for prediction of survival (Strohmaier et al. 1987) and of a critical stage of MOF (> 5). However, high levels of the macrophage marker neopterin are not necessarily related to ARDS (only 60 % in a previous study) (Pacher et al. 1987). In the majority of patients, MOF with poor prognosis is seen, especially when E-alpha$_1$ PI is additionally elevated. Increased

462

neopterin plasma levels were mainly encountered in so-called septic patients and in those who died (Strohmaier et al. 1987; Pacher et al. 1987). According to our criteria (not necessarily positive blood culture or septic focus) these patients might had just a generalized inflammatory reaction without sepsis, as postulated by Goris et al. (1985).

Irrespective of the only weakly defined "sepsis" (according to Montgomery and others), conditions related to MOF seem to depend, at least partly, on activation of phagocytes with the inherent release of destructive proteases, oxygen radicals and eicosanoids. The release of additional potent mediators by phagocytes, such as interleukin 1 and cachectin, with their specific action on the endothelium, places specific emphasis on monitoring the activation stage of the two most important phagocytes, i.e. PMN and macrophages, with the help of Elastase-alpha$_1$ PI and neopterin plasma levels.

REFERENCES

Dittmer H, Jochum H, Fritz M, et al (1986) Freisetzung von granulo-zytärer Elastase und Plasmaproteinveränderungen nach traumatisch hämorrhagischem Schock. Unfallchir 89: 160.
Eiseman B, Beart R, Norton L (1977) Multiple organ failure. Surg. Gynecol. Obstet. 144: 323.
Duswald KH, Jochum M, Schramm W, et al (1985) Released granulocyte elastase: An indicator of pathobiochemical alterations in septicemia after abdominal surgery. Surgery 98: 892.
Goris RJA (1983) The injury severity score. World J Surg 7: 12.
Goris RJA, te Boekhorst TPA, Nuytinck JKS, et al (1985) Multiple organ failure: Generalized autodestructive inflammation? Arch Surg 120: 1109.
Heideman M, Kaijser B, Gelin LE (1978) Complement activation and hematologic, hemodynamic and respiratory reactions early after soft tissue injury. J Trauma 18: 696.
Huber CH, Batchelor JR, Fuchs D, et al (1984) Immune-response associated production of Neopterin release from macrophages primarily under control of Interferon-gamma. J Exp Med 160: 310.
Inthorn D, Szczeponik T, Mühlbayer D, et al (1987) Studies of granulocyte function (chemiluminescence response) in postoperative infection. In: Schlag G, Redl H (eds) Progress in Clinical and Biological Research, Vol. 236 B:First Vienna Shock Forum, Part B: Monitoring and Treatment in Shock, Alan R Liss, New York, p. 51.
Montgomery AB, Stager MA, Carrico CJ, et al (1985) Causes of mortality in patients with the adult respirtory distress syndrome. Amer Rev Resp Dis 132: 485.
Neumann S, Hennrich N, Gunzer N, et al (1984) Enzyme linked immunoassay for human granulocyte elastase in complex with alpha 1 proteinase inhibitor. In: Hörl WA, Heidland A (eds) Proteinases: Potential Role in Health and Disease, Plenum Press, New York, p 379.
Nuytinck JKS, Goris RJA, Redl H, Schlag G, van Munster PJJ (1986) Post-traumatic complications and inflammatory mediators. Arch Surg 121: 886.
Pacher R, Redl H, Frass M, et al (1987) Neopterin and granulocyte elastase in septicemic patients prone to develop multi-organ failure. In: Blair JA, Pfleiderer W, Wachter H (eds) Biochemical and Clinical Aspects of Pteridines, Vol. 5, 6th International Workshop, Walter de Gruyter, in press.
Pepe PE, Potkin TR, Holtman D, et al (1982) Clinical predictors of the adult respiratory distress syndrome. Amer J Surg 144: 124.

Redl H, Dinges HP, Schlag G. (1987) Quantitative estimation of leukostasis in the posttraumatic lung - canine and human autopsy data. In: Schlag G, Redl H (eds.) Progress in Clinical and Biological Research, Vol. 236 A: First Vienna Shock Forum, Part A: Pathophysiological Role of Mediators and Mediator Inhibitors in Shock. Alan R. Liss Inc., New York, p. 43.

Redl H, Pacher R, Woloszczuk W (1987) Acute pulmonary failure - comparison of neopterin and granulocyte elastase in septic and non-septic patients. In Blair JA, Pfleiderer W, Wachter H (eds) Bio-chemical and Clinical Aspects of Pteridines, Vol. 5, 6th International Workshop, Walter de Gruyter, in press.

Redl H, Schlag G, Hammerschmidt DE (1984) Quantitative assessment of leukostasis in experimental hypovolemic-traumatic shock. Acta Chir Scand 150: 113.

Schlag G, Redl H (1983) Posttraumatic ultrastructural changes and the role of granulocytes in the lungs, liver and skeletal muscle. Intens Care Med 9: 148.

Schlag G, Redl H (1985) Morphology of the human lung after traumatic shock. In: Zapol WM, Falke KJ (eds) Acute Respiratory Failure, New York, Basel, Marcel Dekker Inc, p 161.

Schlag G, Voigt WH, Schnells G, et al (1978) Vergleichende Untersuchun-gen der Ultrastruktur von menschlicher Lunge und Skelettmuskulatur im Schock. II. Anaesthesist 26:612-622

Strohmaier W, Redl H, Schlag G, et al (1987) Elevated D-erythro-Neopterin levels in intensive care patients with and without septic complications. In: Schlag G, Redl H (eds.) Progress in Clinical and Biological Research, Vol. 236 A: First Vienna Shock Forum, Part B: Monitoring and Treatment of Shock, Alan R. Liss Inc., New York, p 59.

Zheutlin LM, Thomar EJMA, Jacobs ER, et al (1986) Plasma elastase levels in the adult respiratory distress syndrome. J Crit Care 1: 39.

PMN ELASTASE AND LEUKOCYTE NEUTRAL PROTEINASE INHIBITOR (LNPI) FROM GRANULOCYTES AS INFLAMMATION MARKERS IN EXPERIMENTAL-SEPTICEMIA

R. Geiger[1], S. Sokal[1], G. Trefz[1], M. Siebeck[2] and H. Hoffmann[2]

[1]Department of Clinical Chemistry and Clinical Biochemistry in the [2]Department of Surgery, Univ. of Munich, D-8000 Munich 2, FRG

INTRODUCTION

During severe inflammatory processes, various blood and tissue cells including polymorphonuclear granulocytes release cellular constituents such as lysosomal proteinases and inhibitors from the cytosol extracellularly into the circulation (1 - 4). Such enzymes as well as oxidizing agents produced during phagocytosis enhance the inflammatory response by degrading connective tissue structures, membrane constituents and soluble proteins by proteolysis or oxidation (2, 4).
The physiological function of these inhibitors is of special interest as yet not clearly unterstood. They may, on the one side, act as part of an intracellular defence system to regulate undesired proteolytic activities inside the cells. On the other side, such inhibitors could assist the overall inhibitory potential of the organism, e.g. in inflammatory processes (5), in which leukocytes release large amounts of proteinases by degranulation or during desintegration (6). The inhibitors might function then by elimination of the proteases by complex formation thus preventing degradation of tissues and humoral factors as well as further activation of inflammatory cells.
In this communication we describe the application of recently developed enzyme immunoassays for leukocyte elastase LNPI complexes and LNPI (7) as well. Plasma levels of the elastase-inhibitor complex and of total LNPI (complex and free) have been determined during the development of septicemia induced with live E. coli in pigs. Furthermore, the release of LNPI from polymorphonuclear granulocytes by stimulation with different agents such as endotoxin, zymosan, etc. has been studied.

MATERIAL AND METHODS

Leukocyte neutral proteinase inhibitor (LNPI) has been purified and characterized as described by Trefz et al. (9). Enzyme immunoassays for the determination of leukocyte elastase-LNPI complexes and LNPI have been established recently (7). Septicemia was induced in domestic pigs under pentobarbital anesthesia by infusion of 3×10^{10} live bacteria in 2 hrs (E. coli, ATCC strain 20399). Plasma samples were taken before infusion and at 2 hrs intervals thereafter.

Stimulation of leukocytes in whole blood was performed as follows: 6 ml whole blood and 0.12 ml stimulating reagent solution (6.96 mg zymosan, 0.06 mg endotoxin and 5×10^{7} live E. coli) were incubated at $37^{\circ}C$. Samples of 0.5 ml were taken at different time intervals and centrifuged at 12 000g. The supernatants were used for the determination of LNPI, LNPI complexes and lactate dehydrogenase, respectively. For control measurements, physiological saline was used instead of a stimulant. To stimulate leukocyte suspensions in physiological saline 6 ml of whole blood were centrifuged for 10 min at 4000 g. The leukocyte sediment was gently resuspended in saline, centrifuged again (in total 3 times) and suspended in 6 ml saline. The leukocyte stimulation was performed as described for whole blood.

RESULTS AND DISCUSSION

An intracellular proteinase inhibitor (leukocyte neutral proteinase inhibitor, LNPI; (8, 9)) has been isolated recently from pig granulocytes and characterized in detail. LNPI released after desintegration of leukocytes may appear in plasma of pigs either as a free inhibitor, as leukocyte elastase-LNPI complex, or in complex with other neutral proteinases, such as mast cell chymase.

Leukocyte elastase-LNPI complexes and total LNPI concentrations were measured as shown schemactically in Fig. 1 using recently developed enzyme immunoassays (7). Plasma levels of this LNPI complex were calculated by substraction of the measured concentrations of LNPI (free LNPI and LNPI bound to proteinases other than leukocyte elastase) from the total LNPI concentration (Fig. 2). The plasma concentration of LNPI in non-septic pigs was found to be $150 + 30 \times 10^{-9}$ g/ml blood, and for leukocyte elastase LNPI-complexes of $16 + 11 \times 10^{-9}$ g/ml. In porcine leukocytes a LNPI content of 64×10^{-15} g/PMN granulocyte was measured.

During the development of septicemia induced with live E. coli, the inhibitor was released simultaneously with the lysosomal proteinases. Plasma concentrations of LNPI and leukocyte

elastase LNPI-complexes about 10times above the physiological
levels were observed in septic pigs (n = 8; Fig. 3). Thus LNPI
can serve as a marker of the inflammatory response of the
granulocytes in septic pigs.

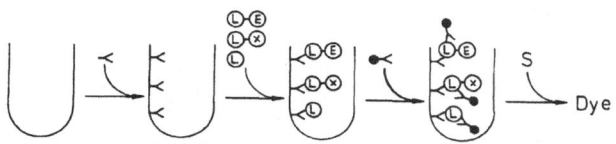

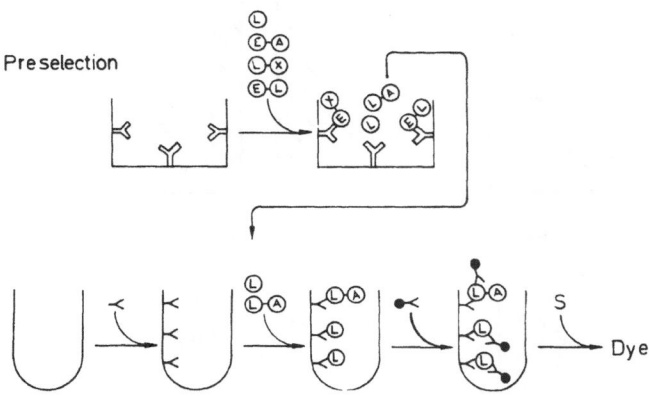

Fig. 1. Scheme of the enzyme immunoassay of (a) total LNPI
(complexed and free LNPI) and of (b) neutral proteinase LNPI-complexes (except leukocyte elastase
LNPI complexes) in porcine plasma.
In enzyme immunoassay typ B (b) a preselection of
leukocyte elastase and leukocyte inhibitor complexes on microtiterplates coated with anti-elastase antibodies was performed. L, LNPI; E, leukocyte elastase; X, neutral proteinase; S, ABTS[R];
⊰ , anti-LNPI antibodies;⊱●, anti-LNPI peroxidase conjugate;⊏ , anti-leukocyte elastase antibodies; A, inhibitor.

In our in vitro studies we stimulated both porcine leukocytes in whole blood and washed leukocytes using zymosan, live
E.coli and endotoxin. Desintegration of leukocytes was monitored by determination of lactate dehydrogenase during incubation. As can be seen from Fig. 4, only zymosan induced a
marked release of LNPI after about 30 min. The release of LNPI
caused by E. coli and endotoxin is mainly due to desintegration of the cells during incubation. This is in contrast
to data obtained for the release of elastase after stimulation
of human leukocytes (10). Obviously, the inhibitor is liberated from the cells only slowly and for the release reaction a
potent stimulus, even desintegration of the cells may be
necessary.

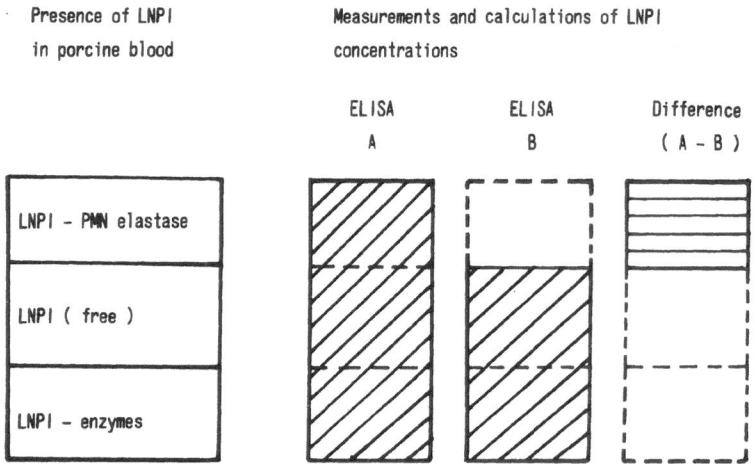

Fig.2. Scheme of LNPI determination in porcine blood.

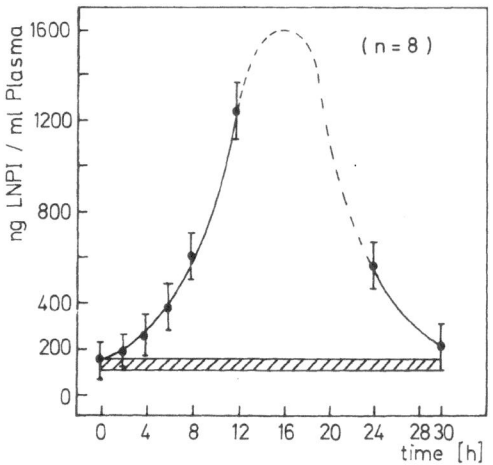

Fig. 3a. Mean plasma levels of LNPI in septic pigs.

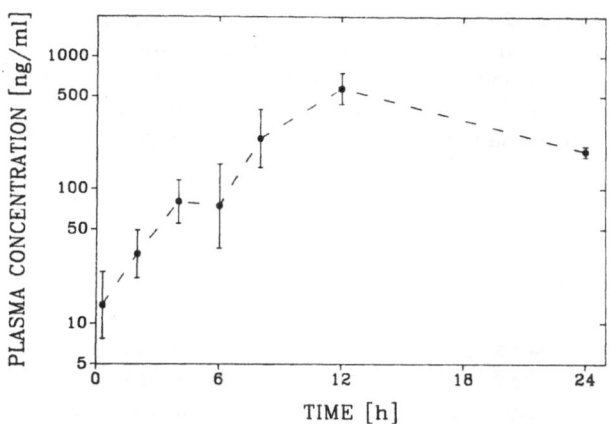

Fig. 3b. Mean plasma levels of leukocyte elastase LNPI-
complexes in septic pigs.

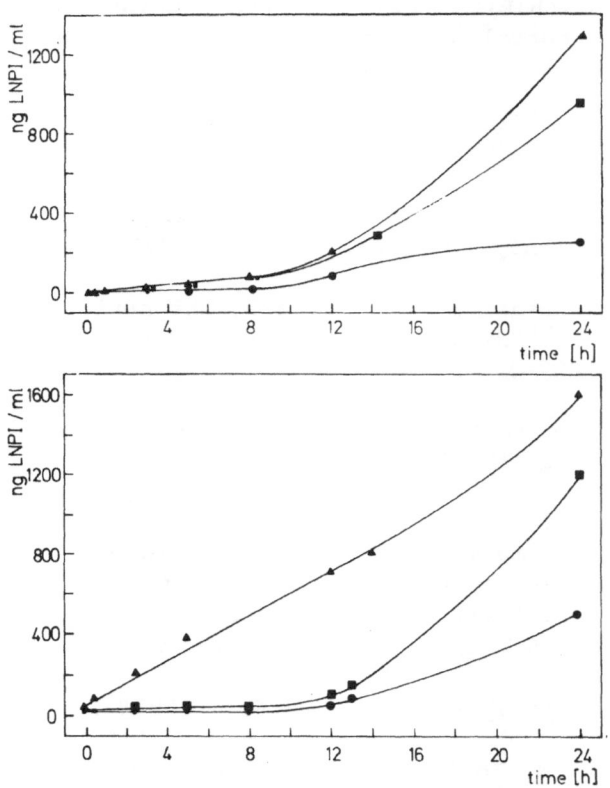

Fig.4. In vitro stimulation of PMN leukocytes.
▲, zymosan (1 mg/(ml x 10^3 cells); ■, E. coli (5 x
10^7); ●, endotoxin (10 ug/ml).

The reason why the inhibitor is not as easily released may be its localization in the leukocytes. Leukocyte elastase is stored in lysosomal granules of the cell, whereas LNPI has been found only in the cytosol. Whether this release reaction is of physiological significance has to be clarified in future studies.

ACKNOWLEDGEMENT

This work was supported by a grant of the Deutsche Forschungsgemeinschaft (We 1002/2-1). We wisk to.thank Prof. Dr. H. Fritz for many helpful discussions during work.

ABBREVIATIONS

LNPI, an inhibitor isolated from porcine leukocytes directed against neutral proteinases

REFERENCES

1. J. T. Dingle, Lysosomes: a laboratory handbook, Elsevier North Holland, Amsterdam, pp.323 (1977).

2. S. J. Klebanoff and R. A. Clark, The neutrophil: function and clinical disorders, Elsevier North Holland, Amsterdam, pp.810 (1978).

3. T. P. Stossel, The mechanism of Phagocytosis, J. Reticulo-enthel. Soc. 19: 237 (1976).

4. H. Fritz, M. Jochum, K. H. Duswald, H. Dittmer, H. Kortmann, S. Neumann and H. Lang, Granulocyte proteinases as mediators of unspecific proeolysis in inflammation: A review, in: "Cysteine Proteinases and their Inhibitors" V. Turk, ed., Walter de Gruyter, Berlin (1986).

5. J. Bieth, Front Matrix Biol. 6: 1 (1978).

6. H. Fritz, M. Jochum, K. K. Duswald, H. Dittmer, H. Kortmann, H. Neumann and H. Lang, H., in: "Seleced Topics in Clinical Enzymology, Vol. 2", D. M. Goldberg and M. Werner, eds., pp. 305, Walter de Gruyter Verlag, Berlin (1986).

7. R. Geiger, S. Sokal, G. Trefz and H. Fritz, Determination of leukocyte elastase LNPI complexes and LNPI by enzyme Immunoassays, leukocyte elastase-inhibitor complexes in porcine blood, III. J. Clin. Chem. Clin. Biochem., in press (1987).

8. M. Kopitar and D. Lebez, Intracellular distribution of neutral proteinases and inhibitors in pig leukocytes, Eur. J. Biochem. 56: 571 (1975).

9. G. Trefz and R. Geiger, Neutral proteinase inhibitors in PMN leukocytes. II. isolation and characterization of a neutral proteinase inhibitor from porcine PMN leukocytes, Biol. Chem. Hoppe Seyler, in press (1987).

10. M. Jochum and A. Dwenger, Sequential release of lysosomal inflammation mediators and chemiluminescence response of stimulated polymorphonuclear leukocytes, Fresenius Z. Anal. Chem. 324: 362 (1986).

PLASMA DERIVATIVE REPLACEMENT THERAPY IN DISS.INTRAVASC.COAG.(DIC) IN-

DUCED BY SEPTIC DISORDERS WITH HIGHLY ELEVATED ELASTASE α1AT-COMPLEXES

Rudolf Egbring, Rainer Seitz, Martin Wolf, Klaus Havemann
Lydia Lerch, Heiner Blanke, and G. Fuchs

Dept. of Internal Medicine/Div. of Haematology/Oncology
and Div. of Cardiology, Baldingerstrasse, 3550 Marburg,
West Germany

In patients with infectious diseases and septic complications as a
result of the inflammatory response a severe bleeding tendency caused by
disseminated intravascular coagulation (DIC) may often be responsible
for lethal outcome. Three proteolytic enzymes are possibly involved in
the platelet coagulation factor and inhibitor turnover, cell membrane
degradation and endothelial cell damage: Thrombin, elastase and plasmin,
which can be determined immunologically as proteinase inhibitor
complexes (PIC) by ELISA technique (Behringwerke, Merck) and rocket
electrophoresis with antisera against the α2antiplasmin-plasmin complexes, (α2AP-PL) as has recently been described[4].

The DIC-fibrinolysis (DIC-F) syndrome, which was first described by
Schneider 1947 and later intensively investigated by several authors,
may arise from extrinsic or intrinsic activation of blood coagulation or
enhancement of the fibrinolytic system and since the seventies a further
mechanism of coagulation factor and inhibitor consumption has been taken
into consideration: the denaturation of plasma factors and inhibitors by
released cell proteinases such as elastase and cathepsin G from mononuclear and polymorphonuclear leukocytes.

Despite the high potential of natural proteinase inhibitors (antithrombin III, α1antitrypsin, α2antiplasmin and α2macroglobulin) in the
plasma, the three enzymes may cause a consumption of coagulation factors
if the coagulation or the fibrinoytic system is pathologically activated
or lysosomal enzymes are released from PMN leukocytes. Elastase release
may be induced by different stimuli (factor XIIa and others) (Table 1).

Thus thrombin and plasmin induce limited proteolysis of fibrinogen
and other coagulation factors, whereas leukocyte elastase may be responsible for unlimited degradation of several clotting proteins in vivo.
The possible interaction of granulocyte neutral proteases during PMN-L-
endothelial cells interaction was summarized by Havemann and Gramse in
1983.

Table 1.

Importance of polymorphonuclear granulocytes for inflammatory

response and protein degradation

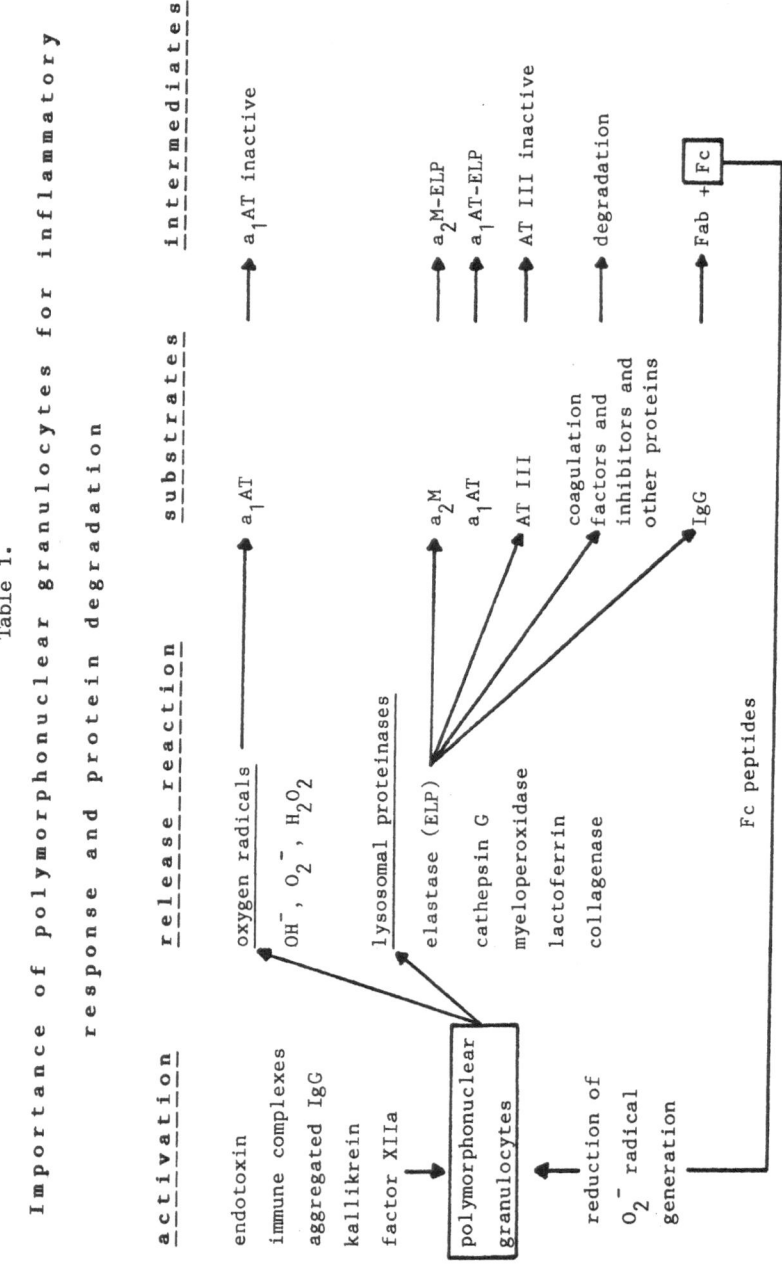

RESULTS

The distribution of the three PICs in infectious diseases is given in Table 2.

Table 2. Proteinase inhibitor complex pattern in 97 patients with infectious disorders, septicemia and shock symptoms

α1AT-ELP	AT III-Thr	α2AP-PI	Number of patients	%
+	+	+	18	18.6
+	+	−	20	20.6
+	−	+	7	7.2
+	−	−	24	24.7
−	−	−	11	11.3
−	−	+	6	6.2
−	+	+	4	4.1
−	+	−	7	7.2

total: 97 patients
+ means α1AT-ELP: > 300 ng/ml
+ means AT III-thrombin: > 5.0 ng/ml
+ means α2AP-PI: > 25%

In more than 60% of all patients αlantitrypsin-elastase (α1AT-ELP) is found to be increased, while thrombin-antithrombin III (TAT) is less often involved. In 18% of patients, all three PICs are increased so that one can assume that all three proteolytic mechanisms are involved.

In septic shock TAT as well as α1AT-ELP was highly increased, whereas in other forms of shock, such as cardiac shock, TAT alone was predominantly increased to high values.

In several patients with infectious diseases only α1-AT-ELP could be found to be elevated. In these cases elastase mediated proteolysis may be the only cause of the consumption of coagulation factors and inhibitors.

Table 3A lists 15 patients with infectious diseases and the behavior of the PICs. In contrast table 3B demonstrates the behavior of PICs in 12 patients with cardiogenic shock (see also [4]). Table 4 shows the PICs pattern in various diseases.

The treatment of the severe disseminated intravascular coagulation syndrome is still somewhat controversial.

To stop the consumption mechanism no other therapeutic treatment other than anticoagulant treatment has yet been applied.
Patients with severe septic disorders induced by either gram positive, gram negative bacterial, viral or other infections can exhibit severe hemostatic disturbances which are caused by an increased consumption of clotting proteins and their particular inhibitors.

In the present study - in addition to heparin treatment - the patients received a plasmaderivative replacement (ATIII concentrates, FFP, and sometimes, if necessary, PBSB and factor XIII).

Table 3A. Behavior of AT III-Thrombin and α1AT-ELP
           Complexes in 15 Patients with Septicemia
           and DIC

Patients	TAT ng/ml	α1AT-ELP ng/ml
A. E.	22.0	3315
B. K.	170.0	311
D. O.	115.0	918
E. H.		1000
D. M.	10.0	300
G. J.	4.0	2040
H. M.	9.0	286
H. G.	19.0	1630
K. K.	85.0	1200
P. E.	9.5	296
R. ±.	95.0	2100
S. H.	9.5	1428
S. E.	40.0	2900
S. C.	10.0	806
S. H.	8.5	1428
S. H.	12.0	520

Table 3B. Proteinase-Inhibitor-Complex Elevation
           (Thrombin-Antithrombin III, ELP α1-Anti-
           trypsin) in 12 Patients with Cardiac Shock

Patients	TAT ng/ml	α1AT-ELP ng/ml
H. W.	16.0	377
C. R.	50.0	230
D. B.	14.0	525
A. R.	11.0	204
L. S.	15.0	41
C. H.	120.0	469
S. H.	34.0	673
H. L.	22.0	515
E. B.	55.0	525
A. S.	14.0	311
F. B.	32.0	214
(complicated by		
septicemia)	(20.0)	(1581)
W. U.	20.0	224

AT III replacement is of importance to increase not only the anti-coagulant effect of heparin, since according to Reeve (1982), the amount of circulating thrombin is dependent on the equation 1:AT III[2] but also to inhibit the factor XIIa induced elastase release from PMN-L[8]. It also influence the elastase mediated proteolysis of platelets and endothelial cells surface proteins[6].

The amount of plasma derivative substitution cannot be standardized, because its effect in limiting the consumption mechanism is dependent on the basic illness.

Recently[4] a young boy suffering from meningococcal septicemia was successfully treated with this replacement therapy. The 4-year-old boy was given 11,900 units ATIII concentrate, 5.05 l FFP and 17,000 units plasminogen.

The time course of PICs and other coagulation and clinical data after plasma-derivative replacemant therapy for this boy and other patients with septic disorders have been published elsewhere [3,4].

In a prospective pilot study 30 patients with septic shock were treated with plasma derivatives, while 15 patients did not receive a replacement therapy. From the first group 13 patients died, the causes of death in three of them were complicated by carcinoma, liver cirrhosis, and chronic renal failure.

In contrast, 13 of 15 patients, who did not undergo replacement therapy died during septic shock.

Subsequent to the replacement therapy the elevated PICs decreased and platelet counts increased, as figure 1. demonstrate. Figure 1 shows the behavior of platelet counts , TAT and α1AT-ELP complexes in 8 patients with septicemia after replacement therapy, whereas platelets increase the PICs values decreases.
So far more than 40 patients have been treated in the above-mentioned way.

As in all previously discussed cases [3,4], the platelets increased within three to 6 days and the PICs decreased shortly after beginning the replacement therapy.

In the case of DIC, it seems likely that after AT III substitution the intravascularly developing thrombin can more easily be eliminated.

According to investigations carried out by Wachtfogel, factor XIIa induces a release of lysosomal enzymes from leukocytes [8]. Therefore AT III replacement may additionally influence the proteinase release, because factor XIIa can be inhibited by the thrombin-inhibitor.

In patients with septic complications, α2macroglobulin is found to be diminished. This is the reason why we combined the ATIII replacement with fresh-frozen-plasma. Other proteins not available as concentrates are present in FFP (Factor V, Plasminogen, α1AT, α2AP, α2M).

In in vitro models bacterial proteinases have been shown to destroy α1AT, activate prothrombin and others. As our in vitro studies have shown so far, α2M may inhibit the proteolytic activity of several bacteria. This could be demonstrated by testing the fibrinolytic activity of several bacteria taking fibrin-plates with or without α2macroglobulin.

Table 4   Thrombin-Antithrombin III (TAT) and α1Antitrypsin-
Elastase (α1AT-ELP) in Patients with Various Diseases
at Onset of Illness

Patients with	n	TAT -- 2 ng/ml*		α1AT-ELP -- 300 ng/ml*
septic infection and/or shock	45	>10	<200**	> 1000**
not septic inf.	52	< 5		< 1000
cardiac shock	19	≈ 45,0		≈ 700
myocardial infarction without shock	65	≈ 8,0		≈ 109
deep vein thrombosis	18	≈ 9,0		≈ 155
deep vein thrombosis and embolism	22	≈12,0		≈ 174
acute hepatic failure	5	≈100		≈ 490
selected patients with lung cancer	8	≈ 50		≈ 250

* normal values, ≈ mean values, ** mean values see Seitz

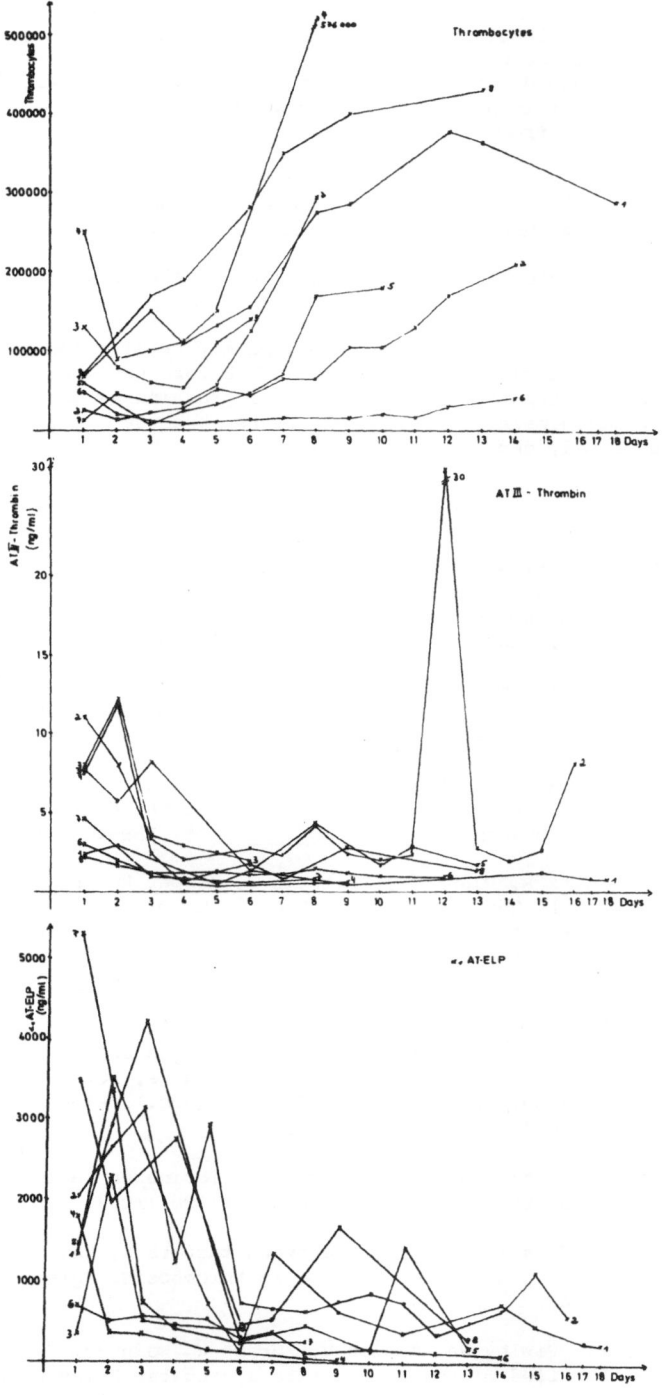

Fig. 1.

CONCLUSION

Elevated proteinase-inhibitor complexes in the plasma of patients with infectious diseases indicate
- activation of blood coagulation and fibrinolysis system with intravascular thrombin and/or plasmin generation or
- elastase release from PMN leukocytes.

The AT III replacement may be necessary to
- stop intravascular coagulation and prevent microcirculatory disturbances, because intravascular thrombin formation leads to a decrease in active inhibitor concentration.
- influence the factor XIIa induced proteinase release reaction from leukocytes.

The replacement of fresh-frozen-plasma may be of value for the substitution of coagulation factors and inhibitors not available as concentrates. The hypothesis that bacterial proteinases may be involved in the consumption mechanism is set up by us, because α2macroglobulin is sometimes extremely decreased in patients with septic infections.

REFERENCES

1. Blanke H, Praetorius G, Leschke M, Seitz R, Egbring R, Strauer BE: Die Bedeutung des Thrombin-Antithrombin III-Komplexes in der Diagnostik der Lungenembolie und der tiefen Venenthrombose - Vergleich mit Fibrinopeptid A, Plättchenfaktor IV und ß-Thromboglobulin. in Druck

2. Egbring R, Schmidt W, Havemann K, Fuchs G: Demonstration of granulocytic proteases in plasma of patients with acute leukemia and septicemia with coagulation defects. Blood 49: 219-231 (1977).

3. Egbring R, Seitz R, Fuchs G, Karges HE: The clinical significance of elastase-induced degradation of clotting factors and inhibitors in patients with septicemia. Symp. Biol. Hun. 25: 389-409 (1983).

4. Egbring R, Seitz R, Blanke H, Leititis J, Kesper HJ, Burghard R, Fuchs G, Lerch L: The Proteinase Inhibitor Complexes (Antithrombin III-Thrombin, α2Antiplasmin-Plasmin and α1Antitrypsin-Elastase) in Septicemia Fulminant Hepatic Failure and Cardiac Shock: Value for Diagnosis and Therapy Control in DIC/F Syndrome. Behring Inst. Mitt. 79: 87-103 (1986).

5. Havemann K, Gramse M: Physiology and pathophysiology of neutral proteinases of human granulocytes. In: Proteases, Potential role in Health and Disease: Hörl WA and Heidland A (eds.); Plenum Publishing Corp., p. 1 (1984).

6. Kornecki E: Elastase-induced exposure of fibrinogen receptors in human platelets. Circulation 70, Suppl. 11: 98 (1984).

7. Reeve EB: Steady state relations between factors X, Xa, wII, IIa, antithrombin III and α2macorglobulin in thrombosis. Thrombos. Res. 18: 19-31 (1980).

8. Wachtfogel YT, Pixley RA, Kucich V, Abrams W, Weinbaum G, Schapira M, Colman RW: Purified plasma factor XIIa aggregates human neutrophils and releases elastase. Circulation 70, Suppl. II: 352 (1984).

NEUTROPHIL ELASTASE, THROMBIN AND PLASMIN IN SEPTIC SHOCK

R. Seitz, M. Wolf, R. Egbring, and K. Havemann

Div. of Internal Medicine/Dept. of Hematology/Oncology
Baldingerstrasse
D-3550 Marburg, West Germany

INTRODUCTION

In septic shock both bleeding and clotting may occur: a hemorrhagic diathesis due to a lack of coagulation factors as well as disseminated intravascular coagulation (DIC) leading to impaired organ perfusion. Proteolytic systems appear to be involved in both the enhanced turnover of hemostatic proteins and the damage of microcirculation. The classical concept of DIC implicates the action of thrombin and plasmin, the key enzymes of coagulation and fibrinolysis, respectively. In the past decade evidence has been accumulated [1,2] that neutral proteases from neutrophil granulocytes may contribute to the above mentioned complications, and thus influence the prognosis of septic shock. The appearance of active proteases in the plasma is immediately followed by the formation of complexes with their specific inhibitors. Such proteinase-inhibitor complexes (PIC) were determined in 43 patients with septic shock.

PATIENTS AND METHODS

In this study 43 patients with septic shock were examined. On occurrence of coagulation defects 29 of these patients received a substitution [3] of hemostatic proteins and inhibitors by means of antithrombin III (ATIII) concentrate and fresh frozen plasma. Out of the 43 patients 17 survived, 15 of them on substitution.

The PIC were determined immunologically: the complex α_1-antitrypsin-elastase (α_1AT-ELP) by ELISA (Merck, Darmstadt), antithrombin III-thrombin (ATIII-Thr) by enzyme immunoassay (EIA), kindly made available for this study by N. Heimburger and H. Pelzer (Behringwerke, Marburg), and α_2-antiplasmin-plasmin (α_2AP-PL) by rocket immunoelectrophoresis, with antibodies donated by H.E. Karges (Behringwerke). The calibration curve for α_2AP-PL was obtained by activation of pooled normal plasma with 1000 iU streptokinase per ml (100%-value). Furthermore the prothrombin time (Quick value), the thrombocyte count and as additional clinical parameter the serum creatinine level were determined. For the statistical analysis (Mann-Whitney U-test and Wilcoxon rank sum W-test, Spearman correlation) the values obtained on three points of time were taken into consideration: on admission, in the early course after 48 to 72 hours and towards the endpoint of the disease.

RESULTS

As shown in table 1, on admission Quick values and thrombocyte counts as well as creatinine levels were not significantly different between survivors and nonsurvivors. These parameters were statistically different at the endpoint which was not a consequence of progressive deterioration in the dying patients, but a result of clear improvement in the survivors. The PIC ATIII-thr and α_2AP-PL were not different between the groups. However, α_1AT-ELP was significantly higher on admission in the survivors, but lower at the endpoint reaching almost the 5% significance level.

While there were no correlations between α_2AP-PL and the other parameters, for both ATIII-Thr (table 2) and α_1AT-ELP (table 3) significant correlations were found.

Table 1. Mean values + SEM of the 17 surviving (upper values) and the 26 dying patients (lower values)

	admission	after 48-72 h	endpoint
Creatinine (mg/dl)	2.8 + 0.7 n.s. 3.7 + 0.6	2.2 + 0.7 p = 0.0077 4.5 + 0.5	1.5 + 0.4 p = 0.0017 4.5 + 0.8
Quick (% of normal)	56.8 + 6.7 n.s. 59.4 + 5.5	68.7 + 5.8 n.s. 61.4 + 5.1	80.6 + 5.2 p = 0.0288 63.8 + 5.0
Thrombocytes ($\times 10^3$/mm^3)	95.3 + 24.9 n.s. 118.9 + 15.1	127.8 + 31.2 n.s. 104.8 + 17.1	313.6 + 47.3 p = 0.0025 133.9 + 25.6
α_1AT-ELP (ng/ml)	2458 + 348 p = 0.0170 1291 + 295	1543 + 267 n.s. 1369 + 366	315 + 54 (p = 0.0599) 652 + 161
ATIII-Thr (ng/ml)	18.8 + 6.3 n.s. 16.5 + 7.2	7.0 + 2.6 n.s. 6.5 + 1.3	2.7 + 0.5 n.s. 3.7 + 0.8
α_2AP-PL*	11.1 + 7.6 n.s. 21.9 + 6.4	10.0 + 7.2 n.s. 20.3 + 6.1	13.9 + 9.1 n.s. 18.1 + 7.1

*unit see methods

Table 2. Correlations of ATIII-Thr with other parameters

	admission	after 48-72 h	endpoint
Creatinine	-	r = 0.3334 p = 0.045	-
Quick	r = -0.4431 p = 0.003	r = -0.4110 p = 0.009	-
Thrombocytes	r = -0.3595 p = 0.024	r = -0.4110 p = 0.009	r = -0.4589 p = 0.005
α_1AT-ELP	-	r = 0.3552 p = 0.027	-

Table 3. Correlations between α_1AT-ELP and other parameters

	admission	after 48-72 h	endpoint
Creatinine	-	r = 0.5279 p = 0.008	r = 0.7854 p = 0.001
Quick	-	-	r = -0.4038 p = 0.020
Thrombocytes	r = -0.3606 p = 0.042	-	r = -0.3402 p = 0.048
ATIII-Thr	-	r = 0.3552 p = 0.027	-

DISCUSSION

In our septic shock patients, the final outcome could be predicted on admission neither by the alterations of coagulation (expressed by Quick value and thrombocyte count, which are still clinically useful parameters), nor by the degree of renal insufficiency.

In our patients (the majority of whom received a substitution of plasma derivatives) α_1AT-ELP was the only parameter which was different on admission. Surprisingly, the initial values were significantly higher in the survivors. This result cannot be explained at present. Nevertheless it indicates that a high initial α_1AT-ELP level ist not necessarily a sign of bad prognosis. The α_1AT-ELP values decreased in both groups during the course, which may be a result of the substitution[4]. The decrease of the α_1AT-ELP levels, however, was much more pronounced in the survivors who had lower levels at the end. The complex ATIII-Thr was elevated on admission and subsequently decreased in both groups; α_2AP-PL was increased throughout the observation period.

In consideration of the correlations between the PIC and the other parameters, the results given in the tables 2 and 3 suggest that ATIII-Thr as an indicator of an activated coagulation is of greater importance in the early course of the disease, while sigingficant correlations of α_1AT-ELP were found rather at the endpoint. It is of particular interest that significant correlations were obtained not only with coagulation parameters, but also with serum creatinine. This suggests that neutrophil elastase may contribute to the microcirculatory failure. This result is in accordance with experimental data showing the role of elastase in endothelial cell injury triggered by endotoxin[5].

In conclusion, the data suggest that at the beginning of septic shock the elastase release may be considerable, but the more important event is the activation of coagulation. An activation of fibrinolysis is found during the whole course without relation to the outcome. In the advanced stage of septic shock the pathogenetic significance of neutrophil release seems to be growing, as the correlations at the endpoint indicate. The plasma levels of neutrophil elastase appear to be of prognostic significance. This might be due to proteolytic effects of elastase on both, hemostatic proteins and endothelial cells.

Acknowledgements: the authors wish to thank Mrs. L. Lerch and Miss I. Rohner for excellent technical assistance, Dr. Zöfel for his help in statistical evaluation and Mrs. S. Harnisch for typing the manuscript.

REFERENCES

1. Egbring R, Schmidt W, Fuchs G, Havemann K: Demonstration of granulocytic proteases in plasma of patients with acute leukemia and septicemia with coagulation defects. Blood 49: 219 (1977).

2. Duswald KH, Jochum M, Schramm W, Fritz H: Released granulocytic elastase: An indicator of pathobiochemical alterations in septicemia after abdominal surgery. Surgery 98: 892 (1985).

3. Seitz R, Egbring R, Radtke KP, Wolf M, Fuchs G, Fischer J, Lerch L, Karges HE: The clinical significance of $\alpha_1$antitrypsin-elastase (α_1AT-ELP) and $\alpha_2$antiplasmin-plasmin (α_2AP-PL) complexes for the differentiation of coagulation protein turnover: Indications for plasma protein substitution in patients with septicemia. Int. J. Tissue React. 7: 321 (1985).

4. Seitz R, Wolf M, Egbring R, Radtke KP, Liesenfeld A, Pittner P, Havemann K: Participation and interactions of neutrophil elastase in hemostatic disorders of patients with severe infections. Eur. J. Haematol. 38: 231 (1987).

5. Smedly LA, Tonnesen MG, Sandhaus RA, Haslett C, Guthrie LA, Johnston RB, Henson PM, Worthen GS: Neutrophil-mediated injury to endothelial cells: Enhancement by endotoxin and essential role of neutrophil elastase. J. Clin. Invest. 77: 1233 (1986).

ELASTASE-α_1-PROTEINASE INHIBITOR: AN EARLY INDICATOR OF SEPTICEMIA AND BACTERIAL MENINGITIS IN CHILDHOOD

Ch.P. Speer*, M. Rethwilm and M. Gahr

Dept. of Pediatrics, University of Göttingen
Humboldtallee 38
D-3400 Göttingen

INTRODUCTION

Neonatal septicemia and meningitis are still a significant cause of morbidity and mortality[1,2,3], in addition, systemic bacterial infections as well as bacterial meningitis in childhood are potentially lifethreatening diseases that require prompt recognition and early and appropriate therapy. Recently elastase-α_1-proteinase inhibitor complex (E-α_1-PI) has been introduced as a valuable indicator of systemic bacterial infections in neonates[4].

In this report we provide more evidence that E-α_1-PI is a sensitive and rapid responsive indicator of septicemia and bacterial meningitis in infancy and childhood.

METHODS

272 preterm and term newborn infants who were admitted to the intensive care unit, the Department of Pediatrics, Göttingen, West-Germany, were included in the study, and were assigned to one of the three groups.

Group one included 19 infants with proven sepsis and/or meningitis; all patients had at least one positive blood and/or CSF culture and clinical course consistent with diagnosis.

Group two consisted of 22 neonates with suspected sepsis or meningitis, and who were given antibiotics. The predisposing risk factors have been recently described[4].

Group three included 231 neonates (102 preterm, 129 term) without history of predisposing factors and without signs of infection, 44 of these infants received prophylactic antibiotics.

---

*Supported by grant Sp 239/2-2 from Deutsche Forschungsgemeinschaft.

In 24 infants and children with bacterial septicemia and/
or meningitis (treated at the University Hospital of Göttingen)
E-α_1-PI concentrations in plasma and CSF were evaluated (sepsis:
n = 7, sepsis and meningitis: n = 6, meningitis: n = 11). All
patients had at least one positive blood and/or CSF culture and
clinical course consistent with diagnosis. Bacterial pathogens
were isolated by standard microbiologic methods. In addition,
15 patients with aseptic meningitis were included in this
study. The criteria for aseptic meningitis were (1) >10 WBC/mm³
CSF, (2) negative blood, CSF and urine bacterial cultures,
(3) no antibiotic therapy prior to diagnosis, (4) a negative
tuberculin skin test and (5) clinical symptoms of meningitis.

Reference values of E-α_1-PI in plasma were obtained from
142 healthy infants and children without any signs of infection
ranging in the age from 5 weeks to 14 years; none of these
patients was treated with antibiotics.

To define the normal range, E-α_1-PI was determined in CSF
of 62 patients (5 weeks to 17 years) undergoing initial evaluation for suspected meningitis or febrile convulsions, none of
these patients had >10 WBC/mm³ and was found to have bacterial
meningitis.

Plasma and CSF specimens were obtained as part of the
routine laboratory analysis; samples were centrifuged at
3 000 rpm for 10 min at 18° C within one hour of collection.
The enzyme-linked immunoassay for E-α_1-PI was performed according to the instructions of Neumann and coworkers[5] and the
instructions of the manufacturer (Merck, Darmstadt, West-
Germany).

RESULTS

E-α_1-PI plasma levels in neonates. As shown in Fig. 1 all
neonates with sepsis and/or meningitis had increased plasma
levels above the range of normal at time of diagnosis. The
gestational age ranged from 27 to 41 weeks.

The reference range for the plasma concentration of E-α_1-
PI during the neonatal period has recently been published[4].

Microorganisms responsible for neonatal septicemia were
β-hemolytic streptococcus group B (GBS, n = 4), Escherichia
coli (n = 7), Klebsiella-Enterobacter species (n = 5), Staph.
epidermidis (n = 1) and others (n = 2).

Only a weak correlation between the total number of neutrophils and the concentration of E-α_1-PI has been found (r = 0.64).
The lowest concentration of E-α_1-PI was observed in a patient
with Enterobacter septicemia, concentration of E-α_1-PI was
247 μg/l (945 total neutrophils/μl) at 20 days of age.

18 of 22 neonates (81%) with suspected systemic bacterial
infection had increased E-α_1-PI levels at diagnosis. After
initiation of antibiotic therapy a clear decrease in E-α_1-PI
plasma concentration was observed in all patients with proven
and suspected septicemia who survived this lifethreatening
disease (Fig. 2). So far, increased plasma levels also have
been observed in 6 neonates with meconium aspiration

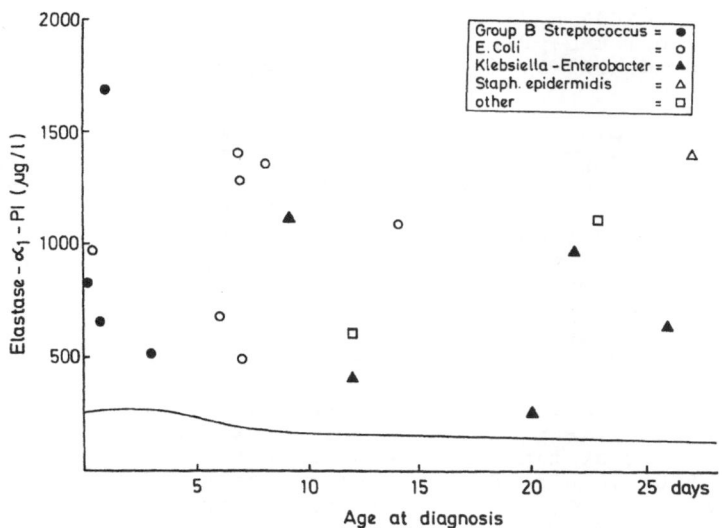

Fig. 1. Plasma concentrations of E-α_1-PI in 19 neonates with septicemia and/or meningitis at age of diagnosis. Solid line: upper range of normal. The symbols represent microbial pathogens responsible for septicemia in individual patients.

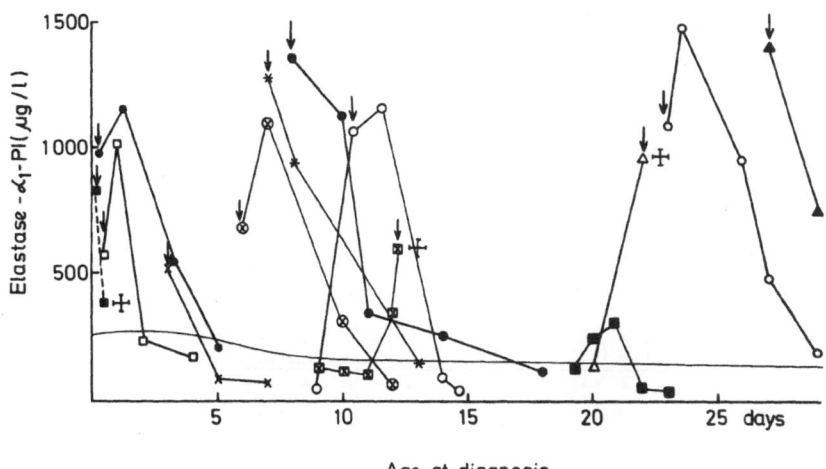

Fig. 2. E-α_1-PI concentrations (plasma) in representative patients with neonatal septicemia. The symbols represent individual cases. Line: upper limit of normal values.

(range: 412-1031 µg/l E-α_1-PI) and neonatal pneumonia (range: 384-1020 µg/l E-α_1-PI). All neonates with hyaline membrane disease grade III-IV (n = 15) had E-α_1-PI levels within the normal range.

E-α_1-PI plasma levels in infants and children. The reference values for the plasma concentrations of E-α_1-PI derived from 142 infants and children (89 boys, 53 girls) without signs of infections are shown in Fig. 3, the range of age was 5 weeks to 14 years. In view of the relatively small number of data points in many time segments, preliminary bounds were drawn by visual inspection. There was no correlation between age of the infants and children and plasma concentration of E-α_1-PI (maximum concentration: 190 µg/l E-α_1-PI, minimum concentration: 25 µg/l E-α_1-PI).

As shown in Fig. 3 all patients (n = 24) with septicemia and/or meningitis exhibited significantly elevated E-α_1-PI plasma levels at time of diagnosis, even those with neutropenia (1137 ± 570; mean ± SD). In a seven month old boy with septicemia caused by N. meningitidis the concentration of E-α_1-PI was 1040 µg/l (720 total neutrophils/mm^3), in a three month old boy with group B streptococcal septicemia E-α_1-PI level was 444 µg/l (506 total neutrophils/mm^3).

There was a weak correlation between the total number of neutrophils and the concentration of E-α_1-PI in plasma (r = 0.48) To date no overlapping has been observed between the plasma values from patients with septicemia and/or meningitis and those with aseptic meningitis (range 25-291 µg/l E-α_1-PI, n = 10).

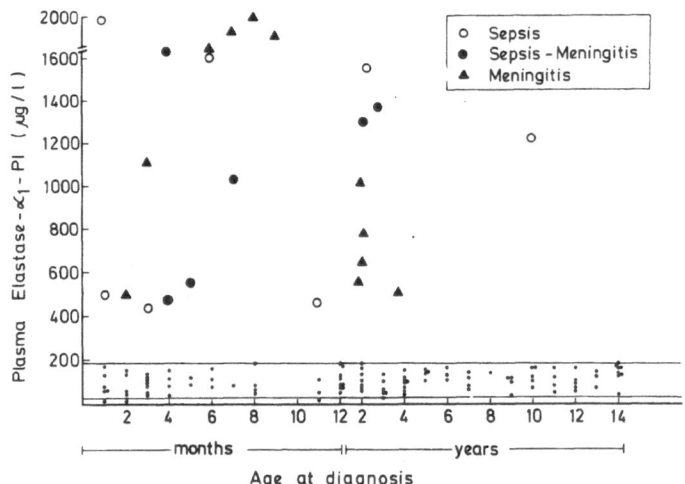

Fig. 3. Plasma concentrations of E-α_1-PI in 24 infants and children with septicemia and/or meningitis at age of diagnosis. Two solid lines, range of normal values.
o Patients with septicemia.
● Patients with septicemia and meningitis.
▲ Patients with meningitis.

A clear decrease of E-α_1-PI plasma concentration was observed in nearly all patients with sepsis and/or meningitis after initiation of antibiotic therapy.

Increased concentrations of E-α_1-PI were observed in children with urinary tract infection (n = 7), pneumonia (n = 16) and bacterial enteritis (n = 7). In contrast normal

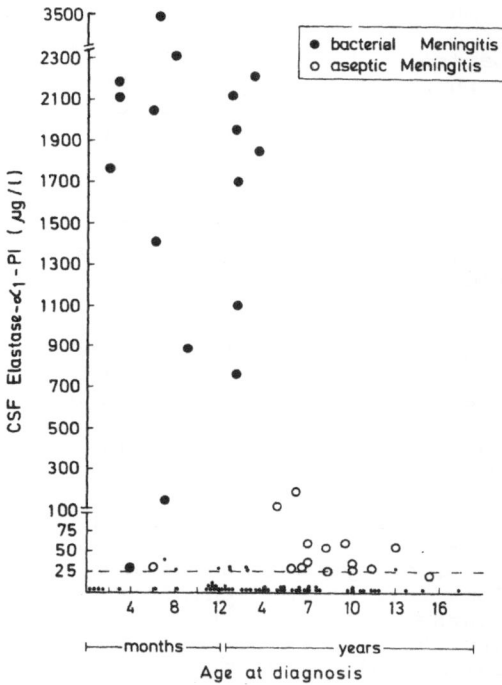

Fig. 4. E-α_1-PI determinations in CSF in patients with septic and aseptic meningitis at age of diagnosis. Solid line, lower limit of detection of the ELISA-assay used. 54 out of 62 reference values of E-α_1-PI were found to be below this limit (<25 μg E-α_1-PI/l).
● Septic meningitis.
o Aseptic meningitis.
· Controls.

to slightly elevated plasma concentrations of E-α_1-PI (maximum value 386 μg/l) could be detected in patients with systemic viral diseases (n = 18) and acute diarrhea caused by Rotavirus (n = 8).

E-α_1-PI CSF levels in infants and children with meningitis. The reference values of E-α_1-PI in CSF are demonstrated in Fig. 4. 16 of 17 patients with septic meningitis had significantly elevated levels of CSF E-α_1-PI (Fig. 4).

The lowest CSF concentrations of $E-\alpha_1-PI$ were observed in two patients with acute meningococcemia; at time of diagnosis 25 cells/mm³ (<10% PMN) and 23 cells/mm³ (50% PMN) were found and CSF cultures were positive for N. meningitidis in either patients. Besides these two patients no overlap was noticed between $E-\alpha_1-PI$ values from patients with bacterial meningitis (1599 ± 898.5; mean ± SD) and aseptic meningitis (range 25-194 µg $E-\alpha_1-PI$/l; 55 ± 48; n = 15); the difference between the groups was statistically significant (p <0.01). There was neither a good correlation between CSF total neutrophil count and CSF $E-\alpha_1-PI$ levels (r = 0.28) nor a correlation between plasma and CSF concentrations of $E-\alpha_1-PI$ (r = 0.16).

In three patients who had exessively increased CSF $E-\alpha_1-PI$, a second lumbar puncture was performed ten days after initiation of treatment; CSF $E-\alpha_1-PI$ concentrations were shown to be normalized in all children (<30 µg/l $E-\alpha_1-PI$).

DISCUSSION

The potential role of $E-\alpha_1-PI$ as an early indicator of neonatal systemic bacterial infections has been recently demonstrated[4,6,7].

In this study we provide more evidence that $E-\alpha_1-PI$ is a rapidly responsive indicator of bacterial infections during and beyond the neonatal period. All neonates and children with sepsis and/or meningitis had increased plasma levels of $E-\alpha_1-PI$ at diagnosis, even those with neutropenia. A normalization of $E-\alpha_1-PI$ levels was observed in all patients who recovered from the illness. In addition, CSF $E-\alpha_1-PI$ concentrations in bacterial meningitis were significantly increased when compared with aseptic meningitis. Our data suggest that $E-\alpha_1-PI$ is a highly sensitive indicator of systemic bacterial infections in childhood. In addition, this enzyme inhibitor complex is helpful in following the course of the illness. Since other local inflammations and bacterial infections cause an increase of $E-\alpha_1-PI$ in plasma (meconium aspiration, pneumonia, urinary tract infections etc.) specificity and predictive value of this test are rather low.

REFERENCES

1. G. H. McCracken, Jr., Perinatal bacterial diseases, in: "Textbook of Pediatric Infectious Diseases", R. D. Feigin, ed., W. B. Saunders, Philadelphia (1981).
2. G. S. Philip, "Neonatal sepsis and meningitis", G. K. Hall, Boston (1985).
3. Ch. P. Speer, D. Hauptmann, P. Stubbe, and M. Gahr, Neonatal septicemia and meningitis in Göttingen, West-Germany, Pediatr. Infect. Dis. 4:36 (1985).
4. Ch. P. Speer, A. Ninjo, and M. Gahr, Elastase-α_1-proteinase inhibitor in early diagnosis of neonatal septicemia, J. Pediatr. 108:987 (1986).
5. S. Neumann, G. Gunzer, N. Hennrich, and H. Lang, "PMN-Elastase Assay": Enzyme immunoassay for human polymorphonuclear elastase complexed with α_1-proteinase inhibitor, J. Clin. Chem. Clin. Biochem. 22:693 (1984).

6. F. K. Tegtmeyer and J. Otte, Early diagnosis of neonatal infections by leukocyte-elastase, Eur. J. Pediatr. 143:249 (1985).
7. Ch. P. Speer and M. Gahr, Leukozytenindizes, C-reaktives Protein und PMN-Elastase/α_1-Proteinaseinhibitorkomplex in der Frühdiagnostik der neonatalen Sepsis, Monatsschr. Kinderheilkd. 133:595 (1985).

SERUM PANCREATIC SECRETORY TRYPSIN INHIBITOR (PSTI) IN

SERIOUSLY INJURED AND SEPTIC PATIENTS

Hiroshi Tanaka, Michio Ogawa, Toshiharu Yoshioka,
and Tsuyoshi Sugimoto

Department of Traumatology
Osaka University Medical School

SUMMARY

Changes of serum PSTI in 49 patients with serious injury or sepsis were investigated. The patients were divided into three groups according to the maximum serum PSTI levels. The mean maximum PSTI was 14.3±5.7 ng/ml (mean ± SD) in Group 1 (17 patients), 61.9±25.6 ng/ml in Group 2 (19 patients) and 678.6±401.3 ng/ml in Group 3 (13 patients). Serum PSTI in Group 1 stayed within the normal range. Serum PSTI in Group 2 rose from the second or third hospital day, reached its maximum within seven days, and returned to normal thereafter. All 13 patients in Group 3 were suffering from either sepsis (8 patients) or irreversible shock (5 patients) and serum PSTI showed a remarkably high level. Judging by multiple organ failure (MOF) index, there was a significant correlation between the maximum serum PSTI and the severity of traumatic stress, tissue destruction or infection.

INTRODUCTION

Pancreatic secretory trypsin inhibitor (PSTI) was isolated from bovine pancreas by Kazal et al.[1] in 1948, and from human pancreatic juice by Haverback et al.[2] in 1960. Using a specific radioimmunoassay for human PSTI, the elevation of serum PSTI was found in patients with acute pancreatitis and chronic relapsing pancreatitis.[3,4] Recently PSTI was shown to be present in various tissues.[5] In addition, serum PSTI was found to rise in various pathological states,[6,7,8] under surgical stress[9] and in serious injuries.[10] However the physiological role of PSTI and the mechanism for the elevation of serum PSTI in various pathological states have not been fully clarified.

In this study, we tried to clarify the mechanism of the change of serum PSTI in seriously injured and septic patients by comparing serum PSTI levels with their clinical symptoms and laboratory data.

MATERIALS AND METHODS

From September 1986 to February 1987, serum PSTI levels were measured in 49 patients at the Department of Traumatology, Osaka University Medical School. Thirty-six patients were seriously injured, 9 patients had burns. Two trauma patients, two burn patients and four patients with panperitonitis after surgery developed septicemia. A total of 8 patients had septicemia. Diagnosis of sepsis was based on our own criteria as shown in Table 1. The patients with pancreatic injury and pancreatitis confirmed by CT scan and/or operation were excluded. Maximum serum PSTI levels were compared with the

Table 1. Criteria for Sepsis

Sepsis is diagnosed by two or more manifestations

1. Fever > 39°C for more than 3 days
2. Leucocyte counts > 30,000/μl
3. Positive limulus test
4. Positive blood culture
5. Positive culture of the same organisms in catheter tip and infectious focus

MOF index calculated at this time, which, as proposed by Goris et al.[11] evaluates the dysfunction of seven affected areas: lung, heart, liver, blood, gastrointestinal tract, central nervous system after multiple injuries and sepsis.

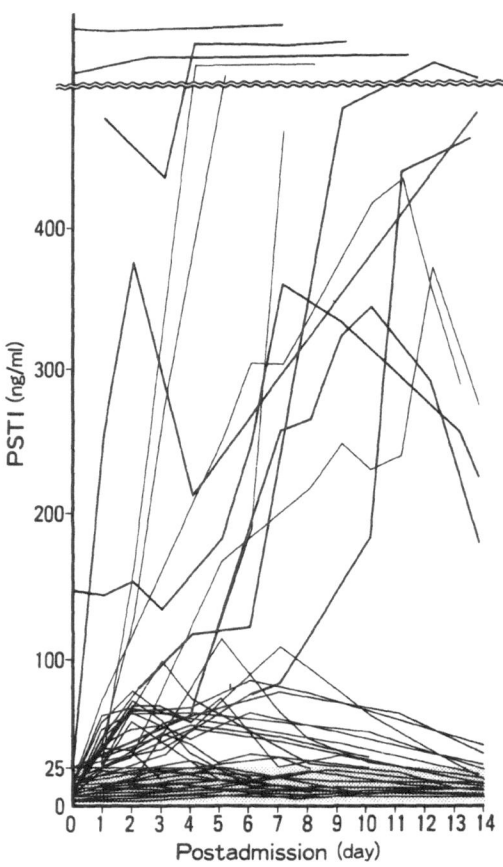

Fig. 1. Serial changes in serum PSTI

Blood samples were obtained on admission and periodically thereafter. The blood samples were immediately separated and stored at -40°C until they were assayed. Serum PSTI content was determined by a RIA Kit (Shionogi, Co. Osaka, Japan). The normal range was from 5.7 to 24.9 ng/ml (mean ± 2SD).

Total bilirubin, sGOT, creatinine, amylase (autoanalyzer), platelet count (microscopic method), α1-antitrypsin, α2-macroglobulin (single radial immunodiffusion) and fibrinogen (thrombin time method) were also measured. Respiratory index was calculated by AaDO2/PaO2.[12] Arterial blood gas was measured with an automatic gas analyzer.

The Wilcoxon rank sum test for comparing data was used for assessing differences among groups and the Separman rank correlation coefficient was used to test the association between the parameters. The level of significance was taken as p < 0.05 in a two-tailed test.

RESULT

Fig. 1 shows the serial changes in serum PSTI of all patients after admission. The patients were divided into three groups according to the maximum serum PSTI levels. The maximum serum PSTI of Group 1 was less than 24.9 ng/ml (17 patients), Group 2 24.9-300 ng/ml (19 patients) and Group 3 more than 300 ng/ml (13 patients). The mean maximum serum PSTI was 14.3±5.7 ng/ml (mean ± SD) in Group 1, 61.9±25.6 ng/ml in Group 2, and 678.6±401.3 ng/ml in Group 3. Serum PSTI was within normal range after admission in Group 1. Serum PSTI of Group 2 rose from the second or third hospital day, reached its maximum within seven days, and returned to normal thereafter. Serum PSTI of Group 3 rose remarkably, and remained at a high level.

Fig. 2 shows the relation of the maximum serum PSTI levels with the causes of traumatic stress. Abnormally high level was seen in Group 3. Five patients were most severly injured and had irreversible shock. The remainder had septicemia.

Fig. 3 shows the serial changes of serum PSTI in patients before and after onset of sepsis. Serum PSTI continued to rise after the onset of sepsis and, when the patients recovered from sepsis, the level of PSTI decreased.

Table 2 shows laboratory data of these patients when serum PSTI reached its maximum. Total bilirubin and the respiratory index rose significantly in Group 3. Creatinine clearance and platelet counts fell significantly in Group 3. However, there was no significant difference among groups in the level of amylase. Fibrinogen, α1-antitrypsin rose in all groups, but α2-macroglobulin was within normal range in all groups.

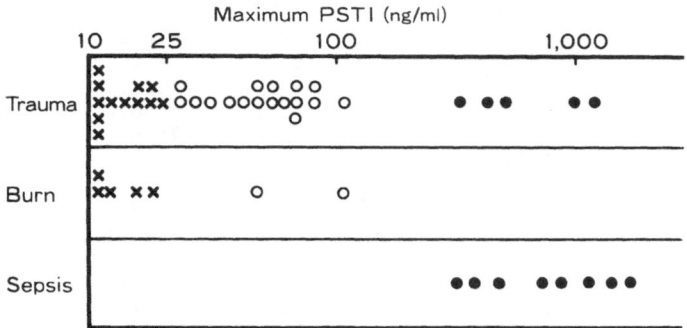

Fig. 2. Maximum levels of serum PSTI
(×) Group 1   (○) Group 2   (●) Group 3

495

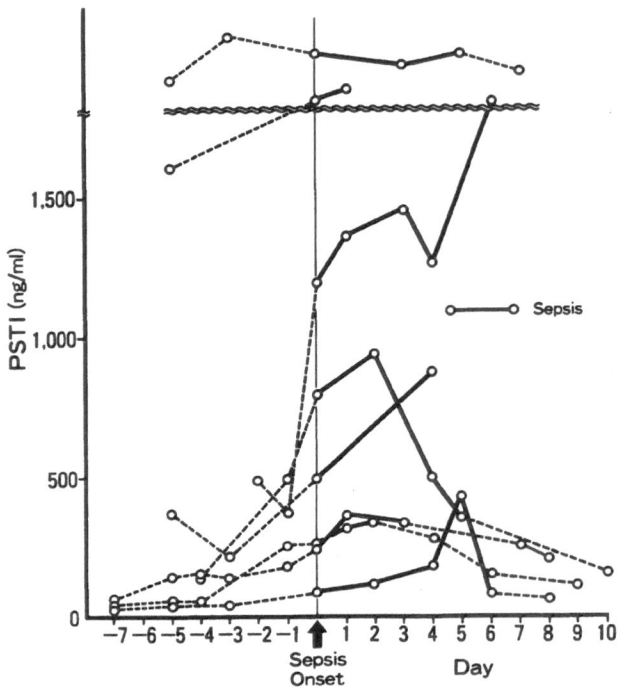

Fig. 3. Serial changes in serum PSTI in septic patients

Table 2. Laboratory Data when Serum PSTI Reached Its Maximum Value

	1 (n=17)	2 (n=19)	3 (n=13)
Total Bilirubin (mg/dl)*	1.1	0.9	5.0
	(0.4–2.0)	(0.3–2.4)	(0.6–22.8)
Creatinine Clearance (ml/min.)**	120	98	46
	(90–172)	(18–180)	(4–102)
Amylase (u/l)	48	46	54
	(10–90)	(18–150)	(18–280)
Respiratory Index*	0.25	0.40	2.05
	(0.05–1.15)	(0.08–0.90)	(0.50–2.80)
Platelet counts (10^4/mm^3)*	17.5	11.5	3.4
	(4.8–38.0)	(5.8–20.0)	(0.5–18.5)
Fibrinogen (mg/dl)	500	570	500
	(360–840)	(400–1000)	(280–1000)
α1-antitrypsin (mg/dl)	350	400	440
	(260–520)	(250–580)	(300–860)
α2-macroglobulin (mg/dl)	160	130	150
	(120–190)	(90–228)	(84–224)

Results are median values.   * p < 0.05   ** p < 0.1

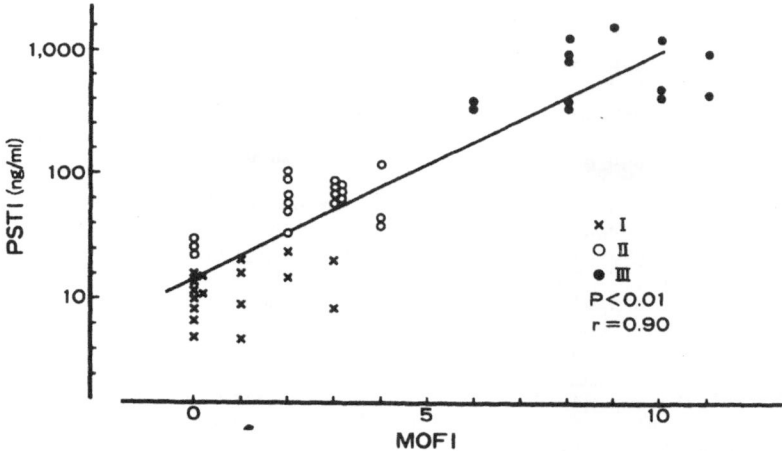

Fig. 4. Correlation between maximum serum PSTI and MOF index

To investigate the relationship between the elevation in serum PSTI and organ failure, the maximum PSTI in each group was compared to the MOF index (Fig. 4). There was a significant correlation between the maximum serum PSTI and the MOF index (r=0.90, p<0.01).

DISCUSSION

The present study demonstrates a high incidence of the elevation of serum PSTI following serious injury and sepsis. In uneventful cases with serious injury (Group 2), serum PSTI rose from the second or third hospital day and reached its maximum within seven days. After that it returned to normal. On the contrary, an abnormal rise was observed in patients under severe stress, such as sepsis or prolonged shock. In septic patients serum PSTI began to rise before the onset of sepsis, and reached its maximum thereafter. When patients recovered from sepsis, the level of serum PSTI decreased. Multiple organ failure is said to be associated with sepsis or severe injury.[11] There was a significant correlation between serum PSTI and the MOF index in the present study. It is suggested that serum PSTI rises according to the severity of traumatic stress, tissue destruction or infection.

Recently several investigators reported[13,14] that PSTI and epidermal growth factor (EGF) were similar in structure and function. EGF[15] is a unique growth promoting hormone and stimulates the growth and differentiation of various epidermal and epithelial cells. Ogawa et al.[16] reported that PSTI stimulated DNA synthesis in human fibroblast. To clarify the mechanism of PSTI in serious injury and sepsis, it is necessary to investigate biological significance of PSTI.

In this study we could not show where PSTI originated in cases of serious injury or sepsis. The measured PSTI was not from the pancreas, as pancreatic injury was excluded and the changes of serum amylase was not correlated with those of serum PSTI. Determination of the origin and the mechanism involved in the elevation of PSTI may shed a light on the healing process of trauma and sepsis at a cellular level.

REFERENCES

1. L. A. Kazal, D. S. Spicer and R. A. Brahinski, Isolation of a crystalline trypsin inhibitor-anticoagulant protein from pancreas, J. Am. Chem. Soc. 70:3034 (1948).

2. B. J. Haverback, B. Dyce, H. Bundy and H. A. Edmondson, Trypsin, Trypsinogen and trypsin inhibitor in human pancreatic juice, Amer. J. Med. 29: 424 (1962).

3. A. Eddeland and K. Ohlsson, A radioimmunoassay for measurement of human pancreatic secretory trypsin inhibitor in different body fluids, Hoppe-Seyler's Z. Physiol. Chem. 359: 671 (1978).

4. T. Kitahara, Y. Takatsuka, K. Fujimoto, S. Tanaka, M. Ogawa and G. Kosaki, Radioimmunoassay for human pancreatic secretory trypsin inhibitor: Measurement of serum pancreatic secretory trypsin inhibitor in normal subjects and subjects with pancreatic diseases, Clin. Chim. Acta 103: 135 (1980).

5. T. Shibata, M. Ogawa, K. Matsuda, K. Miyauchi, T. Yamamoto and T. Mori, Purification and characterization of pancreatic secretory trypsin inhibitor in human gastric mucosa, Clin. Chim. Acta 159: 27 (1986).

6. K. Matsuda, M. Ogawa, A. Murata, T. Kitahara and G. Kosaki, Elevation of serum immunoreactive pancreatic secretory trypsin inhibitor contents in various malignant diseases, Res. Commun. Chem. Pathol. Pharmacol. 40: 301 (1983).

7. A. Murata, M. Ogawa, K. Matsuda, N. Matsuura, G. Kosaki, M. Inoue, G. Ueda and K. Kurachi, Immunoreactive pancreatic secretory trypsin inhibitor in gynecological diseases, Res. Commun. Chem. Pathol. Pharmacol. 41: 493 (1983).

8. A. Lasson, A. Borgstrom and K. Ohlsson, Serum levels of immunoreactive PSTI in acute abdominal disorders, with special reference to a possible extrapancreatic PSTI production, Clin. Chim. Acta 161: 37 (1986).

9. K. Matsuda, M. Ogawa, T. Shibata, S. Nishibe, K. Miyauchi, Y. Matsuda and T. Mori, Postoperative elevation of serum pancreatic secretory trypsin inhibitor, The Am. J. Gastroenterol. 80: 694 (1985).

10. M. Ogawa, K. Matsuda, T. Shibata, Y. Matsuda, T. Ukai, M. Ohta and T. Mori, Elevation of serum pancreatic secretory trypsin inhibitor (PSTI) in patients with serious injury, Res. Commun. Chem. Pathol. Pharmacol. 50: 259 (1985).

11. R. J. Goris, P. A. Boekhorst and K. S. Johannes, Multiple-Organ Failure, Arch. Surg. 120: 1109 (1985).

12. M. A. Goldfarb, T. F. Ciurej, T. C. McAsian, W. J. Sacco, M. A. Weinstein and R. A. Cowley, Tracking respiratory therapy in the trauma patient, Am. J. Surg. 129: 255 (1975).

13. L. A. Scheving, Primary amino acid sequence similarity between human epidermal growth factor-urogastrone, human pancreatic secretory trypsin inhibitor, and members of porcine secretin family, Arch. Biochem. Biophys. 226: 411 (1983).

14. L. T. Hunt, W. C. Barker and M. O. Dayhoff, Epidermal growth factor: Internal duplication and probable relationship to pancreatic secretory trypsin inhibitor, Biochem. Biophys. Res. Commun. 60: 1020 (1974).

15. S. Cohen, Isolation of a mouse submaxillary gland protein accelerating incisor eruption and eyelid opening in the newborn animal, J. Biol. Chem. 237: 1555 (1962).

16. M. Ogawa, T. Tsushima, Y. Ohba, N. Ogawa, S. Tanaka, M. Ishida and T. Mori, Stimulation of DNA synthesis in human fibroblasts by human pancreatic secretory trypsin inhibitor, Res. Commun. Chem. Pathol. Pharmacol. 50: 155 (1985).

CHANGES IN PMN-ELASTASE IN BLOOD AND IN RENAL AND PLASMA KALLIKREIN-KININ SYSTEMS AFTER SEVERE BURN INJURY[*]

Gerd Bönner, W.Niermann, R.Festge, and U.Büchsler

Department of Internal Medicine II, University of Cologne, and Depart. of Plastic Surgery and Burns Treatment, Merheim Hospital, D-5000 Köln 91, FRG

INTRODUCTION

Severe burn injury is often complicated by sepsis, hypotensive shock, adult respiratory distress syndrome (ARDS) and acute renal failure (ARF). The pathogenetic mechanisms of these secondary complications are not at all clarified till today. It is suggested that kinins play a role in hypotensive reactions and in pulmonary edema, that increased PMN-elastase will participate in ARDS, and that an imbalance between renal kallikrein and renin might be promote ARF (1,2).
· This study now was performed in order to investigate the changes in PMN-elastase and in kallikrein-kinin systems in blood as well as the changes in renal kallikrein excretion in man after thermal injury by open fire.

PATIENTS AND METHODS

17 patients suffering from thermal injury by fire were investigated (11 male and 6 female in the age of 16 to 89 years). 8 patients died during the study, 11 patients developed from ARDS, and 3 patients developed ARF. 16 patients had thermal injury of 3rd degree (5 over 40% of skin), 13 of the patients had ventilated the fire smoke for a longer time.
Blood sampling was preformed at 8.oo a.m.; urine was collected over periods of 12 to 24 hours. Excretion rates were calculated per hour. Measurements were made daily untill day 14, then in the follow up the investigations were performed twice a week till normalization of the clinical situation. General therapy was not changed for the study.
For measurement of the given parameters we mainly used the commercial assay kits listed in the following. For further details of major modifications the recent publications are cited. We determined: PMN-elastase-alpha-1-protease-inhibitor complex by ELISA (Merck); plasmaprekallikrein activity by amidolytic assay (3)(S2302, Kabi); renal kallikrein activity by amidolytic assay (4)(S2266, Kabi); kininogen

---

* This study was supported by the "Minister für Wissenschaft und Forschung" of Nordrhein-Westfalen (IVB5-40005886).

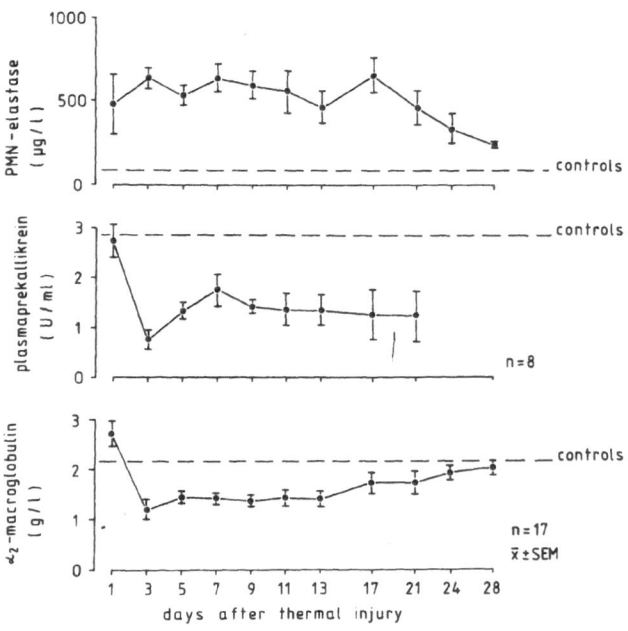

Figure 1. Changes in PMN-elastase, plasmaprekallikrein and alpha-2 macroglobulin by thermal injury.

concentration by bradykinin RIA (5); kininase II activity by radioassay (Ventrex); concentration of alpha-1 protease inhibitor and alpha-2 macroglobulin by radial immunodiffusion (Behring); potassium concentration in urine by flame photometry (Beckman); creatinine concentration according to Jaffe (Boehringer); fibrinogen concentration by coagulometry (Eppendorf); leucocytes by coulter counter technique and hematocrit by microcapillary centrifugation.

Values given represents means $\pm$ SEM. For better evaluation of changes in the patients mean values of healthy volunteers are given as controls.

RESULTS

In all patients there was a marked increase in PMN-elastase accompanied by a parallel increment of alpha-1 protease inhibitor (fig.1, tab.1). Individual courses (fig.3) showed, that PMN-elastase correlated well with leucocytes, either inversely in the acute phase of inflammatory with its local comsumption of leucocytes (all patients: PMN-E from 481+168 to 738+123 µg/l; leucocytes from 20.7+1.8 to 10.6+1.2 1000/mm^3) or directly in the phase of late recovery with parallel decrease of PMN-elastase and leucocytes (PMN-E from 455+80 to 190+23 µg/l; leucocytes from 21.5+2.9 to 10.3+1.9 1000/mm^3). Hematocrit declined during the observation period due to marked loss in blood during and after necrectomy (tab. 1). Plasmaprekallikrein was activated rapidly; in parallel its inhibitor alpha-2 macroglobulin and its substrate HMW-kininogen declined (fig.1, tab.1). Secondary to the activated coagulation pathway fibrinogen in plasma was reduced (tab.1).

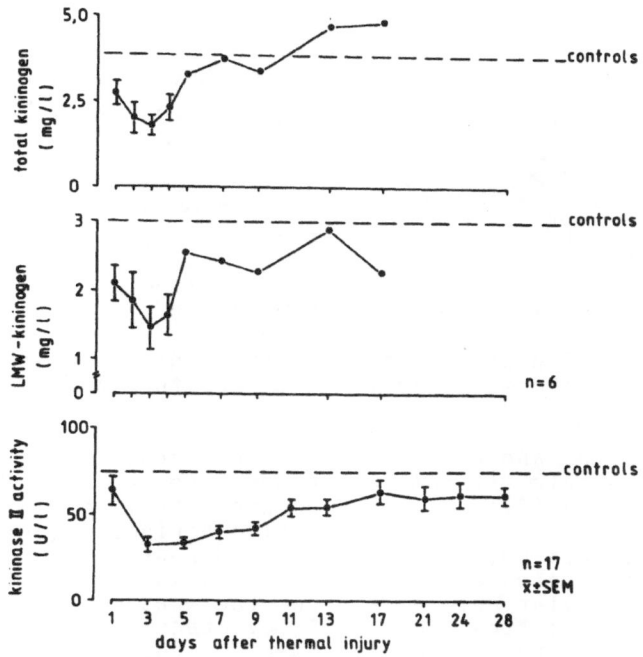

Figure 2. Changes in kininase II and in total resp. low molecular weight (LMW-) kininogen by thermal injury.

Table 1. Changes in creatinine (Cr), fibrinogen (F), alpha-1 protease inhibitor (a1-PI), high molecular weight kininogen (HMWK), leucocytes (L) and hematocrit (HCT) in plasma as well as potassium excretion in urine (UKV) after thermal injury.

Parameter	Day after thermal injury					
	1	3	5	7	13	21
Cr (mg/l)	1.03 ±0.10	1.52 ±0.41	0.98 ±0.08	0.87 ±0.08	1.32 ±0.38	0.78 ±0.14
L (1000/mm^3)	20.7 ± 1.8	9.8 ± 1.4	7.7 ± 1.2	11.7 ± 1.0	21.5 ± 2.9	13.5 ± 1.6
HCT (%)	42.4 ± 2.4	33.9 ± 1.5	34.3 ± 1.1	32.9 ± 1.4	32.1 ± 1.5	31.7 ± 0.8
F (g/l)	2.10 ±0.15	3.31 ±0.25	6.75 ±1.47	6.47 ±0.59	5.76 ±0.77	4.71 ±0.37
a1-PI (g/l)	1.45 ±0.07	2.54 ±0.12	3.97 ±0.12	4.34 ±0.15	3.74 ±0.18	3.03 ±0.27
HMWK (mg/l)	0.62 ±0.13	0.34 ±0.05	0.81 --	1.34 --	1.65 --	--- --
UKV (mmol/h)	5.42 ±1.22	5.02 ±0.76	4.14 ±0.91	4.74 ±0.53	3.95 ±1.19	1.86 ±0.65

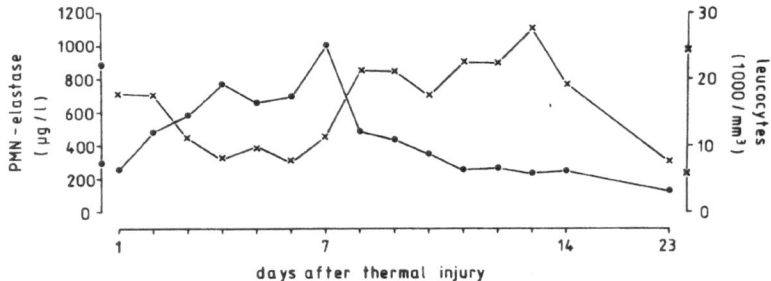

Figure 3. Time course of leucocytes and PMN-elastase in a patient with thermal injury (II°, less than 60% of the skin).

LMW-kininogen and kininase II as components of the glandular kallikrein-kinin systems in blood were reduced too after thermal injury (fig.2). Albumine concentration did not change signifikantly during the observation period. In urine creatinine and potassium excretion remained nearly unchanged, while renal kallikrein excretion was diminished markedly (tab.1, fig.4). Correlations between one the measured parameters and clinical events like ARDS, ARF, hypotension or death are not performed at the moment in regard to the small number of patients in each of these subgroups.

DISCUSSION

In our patients suffering from thermal injury by open fire we found a marked activation of PMN-elastase. Its time course was characterized by two phases, in the acute phase with an inverse correlation to the leucocytes and in contrast to that finding in the late phase of recovery with a direct positive correlation to the leucocytes. This relationship could be of clinical interest since it allows a clear differentiation of the two forms of fall in leucocytes, i.e. peripheral consumption of leucocytes by inflammation resp. reduction in leucocytes in regard to improvement of the disease. In the first case PMN-elastase increased during the fall in leucocytes, in the latter case PMN-elastase decreased in parallel. Whether there is a relationship between PMN-elastase concentration in blood and clinical complications as ARDS, shock or ARF, cannot be clarified from our data, since the special subgroups of patients are still too small. The recently published data are contradictary (6, 7) and therefore it seems still of importance to enlarge the groups of patients in our study furthermore.

In regard to the kallikrein-kinin systems in blood we observed a rapid stimulation. In consequence of the marked degradation of kininogen an increased release in kinins can be suggested. This enhanced kinin activity will probably be potentiated by the reduced kininase II activity in the burned patients which was described never before. The increased kinin release can play a role in the pathogenesis of hypotension and ARDS, since kinins are the most potent vasodilator in man and enhance the vascular permeability in a great extend resulting in edema and protein influx in the perivascular tissue.

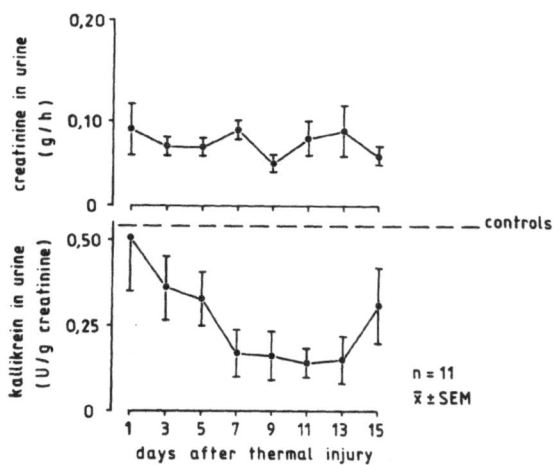

Figure 4. Changes in urinary excretion of creatinine and renal kallikrein after thermal injury.

The renal kallikrein is reduced after burning extensively while creatinine excretion remained unchanged. One possible explanation for these changes could be an enhanced glomerular filtration of kallikrein inhibitors. On the other site, however, decrease in kallikrein could be a very early indicator of an isolated disturbed distal tubular function of the nephron. In the following in vitro studies we will attend to that special problem in order to get further information on the reason of reduced renal kallikrein and its role in the development of ARF after thermal injury.

REFERENCES

1. M.Jochum, K.Duswald, S.Neumann, J.Witte, H.Fritz: Proteinases and their inhibitors in septicemia - basic concepts and clinical implication. Adv.Exp.Med.Biol. 167:391 (1984)
2. W.Hörl, R.Schäfer, K.Scheidhauer, M.Jochum, A.Heidland: Proteolytic activity in patients with hypercatabolic renal failure. Adv.Exp.Med.Biol. 167: 405 (1984)
3. G.Bönner, J.Horn, U.Büchsler, C.Kapp: Measurement of plasmaprekallikrein independent of inhibitors and interfering enzymes. Adv.Exp.Med.Biol. 198 B: 87 (1986)
4. G.Bönner, M.Marin-Grez: Measurement of kallikrein activity in urine of rats and man using a chromogenic tripeptide substrate. J.Clin.Chem.Clin.Biochem. 19: 165 (1981)
5. G.Bönner, D.Iwersen, K.Shimamoto: The analytical value for kinin concentration in blood depends on the antiserum used in the bradykinin radioimmunoassay. J.Clin.Chem.Clin. Biochem. 25: 39 (1987)
6. H.Redl, E.Paul, R.Goris, R.Pacher, W.Woloszczuk, G.Schlag: Plasma levels of elastase-a1-protease inhibitor complex in the monitoring of ARDS and Multiorgan failure - a summary of 3 clinical trials. Adv.Exp.Med.Biol.,this issue (1988)
7. W.Kellermann, M.Jochum: PMN-elastase and activated complement C3a in ARDS after trauma and sepsis. Adv.Exp.Med.Biol., this issue (1988)

SERUM PANCREATIC SECRETORY TRYPSIN INHIBITOR (PSTI) IN PATIENTS WITH INFLAMMATORY DISEASES

Michio Ogawa, Takashi Shibata, Takahiro Niinobu
Ken-ichi Uda, Naoki Takata and Takesada Mori

Second Department of Surgery
Osaka University Medical School
Fukushima-ku, Osaka 553, Japan

INTRODUCTION

Pancreatic secretory trypsin inhibitor (PSTI), first isolated by Kazal et al.[1] in 1948, has been considered to exist only in the pancreas and to be a specific trypsin inhibitor in pancreatic juice. It also enters the circulating blood stream as other pancreatic enzymes do. Using a specific radioimmunoassay for human PSTI, we revealed that PSTI existed in various organs other than the pancreas.[2]

In the earlier clinical investigations, we reported that serum PSTI was significantly elevated in patients with severe acute pancreatitis,[3,4] in patients with various malignancy,[2,5] in patients who received major surgery,[6] and in patients with serious injury.[7] As the elevation of serum PSTI was significantly correlated with that of acute phase reactants, we assumed that the role of increased PSTI in serum might be related to the host defence to surgical stress or invasion.[6,7]

SERUM PSTI IN SEPSIS

As acute inflammation is known to be a potent stimulus for the acute phase response, we studied the changes of serum PSTI in patients with sepsis. Figure 1 shows the serial changes of serum PSTI in patients with multiple injuries who were complicated by sepsis during the posttraumatic course (Group 1), in patients with extended burn who also complicated by sepsis (Group 2) and in patients with non-traumatic sepsis (Group 3). Serum PSTI at admission or onset in most cases remained in the normal range, rose before or soon after the onset of septic complication, and remained elevated till the full recovery.

Figure 2 shows the correlation between PSTI and fibrinogen in these three groups. The magnitude of the elevation of serum PSTI was far greater than that of fibrinogen. The same result was obtained when serum PSTI was compared with serum α_1-antitrypsin.

Changes of serum PSTI together with other laboratory data in a patient with multiple injuries was illustrated in Figure 3. As shown in

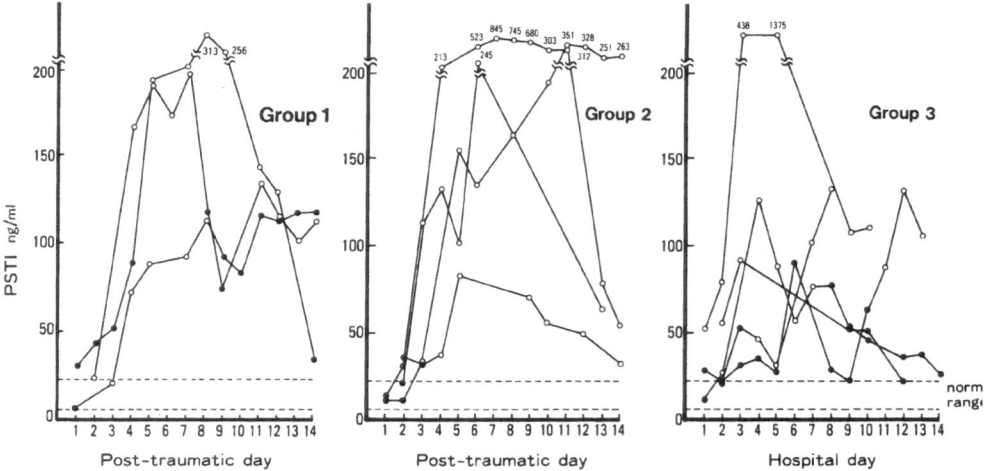

Figure 1.   Changes in serum PSTI in patients with sepsis
            Group 1: Sepsis following multiple injuries
            Group 2: Sepsis following extended burn
            Group 3: Non-traumatic sepsis
            PSTI is determined at septic state (O) or non-septic
            state (●).

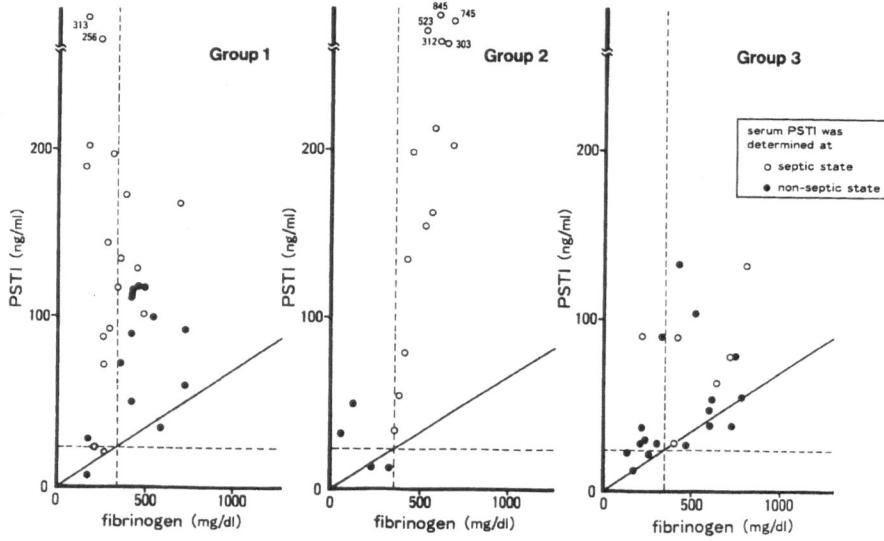

Figure 2.   Correlation between serum PSTI and plasma fibrinogen
            in patients with sepsis
            Group 1: Sepsis following multiple injuries
            Group 2: Sepsis following extended burn
            Group 3: Non-traumatic sepsis
            PSTI and fibrinogen are determined at septic state (O)
            or non-septic state (●). Dotted lines indicate the upper
            limit of the normal range.

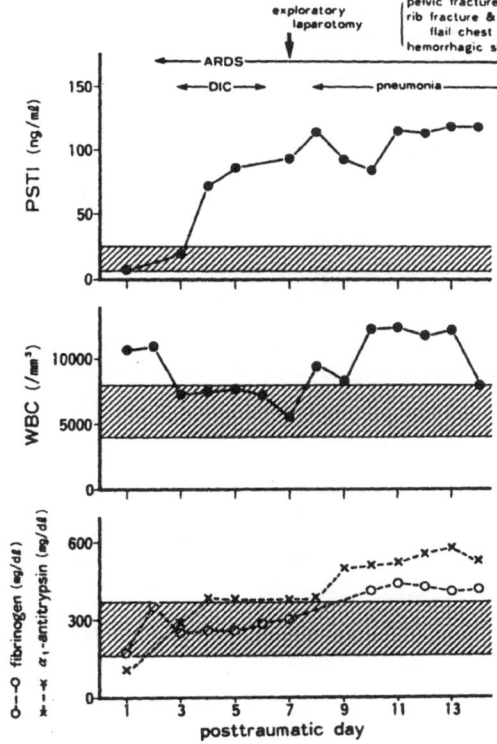

Figure 3.

Changes in serum PSTI and other laboratory data in a patient (44 F) with multiple injuries.

this case, the elevation of serum PSTI was greater and more sensitive than those of white blood cell count, fibrinogen or α_1-antitrypsin.

SERUM PSTI AND C-REACTIVE PROTEIN IN VARIOUS INFLAMMATORY DISEASES

C-Reactive protein (CRP) is the most widely used marker for inflammation. Table shows the results of simultaneous determinations of serum PSTI and CRP in 315 patients with various inflammation. Both the

Table. Relation between serum PSTI and CRP in patients with various inflammatory diseases

Serum CRP (mg/dl)	n	Serum PSTI		
		Mean ± S.E. (ng/ml)	Range (ng/ml)	Incidence of elevation (%)*
0- 0.4	60	13.3± 0.8	6- 39	8.3
0.5- 1.0	53	15.1± 1.4	6- 72	11.3
1.1- 2.5	49	17.7± 1.4	7- 44	26.5
2.6- 3.8	25	34.8± 6.9	8- 126	36.0
3.9- 6.5	36	58.7±17.9	6- 551	44.4
6.6-10.0	32	86.0±35.6	9- 982	43.8
10.1-15.0	42	111.4±35.6	14-1238	64.3
15.1-	18	233.4±62.6	10- 784	83.3

*above upper limit of normal (22.7 ng/ml)

incidence of the elevation and the average serum content of PSTI increased as a whole with the increase of CRP. But, in approximately 10% of the total cases, there was a discrepancy between the levels of serum PSTI and CRP, indicating that the origin or the mechanism of the elevation might be different in a certain pathological condition which is presently unclarified.

CONCLUSION

Serum PSTI was elevated remarkably in patients with sepsis. The change of serum PSTI reflected serious pathological conditions of sepsis. The magnitude of the elevation of serum PSTI was far greater than those of acute phase reactants previously known. A significant correlation between serum PSTI and CRP was also observed in patients with various inflammation. Thus, the determination of serum PSTI can be of value as a useful marker for inflammatory diseases.

ACKNOWLEDGEMENTS

This study was partially supported by Research Grant 60480303 from the Ministry of Education, Science and Culture, and Research Grant for Intractable Diseases of the Pancreas from the Ministry of Health and Welfare of Japan to M.O.

REFERENCES

1. L. A. Kazal, D. S. Spicer, and R. A. Brahinski, Isolation of a crystalline trypsin inhibitor-anticoagulant protein from pancreas. *J. Am. Chem. Soc.*, 70:3034-3040 (1948).
2. K. Matsuda, M. Ogawa, A. Murata, T. Kitahara, G. Kosaki, Elevation of serum immunoreactive pancreatic secretory trypsin inhibitor contents in various malignant diseases. *Res. Commun. Chem. Pathol. Pharmacol.*, 40:301-305 (1983).
3. T. Kitahara, Y. Takatsuka, K. Fujimoto, S. Tanaka, M. Ogawa, and G. Kosaki, Radioimmunoassay for human pancreatic secretory trypsin inhibitor: Measurement of serum pancreatic secretory trypsin inhibitor in normal subjects and subjects with pancreatic diseases. *Clin. Chim. Acta*, 103:135-143 (1980).
4. M. Ogawa, K. Matsuda, Y. Matsuda, K. Miyauchi, J. Nishijima, Y. Horikawa, M. Kurihara, and T. Mori, Evaluation of immunological assays of serum pancreatic enzymes and pancreatic secretory trypsin inhibitor for diagnosis and management of acute pancreatitis. *in*: "Pancreatitis—its pathophysiology and clinical aspects," T. Sato, and H. Yamauchi, eds., University of Tokyo Press, Tokyo, pp. 107-115 (1985).
5. A. Murata, M. Ogawa, K. Matsuda, N. Matsuura, G. Kosaki, M. Inoue, G. Ueda, and K. Kurachi, Immunoreactive pancreatic secretory trypsin inhibitor in gynecological diseases. *Res. Commun. Chem. Pathol. Pharmacol.*, 41:493-499 (1983).
6. K. Matsuda, M. Ogawa, T. Shibata, S. Nishibe, K. Miyauchi, Y. Matsuda, and T. Mori, Postoperative elevation of serum pancreatic secretory trypsin inhibitor. *Am. J. Gastroenterol.*, 80:694-698 (1985).
7. M. Ogawa, K. Matsuda, T. Shibata, Y. Matsuda, T. Ukai, M. Ohta, and T. Mori, Elevation of serum pancreatic secretory trypsin inhibitor (PSTI) in patients with serious injury. *Res. Commun. Chem. Pathol. Pharmacol.* 50:259-266 (1985).

THE EFFECT OF APROTININ ADMINISTRATION ON THE INTRAOPERATIVE HISTAMINE RELEASE AND HAEMOSTATIC DISORDERS

Henning Harke, and Salah Rahman

Department of Anaesthesia
General Hospital of Krefeld
Krefeld, West Germany

INTRODUCTION

Histamine is a biogenic amine which is normally stored in the granula of basophilic leucocytes and tissue mast cells[1,2]. High histamine concentrations are present in the lung, stomach and uterus tissues[4]. Under physiological conditions the plasma histamine level ranges between 0.3-0.7 ng/ml[10]. Excessive histamine liberation can be induced either by specific stimuli such as complement activation, Ig E induced type of allergic reaction or unspecifically via pharmogological interactions with the cell membrane. It can also be liberated mechanically with other vasoactive substances such as serotonin, prostaglandins, leucotrienes, kinins and lysosomal enzymes during tissue injury and shock[9].

All these mediators can induce various pathological reactions, particularly haemostatic disorders and organ insufficiency[9,11]. According to our previous experience, the histamine liberation during blood banking could be diminished by initial addition of 200.000 KIU of aprotinin into blood unit[6]. These observations have encouraged us to investigate the effects of intraoperative aprotinin administration in patients who underwent hysterectomy, an intervention currently associated with excessive histamine release.

MATERIAL AND METHODS

The present study comprises 40 consecutive female patients between 19 and 76 years old who underwent hysterectomy. The patients were allocated by random to one of four groups including 10 patients each. During surgery each patient received a total of 500 ml of the following solutions infused within one hour:

(a) Group 1, the control group, received 500 ml of electrolyte solution.
(b) Group 2 received 200 ml of aprotinin solution including 2 million KIU + 300 ml of electrolyte solution.
(c) Group 3 received 350 ml of aprotinin solution including 3.5 million KIU + 150 ml of electrolyte solution.

(d) Group 4 received 500 ml of aprotinin solution including 5 million KIU.

Blood samples for estimation of histamine concentration[10], blood count, platelet function[3], PTT[12], F VIII[5], plasminogen[7] and alpha-2-antiplasmin[13] were collected immediately after anaesthesia, 1 hour, 2 hours, 3 hours and 19 hours after infusion.

The statistical assessment of the data was performed by the analysis of variance according to Duncan test.

RESULTS

The plasma histamine level of the patients prior to surgery ranged between 0.3-0.7 ng/ml plasma and was in close agreement with values estimated by other investigators.

During surgery, however, the plasma histamine concentration of the control patients increased significantly and reached values between 1.09 and 1.28 ng/ml immediately at the end of surgery. These levels were maintained for a short period and began to decrease again to normal value 1 hour after surgery. In contrast, the plasma histamine level of patients who received aprotinin did not show any significant elevation and remained almost unchanged during surgery (Fig. 1).

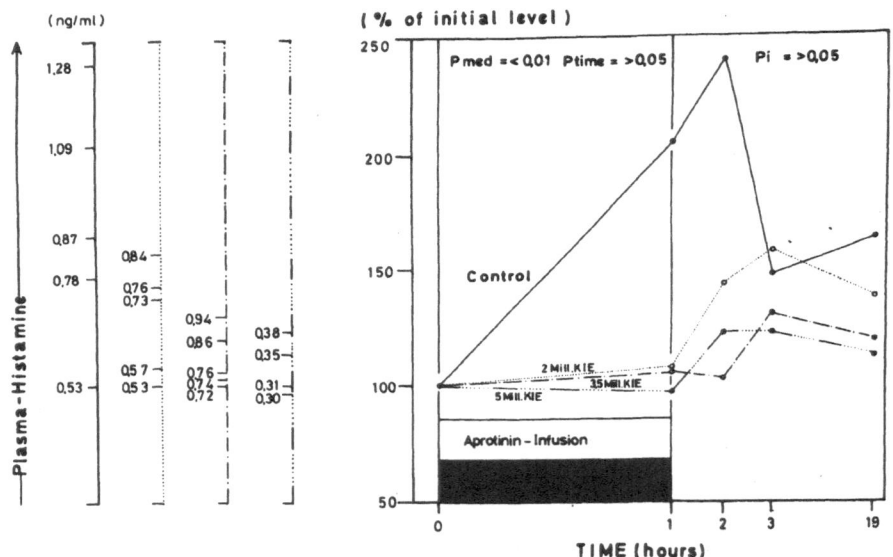

Fig. 1 Changes of the histamine level during and after hysterectomy in patients receiving different dosages of aprotinin.

In addition, the platelet function was significantly reduced by 30 % in the patients of the control group. In contrast, the reduction of platelet function in the patients of the aprotinin groups was less pronounced and reached values of 15 % and 10 % in group 3 and 4 respectively (Fig. 2).

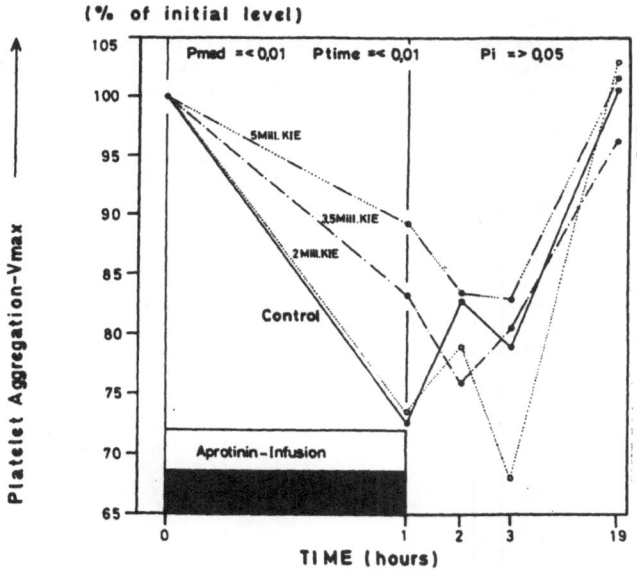

Fig. 2 Alteration of the platelet function during and after hysterectomy in patients receiving different dosages of aprotinin.

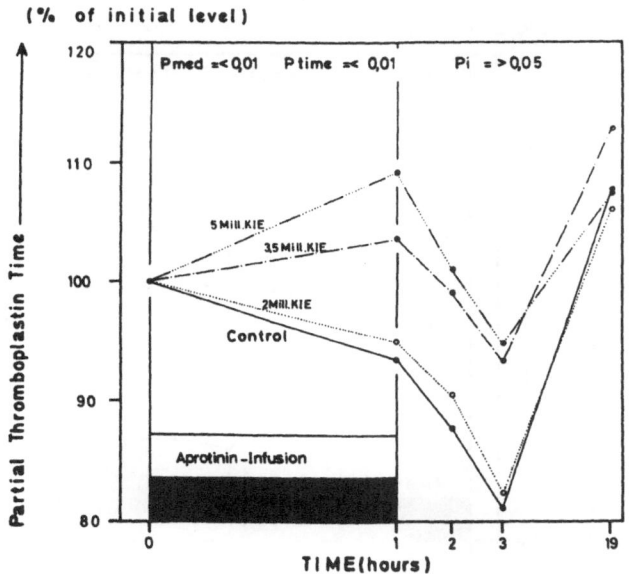

Fig. 3. Changes of the partial thromboplastin time during and after hysterectomy in patients receiving different dosages of aprotinin.

The hypercoagualability normally associated with surgery could not be detected in the patients of the aprotinin groups. For example, the PTT was shortened by 20 % in the controls, while the values in the aprotinin treated patients remained within normal range (Fig.3).

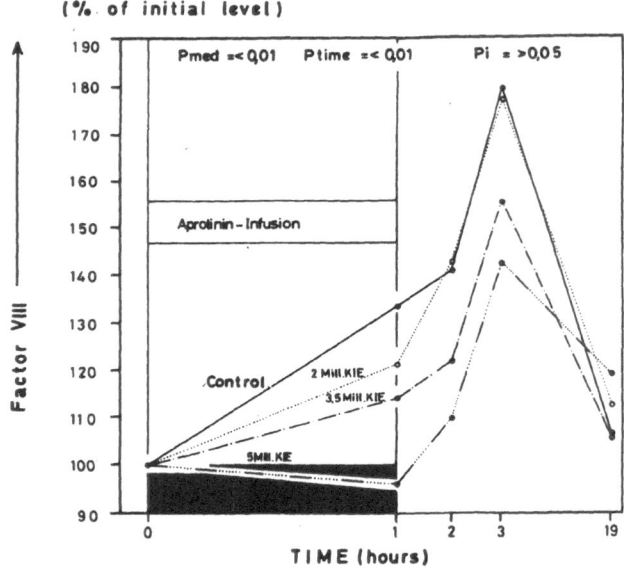

Fig. 4. Changes of F VIII activity during and after hysterectomy in patients receiving different dosages of aprotinin.

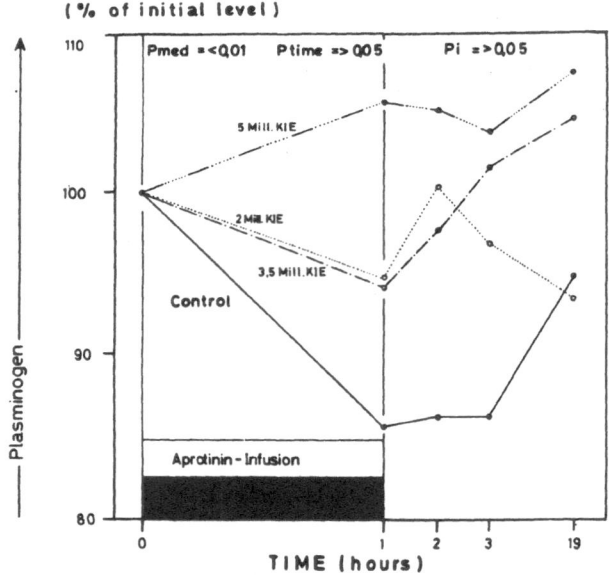

Fig. 5. Changes of the plasminogen level during and after hysterectomy in patients receiving different dosages of aprotinin.

Furthermore, the Factor VIII-activity in the control patients showed a significant elevation of approximately 30 % during surgery and reached 80 % 3 hours thereafter. This increase was considerably lower following aprotinin administration (Fig. 4).

In addition, the exogenous activation of the fibrinolytic system with concomitant consumption of plasminogen following surgery was significantly pronounced in the control patients. However, the circulating plasminogen in the patients of the aprotinin groups remained almost unchanged (Fig. 5).

A surprising reaction pattern could be observed for alpha-2-antiplasmin. Whereas the values in the control group were reduced by 20 %, the alpha-2-antiplasmin-activity after aprotinin infusion showed a dose-dependent elevation by approximately 30 % - 60 % (Fig. 6).

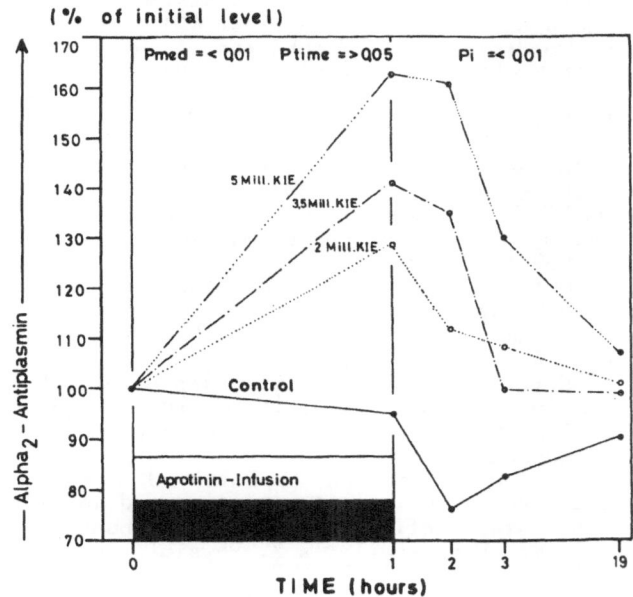

Fig. 6. Changes of the alpha-2-antiplasmin activity during and after hysterectomy in patients receiving different dosages of aprotinin.

CONCLUSION

The effect of aprotinin administration on histamine release was investigated under a typical operative model normally associated with excessive histamine liberation.

The intravenous infusion of various aprotinin dosages showed a dose-dependent interruption of the histamine release during hysterectomy[6].

Furthermore, the haemostatic alterations associated with surgery, particularly the reduction of platelet function, hypercoagulability as well as consumption of plasminogen and alpha-2-antiplasmin were considerably diminished following intraoperative aprotinin administration[8].

Since increased membrane permeability is always associated with a histamine liberation and while the aprotinin administration could prevent this reaction, it can be concluded that aprotinin may have a protective action on cell membrane.

REFERENCES

1. M. A. Beaven, Histamine (First of two Parts), N. Engl. J. Med. 294 : 30 (1976).
2. M. A. Beaven, Histamine (Second of two Parts), N. Engl. J. Med. 294 : 320 (1976).
3. G. V. R. Born, M. J. Cross, The aggregation of blood platelets, J. Physiol. 168 : 178 (1963).
4. G. Günther, W. D. M. Paton, The histamine content of the human uterus, J. Physiol. 154 : 417 (1960).
5. R. M. Hardisty, J. C. Macpherson, A one-stage factor VIII (Antihaemophilic Globulin) assay and its use on venous and capillary plasma, Thromb. Haemostasis 7 : 215 (1962).
6. H. Harke, G. Stienen, S. Rahman, H. Flohr, Aprotinin-ACD-Blood. II. The effect of aprotinin on the release of cellular mediators and enzyms in banked blood, Anaesthesist 31 : 165 (1982).
7. E. Jacobi, H. E. Karges, N. Heimburger, Bestimmung von Plasminogen mit einer Koagulometermethode, Med. Welt 26 : 1969 (1975).
8. G. Kosaki, T. Nomura, J. Kambayashi, The mechanism of the inhibitory effect of proteinase inhibitors on platelet aggregation and cellular synthesis of prostaglandins, Thromb. Res. 20 : 587 (1980).
9. S. G. Laychock, I. W. Putnay, Roles of phospholipid metabolism in secretory cells, in: Cellular regulation of secretion and release, M. P. Conn, ed, Academic Press Inc., London (1982).
10. W. Lorenz, H.-J. Reimann, H. Barth, J. Kusche, R. Meyer, A. Doenicke, M. Hutzel, A sensitive and specific method for the determination of histamine in human whole blood and plasma, Hoppe-Seyler's Z. Physiol. Chem. 353 : 911 (1972).
11. W. Lorenz, A. Doenicke, Histamine release in clinical conditions, M. T. Sinai J. Med. 45 : 357 (1978).
12. R. R. Proctor, S. I. Rapaport, The partial thromboplastin time with kaolin, Amer. J. clin. Path. 36 : 212 (1961).
13. A. C. Teger-Nilsson, P. Friberger, E. Gyzander, Determination of a new rapid plasmin inhibitor in human blood by means of a plasmin specific tripeptide substrate, Scand. J. clin. Lab. Invest. 37 : 403 (1977).

INCREASED MORTALITY IN SEPTIC RATS AFTER LEUPEPTIN APPLICATION

Erwin Kovats, Josef Karner, Günter Ollenschläger[1]
Judith Karner, Annette Simmel and Erich Roth

Ist Surgical Clinic, Metabolic Research Laboratory, University Vienna, Austria
[1] 2nd Medical Department, Univ. of Cologne, FRG

INTRODUCTION

Loss of body weight induced by catabolic stress situation like trauma, burns, operation, sepsis is caused by both enhanced proteolysis and diminished synthesis (1, 2). This catabolic phase is characterized by the net breakdown of body protein, primarily in skeletal muscle. Odessey reported elevated rates of protein degradation in the soleus muscle from a burned limb, while no effects on the rates of protein synthesis were detected (3). The mechanism responsible for increased muscle breakdown is not understood. It has been reported that the rate of proteolysis in muscle is coupled with the activity of cathepsin B and D and also other proteinases (4, 5). Most of the work done in elucidating the role of lysosomal proteinases in mediating protein degradation has been done in cell culture and perfused organ preparation. In this study we investigated the effect of Leupeptin, a thiol proteinase inhibitor on septic rats.

METHODS

Experiments were performed on four groups of male Sprague Dawley rats (Himberg, Austria) with a body weight between 200 and 250 g. For experiments with muscle incubation (soleus) rats with a weight of 220 g and 70 g were used respectively. Sepsis was induced in nonfasting rats by cecal ligation. The animals were allowed water and food after the operation procedures. 5 rats received 40 mg Leupeptin per kg/BW in saline intraperitoneally per day (NL) and a control group (N) received only saline. Another set of rats with cecal ligation received either the same dosage of Leupeptin (SL) or only saline (S). Blood was drawn by heart puncture for measurements of routine chemistry. To study protein degradation with the animal under ether anesthesia the soleus muscles were dissected with intact tendons, weighed and incubated at 37°C in Krebs-Ringer (pH 7.4) bicarbonate buffer containing glucose (10 mM), 0.5 mM cycloheximid and equilibrated with 95 % O_2: 5 % CO_2. Tyrosine release was measured in the supernatant with an automatic amino acid analyser (Liquimat III, Kontron, Zürich, Switzerland).

RESULTS

1. Intake, body and organ weight. Food and water intake was significantly lower in S rats and NL rats than in N rats (Table 1).

TABLE 1. Water and food consumption within 2 days ($\pm$ g)

	Controls n=6	Sepsis n=5	Controls+LP n=5	Sepsis+LP n=4
Food	39 $\pm$ 1.7	5 $\pm$ 2.0*	21 $\pm$ 1.4*	4 $\pm$ 1.2
Water	62 $\pm$ 0.9	36 $\pm$ 1.4*	53 $\pm$ 2.8*	29 $\pm$ 2.3

*$p < 0.05$ vs controls   results given as $\bar{x} \pm$ sem

Weight gain in N rats was 7 g in 2 days whereas sepsis and also the injection of Leupeptin caused a weight loss (Table 2).

TABLE 2. Course of body weight within 2 days

	n	weight gain or loss ($\pm$ g)
Controls	6	+ 7 $\pm$ 1.7
Controls + LP	11	− 9 $\pm$ 2.0*
Sepsis	5	− 17 $\pm$ 4.0*
Sepsis + LP	4	− 24 $\pm$ 1.8

*$p < 0.01$ vs Controls   results given as $\bar{x} \pm$ sem

There were no significant differences in liver, lung and kidney weights between the four groups. Septic rats had lower heart weights than control rats.

2. Blood metabolites and enzymes. Concentrations of creatinine and liver enzymes were increased significantly in septic and also in NL rats. Septic rats receiving Leupeptin had dramatically increased levels of liver enzymes (Table 3).

3. Mortality. Two of 16 rats died after cecum ligation, however, 8 of 11 rats with cecum ligation died during the treatment with Leupeptin ($p < 0.001$).

4. Protein breakdown. Protein breakdown in incubated soleus muscle was increased in septic rats (body weight 70 g) but remained equal in rats receiving Leupeptin (Table 4).

TABLE 3. Plasma concentrations of creatinine and liver enzymes.

	N (n=9)	NL (n=9)	S (n=5)	SL (n=4)
Creatinine [mg/dl]	0.54 ± 0.04	0.8 ± 0.09*	0.93 ± 0.14**	1.86 ± 0.47*
Cholinest. [U/l]	480 ± 70	221 ± 24*	164 ± 21*	150 ± 69*
Alc.Phosph. [U/l]	274 ± 61	203 ± 36	494 ± 100*	1097 ± 90***
GOT [U/l]	37 ± 3	110 ± 32*	98 ± 18**	824 ± 554***
LDH [U/l]	213 ± 41	428 ± 110	219 ± 27	3626 ± 2862***

*p < 0.05 vs Controls, **p < 0.01 vs Controls, ***p < 0.05 vs Sepsis

TABLE 4. Tyrosine release° from the M.soleus of the rat.

	n	body weight (60-80g)	n	body weight (200-230g)
Controls	8	655 ± 18	10	180 ± 9
Sepsis	8	758 ± 41*	10	179 ± 8
Controls + LP	8	656 ± 21	10	174 ± 11
Sepsis + LP	6	801 ± 18**	4	235 ± 6

*p <0.05 vs Controls, **p <0.01 vs Controls+LP   results given as $\bar{x}$ ± sem
°$\mu mol \times g^{ww^{-1}} \times 2h^{-1}$

DISCUSSION

    Leupeptin, a specific inhibitor of thiol and some serine proteases, is known to inhibit intracellular degradation of various proteins, presumably by inhibiting proteases in lyso-somal and prelysosomal compartments. Furthermore it has been proved that Leupeptin inhibits coagulation of human and rabbit blood, the stimulation of [^{3}H]thymidine incorporation by phytohemagglutinin in peripheral blood lymphocytes, the chemicaltumorigenesis in mouse skin and colon, the vascular metastasis of hepatoma to the lung (6). Moreover, Leupeptin causes a relative increase in the weights of livers and diaphragms after injury and causes a marked increase in the RNA and DNA content of diaphragms (7).

    In our study the intraperitoneal injection of Leupeptin dramatically increase the mortality in septic rats. This increase in mortality may be due to a liver damage evoked by Leupeptin. The administration of Leupeptin significantly reduced the body weight of control and septic rats. This weight loss was accompanied by a reduced food intake.

    Control and septic rats receiving Leupeptin had increased plasma levels of creatinine and liver enzymes indicating a toxic effect of Leupeptin. According to our knowledge no hepatotoxic effect of Leupeptin has been described.

The incubation of soleus muscle and measurement of tyrosine
released from the muscle is a standardized model for the determination of the proteolytical activity of tissue. Our results show that the septic progress enhance the proteolytical
capacity of the soleus muscle. This proteolytical activity is
not inhibited by the in vivo administration of Leupeptin.

REFERENCES

1. G. H. Clowes, B. C. George, C. A. Villee, C. A. Sara-
   fis, Muscle proteolysis induced by a circulating
   peptide in patients with sepsis or trauma, New
   Engl.J.Med.308:545 (1983)
2. M. E. Tischler, J. Fagan, Response to trauma of protein
   amino acid, and carbohydrate metabolism in injured
   and uninjured rat skeletal muscles, Metabolism
   32:853 (1983)
3. R. Odessey, B. Paar, Effect of insulin and leucin on
   protein turnover in rat soleus muscle after burn
   injury, Metabolism 31:82 (1982)
4. G. F. Guarnieri, G. Toigo, R. Situlin, Muscle biopsy
   studies on protein metabolism in traumatized patients, in: "Clinical Nutrition and Metabolic Research," G.Dietze, A.Grünert, G. Kleinberger,
   G. Wolfram, ed., Karger, München (1986)
5. J. Newman, D.R. Strome, C.W. Goodwing, A.D. Mason,
   B.A. Pruitt, Neutral proteinase activity in skeletal muscle from thermally injured rats, J.Surg.
   Res. 35:515 (1983)

6. U. H. Umezawa, T. Aoyagi, Activities of proteinase inhibitors of microbial origin, in: "Proteinases in
   Mammalia and Tissues," A.J. Barrett, ed., North-
   Holland Publishing Company, New York (1977)
7. B. M. Miller, B. Hoxworth, R. Buckspan, L. Nanny,
   W.W. Lacy, N.N. Abumrad, Leupeptin's effect on
   organ weight, RNA, DNA, and protein content after
   long bone fracture in the rat, J.Surg.Res. 36:453
   (1984)

LYSOSOMAL ENZYMES AND GRANULOCYTE ELASTASE IN SYNOVIAL FLUID AFTER MULTIPLE TRAUMATIC INJURIES

Maximilian Hörl and Hans-Peter Bruch

Department of Surgery

University of Würzburg, 8700 Würzburg, FRG

INTRODUCTION

Polytrauma is often associated with soft tissue trauma and knee injury. The treatment of open and closed fractures requires an exact classification for adequate management. The presence of a hemarthrosis, generally, is a sign of particularly severe joint lesions. A direct enzymatic damage of the joint cartilage or in combination with trophical alterations appearing later on can lead to a formation and aggravation of arthrosis[1,2].

The present study was designed to investigate the following questions:
1. Is there a dependence of the activity of lysosomal enzymes on the extent of knee injury ?
2. Is there a dependence of the elastase-alpha-1-proteinase inhibitor complex (E-alpha-1 PI) on the age of hemarthrosis ?
3. Is there a correlation of E-alpha-1 PI with the amount of soft tissue damage ?

PATIENTS AND METHODS

Eightynine synovial fluid samples of patients after knee injury were studied. Samples were obtained from 10 patients with serous aspirates and 79 samples from patients with hemarthrosis (meniscal lesions n=12; stretched or sprained ligaments n=26; ruptured ligaments n=27; fractures into the joint n=24).
Determination of RNAse was performed according to the method of Leubner[3]. Arylsulfatase and beta-glucuronidase were determined according to the method of Bergmeyer[4]. Cathepsin D was analyzed as modified according to Anson[5]. Plasma and synovial fluid levels of granulocyte elastase in complex with alpha-1-proteinase inhibitor were measured according to Neumann and coworkers[6].

RESULTS

Table 1 shows the activity of arylsulfatase in dependence on

Table 1 : Effect of traumatic knee joint lesions on different lysosomal
enzyme activities, E-alpha-1 PI concentration and
granulocyte count in the synovial fluid

Parameter	Serous aspirates (n=10)	Hemarthrosis			
		Meniscal lesions (n=12)	Stretched/ sprained ligaments (n=26)	Ruptured ligaments (n=27)	Fractures into the joint (n=24)
RNAse E260/30min/50μl	0.47 ±0.03	0.60* ±0.02	0.63** ±0.02	0.60* ±0.02	0.64** ±0.03
Arylsulfatase mU/ml	0.22 ±0.03	0.26 ±0.04	0.45* ±0.07	0.48** ±0.06	0.88** ±0.10
Beta-Glucuronidase mU/ml	5.7 ±1.3	4.8 ±1.4	6.2 ±1.0	6.4 ±1.2	10.0* ±1.8
Cathepsin D mU/ml	0.06 ±0.03	0.10 ±0.03	0.12 ±0.02	0.23* ±0.05	0.30** ±0.05
E-alpha-1 PI ng/ml	54 ±9	n.d.	3940 ±1009	n.d.	6382 ±2687
Leukocytes cells/μl	0-81	n.d.	650-23550	n.d.	1100-39600

Mean values ± SEM;
* $p < 0.05$ versus serous aspirates
** $p < 0.01$ versus serous aspirates
n.d. = not done

the extent of trauma (meniscal lesions, strechted/sprained ligaments, ruptured ligaments and fractures into the joint). After meniscal lesions the activity in hemarthrosis was 0.26 mU/ml (serous aspirates 0.22 $\pm$ 0.032 mU/ml). In knee injuries with fractures into the joint the activity was significantly elevated to 0.88 mU/ml. The activity of plasma samples obtained from healthy subjects was 1.33 mU/ml. The activity of RNAse in dependence on the extent of knee injury is also shown in this table. The activity was significantly elevated in all four groups of patients.

The correlation of the activity of beta-glucuronidase with the extent of trauma is comparable with the results demonstrated before. In patients with hemarthrosis and sprained ligaments an activity of 6.2 mU/ml was measured. After fractures into the joint the beta-glucuronidase values increased to 10.0 mU/ml. Beta-glucuronidase activity was 4.8 $\pm$ 1.4 mU/ml in samples of patients with hemarthrosis showing not any further lesions. We finally report about the results of cathepsin D. While the activity after meniscal lesions is comparable with values in serous aspirates, there were high aspirate levels after ruptured ligaments of fractures into the joint (table 1).

There was also a dependence of elastase-alpha-1-proteinase inhibitor complex in synovial fluid on the time of aspiration. The highest aspirate levels were measured on the second and third day after trauma (data not shown).

Monitoring of plasma elastase-alpha-1-proteinase inhibitor is helpful in evaluating patients with multiple trauma. Soft tissue trauma was followed with a 10 to 20-fold increase (from 70.3 $\pm$ 4.4 ng/ml to 1109 $\pm$ 50 ng/ml. After surgical therapy (for example amputation or muscle debridement) a rapid decrease of the plasma E-alpha-1 PI levels was observed.

DISCUSSION

A recent study demonstrated large amounts of E-alpha-1 PI in the inflammatory synovial fluid (9947 $\pm$ 124 µg/1) compared with 41 $\pm$ 0.5 µg/1 in non-inflammatory synovial fluids[7]. There was no difference between the plasma E-alpha-1 PI values of both groups of patients. However, a close relation to the number of granulocytes in the synovial fluid could be observed[7]. The present study documents similar results in synovial fluid samples obtained from patients with multiple traumatic injuries. E-alpha-1 PI concentration, leukocyte count and the activity of several lysosomal enzymes increase with the severity of joint lesions.

Approximately 90% of the elastase released from human granulocytes is immediately complexed by alpha-1-proteinase inhibitor, 10% by alpha-2-macroglobulin[8]. Exposure of cultured synovial fibroblasts to elastase-alpha-2-macroglobulin complexes caused an increase in the synthesis of hyaluronate, chondroitin sulfate and dermatan sulfate. Elastase-alpha-2-macroglobulin complexes also stimulated synovial cells to release prostaglandin E2[9]. On the other hand, PGE2 may stimulate lysosomal enzymes[10].

Septicemia is responsible for increased plasma E-alpha-1 PI values[10]. After operation this parameter returned to normal in patients without further complications[11]. Similar observations were made in the present study.

In summary after knee injury the activity of lysosomal enzymes increases with the extent of trauma. There were also high levels of E-alpha-1 PI in hemarthrosis with a maximum two and three days after cruciate ligament rupture. Soft tissue damage, amputation or muscle debridement leads to a rapid decrease of the 20-fold increased plasma levels of E-alpha-1 PI.

REFERENCES

1. H.Cotta, and F.U.Niethard, Biomechanische und biochemische Grundlagen der Entstehung einer posttraumatischen Arthrose, Chirurg 50: 595 (1979).
2. H.Cotta, W.Puhl, and F.U.Niethard, Der Einfluß des Hämarthros auf den Knorpel der Gelenke, Unfallchirurgie 8: 145 (1982).
3. H.Leubner, Aktivitätsbestimmung von Pankreasenzymen, einschließlich der Nukleasen, in der Klinik, Ärztl. Lab. 1: 2 (1963).
4. H.Bergmeyer, Methoden der enzymatischen Analyse, Verlag Chemie, Weinheim (1974).
5. M.L.Anson, The estimation of cathepsin with hemoglobin and the partial purification of cathepsin, J.Gen.Physiol. 20: 565 (1937).
6. S.Neumann, N.Heinrich, G.Gunzer, and H.Lang, Enzyme-linked immunoassay for human granulocyte elastase in complex with -1-proteinase inhibitor, in: Proteases: potential role in health and disease, W.H.Hörl, and A.Heidland, eds., Plenum, London, p. 379 (1984).
7. K.Kleesiek, and H.Greiling, Pathobiochemical mechanisms during the acute phase response, Int.J.Microcirc.Clin.Exp. 3: 131 (1984).
8. K.Ohlsson, and C.B.Laurell, The disappearance of enzyme-inhibitor complexes from the circulation of man, Clin.Sci. Mol.Med. 51: 87 (1976).
9. K.Kleesiek, R.Reinards, and H.Greiling, Effect of granulocyte elastase on the metabolism of proteoglycans in inflammatory joint diseases, in: Proteinases in inflammation and tumor invasion, H.Tschesche, ed., Walter de Gruyter, Berlin - New York, p.25 (1986).
10. V.Baracos, H.P.Rodemann, C.A.Dinarello, A.L.Goldberg, Stimulation of muscle protein degradation and prostaglandin E2 release by leukocytic pyrogen (interleukin-1). N.Engl.J.Med. 308: 553 (1983).
11. M.Jochum, K.H.Duswald, S.Neumann, J.Witte, and H.Fritz, Proteinases and their inhibitors in septicemia - basic concepts and clinical implications, in: Proteases: Potential role in health and disease, W.H.Hörl, and A. Heidland, eds., Plenum, London, p. 391 (1984).

A SERINE PROTEINASE INHIBITOR IN HUMAN ARTICULAR CARTILAGE

POSSIBLE ROLE IN THE PATHOGENESIS OF INFLAMMATORY JOINT DISEAESES

Harald Burkhardt[1], Michael Kasten[2] and Silke Rauls[2]

Medizinische Hochschule Hannover,Abteilung für Krankheiten der
Bewegungsorgane und des Stoffwechsels[1], Universität Bielefeld
Fakultät für Chemie, Lehrstuhl für Biochemie[2]

INTRODUCTION

Inflammatory joint diseases are characterized by a progressive loss
of intact cartilage matrix, resulting from an increase in catabolic
activities of proteolytic enzymes. At present a variety of cells supplied
with cartilage degrading enzymes have to be taken into account as effectors
of matrix degradation. Amoung the different enzymes those with neutral
pH-optimum seem to be of major importance for the extracellular matrix-
breakdown and especially the serine proteinase elastase of polymorphonuclear
leukocytes has been made responsible for the initiation of this process,
because of its ability to attack the proteoglycans as well as the collagen-
meshwork [1,2] . ALthough the proteolytic activity of the enzymes released
into the joint space is antagonized by the serum proteinase inhibitors of
the synovial fluid (α_1-proteinaseinhibitor (α_1-PI), β_2-macroglobulin
(β_2-M), β_1-anticollagenase (β_1-AC)) a local disturbance of the enzyme-
inhibitor balance may occur and promote tissue destruction This incomplete
protection of cartilage by the enzyme inhibitors of synovial fluid led to the
question, whether an additional antiproteinase potential is integrated
within the tissue matrix itself as a second barrier against proteolytic
attack.Recent work of several investigators gave indications of low
molecular weight inhibitors in connective tissues [3] . Whereas an inhibitor
of metalloproteinases with a molecular weight of 28000 is already well
characterized including amino acid sequence [4] the situation is far less
clear concerning the serine proteinase inhibitory activity as indicated by
diversities of molecular weight estimations.Therefore in this paper we
will trie to characterize a low molecular weight inhibitor of serine
proteinases purified from human articular cartilage and discuss its possible
role in the regulation of proteolytic activity with respect to the pathogenesis of rheumatic joint destruction.

EXPERIMENTAL PROCEDURES

Cartilage Extraction

Human articular cartilage was taken from the tibia plateaus and
femoral condyli of adolescent trauma victims post mortem and by microscopic
staining for proteoglycans significant autolysis of the tissue specimens
was excluded prior to the application of further experimental procedures.

After mechanical fragmentation of the cartilage the specimens were subjected to extraction of matrix components in a 200 mM Tris-HCl-buffer (pH 7.4) containing 0.5 M NaCl, 50 mM $CaCl_2$, 2 M guanidinium chloride and 0.05% (v/v) Brij 35 in a ratio of 1:5 and stirred at 4° C for 48 h. Following centrifugation at 48000 x g for 30 min. at 4° C the matrix molecules dissolved in the supernatant were ultrafiltrated through an Amicon XM100-membrane, concentrated on an Amicon UM2-membrane and equilibrated with 0.05 M sodium acetate buffer (pH 5.5) at 4° C.

Ion Exchange Chromatography

The cartilage extract was applied to a column of CM-Sephadex C-50 (2.8 x 45 cm), eluted with a linear increasing gradient of 0 – 1 M NaCl and collected in fractions of 7 ml. A second run on CM-Sephadex C-50 (column: 1.8 x 15 cm) was performed and a gradient from 0.3 – 0.7 M NaCl in sodium acetate buffer chosen. The eluate was collected in fractions of 2 ml volume. All gel-chromatographic procedures were performed at 4° C.

Gel Electrophoresis

Sodium dodecyl sulphate (SDS) electrophoresis was carried out according to the method of Weber and Osborn [5] in a Pharmacia slab gel apparatus. For this purpose 12% gels of 3mm thickness were used. Standard proprotein mixtures (from 6.5 kDa up to 92 kDa) were applied for relative molecular mass calibration and staining was performed with 0.1% Coomassie Brilliant Blue R for 2 h.

Isoelectric Focussing

Estimation of the isoelectric point (pI) was performed by a run on Servalyt-Precotes, pH 3-10 (Serva, Heidelberg, FRG), following the instructions of the manufacturer. A protein test mixture (pI 3.5 – 10.65) was run simultaneously.

Enzymes

Granulocytic enzymes were purified from buffy coats of healthy blood donors, and in the case of elastase (EC 3.4.21.37) and cathepsin G (EC 3.4.21.20) the method of Engelbrecht et al. [6] was applied with slight modifications. The purification of granulocytic collagenase (EC 3.4.24.7) was performed in accordance with previously described methods [7,8]. Trypsin (bovine pancreas) (EC 3.4.21.4) was purchased from Serva (Heidelberg, FRG).

Assays

Elastase inhibitory activity was determined using Suc-(Ala)$_3$-4NA (SAPA) [9] and MeO-Suc-Ala-Ala-Pro-Val-4NA [10] as substrates on an Eppendorf 1101M spectrophotometer.
Suc-Phe-Pro-Phe-4NA was employed to measure cathepsin G activity [11,12]. Trypsin activity was assessed by the method of Erlanger et al. [13] with benzoyl-DL-Arg-4NA.
Collagenase inhibition was assayed by the fluorescamine method of Tscheshe and Macartney [14] on a Shimadzu RF-510 spectrofluorophotometer.
All assays for inhibitory activity were run after a preincubation period 15 min.) of the probe with the enzyme.
Dissociation experiments of the enzyme –inhibitor complex were performed according to the method of Aubry and Bieth [15], using a 10-fold molar excess of α_2-macroglobulin.
Protein concentration was measured by the microtiter method of the BCA-assay [16] from Redingbaugh and Turley [17], using bovine serum albumin as standard.

Concentration of uronic acids was determined in accordance with the method described by Blumenkrantz and Asboe-Hansen [18] after standardization with commercially available hyaluronic acid. Sugar content was assessed by the colorimetric method of Dubois et al. [19], using D-galactose as a standard.

RESULTS

In the ultrafiltrate of the matrix components extracted from the articular cartilage no contamination of proteoglycans was detectable. According to the further experimental procedure the filtrate was subjected to ion exchange chromatography on CM-Sephadex C-50. By an elution with a linear increasing NaCl-gradient between 0 - 1 M a main inhibitory peak with activity towards trypsin, cathepsin G and elastase was obtained. The assays for quantification of the inhibitor capacity revealed an 80% inhibiton of trypsin (67 mU/ml), 85% inhibition of elastase (2 mU/ml, substrate: SAPA) and 90% inhibition of cathepsin G (6 mU/ml). A second peak with inhibitory activity towards cathepsin G up to a maximum of 60% was found at lower ionic strength (fig. 1).
The fractions exhibiting an inhibitory activity towards all 3 serine proteinases were loaded again onto the column and eluted with a closely selected gradient (0.3 - 0.7 M NaCl). A sharp and single inhibitory peak was found and the pooled fractions after concentration in an ultrafiltration chamber (cut off: 1 kDa) showed a protein content of 24 μg/ml and a sugar concentration of 2.5 μg/ml. The specific activity of the inhibitor pool was determined to 7.3 mU/1 μg for trypsin, 0.23 mU/1 μg for elastase and 0.034 mU/1 μg for cathepsin G.
To examine the purity of the inhibitory fractions SDS gel electrophoresis was carried out and revealed a single band with an apparent molecular weight of 16.5×10^3 after staining with Coomassie Blue (fig. 2). For determination of pI the inhibitor was applied to Servalyt Precotes. After 2 h (start 200 V, limited power 4 W), the band was run out of the gel, indicating a pI greater than pH 10.
The titration of elastase with the inhibitor revealed a strong dosedependency of their interaction and the minimum amount of inhibitor resulting in complete inhibition of enzyme (1.7×10^{-8} M) was determined to 39 μl

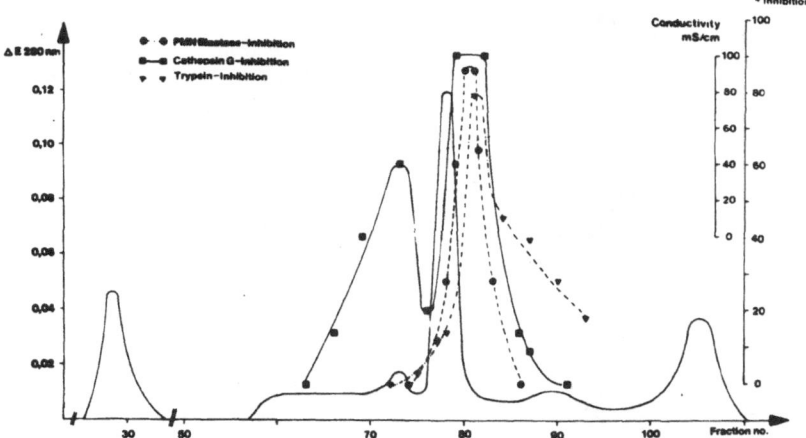

Fig. 1. Ion exchange chromatography on CM-Sephadex C-50 elution profile of the protein ($\Delta E_{280\ nm}$) and inhibitory activity towards PMN-elastase (●), cathepsin G (▪) and trypsin (▼).

equivalent to 9.3 x 10^{-8} M. The attempt to displace leucocyte elastase from its binding to the inhibitor by adding α_2-macroglobulin in a 10-fold molar excess did not reveal a recovery of enzymatic activity towards MeO-Suc-Ala-Ala-Pro-Val-4NA during an incubation period of 32 h, while controls of α_2-macroglobulin-elastase complexes retained their activities. The time dependent inhibition of elastase was assayed by using a recorder equipped photometer and the association rate constant determined to 2.2 x 10^6 M^{-1} sec^{-1}.

DISCUSSION

In the pathogenesis of rheumatoid joint destruction proteolytic enzymes from various sources seem to play a major role in cartilage matrix catabolism. Especially polymorphonuclear leukocytes, abundant in effusions of inflamed joints contain three proteinases with neutral pH-optimum : elastase, cathepsin G and collagenase [1], capable of degrading the major components of cartilage matrix the proteoglycans as well as the collagen molecule [3]. Whereas under physiological conditions the proteinase inhibitors of synovial fluid supervene proteolytic activity in the joint space [20], in disease states such as arthritis some mechanisms have been described, which may disturb the local enzyme-inhibitor balance. Amoung these mechanisms the enzyme release by the so called "frustrated phagocytosis" [21] and the oxidative inactivation of the major serine proteinaeinhibitor α_1-PI by oxygen radicals generated in the cell membrane of activated phagocytes [22] are of special importance. In addition the relative high molecular weight of the serum antiproteinases (α_1-PI = 54000 , α_2-M = 725000) above the upper limit of matrix penetration seems to be disadvantageous for protection of cartilage against the proteolytic attack of the smaller enzymes (elastase = 30000) as indicated by recent work [23]. Therefore the question has been raised, whether the cartilage itself may contain an antiprotease potential constituting a second barrier in the defence against proteolytic attack in order to maintain its structural integrity. The investigations of several laboratories have contributed to an extended knowledge about the presence of low molecular weight inhibitors in connective tissue during the last decade.

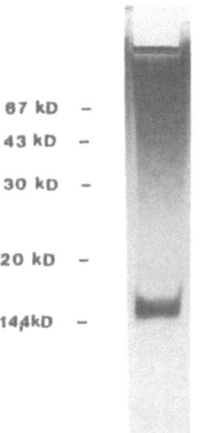

67 kD –
43 kD –
30 kD –
20 kD –
14,4kD –

Fig. 2. SDS gel electrophoresis of cartilage inhibitor. By the marks on the left the electrophoretic mobility of marker proteins with the respective MW is presented.

An inhibitor of the metalloproteinases collagenase, proteoglycanase and gelatinase was purified from rabbit bone by Cawston et al [24] and the same group succeeded in determination of amino acid sequence and in vitro clo-

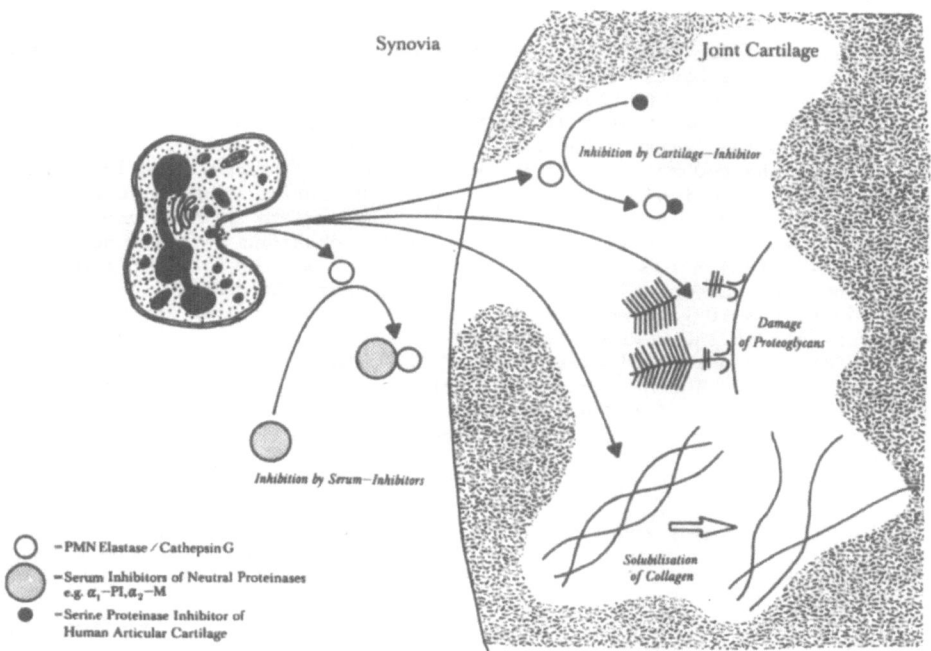

Synovia

Joint Cartilage

Inhibition by Cartilage–Inhibitor

Damage
of Proteoglycans

Inhibition by Serum–Inhibitors

Solubilisation
of Collagen

○ = PMN Elastase / Cathepsin G

◐ = Serum Inhibitors of Neutral Proteinases
e.g. α₁–PI, α₂–M

● = Serine Proteinase Inhibitor of
Human Articular Cartilage

Fig. 3. Hypothesis concerning the possible pathophysiological role of the cartilage inhibitor of
serine proteinases : while the proteolytic enzymes secreted by granulocytes may bypass the
antiprotease potential of synovial fluid,
their action on matrix components and chondro-
cytes is antagonized by the cartilage inhibitor as a second barrier of chondroprotection.

ning of this so-called "tisssue-inhibitor of metalloproteinases (TIMP)" [4],
which has also been identified in connective tissue of human origin [25]
Recently by the work of Cawston et al. [26] the identity of this inhibitor with the serum component β_1-AC was indicated. In contrast to this
clear situation there has been described a variety of fractions with inhibitory activities towards serine proteinases in articular cartilage. While
Knight et al. [27] eluated from a gel filtration column a fraction with an
approximate molecular weight of 14×10^3 exhibiting inhibitory activity
towards extracts from PMN – granules, Killakey et al. [3] revealed distinct inhibitor capacities, differing in molecular size. They found that
the inhibitory activity towards serine proteinases was resolved into two
components of apparent molecular weight 62×10^3 and 12×10^3. Lesjak and
Gosh [28] extracted a serine proteinase inhibitor from human knee and hip
joints displaying a molecular weight of 7×10^3 and consisting of two
subunits.

By the purification procedure presented in this paper we succeeded in the
isolation of an inhibitor with activity towards the three serine proteinases tested :elastase, cathepsin G and trypsin. A homogeneity of the
fraction was evidenced by running as a single band in SDS-gel electrophoresis. As the results of the electrophoretic procedure were the same under
reducing as well as unreducing conditions no dissociation into subunits
was observed in contrast to the data presented by Lesjak and Gosh [28]
for their inhibitor with a lower molecular weight and a broader inhibitor
spectrum including a collagenase of bacterial origin. On the other hand

527

the molecular weight estimation from the electrophoretic mobility of our purified inhibitor fits well into the range given by others for inhibitory fractions on the basis of gelchromatographic procedures [27,3]. The cationic nature of our inhibitor revealed by isoelectric focussing with a pI > 10 seems to be of special importance since protein – cartilage–matrix interactions are charge dependent [29] with a preferential retainance of negatively charged molecules. In addition the preliminary kinetic studies presented here indicate a high affinity of the serine proteinase elastase to the inhibitor molecule as the association constant is high and with 2.2×10^6 M^{-1} sec^{-1} in the range of the value obtained by elastase $- \alpha_2$-M interaction (3.4×10^6 M^{-1} sec^{-1}), while no enzyme dissociation from inhibitor binding was observed during 32 h even under competetive conditions with α_2-M in a 10-fold molar excess. On the basis of these kinetic parameters and refering to the review article of Bieth [30] it seems to be justified to hypothesize a physiological role of the inhibitor molecule in vivo as shown in fig. 3. Whether the serineproteinase inhibitor is of importance for the regulation of proteolytic activity in the joint under pathophysiological conditions such as arthritis remains to be elucidated.

ACKNOWLEDGEMENTS

This work was supported by Deutsche Forschungsgemeinschaft, SFB 223

REFERENCES

1. M. Baggiolini, U. Bretz, and B. Dewald, Subcellular localization of granulocyte enzymes, in : Neutral proteases of human polymorphonuclear leukocytes , K. Havemann, A. Janoff, ed., Urban & Schwarzenberg, Baltimore (1978)

2. P. M. Starkey, A. J. Barrett, and M. C. Burleigh, The degradation of articular collagen by neutrophil proteinases, Biochim Biophys Acta 483 : 386 (1977)

3. J. J. Killakey, P. J. Roughley, and J.S. Mort, Proteinase inhibitors of human articular cartilage, Collagen Rel Res 3 : 419 (1983)

4. A. J. P. Docherty, A. Lyons, B. J. Smith, E. M. Wright, P. E. Stephens, T. J. R. Harris, G. Murphy, and J. J. Reynolds, Sequence of human tissue inhibitor of metalloproteinases and its identity to erythroid-potentiating activity, Nature 318 : 66 (1985)

5. K. Weber and M. Osborne, The reliability of molecular weight determinations by duodecylsulfate-polyacrylamid gel electrophoresis, J. Biol. Chem. 244 : 4406 (1969)

6. S. Engelbrecht, E. Pieper, H. W. Macartney, W. Rautenberg, H. R. Wenzel, and H. Tschesche, Separation of the human leukocyte enzymes alanine aminopeptidase, cathepsin G, collagenase, elastase and myeloperoxidase, Hoppe-Seyler's Z Physiol Chem 363 : 305 (1981)

7. H. W. Macartney and H. Tschesche, Latent and active human polymorphonuclear leukocyte collagenase. Isolation, Purification and characterization, Eur J Biochem 130 : 71 (1983)

8. K. Ohlsson and I. Ohlsson, The neutral proteinases of human granulocytes. Isolation and partial characterization of granulocyte elastase, Eur J Biochem 36 : 473 (1973)

9. H. Tschesche, S. Engelbrecht, H. R. Wenzel, Leukocyte elastase, in : Methods of enzymatic analysis, Vol. 5, H . U. Bergmeyer, ed., Verlag Chemie, Weinheim FRG (1979)

10. K. Nakajima , J. C. Powers, M. J. Castillo, B. A. Ashe, and M. Zimmermann, Mapping the extended substrate binding site of cathepsin G and human leukocyte elastase, J Biol Chem 254 : 4027 (1979)

11. R. G. Woodbury and H. Neurath, Structure, specifity and localization of the serine proteinases of connective tissue, FEBS Lett 114 : 189 (1980)

12. N. Yoshida, M. T. Everitt, H. Neurath, R. G. Woodbury, and J. C. Powers, Substrate specifity of two chymotrypsin-like proteases from rat mast cells. Studies with peptide 4-nitroanilides and comparison with cathepsin G. Biochemistry 19 : 5799 (1980)

13. F. E. Erlanger, N. Kokowski, W. Cohen, The preparation and properties of two new chromogenic substrates of trypsin, Arch of Biochem and Biophys 95 : 271 (1961)

14. H. Tschesche and H. W. Macartney, Collagenase, in : Methods of enzymatic analysis, Vol. 5, H. U. Bergmeyer, ed., Verlag Chemie, Weinheim FRG (1979)

15. M. Aubry and J. Bieth, A kinetic study of the inhibition of human and bovine trypsin and chymotrypsin by the inter-alpha-inhibitor from human plasma. Biochim Biophys Acta 438 : 221 (1976)

16. P. K. Smith, R. I. Krohn, G. T. Hermanson, A. K. Mallia, F. H. Gartner, M. D. Provenzano, E. K. Fujimoto, N. M. Goeke, B. J. Olson, and D.C. Klenk, Measurement of protein using bicinchoninic acid, Anal Biochem 150 : 76 (1985)

17. M. G. Redingbaugh and R. B. Turley, Adaptation of the bicinchoninic acid protein assay for use with microtiter plates and sucrose gradient fractions, Anal Biochem 153 : 267 (1986)

18. N. Blumenkrantz and G. Asboe-Hansen, New method for quantitative determination of uronic acids, Anal Biochem 54 : 484 (1973)

19. M. Dubois, K. A. Gilles, J. K. Hamilton, P. A. Rebers, and F. Smith, Colorimetric method for determination of sugars and related substances, Anal Chem 28 : 350 (1956)

20. A. J. Barrett, The possible role of neutrophil proteinases in damage to articular cartilage, Agents Actions 8 : 11 (1978)

21. P. M. Henson, The immunologic release of constituents from neutrophil leukocytes : II. Mechanism of release during phagocytosis and adherence to nonphagocytosable surfaces, J Immunol 107 : 1535 (1971)

22. H. Carp, A. Janoff, In vitro suppression of serum elastase inhibitory capacity by reactive oxygen species generated by phagocytosing polymorphonuclear leukocytes, J Clin Invest 63 : 793 (1979)

23. M. Kasten, H. Burkhardt, S. Rauls, Alpha₁-Proteinaseinhibitor - ein wirksamer Schutz der Knorpelmatrix gegen die protelytische Aktivität der Granulozytenelastase? Z Rheumatol 45 : 183 (1986)

24. T. E. Cawston, W. A. Galloway, E. Mercer, G. Murphy, and J. J. Reynolds, Purification of rabbit bone inhibitor of collagenase, Biochem J 195 : 159 (1981)

25. R. M. Hembry, G. Murphy, J. J. Reynolds, Immunolocalization of tissue inhibitor of metalloproteinases (TIMP) in human cells. Characterization and use of a specific antiserum, J Cell Sci 73 : 105 (1985)

26. T. E. Cawston, D. N. Noble, G. Murphy, A. J. Smith, C. Woodley, B. L. Hazleman, Rapid purification of the tissue inhibitor of metalloproteinases (TIMP) from human plasma and identification as a γ-serum protein, Biochem J 238 : 677 (1986)

27. J. A. Knight, R. W. Stephens, G. R. Bushell, P. Gosh, and K. F. Taylor, Neutral protease inhibitors from human intervertebral disc and femoral head articular cartilage, Biochim Biophys Acta 584 : 304 (1979)

28. M. S. Lesjak and P. Gosh, Polypeptide proteinase inhibitor from human articular cartilage, Biochim Biophys Acta 789 : 266 (1984)

29. A. Maroudas, Physicochemical properties of articular cartilage, in : Adult articular cartilage, M. A. R. Freeman, ed., Pitman, London (1973)

30. J. Bieth, Pathophysiological interpretation of kinetic constants of proteinase inhibitors, Bull Europ Physiopath Resp 16 (Suppl) : 183 (1980)

DETECTION OF GRANULOCYTE ELASTASE SPECIFIC IgG SPLIT PRODUCTS

IN RHEUMATOID SYNOVIAL FLUID

Ilsebill Eckle*, Gerald Kolb*, Friedrich Neurath°, and Klaus Havemann*

*Division of Hematology/Oncology, Baldingerstrasse, D-3550 Marburg, °Dept. of Operative Medicine, Baldingerstrasse D-3550 Marburg, F.R.Germany

ABSTRACT

In vitro granulocyte elastase is known to cleave a large number of substrates e.g. complement components, fibrinogen, collagen and IgG. In vivo the enzyme is rapidly complexed by the plasma inhibitors a_1 proteinase inhibitor (a_1PI) and a_2 macroglobulin. Therefore in biological fluids elastase is measured as the inactive a_1PI-complex.

We present a radioimmunoassay for elastase specific IgG split products as marker for the elastase activity in vivo.

Elastase splits human IgG_1 in Fab and Fc fragments and low molecular weight peptides. We produced specific antibodies against the elastase induced Fc fragment by immunization with an elastase generated peptide. After purification of the antibodies there is no crossreactivity with native IgG nor with similar Fc fragments produced by plasmin or papain.

The elastase specific IgG split products are detected in synovial fluid samples of patients with rheumatoid arthritis. The measured concentrations are higher in the RA group than in control groups of patients with other inflammatory joint diseases.

INTRODUCTION

Human leukocyte elastase (HLE) is an unspecific proteinase which splits a number of physiological substrates like clotting factors, complement components, fibrinogen, different collagen types and IgG (1-4). IgG is cleaved by elastase into Fab and Fc fragments and additionally low molecular weight peptides, other substrates (s.above) are degraded completely. In vivo elastase is complexed by a_1PI and a_2 macroglobulin The commercially available test measures only the inactive HLE-a_1PI-complex[5]. In synovial fluid of patients suffering from rheumatoid arthritis (RA) IgG is split into Fab and Fc fragments. We present an assay for the HLE induced Fc fragments to get information about the actual proteolytic activity in vivo.

MATERIAL AND METHODS

IgG$_1$ was cleaved by isolated leukocyte elastase (enzyme-to-substrate-ratio 1:1000, w/w, incubation 80 h, 37°C) into Fab and Fc fragments and low molecular peptides. Fc was purified by affinity chromatography on protein A sepharose and gelfiltration (Sephadex G 100). Isolated Fc was labelled with ^{125}J according to the method of Bolton and Hunter[6].

A peptide fraction from the hinge region of IgG$_1$ was purified by anion exchange chromatography (DEAE-Sepharose CL6B) and gelfiltration (Sephadex G 25). The isolated peptide fraction was coupled to bovine serum albumin with glutaraldehyd[7], and rabbits were immunized with the conjugated peptide. Antisera were purified by affinity chromatography on IgG$_1$ Sepharose.

For the radioimmunoassay sample or standard and antisera were diluted in 50 mM sodiumphosphatbuffer pH 7.0 with 0.1% human serum albumin. Incubation was performed for 48 h at 4°C, label was added after 24 h. The bound antigen was precipitated with polyethylenglycol[8] and centrifugated. The supernatants were aspirated and the radioactivity of the residues measured.

Synovial fluid was aspirated from the knee joints and collected in heparine (75 U/ml). After centrifugation supernatants were stored frozen at -20°C.

RESULTS AND DISCUSSION

Radioimmunoassay for elastase specific Fc

Incubation of IgG$_1$ with elastase (enzyme-to-substrate-ratio 1:1000) for 80 h results in 95% proteolysis of IgG. About 91% are Fab and Fc fragments, 4% are low molecular weight peptides. Separation of the peptides by aionexchange and gelfiltration gives a peptide from the hinge region of IgG which bears the same neoantigen as the Fc fragment. Immunization with the peptide rather than with the Fc fragment gives more specific antibodies against the neoantigen. After adsorption of the antibodies on native IgG$_1$ they show no more crossreactivity with indigested IgG.

The calibration curve of the radioimmunoassay is linear in the range of 100 to 500 ng HLE-Fc/ml. The interassay and intraassay variation coefficient of 6.5% and 11% respectively show good reproducibility. Recovery of Fc after incubation in plasma or serum for 1 h at room temperature is more than 92%, which shows that it is not further degraded in biological fluids. The Fc fragment is specific for HLE, there is no crossreactivity with similar Fc fragments produced by plasmin or papain.

IgG incubation with HLE (enzyme-to-substrate-ratio 1:200, 37°C) gives first detectable cleavage products after 30 min. In contrast first antibody binding in our assay is after 6 h HLE-digestion. This suggests, that only further proteolysis reveals the neoantigenic structures.

Elastase specific Fc fragments in rheumatoid arthritis

Patients were divided into three groups according to the clinical status and the ARA criteria[9].

Group I   Rheumatoid arthritis (RA) (n=23)
Group II   Inflammatory joint disease except RA, i.e. arthritis
urica (gout arthritis), Morbus Reiter, septic syno-
vitis, seronegative spondylarthritis (n=23)
Group III   Joint effusions after trauma or in osteoarthrosis, i.e.
post operative (meniscotomy), posttraumatic hemarthrosis, osteoarthrosis with gonarthrosis (n=19)

In group I 15 synovial fluid samples of 23 (=65%) are positive for
Fc, in group II 11 of 23 (=48%) and in group III 7 of 19 (=37%). The median values and range of the Fc concentration and the corresponding concentrations of elastase-a_1PI-complex are shown in table 1.

Table 1.   Median concentration of elastase specific Fc fragment
and elastase-a_1PI-complex in synovial fluid samples of
patient groups I to III.

Patient group	n		Fc (μg/ml SF)	ELP-a_1PI (ng/ml SF)
I	23	median	0.62	10,600
		range	(0-3.3)	(0-114,000)
II	23	median	0	5,200
		range	(0-2.8)	(255-78,500)
III	19	median	0	66
		range	(0-2.3)	(0-4,730)

The concentration of the elastase specific split products does not
correlate with the concentration on the elastase-a_1PI-complex.

The measurement of the elastase specific Fc fragment and of the
elastase a_1PI-complex give different informations: elevated elastase-a_1
PI-levels show that elastase has been released but cannot show whether
the enzyme has acted on proteinsubstrates. The detection of elastase
specific IgG fragments shows that elastase has split IgG in vivo. As the
Fc fragment which is detected in the RIA seems to be a late digestion
product, the action of elastase has to be rather long.

This explains why the Fc fragment is more often and in high concentrations found in synovial fluids of the RA-group: it can serve as a
marker of the chronic inflammatory process.

REFERENCES

1. W. Schmidt, R. Egbring, and K. Havemann, Effect of ELP and CLP from
human granulocytes on isolated clotting factors, Thromb. Res. 6:315-
326 (1975).
2. K. Johnson, K. Ohlson, and J. Olsson, Effects of granulocyte neutral
proteases on complement components, Scand. J. Immunol. 5:421-427
(1976).
3. M. Gramse, C. Bingenheimer, R. Egbring, and K. Havemann, Degradation
products of fibrinogen by elastase-like neutral protease from human
granulocytes, J. Clin. Invest. 61:1027-1033 (1978).
4. A. Baici, M. Knöpfel, K. Fehr, F. Skvaril, and A. Böni, Kinetics of
the different susceptibilities of the four human immunoglobulin G
classes to proteolysis by human lysosomal elastase, Scand. J.
Immunol. 12:41-50 (1980).

5. S. Neumann, G. Gunzer, N. Hennrich, and H. Lang, "PMN-elastase assay" enzyme immunoassay for human polymorphonuclear elastase complexed with alpha-1-proteinase inhibitor, J. Clin. Chem. Clin. Biochem. 22: 693-697 (1984).

6. A.E. Bolton, and W.M. Hunter, The labelling of proteins to high specific radioactivities by conjugation to a ^{125}J-containing acylating agent, Biochem. J. 133:529-539 (1973).

7. D.J. O'Shaughnessy, Antibodies, in: "Radioimmunoassay of Gut Regulatory Peptides", S.R. Bloom and R.G. Chong, ed., WB Sauners Company Ltd., London, Philadelphia, Toronto, pp. 11-20 (1982).

8. J.H. Livesey, and R.A. Donald, Prevention of adsorption losses during radioimmunoassay of polypeptide hormones: effectivness of albumin, gelatin, caseins, Tween 20 and plasma, Clin. Chim. Acta 123:193-198 (1983).

9. M.W. Ropes, E.A. Bennet, S. Lobb, R. Jacox, and R. Jessar, Revision of diagnostic criteria for rheumatoid arthritis, Bull. Rheum. Dis. 9: 175-183 (1958).

THE T CELL SPECIFIC SERINE PROTEINASE TSP-1: BIOCHEMICAL CHARACTERIZATION, GENETIC ANALYSIS, AND FUNCTIONAL ROLE

H.-G. Simon, U. Fruth, R. Geiger, M.D. Kramer* and M.M. Simon

Max-Planck-Institut für Immunbiologie, *Ernst Rodenwaldt Institut Koblenz, FRG, and Abteilung für Klinische Chemie und Klinische Biochemie in der Chirurgischen Klinik der Universität München

INTRODUCTION

Activation of T lymphocytes by antigen leads to the production of a variety of hormone like factors - i.e. lymphokines (1) - and of effector molecules - i.e. leukolysins (2) - that either control cellular and humoral immune responses or exert effector functions. More recently a number of independent groups have identified several serine esterase(s) (3-5)/serine proteinase(s) (6,7) respectively, in T effector cells. In these and previous reports the involvement of proteolytic enzymes in T cell mediated processes such as cytolysis (3,8,9), induction of lymphocyte proliferation (7,10-12) and cellular migration (13,14) have been suggested. Here we describe the biochemical characterization and genetic analysis of a T cell specific serine proteinase with highly restricted substrate specificity, termed TSP-1, which was isolated from cloned murine cytolytic T lymphocytes. We demonstrate that TSP-1 is packaged into cytoplasmic granules of T effector cells and secreted into the extracellular space as a result of antigenic stimulation. Our functional studies indicate that TSP-1 is involved in the process of cell mediated lympholysis (CML) and in the T cell control of virus replication. Moreover the data suggest that the same enzyme regulates cellular proliferation and migration of lymphocytes.

RESULTS AND DISCUSSION
Isolation and Characterization of TSP-1

TSP-1 was purified from postnuclear supernatant of the cytolytic T lymphocyte line (CTLL) 1.3E6 by ammonium sulfate precipitation followed by affinity - and ion-exchange - chromatography (7). The characteristics of TSP-1 are outlined in Table 1. The enzyme has proteolytic activity on high molecular weight proteins (14) and is a serine type endopeptidase as revealed by its exclusive sensitivity to inhibitors of serine -, but not to those of either carboxyl-, thiol- or metalloproteinases (7). TSP-1 is a disulfide linked homodimer with a mol.wt. $\sim$ 60 kD (7). The most prominent feature of the enzyme is its highly restricted specificity for the amino acid (AA) sequences Phe-Arg and Pro-Phe-Arg as determined on a panel of 28 chromogenic model peptide substrates (7). The pH optimum of proteolytic and amidolytic activity of TSP-1 is around pH 8.0. The kinetic constants (K_m, V_{max}) for the TSP-1 catalyzed hydrolysis of Ac-Phe-Arg-OEt revealed a cleavage rate in the order of magnitude similar to that observed with a comparable tissue type serine proteinase from human, i.e. urinary kallikrein (15).

Table 1.   Biochemical characterization and kinetic data of TSP-1

Cell source	:	CTLL 1.3E6; long-term cultured murine cytolytic T lymphocyte
Preparation for TSP-1 purification	:	Post nuclear supernatant (PNSN)
Characterization of TSP-1		
Enzyme class	:	Proteinase
Subgroup	:	Serine endopeptidase
Substrate specificity	:	Phe-Arg / Pro-Phe-Arg*
pH-optimum	:	8.0 - 8.5
Mol.wt.	:	~60 kD (SDS-PAGE, nonreducing) ~30 kD (SDS-PAGE, reducing)
Kinetic data		
K_m [mol/l]	:	1.6×10^{-4} (on Ac-Phe-Arg OEt)
V_{max} [μmol $\times$ min^{-1} $\times$ mg^{-1}]	:	152 (on Ac-Phe-Arg OEt)

* Cleavage of the chromogenic substrates Phe-Arg-nitro-anilide (NA) and H-D-Pro-Phe-Arg-NA occurs at the carboxy-terminus of the basic amino acid L-arginine.

Subcellular Localization and Secretion of TSP-1

Subcellular fractionation analysis was accomplished by Percoll density gradient centrifugation of CTLL cell lysates (16) according to Millard et al. (17). Fractions were then assayed for the distribution of subcellular enzyme markers and for the granule constituent cytolysin (18,19) to assess for the enrichment of cytoplasmic organelles. As shown in Fig. 1a, fraction 3 of the Percoll gradient contained the main peak of TSP-1 activity (80-90%) and was also enriched for cytolysin activity. Morphological examination of this fraction revealed the characteristic features of granules (Fig.1b) thus demonstrating that TSP-1 is associated with these cytoplasmic organelles. The identity of the granule associated proteinase in enriched granule preparations with that in total cell lysates of independent CTLL was indicated by a) its specificity for the chromogenic substrate H-D-Pro-Phe-Arg-NA, b) its sensitivity to class-specific as well as TSP-1 specific enzyme inhibitors, c) by its binding of the serine proteinase-specific affinity ligand tritiated diisopropyl fluorophosphate ([^{3}H] DFP) and d) by its reactivity with a polyvalent TSP-1 specific rabbit antiserum (16).

Previous studies have shown that CTL mediated target cell lysis is accompanied by antigen induced exocytosis of the granule component cytolysin into the extracellular space (18,19). It was therefore to be expected that TSP-1 contained in the same cytoplasmic granules is simultaneously secreted during effector-target cell interaction. Using a panel of antigen-specific CTL it was indeed found that secretion of TSP-1 was induced by exposure of CTL to those tumor cells which are specifically lysed by CTL but not or only marginally by third party tumor cells. One example is depicted in Fig. 2 in which CTLL 1.D.9 (specific for the H-2^d haplotype) was incubated with either H-2^d tumor cell (P815) or with tumor cells of a different haplotype (EL-4, H-2^b). Interestingly, it was found that TSP-1 was also released by those cloned T helper cells (THL) which were shown before to express this enzyme (6, and see below) upon interaction with appropriate target cells. Thus the data demonstrate that effector cells of both major T cell subsets can be specifically induced to release TSP-1 into the extracellular space.

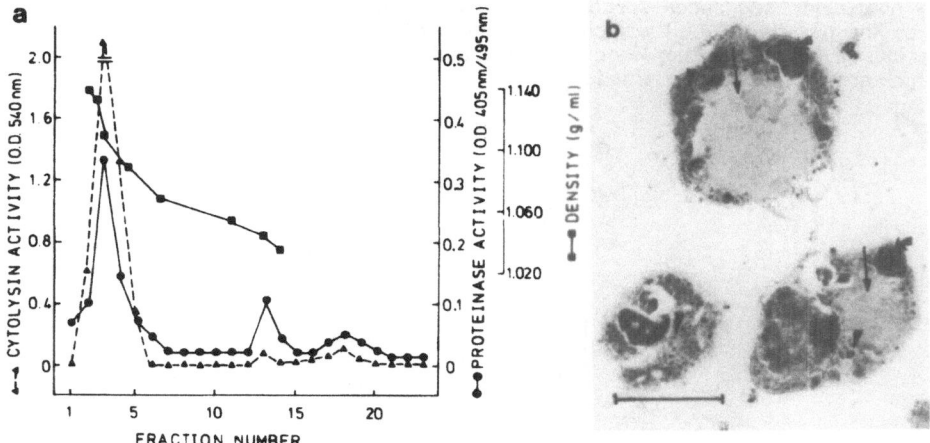

Fig. 1.   Biochemical and ultrastructural analysis of Percoll gradient fractionation of postnuclear supernatant (PNSN) from CTLL HY3-Ag3 (16).
CTLL HY3-Ag3 T cells were disrupted by nitrogen cavitation and
subcellular compartments were separated by centrifugation of the resulting PNSN through a self generating Percoll gradient (48% ⌈v/v⌉ Percoll
stock solution, 60.000 x g, 30 min, 0°C) (17). Fractions of 1 ml were
collected from bottom (Fr. 1) to top (Fr. 24) of the ultracentrifuge tubes.
Percoll was removed by subsequent ultracentrifugation (105.000 x g, 120
min, 0°C). The quality of separation was routinely controlled by marker
enzyme analysis (not shown).

a) TSP-1 activity was measured on the peptide substrate H-D-Pro-Phe-
Arg-pNA. 10 µl of test samples were mixed with 100 µl (0.3 M) of
chromogenic substrate H-D-Pro-Phe-Arg-NA in 100mM Tris-HCl, pH
8.5 (on 0.2 ml final). O.D. was tested at 405 nm after 1h incubation at
37°C. Cytolysin was assayed for lytic activity on sheep erythrocytes
as described (16) and the density profile was determined by the use of
density marker beads.

b) Electron micrograph of an ultrathin section of fraction 3 from the

Percoll gradient containing the peaks of TSP-1/cytolysin activities
(bar = 1 µm). Big arrow: typical T cell cytoplasmic granule with darkly
staining matrix (thin arrow) and surrounding small vesicles (arrow-
head). Bended arrow: structures of amorphous, very electron dense
material.

Cloning of the TSP-1-specific gene and its induction in T lymphocyte subsets

For genetic analysis of the TSP-1 specific gene we have established a cDNA expression library in gt11 phages from poly (A)$^+$ mRNA of CTLL 1.3E6 which has been used before as a cell source for the purification of the enzyme. Clones specific for TSP-1 and related serine proteinases were identified by applying a combination of screening techniques using a polyclonal TSP-1 specific rabbit antiserum and a set of oligonucleotide probes which have been designed according to nucleotide sequences specific for each of three recently published cDNA's encoding putative T cell associated serine proteinases (20-22). We have isolated various cDNA clones that code for TSP-1 and similar proteinases the sequences of which are presently determined.

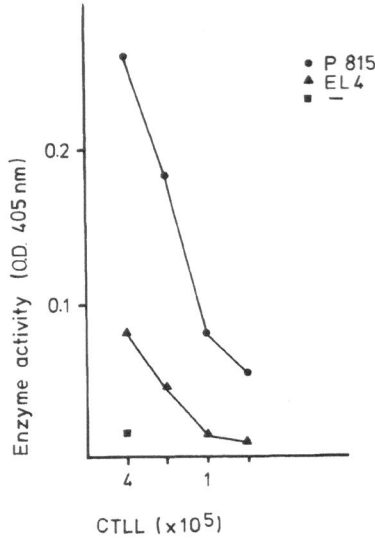

Fig. 2. Antigen induced TSP-1 secretion from CTLL 1.D.9.
CTLL 1.D.9 (4x10^5/well) were titrated in several dilutions to either P815 (H-2^d) or EL-4 (H-2^b) tumor cells (4x10^5/well) and incubated in RPMI medium (w.o. phenol-red) supplemented with 1 mg/ml bovine serum albumin (Boehringer, Mannheim) and 0.01 M HEPES buffer. After centrifugation at 500 rpm the plates were incubated at 37°C for 3h. Supernatant (SN, 150 µl) was removed from replicate wells. Proteinase activity was tested in 50 µl of pooled SN on the chromogenic substrate H-D-Pro-Phe-Arg-NA. O.D. was tested after 1h incubation (37°C) at 405 nm.

In order to investigate the induction and regulation of TSP-1 expression in T cell subsets we used an oligonucleotide probe specific for the H-factor gene (20) which was shown to be identical with the TSP-1 gene (H.G. Simon et al., submitted for publication). For the design of the oligonucleotide we chose a sequence (corresponding to nucleotides 560-583 of the H-factor) that has less than 31% homology with any sequence found in the two other serine proteinase encoding cDNAs. It is seen from Table 2 that both T cell subsets - Lyt-2$^+$, L3T4$^-$ and L3T4$^+$, Lyt-2$^-$ - previously separated by flow cytofluorometry and activated by concavalin A (Con A) and Interleukin-2 (IL-2), expressed cytolytic activity in the lectin facilitated lympho-lysis assay which - as expected - was much more pronounced in the Lyt-2$^+$ subset. Most importantly both populations expressed TSP-1-specific mRNA after activation by lectin and IL-2 in vitro. The probe hybridized to mRNA of both activated T cell subsets with a single band (about 1 kb in Northern blots) and does not give any detectable signal when hybridized to mRNA from resting T cells (Table 2). Consistently, higher amounts of TSP-1-specific mRNA were found in L3T4$^+$ as compared to Lyt-2$^+$ populations. Together with the finding that both of the purified activated cell populations were homogeneous (> 97%) with respect to the respective surface markers Lyt-2$^+$ and L3T4$^+$ these results clearly demonstrate that the expression of TSP-1 is not restricted to a particular T cell subset in mice. This is also supported by the observation that all CTLL (Lyt-2$^+$, L3T4$^-$) and at least some of the cloned T helper cell lines (THL; L3T4$^+$, Lyt-2$^-$) tested by us so far express TSP-1 activity (6) and show comparable signals with the oligonucleotide probe (data not shown).

Table 2. Expression of TSP-1 specific mRNA in resting T cells and in unselected as well as selected Lyt-2$^+$, L3T4$^-$ and L3T4$^+$, Lyt2$^-$ T lymphocytes activated by ConA and IL-2.

		T unselected resting	T unselected activated	Lyt-2$^+$ activated	L3T4$^+$ activated
Phenotype	Lyt-2$^+$	39.2%	n.d.	> 98%	< 2%
	L3T4$^+$	58.3%	n.d.	< 2%	> 97%
Cytolytic activity (P815 / PHA)		−	++	+++	+
Expression of mRNA specific for TSP-1	dil 1:2				
	1:4				
	1:8				

Nylon wool enriched C57BL/6 (B6) lymph node T cells were separated into Lyt-2$^+$, L3T4$^-$ and L3T4$^+$, Lyt-2$^-$ lymphocytes by flow cytofluorometry (FCF) as described before (23). Enriched unselected B6 T cells or FCF-sorted Lyt-2$^+$ and L3T4$^+$ lymphocytes (1x10^5 cells/well) were incubated with ConA (3 µg/ml), irradiated syngeneic spleen feeder cells (5x10^6/well; 3000 rad) and recombinant human IL-2 (rec.hIL-2) as lymphokine source. On day 5, activated lymphocytes were again analysed for their phenotype by FCF and were tested for cytolytic activity on ^{51}Cr-labeled P815 tumor target cells (in the presence of phytohemagglutinin, PHA) as described (23). For the analysis of mRNA from T cells (1x10^6) the individual populations were washed in PBS once and pelleted in a 1.5 ml tube (15000 x g, 15s). After resuspension in 45 µl of ice-cold 10 mM Tris (pH 7.0), 1 mM EDTA, cells were lysed by addition of two 5 µl aliquots of 5% Nonidet P-40 with 5 min on ice in between. The nuclei were pelleted (15.000 x g , 2.5 min) and 50 µl of the supernatant were mixed with 30 µl of 20 x SSC (standard saline citrate) plus 20 µl of 37% (w/w) formaldehyde. The mixture was then incubated at 60°C for 15 min, and stored at -70°C. For analysis, the samples were serially diluted with 15 x SSC in a final volume of 150 µl and were applied to a nitrocellulose membrane using a Minifold II apparatus (Schleicher and Schuell). The nitrocellulose sheet was baked (80°C, 2h) in vacuo to fix cytoplasmic macromolecules (24).

The hybridization of the filters with the ^{32}P-labeled gel-purified oligonucleotides was done as outlined by Miyada (25). After the filters were washed, they were exposed to Kodak X-Omat film.

Role of TSP-1 in Cytolysis

The assumption that proteolytic enzymes are involved in CTL or natural killer cell (NK) mediated cytolysis is derived from a variety of studies showing that low molecular weight inhibitors of proteinases are able to interfere with the cytolytic potential of intact cells (8,9). The fact that cytoplasmic granules contain the lytic machinery of CTL (18,19,26) and the finding that granule associated TSP-1 is exocytosed during cytolysis (27) strongly suggested that this enzyme might play a role in the process of target cell lysis. It was found, however, that purified TSP-1 does not express cytolytic activity on its own when tested on a panel of tumor target cells or on sheep red blood cells (7). Thus we reasoned that TSP-1 might play, if at all, an indirect role in the lytic event. To elucidate the involvement of TSP-1 in the lytic process we have employed a highly specific and irreversible inhibitor molecule, i.e. H-D-Pro-Phe-Arg-chloromethyl ketone (PFR-CK), which was synthesized according to the known AA target sequence recognized by TSP-1 (27). The inhibitory effect of PFR-CK on cytolytic activity was studied using either intact CTLL or preparations of enriched cytoplasmic granules which have been shown before to lyse erythrocyte or tumor target cells (18,19). A schematic representation of the data is given in Fig.3.

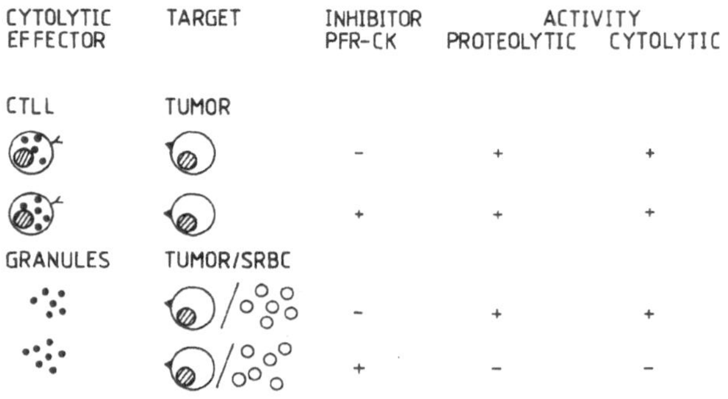

CYTOLYTIC EFFECTOR	TARGET	INHIBITOR PFR-CK	ACTIVITY PROTEOLYTIC	CYTOLYTIC
CTLL	TUMOR			
		–	+	+
		+	+	+
GRANULES	TUMOR/SRBC			
		–	+	+
		+	–	–

Fig. 3. Schematic representation of the effect of the proteinase inhibitor PFR-CK on cytolytic potential of intact CTL and of enriched cytoplasmic granules (27).

The cytolytic potential of intact CTLL was neither influenced by preincubation of effector cells with PFR-CK (1-4 h) nor when the inhibitor was present during the entire time period of CML assays (4 h). Furthermore, pretreatment of CTLL with PFR-CK did not inhibit T cell associated TSP-1 activity, indicating that the inhibitor does not penetrate CTL membranes under these conditions. In contrast, PFR-CK was able to inhibit in a dose dependent way both TSP-1 activity and the cytolytic potential of granules on erythrocytes and tumor cells. Since TSP-1 has a pH optimum around 8.0 and has only little activity at pH 5-6, a pH range suggested for the granule compartment (5), the data suggest that the enzyme performs its function in later stages of CML, i.e. during exocytosis of granule structures. Thus, TSP-1 seems to be an important element of the lytic machinery of CTL. However, the physiological substrate(s) of TSP-1, which may be either a structure(s) associated with cytoplasmic granules or a molecule(s) expressed on the membrane of target cells, are yet to be elucidated.

Role of TSP-1 in Humoral Immune Response

In the past, serine proteinases such as trypsin, chymotrypsin or thrombin have been shown to induce proliferation in murine spleen cells (10-12). Moreover it was found that this induction process can be suppressed by proteinase inhibitors (28). On the basis of these observation we have performed studies to elucidate whether TSP-1 provides similar signals for lymphocyte populations. For this matter unselected spleen cells or enriched B-lymphocyte populations were cultured together with various concentrations of the purified enzyme (7). As can be seen from Table 3, TSP-1, like pancreatic trypsin, is able to induce DNA synthesis in B lympocytes in an antigen independent way. The fact that spleen cells of nu/nu mice mainly consists of B lymphocytes and that proliferation was only induced by the B cell mitogen LPS but not by the T cell mitogen Con A under these conditions indicate that TSP-1 exerted its function on B lymphocytes. Pretreatment of TSP-1 with the proteinase inhibitor phenylmethane sulfonylfluoride (PMSF) abolished this effect. We are presently investigating, whether TSP-1 acts directly on B cells or whether it induces their proliferation indirectly, i.e. via accessory cells.

Table 3. Induction of DNA synthesis in spleen cell population from C3H nu/nu mice by TSP-1

Addition to culture	Incorporation of $[^3H]$ dThd in C3H nu/nu spleen cells (c.p.m.)
None	4,288 + 73
LPS 1 μg/ml	36,939 + 1,039
ConA 3 μg/ml	390 + 83
TSP-1 750 ng/ml	13,054 + 932
" , pretreated with PMSF	4,083 + 260
Pancreatic trypsin 500 ng/ml	27,224 + 1,305
" , pretreated with PMSF	4,178 + 291

Responder cells (5×10^5/well) were incubated in microcultures (0.2 ml; Triplicates) in RPMI medium (without fetal calf serum) with either ConA (3 μg/ml), LPS (1μg/ml), TSP-1 or pancreatic trypsin (untreated or pretreated with the proteinase inhibitor PMSF. On day 3,12-14h before harvest, 1.25μCi[^{3}H]dThd were added to each micro-well. Cells were harvested onto filter strips and counted for ß emission (7).

Role of TSP-1 in migration/invasiveness

Many reports in the past have suggested that proteolytic enzymes may be important elements involved in the migration of lymphocytes through endothelial cell walls and in invasion of tumor cells into distant organs (29). During our studies on cloned murine T effector cells we have derived a variety of in vitro variants which were shown to be tumorgenic when injected into syngeneic recipients (30). It was found that TSP-1 was highly expressed in both normal CTLL and malignant variants and that the enzyme was produced and secreted constitutively by the latter but not by the former population. These findings promted us to investigate whether TSP-1 is able to attack those structures that constitute pyhsiological tissue barriers (14). To this end we tested the proteolytic activity of intact CTLL and of purified TSP-1 on in vitro produced subendothelial extracellular matrix (ECM). These ECM preparations have been shown to mimic the basal laminae of the vascular endothelium in vivo (31) and to be similar in organization and supramolecular composition to naturally occuring ECM. This study showed that both, cell lysates of CTLL as well as purified TSP-1, were able to release high molecular split products from sulphated radiolabeled proteoglycans in ECM (Fig. 4).

Whereas growth and expression of TSP-1 in T-effector cells are regulated by stimuli provided by antigens and/or lymphokines the transformed variants are growing autonomously and produce the enzyme constitutively. The latter properties may be one reason for the spread of T-lymphoma cells to distant organ sites. On the other hand normal activated T cells might use TSP-1 during extravasation in response to an activation signal from an affected organ.

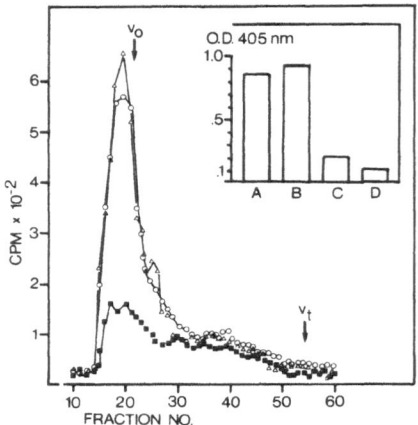

Fig. 4.   Degradation of ECM associated sulphated proteoglycans by Triton X-100 cell lysate or by TSP-1. Aliquots of either Triton X-100 cell lysate (CTLL 1.3E6K1, o) or of the enriched TSP-1 preparation ($\triangle$) were incubated with ECM from bovine corneal endothelial cells labeled with $Na_2 [^{35}SO_4]$ (32). As control, TSP-1 was incubated with ECM in the presence of the serine proteinase inhibitor aprotinin [■]. After 50hr at 37°, the soluble radioactive split products were analysed via molecular sieve chromatography on Sepharose 4B as described (14). Inset: Amidolytic activities of the same samples tested in parallel: CTLL cell lysate (A), TSP-1 (B), TSP-1 + 1 mg/ml aprotinin (C), chromogenic substrate H-D-Pro-Phe-Arg-NA alone (D).

Role of TSP-1 in virus control

The model, according to which CTL inhibit virus replication in vivo, purports that CTL recognize viral antigens on the surface of infected cells and kill them before the virus begins to replicate (33). However, in addition, studies describing the production and secretion of CTL associated molecules such as interferon γ (34,35), and TSP-1 (7,27) suggest that soluble factors might also play a role in antiviral responses. Moreover, in an approach to elucidate physiological substrates for TSP-1 we have screened data banks for proteins containing peptide sequences (Phe-Arg, Pro-Phe-Arg) optimally recognized by TSP-1 and we have found that a large number of structures containing these or very similar sequences were encoded by DNA and RNA viruses. We therefore investigated whether viral proteins might be substrates for the T cell derived proteinases.

For this purpose we incubated a preparation of recombinant reverse transcriptase (RT) from Moloney murine leukemia virus (MoMuLV) with TSP-1. Residual activity of RT(MoMuLV) was tested in a reverse transcriptase assay system containing rabbit globin mRNA and radioactively labeled nucleotides. First strand cDNA molecules of the globin mRNA template were analyzed by urea-polyacryla mide gelelectrophoresis and autoradiography of the gel. In parallel aliquots of untreated and treated RT(MoMuLV) were separated on SDS-PAGE under reducing conditions and analysed by protein staining. The data revealed that TSP-1 is able to reduce the polymerase activity of RT(MoMuLV) in a dose dependent manner. The inactivation of RT(MoMuLV) was accompanied by limited proteolysis of the polymerase, resulting in three distinct degradation products. Furthermore, preliminary experiments have shown that preincubation of intact MoMuLV with TSP-1 resulted in considerable reduction of the capacity of the viruses to infect 3T3 target cells (H.G. Simon et al., submitted for publication). These results indicate that TSP-1 might also degrade protein structures of the viral envelope. Together these studies strongly suggests that in addition to their cytolytic potential, CTL might be able to inhibit virus replication by proteolytic cleavage of virus specific proteins via their secreted TSP-1.

CONCLUSION

This study demonstrates that T lymphocyte effector cells - CTL and TH - but not their precursor cells express a serine proteinase, termed TSP-1, with a highly restricted specificity. This enzyme is allocated to cytoplasmic granules of CTL and THL and is released together with other granule components into the extracellular space in response to antigenic stimulation. Recently, serine esterases with similar properties have been described independently by other groups (3,4,5).

In the immune system, T lymphocytes play a central role as regulator cells for both humoral and cellular immune responses and as effector cells which destroy virus infected and tumorigenic syngenic target cells. There is good evidence that all T-cell mediated activities are associated with regulated secretion of functional molecules such as IL-2, IL-3, IL-4, and Interferon γ (1) and of leukolysins such as, lymphotoxin, cytolysin (2). The new finding reported in this paper that TSP-1 is inducible in resting T cells and is specifically released as a result of interaction between T effector cells and appropriate target cells suggests its involvement in T cell associated processes (36). In fact, our data strongly indicate that TSP-1 participates in the lytic event of CTL and interferes with virus replication in vivo most probably by proteolytic degradation of virus proteins. The possibility that T effector cells might also regulate humoral immune responses via TSP-1 is suggested by the finding that the enzyme is able to induce B lymphocytes for proliferation. Moreover, the fact that both intact CTLL, including malignant variants, and purified TSP-1 are capable to degrade sulphated proteoglycan structures of subendothelial ECM indicates that this enzyme allows activated T lymphocytes or malignant T cells to traverse membrane barriers.

Taken together, our studies indicate that TSP-1 and possibly other comparable serine proteinases constitute a new set of functional molecules of T cells with rather restricted substrate specificity which are used together with other soluble mediators by these lymphocytes to carry out their tasks in vivo. Needless to say that the revelation and characterization of physiological substrates of TSP-1 will further contribute to elucidate the various mechanisms by which T cells regulate immune response.

ACKNOWLEDGEMENT

This work was in part supported by the grant Si 214/5-4 from the Deutsche Forschungsgemeinschaft.

REFERENCES

1. T. Kishimoto, Factors affecting B-cell growth and differentiation, Ann. Rev. Immunol. 3: 133 (1985).
2. H.S. Koren, Proposed classification of leukocyte-associated cytolytic molecules, Immunol. 8: 3 (1987).
3. M.S. Pasternack, C. R. Verret, M.A. Liu, and H.N. Eisen, Serine esterase in cytolytic T lymphocytes, Nature 322: 740 (1986).
4. D. Masson, M. Nabholz, C. Estrade, and J. Tschopp, Granules of cytolytic T-lymphocytes contain two serine esterases, EMBO J. 5:1595 (1986).
5. J.D.E. Young, L.G. Leong, C.C. Liu, A. Damiano, D.A. Wall, and Z.A. Cohn, Isolation and characterization of a serine esterase from cytolytic T cell granules, Cell 47: 183 (1986).
6. M.D. Kramer, L. Binninger, V. Schirrmacher, H. Moll, M. Prester, G. Nerz, and M.M. Simon, Characterization and isolation of a trypsin-like serine protease from a long-term culture cytolytic T cell line and its expression by functionally distinct T cells, J. Immunol. 136: 4644 (1986).
7. M.M. Simon, H. Hoschützky, U. Fruth, H.G. Simon, and M.D. Kramer, Purification and characterization of a T cell specific serine proteinase (TSP-1) from cloned cytolytic T lymphocytes, EMBO J. 5: 3267 (1986).
8. T.W. Chang, and H.N. Eisen, Effects of N α-Tosyl-L-Lysyl-Chloromethylketone on the activity of cytotoxic T lymphocytes, J. Immunol. 124: 1028 (1980).
9. D. Redelman, and D. Hudig, The mechanism of cell-mediated cytotoxicity. I. Killing by murine cytotoxic T lymphocytes requires cell surface thiols and activated proteases, J. Immunol. 124: 870 (1980).
10. T.L. Vischer, Stimulation of mouse B lmyphocytes by trypsin, J. Immunol. 113: 58 (1974).
11. J.G. Kaplan, and C. Bona, Proteases as mitogens, Exp. Cell Res. 88: 388 (1974).
12. L.B. Chen, N.N.H. Teng, and J.M. Buchanan, Mitogenicity of thrombin and surface alterations on mouse splenocytes, Exp. Cell Res. 101: 41 (1976).
13. M.D. Kramer, P. Robinson, I. Vlodavsky, D. Barz, P. Friberger, Z. Fuks, and V. Schirrmacher, Characterization of an extracellular matrix-degrading protease derived from a highly metastatic tumor cell line. Eur. J. Cancer Clin. Oncol. 21: 307 (1985).
14. M.M. Simon, H.G. Simon, U. Fruth, J. Epplen, H.K. Müller-Hermelink, and M.D. Kramer, Cloned cytolytic T-effector cells and their malignant variants produce an extracellular matrix degrading trypsin-like serine proteinase. Immunol. 60: 219 (1987).
15. R. Geiger, U. Stuckstedte, and H. Fritz, Isolation and characterization of human urinary kallikrein, Hoppe-Seyler's Z. Physiol. Chem. 361: 1003 (1980).
16. U. Fruth, M. Prester, J.R. Golecki, H. Hengartner, H.G. Simon, M.D. Kramer, and M.M. Simon, The T cell specific serine proteinase TSP-1 is associated with cytoplasmic granules of cytolytic T lymphocytes, Eur. J. Immunol. in press (1987).
17. P.J. Millard, M.P. Henkart, C.W. Reynolds, and P.A. Henkart, Purification and properties of cytoplasmic granules from cytotoxic rat LGL tumors, J. Immunol. 132: 3197 (1984).
18. P.A. Henkart, Mechanism of lymphocyte-mediated cytotoxicity, Ann. Rev. Immunol. 3: 31 (1985).
19. E.R. Podack, The molecular mechanism of lymphocyte-mediated tumor cell lysis, Immunol. Today 6: 21 (1985).
20. H.K. Gershenfeld, and I.L. Weissman, Cloning of a cDNA for a T cell-specific serine protease from a cytotoxic T lymphocyte, Science, 232: 854 (1986).
21. C.G. Lobe, B.B. Finlay, W. Paranchych, V.H. Paetkau, R.C. Bleackley, Novel serine proteinases encoded by two cyotctoxic T lymphocyte-specific genes, Science 232: 858 (1986).

22. J.F. Brunet, M. Dosseto, F. Denizot, M.G. Mattei, W.R. Clark, T.M. Haqqi, P. Ferrier, M. Nabholz, A.M. Schmitt-Verhulst, M.F. Luciani, and P. Golstein, The inducible cytotoxic T-lymphocyte-associated gene transcript CTLA-1 sequence and gene localization to mouse chromosome 14, Nature 322: 268 (1986).

23. M.M. Simon, U. Fruth, H.G. Simon, and M.D. Kramer, A specific serine proteinase is inducible in Lyt-2$^+$, L3T4$^-$ and Lyt-2$^-$, L3T4$^+$ T cells in vitro but is mainly associated with Lyt-2$^+$, L3T4$^-$ effector cells in vivo, Eur. J. Immunol. 16: 1559 (1986).

24. B.A. White, and F. C. Bancroft, Cytoplasmic dot hybridization, J. Biol. Chem. 257: 8569 (1982).

25. C.G. Miyada, C. Klofelt, A.A. Reyes, E. McLaughlin-Taylor, and R.B. Wallace, Evidence that polymorphism in the murine major histocompatibility complex may be generated by the assortment of subgene sequences, Proc. Natl. Acad. Sci. USA 82: 2890 (1985).

26. D. Masson, and J. Tschopp, Appearance of a lytic, pore-forming protein (perforin) from cytolytic T lymphocytes, J. Biol. Chem. 260: 9096 (1985).

27. M.M. Simon, U. Fruth, H.G. Simon, and M.D. Kramer, Evidence for the involvement of a T cell associated serine proteinase (TSP-1) in cell killing, 17th Forum in Immunol. Ann. Inst. Pasteur, 138: 285 (1987).

28. G.S.B. Ku, J.P. Quigley, and B.M. Sultzer, The inhibition of the mitogenic stimulation of B lymphocytes by a serine proteinase inhibitor: commitment to proliferation correlates with an enhanced expression of a cell-associated arginine-specific serine enzyme, J. Immunol. 131: 2494 (1983).

29. D.E. Mullins, and S.T. Rohrlich, The role of proteinases in cellular invasiveness, Biochim. Biophys. Acta 695: 177 (1983).

30. M.M. Simon, S. Ali, R. Tewari, H.G. Simon, H.K. Müller-Hermelink, and J.T. Epplen, Leukemic cells arise from cloned cytotoxic lymphocytes during cell culture, Eur. J. Immunol. 16: 1269 (1986).

31. D. Gospodarowicz, G. Greenburg, J.M. Foidart, and N. Savion, The production and localization of laminin in cultured vascular and corneal endothelial cells, J. Cell. Physiol. 107: 171 (1983).

32. D. Gospodarowicz, A.R. Mescher, and C.R. Birdwell, Stimulation of corneal endothelial cell proliferation in vitro by fibroblast and epidermal growth factors, Exp. Eye Res. 25: 75 (1977).

33. R.M. Zinkernagel, and P.C. Doherty, MHC-restricted cytotoxic T cells: Studies on the biological role of polymorphic major transplantation antigens determining T-cell restriction-specificity, function, and responsiveness, Adv. Immunol. 27: 51 (1979).

34. A.G. Morris, J.L. Lin, and B.A. Askonas, Immune interferon release when a cloned cytotoxic T-cell line meets its correct influenza-infected target cell, Nature 295: 150 (1982).

35. J.R. Klein, D.H. Raulet, M.S. Pasternack, and M.J. Bevan, Cytotoxic T lymphocytes produce immune interferon in response to antigen or mitogen, J. Exp. Med. 155: 1198 (1982).

36. M.D. Kramer and M.M. Simon, Are proteinases functional molecules of T lymphocytes> Immunol. Today, 8: 140 (1987).

PANCREATIC SECRETORY TRYPSIN INHIBITOR IN CANCER

Michio Ogawa, Naohiro Tomita, Akira Horii
Nariaki Matsuura, Masahiko Higashiyama, Tatsuo Yamamoto
Atsuo Murata, Takesada Mori and Kenichi Matsubara*

Second Department of Surgery, Osaka University Medical
School, Osaka 553, Japan, and *Institute for Molecular
and Cellular Biology, Osaka University, Suita 565, Japan

INTRODUCTION

In normal pancreatic juice, trypsinogen is protected from activation by a certain amount of a protease inhibitor known as Kazal inhibitor (Kazal et al., 1948), which was named as pancreatic secretory trypsin inhibitor (PSTI) (Greene and Giordano, 1969). Until recently PSTI has been considered to exist solely in the pancreas and its role has been regarded only as a specific trypsin inhibitor in pancreatic juice.

In 1978, Eddeland and Ohlsson developed a specific radioimmunoassay (RIA) for human PSTI, and later we also developed it (Kitahara et al., 1979; 1980). Using the RIA we demonstrated that immunoreactive PSTI existed in various tissues other than the pancreas (Matsuda et al., 1983a). Moreover, we found that serum PSTI was frequently elevated in patients with various malignant diseases (Matsuda et al., 1983a; Murata et al., 1983). These findings prompted us to investigated a possibility that PSTI has a previously unrecognized physiological function.

In this paper, we reviews recent clinical and fundamental studies on PSTI in cancer.

SERUM PSTI IN VARIOUS MALIGNANCIES

Table 1 is the first clinical data showing that serum PSTI was frequently elevated in patients with various malignancies (Matsuda et al., 1983a). As we found later that serum PSTI increased with age (Matsuda et al., 1983b), we extended the normal range to include the normal content in the aged. A little lower but still a high incidence of elevated serum PSTI was observed (Matsuda et al., 1983b).

In relation to PSTI, a very similar or identical polypeptide, named tumor-associated trypsin inhibitor (TATI), has been reported (Stenman et al., 1982). It was a 6,000 dalton trypsin inhibitor from urine from a patient with ovarian cancer. Analysis of the first 23 N-terminal amino acid sequence and the amino acid composition showed that TATI was closely related or identical to PSTI (Huhtala et al., 1982). Recently Hagland et al. (1986) reported that elevation of serum TATI was seen in 85% of patients with pancreatic cancer.

Table 1. Serum PSTI contents in patients with various malignant
diseases

diseases	n	serum PSTI** (ng/ml)	range (ng/ml)	incidence of elevated PSTI[†]
normal control	28	11.5 ± 2.5	6.0 - 16.3	
pancreatic cancer	19	36.1 ± 41.9*	8.7 - 194.6	74%
esophageal cancer	14	27.9 ± 30.0*	9.1 - 120.6	57%
gastric cancer	41	32.7 ± 36.1*	8.5 - 164.5	66%
hepatoma	5	45.9 ± 45.0*	11.5 - 130.8	60%
breast cancer	17	19.3 ± 13.7*	9.8 - 66.0	24%
thyroid cancer	7	18.0 ± 6.3*	9.6 - 30.5	57%

* : significantly different from the normal control (p<0.01)
** : mean ± S.D.
† : above control mean plus 2 S.D. (16.5 ng/ml)
Matsuda et al. (1983a) with permission of the publishers.

Table 2. Incidence of the presence of PSTI-immunoreactive cells in
various cancer tissues

cancer tissues	No. of tissues examined	No. of tissues with PSTI-immunoreactive cells	incidence (%)
esophagus	6	0	0
stomach	50	40	80
colon	7	2	29
liver	15	5	33
choledochus	5	3	60
pancreas	9	5	56
lung	30	9	30
breast	9	1	11
thyroid	7	3	43
total	138	68	49

Ogawa et al. (1987) with permission of the publishers.

PRESENCE OF PSTI IN VARIOUS CANCER TISSUES

In patients with advanced or metastasized malignancies, the incidence and magnitude of the elevation of serum PSTI were far greater than
in those in early stage. Since there is a possibility that PSTI could
be produced by cancer tissues, we carried out immunohistochemical staining using the peroxidase-antiperoxidase method on formaline-fixed paraffin-embedded sections (Matsuura et al., 1984; Ogawa et al., 1987).

Table 2 shows the incidence of the presence of PSTI immunoreactive cells in various cancer tissues (Ogawa et al., 1987). The highest incidence was observed in gastric cancer, followed by choledochus cancer, pancreatic cancer and thyroid cancer. The presence of PSTI was observed in cancers of tissues in which no PSTI-immunoreactive cells were detected under normal conditions. Moreover, immunoreactivity of PSTI was detected even in cancer tissues in early stage, or whose serum PSTI remained in normal range. Therefore, it is unlikely that PSTI in cancer tissues is solely responsible for the elevation of serum PSTI in cancer patients.

EXPRESSION OF PSTI IN CANCER CELLS

To clarify whether PSTI in cancer tissues detected by immunoperoxidase staining was produced by cancer cells or PSTI from other organs bound to cancer cells, we did two experiments.

Table 3 shows the result of immunoperoxidase staining of several cell lines derived from various human cancer tissues (Ogawa et al., 1987). Most of the adenocarcinoma-derived cell lines expressed PSTI. The expression of PSTI in cancer cells was also confirmed by Northern blotting analysis of PSTI-positive lung cancer and sigmoid colon polyp tissues using PSTI cDNA (Yamamoto et al., 1985) as a probe (Figure 1). mRNAs with the same size as pancreatic PSTI mRNA were detected in lung cancer and sigmoid colon polyp tissues.

Table 3. PSTI-immunoreactivity in various human cell lines

cell lines	source	PSTI-immunoreactivity
BxPC-3	pancreatic cancer	+
MIA PaCa-2	pancreatic cancer	+*
PANC-1	pancreatic cancer	+
KATO-III	gastric cancer	+
COLO 201	colon cancer	+
SW 403	colon cancer	+*
PA-1	ovarian cancer	−
WIHOVCAR 3	ovarian cancer	+
HeLa Ohio	cervical cancer	+*
KB	nasopharyngeal cancer	−
HEp-2	laryngeal cancer	−
A549	lung cancer	+*
SK·HEp-1	liver cancer	+
G-361	melanoma	−
G-292	osteosarcoma	−

+ : All cells are positive
+* : Many but not all cells are distinctly positive
Ogawa et al. (1987) with permission of the publishers.

These results on PSTI in cancer are still controversial. Although
PSTI was produced in certain cancer tissues, there was no correlation
between the presence of PSTI in cancer tissues and the elevation of
serum PSTI.

Recently we demonstrated that serum PSTI was remarkably elevated in
patients with severe acute pancreatitis, in patients who received major
surgery or in patients with serious injury, and that the elevation of
serum PSTI was significantly correlated with those of acute phase reactants (Ogawa et al., 1985a; Matsuda et al., 1985; Ogawa et al., 1985b).
The elevation was remarkable when those patients were complicated by

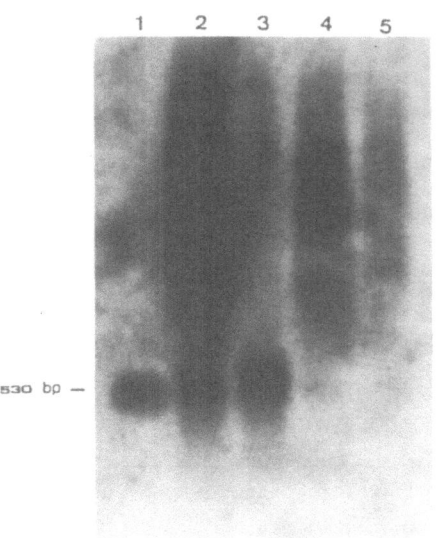

Figure 1. Northern blotting of poly(A) RNAs from human pancreas,
            lung cancer, sigmoid colon polyp, normal colon mucosa
            and placenta using nick translated PSTI cDNA as a probe.
            Lane 1: 0.1 μg of pancreatic poly(A) RNA
            Lane 2: 5 μg of lung cancer poly(A) RNA
            Lane 3: 5 μg of sigmoid colon polyp poly(A) RNA
            Lane 4: 5 μg of normal colon mucosa poly(A) RNA
            Lane 5: 5 μg of placenta poly(A) RNA
            The RNAs were denatured, electrophoresed and subjected
            to Northern blotting analysis.

serious infection. These findings together with the fact that patients
with an advanced or metastasized cancer which is known to be a stimulus
for acute phase response shows a higher level of serum PSTI indicated
that serum PSTI can be one of acute phase reactants.

In contrast, PSTI in cancer tissues was detected independently of
the level of serum PSTI. Therefore we think that serum PSTI and tissue
PSTI should be concerned with separately. Figure 2 shows our present
working hypothesis to investigate the role of PSTI in cancer.

PHYSIOLOGICAL FUNCTION OF PSTI IN CANCER

Previously Hunt et al. (1974) and Scheving (1983) reported that amino acid sequence of PSTI was similar to that of epidermal growth factor (EGF). We also demonstrated a high homology between human PSTI mRNA and mouse PSTI mRNA (Yamamoto et al., 1985). EGF is a growth-stimulating hormone and stimulates the growth and differentiation of various epidermal and epithelial cells (Cohen, 1962; Carpenter and Cohen, 1979). EGF-binding protein has been reported to be capable of autodigestion (Green and Moore 1980), suggesting that the function of PSTI and EGF also resemble each other. We recently demonstrated that human PSTI stimulated [^{3}H]-thymidine incorporation into DNA in human fibroblast (Ogawa et al., 1985c), and that PSTI bound specifically to

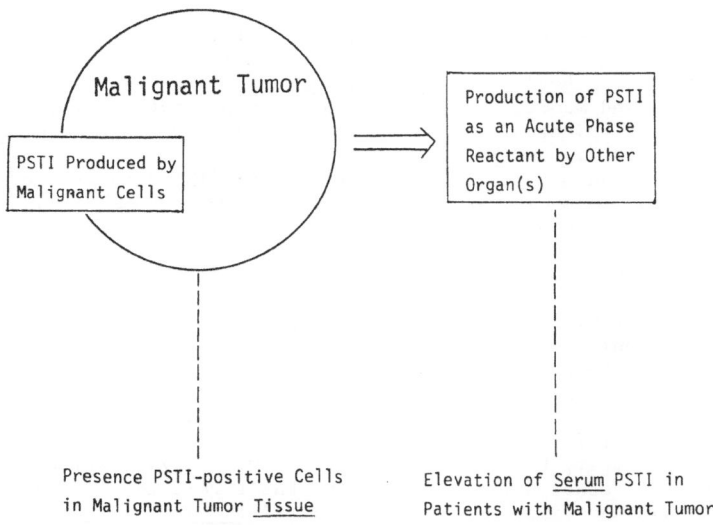

Figure 2. PSTI in cancer — working hypothesis

various cultured cells (Niinobu et al., 1986). The binding sites for PSTI were distinct from the receptor for EGF. These findings strongly suggested that PSTI is also a growth-stimulating factor. In this con-nection, McKeehan et al. (1986) recently reported that one of endothe-lial cell growth factors they purified from the medium of human hepatic cancer cell line was very similar or identical to PSTI.

CONCLUSION

High incidence of the elevation of serum PSTI in patients with various malignancies and the expression of PSTI in cancer tissues were demonstrated. PSTI in cancer tissue may possess autocrine or paracrine function for the proliferation of cancer tissues, whereas PSTI in sera of advanced cancer patients may act as an acute phase reactant for the host defence or the repair of destructed tissues. We concluded that PSTIs in serum and tissue should be evaluated separately, even though both could play a similar role.

ACKNOWLEDGEMENTS

This investigation was partially supported by Research Grant 60480303 from the Ministry of Education, Science and Culture, Grant-In-Aid for Cancer Research 60-5 and Research Grant for Intractable Diseases of the Pancreas from the Ministry of Health and Welfare of Japan to M.O.

REFERENCES

Carpenter, G., and Cohen, S., 1979, Epidermal growth factor, *Ann. Rev. Biochem.*, 48:193-216.

Cohen, S., 1962, Isolation of a mouse submaxillary gland protein accelrating incisor eruption and eyelid opening in the newborn animal. *J. Biol. Chem.*, 237:1555-1562.

Eddeland, A., and Ohlsson, K., 1978, A radioimmunoassay for measurement of human pancreatic secretory trypsin inhibitor in different body fluids. *Hoppe-Seyler's Z. Physiol. Chem.*, 359:671-675.

Green, D. A., and Moore, J. B. Jr., 1980, Demonstration of protease activity for epidermal growth factor-binding protein. *Arch. Biochem. Biophys.*, 202:201-209.

Greene, L. J., and Giordano, J. S., Jr., 1969, The structure of the bovine pancreatic secretory trypsin inhibitor—Kazal's inhibitor, I. The isolation and amino acid sequences of the tryptic peptides from reduced aminoethylated inhibitor, *J. Biol. Chem.*, 244:285-298.

Haglund, C., Huhtala, M.-L., Halia, H., Nordling, S., Roberts, P. J., Scheinin, T. M., and Stenman, U.-H., 1986, Tumour-associated trypsin inhibitor, TATI, in patients with pancreatic cancer, pancreatitis and benign biliary diseases, *Br. J. Cancer*, 54:297-303.

Huhtala, M.-L., Pesonen, K., Kalkkinen, N., and Stenman, U.-H., 1982, Purification and characterization of a tumor-associated trypsin inhibitor from the urine of a patient with ovarian cancer. *J. Biol. Chem.*, 257:13713-13716.

Hunt, L. T., Barker, W. C., and Dayhoff, M. O., 1974, Epidermal growth factor: internal duplication and probable relationship to pancreatic secretory trypsin inhibitor. *Biochem. Biophys. Res. Commun.*, 60:1020-1028.

Kazal, L. A., Spicer, D. S., and Brahinski, R. A., 1948, Isolation of a crystalline trypsin inhibitor-anticoagulant protein from pancreas. *J. Am. Chem. Soc.*, 70:3034-3040.

Kitahara, T., Takatsuka, Y., Fujimoto, K., Tanaka, S., Ogawa, M., and Kosaki, G., 1979, Diagnostic significance of measurement of serum immunoreactive pancreatic secretory trypsin inhibitor in pancreatic diseases. *Proc. Symp. Chem. Physiol. Pathol.*, 19:35-40 (in Japanese).

Kitahara, T., Takatsuka, Y., Fujimoto, K., Tanaka, S., Ogawa, M., and Kosaki, G., 1980, Radioimmunoassay for human pancreatic secretory trypsin inhibitor: measurement of serum pancreatic secretory trypsin inhibitor in normal subjects and subjects with pancreatic diseases. *Clin. Chim. Acta*, 103:135-143.

Matsuda, K., Ogawa, M., Murata, A., Kitahara, T., and Kosaki, G., 1983a, Elevation of serum immunoreactive pancreatic secretory trypsin inhibitor contents in various malignant diseases. *Res. Commun. Chem. Pathol. Pharmacol.*, 40:301-305.

Matsuda, K., Ogawa, M., Murata, A., Miyauchi, K., Kurihara, M., and Kosaki, G., 1983b, Studies on human pancreatic secretory trypsin inhibitor (V): change in serum PSTI content with age and its elevation in patients with various malignant diseases. *Gastroenterol. Surg.*, 6:1883-1886 (in Japanese).

Matsuda, K., Ogawa, M., Shibata, T., Nishibe, S., Miyauchi, K., Matsuda, Y., and Mori, T., 1985, Postoperative elevation of serum pancreatic secretory trypsin inhibitor. *Am. J. Gastroenterol.*, 80:694-698.

Matsuura, N., Matsuda, K., Ogawa, M., Murata, A., Mori, T., Sakurai, M., and Kosaki, G., 1984, Immunohistological study of pancreatic secretory trypsin inhibitor in cancerous and non-cancerous tissues. *Saishin-Igaku*, 39:617-618 (in Japanese).

McKeehan, W. L., Sakagami, Y., Hoshi, H., and McKeehan, K. A., 1986, Two apparent human endothelial cell growth factors from human hepatoma cells are tumor-associated proteinase inhibitors. *J. Biol. Chem.*, 261:5378-5383.

Murata, A., Ogawa, M., Matsuda, K., Matsuura, N., Kosaki, G., Inoue, M., Ueda, G., and Kurachi, K., 1983, Immunoreactive pancreatic secretory trypsin inhibitor in gynecological diseases. *Res. Commun. Chem. Pathol. Pharmacol.*, 41:493-499.

Niinobu, T., Ogawa, M., Shibata, T., Nishibe, S., Murata, A., Mori, T., and Ogata, N., 1986, Specific binding of human pancreatic secretory trypsin inhibitor to various cultured cells. *Res. Commun. Chem. Pathol. Pharmacol.*, 53:245-248.

Ogawa, M., Matsuda, K., Matsuda, Y., Miyauchi, K., Nishijima, J., Horikawa, Y., Kurihara, M., and Mori, T., 1985a, Evaluation of immunological assays of serum pancreatic enzymes and pancreatic secretory trypsin inhibitor for diagnosis and management of acute pancreatitis, *in*: Pancreatitis—its pathophysiology and clinical aspects. T. Sato, and H. Yamauchi, eds., University of Tokyo Press, Tokyo, pp. 107-115

Ogawa, M., Matsuda, K., Shibata, T., Matsuda, Y., Ukai, T., Ohta, M., and Mori, T., 1985b, Elevation of serum pancreatic secretory trypsin inhibitor (PSTI) in patients with serious injury. *Res. Commun. Chem. Pathol. Pharmacol.*, 50:259-266.

Ogawa, M., Tsushima, T., Ohba, Y., Ogawa, N., Tanaka, S., Ishida, M., and Mori, T., 1985c, Stimulation of DNA synthesis in human fibroblasts by human pancreatic secretory trypsin inhibitor. *Res. Commun. Chem. Pathol. Pharmacol.*, 50:155-158.

Ogawa, M., Matsuura, N., Higashiyama, K., and Mori, T., 1987, Expression of pancreatic secretory trypsin inhibitor in various cancer cells, *Res. Commun. Chem. Pathol. Pharmacol.*, 55:137-140.

Scheving, L. A., 1983, Primary amino acid sequence similarity between human epidermal growth factor-urogastrone, human pancreatic secretory trypsin inhibitor, and members of porcine secretin family. *Arch. Biochem. Biophys.*, 226:411-413.

Stenman, U.-H., Huhtala, M.-L., Koitinen, R., and Seppälä, M., 1982, Immunochemical demonstration of an ovarian cancer-associated urinary peptide. *Int. J. Cancer*, 30:53-57.

Yamamoto, T., Nakamura, Y., Nishide, T., Emi, M., Ogawa, M., Mori, T., and Matsubara, K., 1985, Molecular cloning and nucleotide sequence of human pancreatic secretory trypsin inhibitor (PSTI) cDNA. *Biochem. Biophys. Res. Commun.*, 132:605-612.

PROTEASES AND ANTIPROTEASES IN ASCITES – DIFFERENTIATION OF MALIGNANT AND NONMALIGNANT ASCITES AND PREDICTION OF COAGULOPATHY IN ASCITES RETRANSFUSION

Jürgen Schölmerich, Eckart Köttgen, Brigitte A. Volk,
and Wolfgang Gerok

Department of Internal Medicine
University of Freiburg
FRG

INTRODUCTION

Ascites is a symptom of a severe disease. In most cases prognosis is poor. However, the differentiation of malignant and nonmalignant ascites is of clinical significance since it leads to different treatment. Thus far, most attempts using laboratory parameters (protein, pH, LDH, lactate, WBC) did not reveal sufficient separating potential to be useful in clinical routine. Recently fibronectin[1] and cholesterol[2] have been introduced having a sensitivity and specifity between 90 and 100 %. However, it is not clear if these parameters can differentiate all forms of ascites.

Refractory ascites can be treated effectively by ascites retransfusion or peritoneovenous shunting[3,4]. However, life threatening coagulation disorders and bleeding have been reported as most predominant complication of these procedures[4,5]. We have previously proposed that the plasminogen activity in ascites may be a reliable predictive parameter of coagulopathy in ascites retransfusion[4,5].

It has not been studied if the determination of other proteases is helpful in the differential diagnosis of ascites or can predict coagulation disorders in ascites retransfusion procedures. We therefore analyzed the activity or concentration of several proteases, antiproteases or proteins (plasminogen, α_2-antiplasmin (α_2-AP), α_1-protease inhibitor (α_1-PI), α_2-macroglobulin (α_2-MG), antithrombin III (AT III), fibronectin) in ascites and plasma of patients undergoing ascites retransfusion in order to assess their practical value with regard to these questions.

Since the intraperitoneal injection of dexamethasone has been shown to prevent coagulation disorder in ascites retransfusion procedures the effect of dexamethasone injection on activity or concentration of proteases was studied, in addition.

METHODS

54 patients with ascites (17 malignant, 37 nonmalignant) were studied. In 17 patients ascites retransfusion procedures[3,4] were performed. The activity or concentration of plasminogen, α_2-AP, α_1-PI, α_2-MG, AT III,

and fibronectin was assessed as described[6]. In 13 patients at 15 occassions 16 mg dexamethasone was injected intraperitoneally. Proteases were assessed before injection and at daily intervals for four days.

RESULTS

Malignant versus nonmalignant ascites: In plasma of patients with malignant ascites higher activities or concentrations of most of the proteins were found as compared to patients with liver cirrhosis. The difference was statistically significant for plasminogen, α_2-AP, α_1-PI, and AT III. However, due to a large overlap no good differentiation was achieved. In ascites significant differences ($p < 0.001$) were found for fibronectin, plasminogen, α_2-MG, α_1-PI, and AT III (figure 1). However, only fibronectin, α_1-PI, and AT III revealed a sufficient differentiation (figure 2, table 1). If the ratio protease or protein to albumin was used only the fibronectin/albumin ratio was better than the value without relation to albumin[6].

Table 1. Accuracy of proteases for detection of malignant ascites

	Sensitivity (%)	Specifity (%)	Cutoff
Fibronectin	100	96	75 mg/l
Antithrombin III	94	94	3.1 IU/l
α_2-Protease inhibitor	94	89	1.95 g/l
Albumin	90	87	15 g/l
α_2-Macroglobulin	79	71	57 mg/l
Plasminogen	59	76	0.4 CTA U ml

Prediction of coagulopathy: None of the parameters in plasma (including coagulation assessment) revealed a significant difference between patients developing coagulation disorder and those who did not. Plasminogen, α_2-AP and AT III activities in ascites were significantly ($p < 0.01$) lower in patients developing coagulopathy. But only plasminogen and α_2-AP were able to predict coagulation disorders properly (figure 3). Those two proteases were significantly correlated ($r = 0.867$, $p < 0.001$) (figure 4). Both had a predictive value for the occurence of coagulopathy of 100 % (table 2).

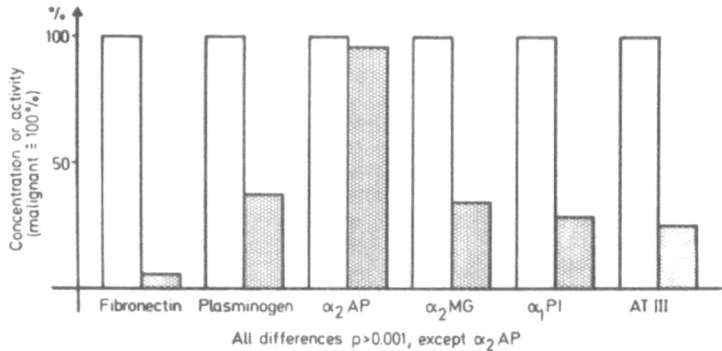

Figure 1. Mean values for malignant (□) and nonmalignant (▨) ascites

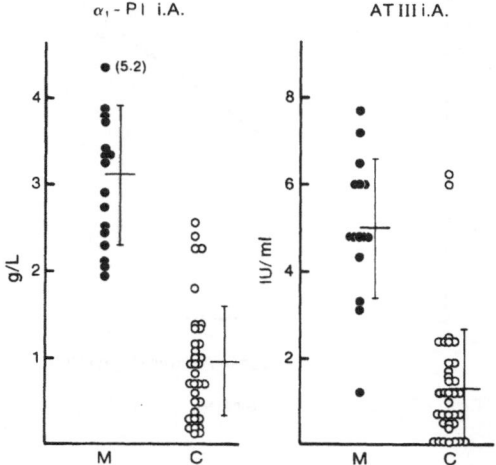

Figure 2. Differentiation between malignant (●) and nonmalignant ascites by protease determination

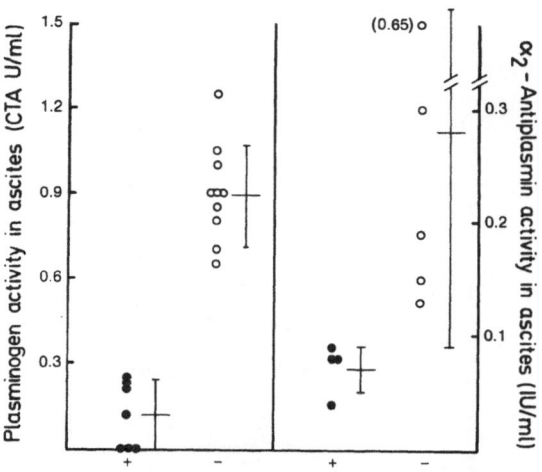

Figure 3. Prediction of coagulopathy (●) during ascites retransfusion

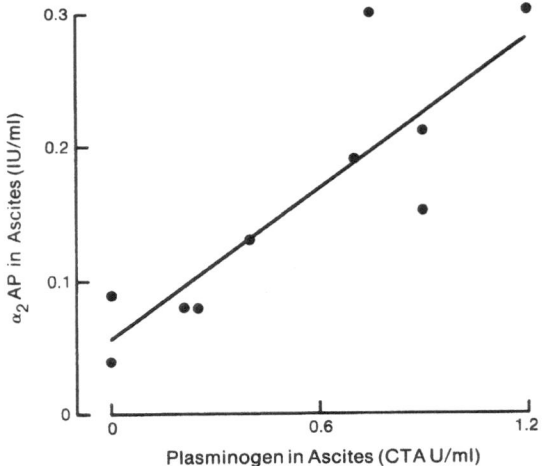

Figure 4. Correlation of plasminogen and α_2-AP in ascites

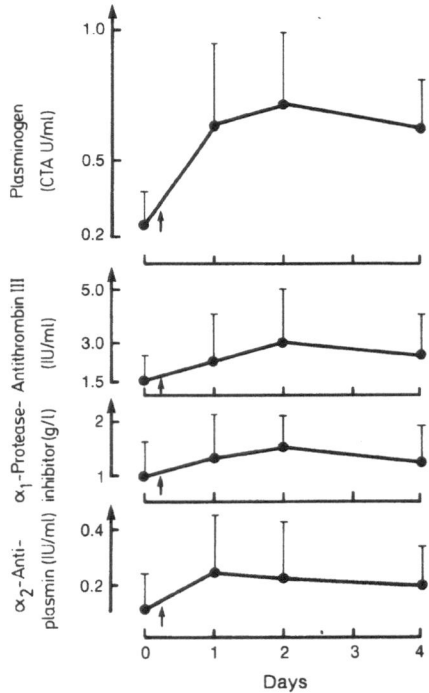

Figure 5. Influence of dexamethasone
injektion (↑) on proteases
in ascitic fluid

Table 2. Predictive value for the occurence of
coagulopathy during ascites retransfusion

	Cutoff	PPV (%)	NPV (%)
plasminogen (CTA U/ml)	< 0.6	100	100
α_2- AP (U/ml)	< 0.1	100	100

Effect of dexamethasone: The injection of dexamethasone (16 mg) into
the ascitic fluid resulted in an increase of the activity of plasminogen,
α_2-AP, and AT III, and of the concentration of α_1-PI (figure 5). However,
the increase of plasminogen activity was significantly correlated with
that of α_2-AP (r = 0.751, p < 0.01) while the increase of α_2-AP activity
was correlated with that of α_1-PI (r = 0.792, p < 0.01). Both (plasminogen and α_2-AP) increases were inversely correlated with the increase of AT
III (r = -0.772 and -0.539).

DISCUSSION

The determination of fibronectin, α_1-PI, and AT III in ascites is of
practical value in the differentiation of malignant and nonmalignant ascites. This confirms earlier observations with regard to fibronectin and AT
III[7]. It has been shown that the cause for the increase of these proteins
is different from that for the well known increase of total protein or albumin . A dissociation of fibronectin from the surface of malignant transformed cells has been demonstrated[1]. The reason for the increased activity
or concentration of the other proteases is not definitively known.

The assessment of plasminogen and α_2-AP can reliably predict the occurence of a coagulation disorder following ascites retransfusion or peritoneovenous shunting. These and other data[3,4,5] led us to conclude that
the coagulopathy is a primary fibrinolysis and not a disseminated intravascular coagulation. Both parameters are equally effective and can thus
be used according to local possibilities. The intraperitoneal injection of
dexamethasone prevents coagulopathy after ascites retransfusion. The effects on the activity or concentration of proteases in ascites observed
suggest an inhibition of the release of plasminogen activators from peritoneal macrophages as active mechanism[8].

In summary, the assessment of activity or concentration of proteases
in ascites can help to differentiate between malignant and nonmalignant
ascites. It can furthermore predict the occurence of coagulation disorders
in ascites retransfusion. The intraperitoneal injection of dexamethasone
influences protease activity in ascites and can prevent coagulopathy in
ascites retransfusion.

REFERENCES

1. J. Schölmerich, B. A. Volk, E. Köttgen, S. Ehlers, and W. Gerok, Fibronectin concentration in ascites differentiates between malignant
   and nonmalignant ascites, Gastroenterology 87:1160 (1984).
2. D. Jüngst, A. L. Gerbes, R. Martin, and G. Paumgartner, Value of ascitic lipids in the differentiation between cirrhotic and malignant ascites, Hepatology 6:239 (1986).
3. J. Schölmerich, W. Diener, E. Köttgen, K.-P. Maier, U. Costabel, and
   W. Gerok, Die extrakorporale Ascites-Reinfusion zur raschen Ascites-
   Elimination bei gastroenterologischen Notfallpatienten, Intensivmedi-
   zin 22:308 (1985).

4. B. A. Volk, J. Schölmerich, H. Wilms, K. Hasler, E. Köttgen, E. H. Farthmann, and W. Gerok, Treatment of refractory ascites by transfusion and peritoneovenous shunting, Digest. Surg. 2:93 (1985).
5. B. A. Volk, J. Schölmerich, H. Wilms, E. Köttgen, G. Witz, P. Billmann, P. Hoppe–Seyler, and W. Gerok, Peritoneo–venöser Shunt in der Aszitestherapie: Komplikationen der Behandlung, Dtsch. med. Wschr. 110:1685 (1985).
6. J. Schölmerich, U. Zimmermann, E. Köttgen, B. A. Volk, S. Ehlers, and W. Gerok, Proteases and antiproteases related to the coagulation system in plasma and ascites. An approach to differentiate between malignant and cirrhotic ascites, Klin. Wschr. in press (1987)
7. V. Chan, C. L. Lai, and T. K. Chan, Metabolism of antithrombin III in cirrhosis and carcinoma of the liver, Clin. Sci. 60:681 (1981).
8. J. D. Vassalli, J. Hamilton , and E. Reich, Macrophage plasminogen activator: modulation of enzyme production by antiinflammatory steroids, mitotic inhibitors, and cyclic nucleotides, Cell 8:271 (1976).

ALPHA$_1$-ANTITRYPSIN AND ALPHA$_1$-ANTICHYMOTRYPSIN SERUM LEVEL IN RELATION TO STAGING AND POSTOPERATIVE CLINICAL COURSE OF HUMAN COLORECTAL CANCER

Anna Kuryliszyn-Moskal, Krystyna Bernacka,
and Stanislaw Sierakowski

Department of Rheumatology, Medical Academy, Bialystok
Poland

INTRODUCTION

During the past years, increasing attention has been devoted to the connection between the neoplasmatic process and proteolytic activity mediated by proteases-antiproteases balance.[1,2] A possible influence of antiproteases on the course of neoplasmatic processes [3] made us investigate the relationship between the serum alpha$_1$-antitrypsin (A$_1$AT) and alpha$_1$-antichymotrypsin (A$_1$AChy) levels and the clinico-pathological staging of colorectal cancer. Besides it would be important from the clinical point of view, to explain to what extent measurement of the serum antiprotease concentrations may indicate the effectiveness of the surgery during postoperative follow up.

MATERIALS AND METHODS

The study included 61 patients (25 men and 36 women) aged 22-79 (mean 55.8) with colorectal cancer. In all cases the stage of disease was determined according to the International TNM Classification [4] on the basis of surgical findings and pathologic examinations of the resected specimens. Histologically 60 cases (98.4%) were defined as adenocarcinomas and one case (1.6%) was termed as a basocellular carcinoma.

Radical surgery i.e. total resection of the mobile tumor including regional lymph nodes and surrounding fatty tissue was performed in 54 patients. In case of the metastatic spread the curative treatment included resection of the visceral or liver metastases. The palliative surgery i.e. laparotomy or bypass of unresectable disease was performed in 7 cases.
Patients who had the radical surgery were divided into two subgroups, according to the response to the postoperative chemotherapy. Total disappearance of clinical radiographic and biologic evidence of the disease was defined as a complete remission. Clinical radiographic or biologic evidence of tumor growth or metastases was defined as a progression.

All patients received postoperative adjuvant chemotherapy. Two treatment programms were used i.e. CCNU (Lomustine) 80 mg/M^2 orally on day 1, Vincristin 1.5 mg/M^2 i.v. on day 1, 5-fluorouracil (5-FU) 500 mg/M^2 i.v. on days 1 and 5 as a 28 day cycle or CCNU 100 mg/M^2

orally on day 1 Adriamycin (doxorubicin) 40 mg/M^2 i.v. on days 1,36, 5-FU 350mg/M^2 i.v. on days 1-5 and 36-40 as a cycle repeated every 10 weeks. The procedure was repeated for 1 year or longer in cases of disease progression.
The control group consisted of 40 healthy blood donors of similar age.

The study of serum A_1AT and A_1AChy was based on the venous blood samples drawn on the day prior to the operation and further samples taken at regular intervals in the postoperative period during the follow-up time of between 1-5 years. The serum samples were stored frozen at -20°C until investigation.
The serum concentration of A_1AT and A_1AChy were measured according to the Laurell method using antisera supplied by Behringwerke (West Germany). [5]
The results of the above determinations were analysed statistically by the Student`s t-test at the significance level p<0.05.

RESULTS

The analysis of the mean A_1AT and A_1AChy levels in different stages of colorectal cancer compared with the control group are shown in Table 1.

Table 1. Serum A_1AT and A_1AChy in relation to different stages of colorectal cancers

Group	Stage	n	A_1AT (g/l) mean±SEM	A_1AChy (g/l) mean±SEM
Control (c)	–	40	1.61±0.05	0.28±0.01
Colo-rectal cancer	I	11	2.47±0.22	0.53±0.04
	II	15	2.76±0.19	0.67±0.05
	III	21	3.48±0.22	0.88±0.06
	IV	14	3.37±0.23	0.72±0.05
	Total(T)	61	3.19±0.08	0.75±0.03

1/p≤0.05
2/p≤0.01
3/p<0.001
n=number of patients

C and T,I,II,III,IV[3]
I and III[2]
I and IV[1]
II and III[1]

C and T,I,II,III,IV[3]
I and II[1]
I and IV[2]
III and I[3],II[2],IV[1]

Both antiproteases serum levels in the cancer groups, independently of the stage, were significantly higher than the mean value of the healthy subjects (p<0.001). The lowest concentration of A_1AT and A_1AChy was found in the patients of Stage I. The A_1AT values in this subgroup were significantly lower than in subgroups III and IV and concerning the A_1AChy level the differences were significant in comparison to the remaining subgroups. The highest levels of A_1AT and A_1AChy were found in the patients of Stage III.
A significant correlation between the level of the both antiproteases and the colorectal cancer stages was observed. The correlation coefficient r for A_1AT was 0.353 and for A_1AChy was 0.338 (P_r<0.01).

In view of the above data, it is interesting to study a possible relationship between the clinical course of the disease during postoperative follow-up and both antiprotease serum levels. As can be seen from Fig.1, the pretreatment values of A_1AT and A_1AChy were similar in the remission and progression subgroups. But, during the follow-up, the levels of both antiproteases in patients with remission were lower and still significantly decresed in comparison to the progression subgroup and the initial values. In the patients with the progressive disease an increase of A_1AT and A_1AChy was found. Significant differences between both subgroups concerning A_1AT level were observed in 3rd, 6th and 9th months and concerning the A_1AChy in 6th and 9th months of the follow up.

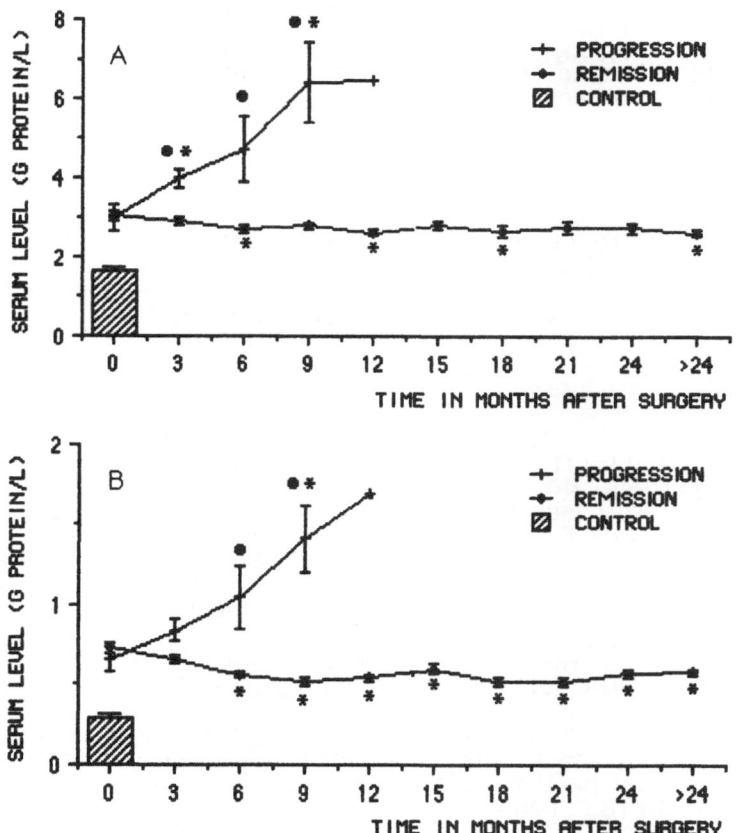

Fig. 1. Serum level (mean±SEM) of A_1AT (A) and A_1AChy (B) in relation to the clinical course of colorectal cancer after radical surgery * $P<0.05$ – differences between pretreatment value and the values in the particular months after surgery
● $P<0.05$ – differences between values in the remission and progression subgroups.

DISCUSSION

The enhanced levels of A_1AT and A_1AChy demonstrated in this study were also observed by many investigators in other malignant diseases.[6,7] However, a relation of these antiprotease serum levels to the cancer stage and clinical course of disease after surgical treatment is still unknown.

In this study we observed a correlation between A_1AT and A_1AChy levels and the cancer stage. It may be connected with enhanced production of these antiproteases by tumor cells [3], mononuclear cells of tumor infiltration [8] or by the host liver cells. [9] However, the values both of these antiproteases in serum increased with the severity of the disease, the value in Stage III was higher than in Stage IV. It may suggest its limited biosynthesis in advanced disease.

The analysis of A_1AT and A_1AChy serum levels during postoperative chemotherapy showed a decrease of those antiproteases in sera of the patients with remission in comparison to the progression patients. The tendency to normalisation of serum antiproteases in the patients whose response to chemotherapy was good may reflect a returning to proteolytic-antiproteolytic balance which may play a potential role in tumor invasiveness. [2]

Our data may suggest, that the increase of A_1AT and A_1AChy levels in patients with colorectal cancer may provide negative prognostic information, if other pathological conditions increasing these inhibitors are excluded.

AKNOWLEDGMENT

We are grateful to (1) Professor Dr. J. Zalewski for providing us with some patients, (2) Professor Dr. A. Gabryelewicz and (3) Doc. Dr. B. Musiatowicz for efficient assistance. This work was supported by grands from the Polish National Cancer Program (PR-6/0304).

REFERENCES

1. R. K. Chawla, D. J. Rausch, F. W. Miller, W. R. Vogler, and D. H. Lawson, Abnormal profile of serum proteinase inhibitors in cancer patients Cancer Res., 44:2718 (1984).
2. H. Heine, Proteases-proteases inhibitors; a local cellular information system, Adv. Exp. Med. Biol., 167:121 (1984).
3. E. Tahara, H. Ito, K. Taniyama, H. Yokozaki, and J. Hata, Alpha$_1$-antitrypsin, alpha$_1$-chymotrypsin and alpha$_2$-macroglobulin in human gastric carcinomas:a retrospective immunohistochemical study. Human Pathol., 15:957 (1984).
4. D. A.Wood, G. F. Robbins, C. Zippin, D. Lum and M. Stearns, Staging of cancer of the colon and rectum, Cancer, 43:961 (1979).
5. C. B. Laurell, Quantitive estimation of proteins by electrophoresis in agarose gel containing antobodies, Anal. Biochem., 15:45 (1966).
6. H. Hasegawa, and K. Aiba, Significance of cancer - related serum proteins in patients with gynecologic neoplasms, Yokohama Med. Bull., 34:185 (1983)
7. A. L. Latner, G. A. Turner, and M. M. Lamin, Alpha$_1$-antitrypsin levels in early and late carcinoma of the cervix, Oncology, 33:12 (1976)
8. G. Wiessmann, R. B. Zurier, and S. Haffstein, Leukocytic proteases and the immunological release of lysosomal enzymes, Am. J. Path., 68:539 (1970)
9. J. Travis, and G. S. Salvesen, Human plasma proteinase inhibitors, Ann. Res. Biochem., 52:655 (1983)

INHIBITION OF PROTEASES DURING EXTRACORPOREAL EXTREMITY PERFUSION

EXPERIMENTAL AND CLINICAL RESULTS

H. Walther, H. Müller, and K.R. Aigner

Department of General and Oncologic Surgery

Trostberg Hospital, Siegerthöhe 1, D-8223 Trostberg

Since 1957 the isolated hyperthermic limb perfusion with cytotoxic drugs became a standarized method in the treatment of malignant melanomas of the extremities.

Meanwhile operation techniques have been standarized and cytostatics like cis-DDP, Vindesine, DTIC and Mitomycin C were introduced.

Simultaneous hyperthermia of the extremities during cytostatic perfusion enhances tumor eoxicity of most drugs. The incidence of complications like edema and peripheral nerve damage is considerably high in drugs introduced recently for isolation perfusion like Vindesine and cis-DDP and increases with perfusion temperatures above 40° C. We took into consideration that occurence of edema is related to the increase of proteases during extracorporeal circulation. In an experimental pilot study we found a significant reduction of the postoperative edema when Aprotinin (Trasylol, Bayer Pharma) was applied perioperatively. Before transferring these findings into clinical management we performed a prospective randomized trial in beagle dogs and tested the interference of Aprotinin with cytostatic drugs in a human melanoma cell line.

1. PROSPECTIVE RANDOMIZED TRIAL IN 18 BEAGLE DOGS

In all dogs the lower extremity was perfused in an isolated extracorporeal circuit for 60 minutes up to 40° C. The arterial pressure did not exceed 120 mmHg at flow rates between 150 and 200 ml/min.. All dogs received 2.0 mg/kg limb weight Vindesine. Half of the dogs were perfused on the addition of 200.000 KIU/kg limb weight Aprotinin (Trasylol, Bayer Pharma) into the perfusate and the rinsing solution.

The histological examinations taken right after the perfusion of the control group showed a distinct inter- and extrafibrillar edema, while in the Aprotinin group only a slight extrafibrillar edema was found. Control biopsies performed four weeks later showed regular muscle structures in both groups.

Granulocytes of the control group showed a maximum increase of 90% while the maximum increase of the Aprotinin group ranged at 28%.

LDH: The increase of the lactatdehydrogenase in the group treated with Aprotinin was 350% in comparison to 1.000% of the untreated control group.

Creatinine Kinase: Follow up of the creatinine kinase levels in the control group showed significantly higher levels up to 67% at the 6th postoperative day in comparison to the Aprotinin group. Maximum levels in this group reached 18%.

Limb volume: The maximum increase of the limb volume calculated in % is related directly to the tissue levels during perfusion. At tissue levels between 1.5 and 3.5 KIU/ml the edema of the upper tigh, knee and knuckle is lowered significantly as compared with the untreated control group.

2. TUMOR CELL COLONIES

Human melanoma cell line M 37/7 was used to check a positive interference of Aprotinin and drug action in vitro.

The cells were treated with Aprotinin and cytostatics in a concentration range representative for clinical use in extremity perfusion. The cytostatics were cis-DDP and Vindesine. Aprotinin alone has a positive effect on tumor cell growing. In case of cis-DDP and Vindesine the obvious cytostatic effect is only merely inhibited. These results have to be confirmed in a larger series of experiments but they encouraged us to apply Aprotinin in the same dosage in isolated perfusion of the extremities in man.

3. CLINICAL STUDY

In 30 patients the iliac artery and vein were exposed and cannulated above the inguinal ligament. After connecting the cannulas to an extracorporeal circuit, consisting of a roller pump and an oxygenator containing a heatexchanger, the isolated perfusion was maintained over 60 minutes at a hyperthermia of 40° C tissue temperature. The tubing system was primed with 500 ml of blood and 500 ml of 5 % glucose. A tourniquet is placed around the proximal upper tigh in order to avoid leakage of the perfusate into the systemic circulation. Flow rates beween 300 - 600 ml/min. at an arterial pressure of 120 - 140 mmHg.

All patients received 20 mg cis-DDP/l limb volume, measured by water displacement.

In 20 patients receiving isolated perfusion under addition of 200.000 KIU per kg limb weight no particular side effects were noted. 10 of these patients received the agent in the perfusate only, in the other 10 patients Aprotinin was added to the perfusate as well as to the rinsing solution. The control group, consisting of 10 patients, was perfused at the same concentrations of cytostatics without Aproinin.

Results: By means of Aprotinin in the perfusate the expected 13 % postoperative edema was reduced to 3 %. In case of rinsing with Aprotinin additionally, only a. 1 % increase of limb volume was observed. In contrast to the control group patients, who required repeated analgetic injections via a peridural catheter, the Aprotinin-treated patients did not require any analgetic therapy at all.

4. SUMMARY

In an experimental and clinical study a significant reduction of the postoperative edema after extracorporeal extremity perfusion under addition of Aprotinin was observed. Together with reduction of edema usually expected complications like pain and peripheral nerve damage could be reduced significantly.

Preliminary, yet unpublished data on cytostatic infusion of the panreas also indicate that intraarterial Aprotinin when given simultaneously may prevent pancreatitis.

It is mandatory Aprotinin to be given early enough in sufficiently high dose and there seems to be an advantage when the arterial route is used.

INDEX

Inter-alpha-trypsin-
  inhibitor, 75, 89

Kallikrein, 172, 453, 499
Kidney, 275, 295
  sections, 277
  transplantation, 309
Kininase, 13
Kininogens, 175, 361, 501
Kupffer cells, 193
Kwashiorkor, 413

Laminin, 268
Leukocyte elastase, 107, 115,
  128, 435
Leupeptin, 324, 515
Long chain fatty acids, 216
Lung elastin, 107, 115
Lysosomal
  enzymes, 519
  hydrolases, 226
  proteases, 283, 305
Lysosomes, 160, 284, 305
Lysozyme, 226

Macrophages, 193, 425
Magnesium, 299
Malignant tumor, 551
Malnutrition, 411
Marasmus, 413
Mast cells, 133
Melanoma, 566
Meningitis, 485
Meprin, 293
Metabolic acidosis, 318
Metalloendopeptidase, 13
Metalloproteinase, 33, 293
Methylcasein, 217
Monoclonal antibodies, 207
Monocytes, 193, 425

Monocyte-derived factor, 191
Monokines, 226
Mononuclear phagocytes, 225
Multicatalytic proteinase,
  216
Muscle protein turnover, 225
Muscle proteolysis, 318
Myofibrillar proteinase, 263

N-acetylglucosaminidase, 402,
  413, 450
N-end rule, 161
Neopterin, 461
Nephrotoxic nephritis, 267
Neurotensin, 15
Neutral proteases, 226
Neutrophil elastase, 27, 28,
  65, 75, 123
Nitrogen balance, 323
Non-lysosomal proteases, 235
N^t-methyl-histidine, 262,
  328, 336
Nutrition, 411

Oleic acid, 219
Oleoyl-coenzyme A, 220
Opsonins, 441
Ouchterlony double diffusion
  assay, 185
Overnutrition, 415
Oxygen radical, 474
Oxytocin, 15

Pancreatic elastase, 27
Pancreatic elastase 2, 97
Pancreatic secretory trypsin
  inhibitor, 101, 493, 505,
  547
Pancreatitis, 245
Papain, 161, 175